FRONTIERS IN QUANTUM OPTICS

MALVERN PHYSICS SERIES

General Series Editor: **Professor E R Pike** FRS

FRONTIERS IN QUANTUM OPTICS

Edited by **E R Pike** and **Sarben Sarkar**
Royal Signals and Radar Establishment, Malvern

Adam Hilger, Bristol and Boston

British Library Cataloguing in Publication Data

Frontiers in quantum optics.—(Malvern physics series; 2)
1. Quantum optics
I. Pike, E.R. II. Sarkar, Sarben
III. Series
535′.15 QC446.2

ISBN 0-85274-577-X

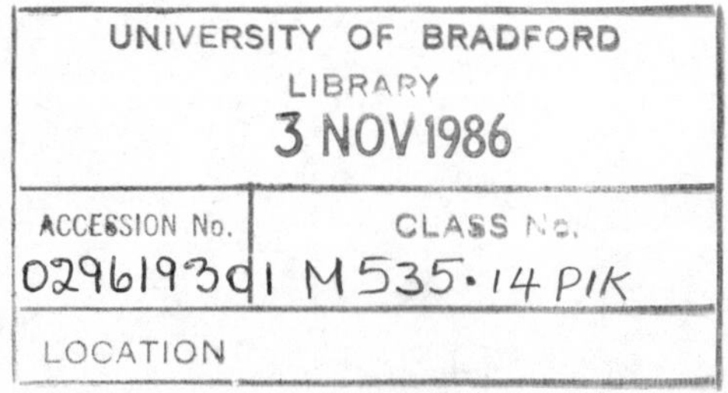

Published under the Adam Hilger imprint by IOP Publishing Limited
Techno House, Redcliffe Way, Bristol BS1 6NX, England
PO Box 230, Accord, MA 02018, USA

Printed in Great Britain by J W Arrowsmith Ltd, Bristol

CONTENTS

PREFACE

We were fortunate in being able to enjoy the participation of a number of leading figures in quantum optics at the Malvern Symposium of December 1985. Moreover, the speakers have been very prompt in providing written versions of their talks. The topics covered are non-classical states of light fields, quantum fluctuations in optical bistability and in the laser and instabilities and chaos in quantum optics, as well as quantum measurement theory with special reference to quantum optics.

We hope that the timely production of this volume will bring many more people some of the pleasure which we had at first hand listening to the contributions, although, unfortunately the stimulating and friendly atmosphere in which they were presented cannot be conveyed.

The success of the meeting was helped in no small part by the excellent organisation of Mrs D Costanzo as well as of Mrs M F Allso and Mrs D J Harris. We thank them warmly. In addition, valuable help has been given by J Zeuli, E M Flanagan, D G Jackson, J L Davies, and M A Williams. We would like also to express our thanks to the Director and Board of Management of RSRE whose valuable support makes this series of symposia possible.

E R Pike
Sarben Sarkar

Malvern
January 1986

LIST OF CONTRIBUTORS

S M BARNETT
Optics Section
Blackett Laboratory
Imperial College
Prince Consort Road
London SW7 2BZ
UK

G BROGGI
Physik Institut der Universitat Zurich
Schonberggasse 9
8001 Zurich
Switzerland

H J CARMICHAEL
Department of Physics
104 Physics Building
University of Arkansas
Fayetteville
Arkansas 72701
USA

A COLOMBO
Universita Degli Studi di Milano
Dipartimento di Fisica
Via Celoria 16
I-20133 Milano
Italy

P FILIPOWICZ
Max-Planck Institut für Quantenoptik
D-8046 Garching bei München
FRG

R J GLAUBER
Lyman Laboratory of Physics
Harvard University
Cambridge
MA 02138
USA

H HAKEN
Institut für Theoretische Physik
Universität Stuttgart
Pfaffenwaldring 57/IV
S-7000 Stuttgart 80
FRG

M W HAMILTON
Max-Planck Institut für Quantenoptik
D-8046 Garching bei München
FRG

E JAKEMAN
Centre for Theoretical Studies
Royal Signals and Radar Establishment
St Andrews Road
Great Malvern
Worcestershire WR14 3PS
UK

J JAVANAINEN
Department of Physics and Astronomy
University of Rochester
Rochester
New York 14627
USA

H J KIMBLE
Department of Physics
University of Texas at Austin
Austin
Texas 78712
USA

P L KNIGHT
Optics Section
Blackett Laboratory
Imperial College
Prince Consort Road
London SW7 2BZ
UK

G LEUCHS
Max-Planck Institut für Quantenoptik
D-8046 Garching bei München
FRG

R LOUDON
Physics Department
Essex University
Colchester CO4 3SQ
UK

L A LUGIATO
Università Degli Studi di Milano
Dipartimento di Fisica
Via Celoria 16
I-20133 Milano
Italy

L MANDEL

Department of Physics and Astronomy
University of Rochester
Rochester
New York 14627
USA

P MANDEL

Service de Chimie Physique II
Code Postal No 231
Campus Plaine ULB
Boulevard du Triomphe
1050 Bruxelles
Belgium

P MEYSTRE

Optical Sciences Center
University of Arizona
Tucson
Arizona 85721
USA

E R PIKE

Centre for Theoretical Studies
Royal Signals and Radar Establishment
St Andrews Road
Great Malvern
Worcestershire WR14 3PS
UK

M D REID

Physics Department
University of Waikato
Hamilton
New Zealand

S SARKAR

Centre for Theoretical Studies
Royal Signals and Radar Establishment
St Andrews Road
Great Malvern
Worcestershire WR14 3PS
UK

J S SATCHELL

Centre for Theoretical Studies
Royal Signals and Radar Establishment
St Andrews Road
Great Malvern
Worcestershire WR14 3PS
UK

and

Clarendon Laboratory
University of Oxford
Parks Road
Oxford OX1 3PU
UK

R E SLUSHER
AT & T Bell Laboratories
Research, Physics Division
600 Mountain Avenue
Murray Hill
NJ 07974
USA

D F WALLS
Physics Department
University of Waikato
Hamilton
New Zealand

B YURKE
AT & T Bell Laboratories
Research, Physics Division
600 Mountain Avenue
Murray Hill
NJ 07974
USA

SQUEEZED STATE GENERATION EXPERIMENTS IN AN OPTICAL CAVITY

R E SLUSHER AND B YURKE

1. INTRODUCTION

A broad class of states of the electromagnetic field where the quantum noise associated with the field is nonuniformly distributed in phase has recently attracted both theoretical and experimental interest as described in a recent review by Walls (1984). The electric field of a single mode of light at frequency $\omega/2\pi$ can be written as

$$E = E_0(X_1 \cos\omega t + X_2 \sin\omega t) \tag{1.1}$$

where E_0 is a constant containing the amplitude of the field and X_1 and X_2 are the field quadrature operators (in the Heisenberg picture) which obey the commutator relation

$$[X_1, X_2] = i/2 \tag{1.2}$$

Even if all other noise sources are eliminated, the field amplitudes are still "noisy" due to the zero-point fluctuations of the light field. In this case the product of the variances of the quadrature operators are related by the minimum uncertainty relationship

$$\Delta X_1 \cdot \Delta X_2 = 1/4 \tag{1.3}$$

Coherent light generated by a single mode laser is a member of this class where the variances of the two quadratures are *equal*. This state of the light field is represented by a "coherent state" which is a linear superposition of the energy eigenstates of the light. The coherent state has noise uncertainty in both amplitude and phase of its field. In a statistical sense, the photons in a coherent state are correlated by Poisson statistics.

The minimum uncertainty class defined by eq. 1.3 also includes states where the variances of the two quadrature operators are *not equal.* These states are called "squeezed coherent states." These squeezed states have noise which can be nonuniformly distributed over the period of the field oscillation so that the quantum noise in either the phase or amplitude of the wave can be reduced. In the statistical picture, the photons in a squeezed coherent state are not Poisson correlated as in the coherent state, i.e. they exhibit non-Possonian statistics. In particular, the vacuum state can be squeezed and the vacuum noise can be periodically reduced below the normal vacuum fluctuation level (and correspondly increased at another point in the field period). This vacuum noise redistribution is the subject of the present article. It results in experimental effects which can only be explained using the fully quantum model. For example, as observed in the present experiments, the "shot noise" in a homodyne detector can be reduced for a particular phase of the detected field. In a semiclassical model where only the electron states used for detection are quantized and the electromagnetic field is treated classically, this "shot noise" is due to random excitation of the electrons and cannot be reduced. The comparison of semiclassical and fully quantum descriptions of homodyne detection are treated by Shapiro (1985).

Squeezing the vacuum noise in practice is limited by excess noise originating in both the generation and detection processes. The experiments described here use a simple atomic Na beam pumped near resonance by a dye laser to generate the squeezed states and a balanced homodyne detector to detect them. Even for this relatively simple system a number of excess noise processes were found which nearly mask the modifications of the basic quantum noise processes. These excess noise mechanisms include spontaneous emission from the pumped Na atoms, "phase jitter" noise due primarily to the pump dye laser and optical inefficiencies due to the optical mode quality and the detector efficiency. These processes will be described in detail.

In most cases this excess noise can in principle be eliminated. Thus, there is hope that the noise squeezing can be a large (x10 or better) effect. Such large squeezing effects may have important applications, especially in the field of precision measurement using interferometers. The accuracy of interferometric phase measurement in interferometers at present is limited by

$$\Delta\phi \geqslant 1/\sqrt{N} \qquad [1.4]$$

where $\Delta\phi$ is the variance in the optical phase and N is the mean number of photons used in the measurement. For example, this limit is obtained in ring laser gyroscopes described by Davis and Ezekiel (1981). The quantum uncertainty relation requires

$$\Delta\phi \gtrsim 1/N \, . \qquad [1.5]$$

For typical experiments, N is large (e.g. $N \sim 10^{12}$) so that the difference between eqs. 1.4 and 1.5 is large and important. Squeezed light, used at one or both entrance ports of an interferometer, allow the limit in eq. 1.4 to be surpassed. This application of squeezed states is discussed in articles by Caves (1981), Bondourant and Shapiro (1984) and Yurke, McCall and Klauder (1986).

This discussion of squeezed state generation experiments will first describe the four-wave mixing in a Na atomic beam used to form the squeezed states. An optical cavity is used to coherently build-up the squeezing effect in a particular mode of the light field. A simple phenomenological model is used to describe this process. Next in Section 3 the homodyne detection of the squeezed states is discussed. The experimental apparatus is described in Section 4 with emphasis on the dominant excess noise sources. Finally, in Section 5 the initial experimental results are described which show a decrease in the vacuum fluctuation noise level for one field quadrature. Variation of the pump intensity is also discussed in a qualitative way. It is shown that the data indicate that the simple noise sources used initially in the phenomenological model are inadequate. More complex (and realistic) noise sources described by Klauder, McCall and Yurke (1986) are in better agreement with the data although phase jitter in the experiments makes direct comparison with the theory difficult at present.

2. *FOUR-WAVE MIXING IN AN OPTICAL CAVITY*

Many phase-dependent nonlinear phenomena have been suggested for squeezed state generation as described by Walls (1984). Yuen and Shapiro (1979) suggested four-wave mixing as a generation mechanism. The experiments described here use backward nondegenerate four-wave mixing. Backward refers to the two pump beams which are incident on the nonlinear medium (Na atoms) in opposing directions as shown in Fig. 1.

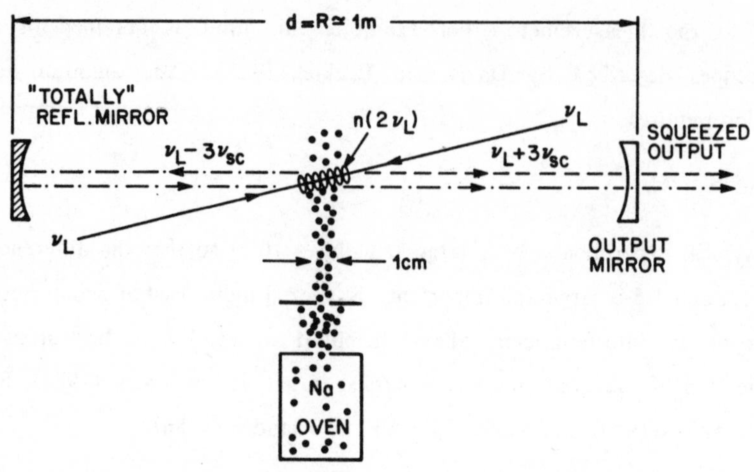

Fig. 1 A schematic diagram shows the cavity, Na beam and laser pump configuration used to generate squeezed light by four-wave mixing. The output and "totally" reflecting mirrors form a confocal cavity at right angles to the Na atomic beam. Photon pairs are generated by the standing-wave pump and are indicated here by the dashed line for the component at frequency $\nu_L + 3\nu_{sc}$ and the dash-dot line for the component at frequency $\nu_L - 3\nu_{sc}$. The index of refraction $n(2\nu_L)$ varies at a frequency equal to twice the pump frequency.

The standing wave generated in the Na atom beam by these opposing pump beams causes a time varying index of refraction at twice the pump frequency, ν_L. This leads to the generation of highly correlated photon pairs at frequencies symmetrically located about the pump frequency,

$$\nu_{1,2} = \nu_L \pm n\nu_{sc} \qquad [2.1]$$

where n is an integer and ν_{sc} is the cavity mode-spacing frequency. The photons in this pair also propagate in opposite directions at a small angle to the pump beam as shown in Fig. 1. Both energy and momentum are conserved for this four-wave process over the interaction region where pump and generated pair beams overlap. As suggested by Yurke (1984), this four-wave process can be built-up coherently by using a cavity to reflect the generated pairs back into the interaction region. The cavity is shown in Fig. 1 and has one "totally" reflecting mirror (reflectivity of 0.995) and one output mirror with a transmission of 2%. The squeezed state is formed by a superposition of the highly correlated photon pairs which collinearly exit the output mirror. It is important that the mirror reflectivities be asymmetric, otherwise uncorrelated vacuum noise will enter the cavity at the same rate the correlated squeezed light is generated thus limiting the degree of squeezing. The experimental cavity is a confocal cavity with a mirror separation near 107 cm. The squeezing cavity mode-spacing frequency ν_{sc} is 140.5 MHz. The squeezed noise effects are coherently built-up at the cavity resonant frequencies while the pump frequency is tuned to one of the cavity resonances. Further experimental details of the squeezing cavity are described in Section 4.

The frequencies ν_L and $\nu_{1,2}$ are fixed near the Na D_2 atomic resonance with a wavelength of 589 nm. The frequencies of interest are shown in Fig. 2.

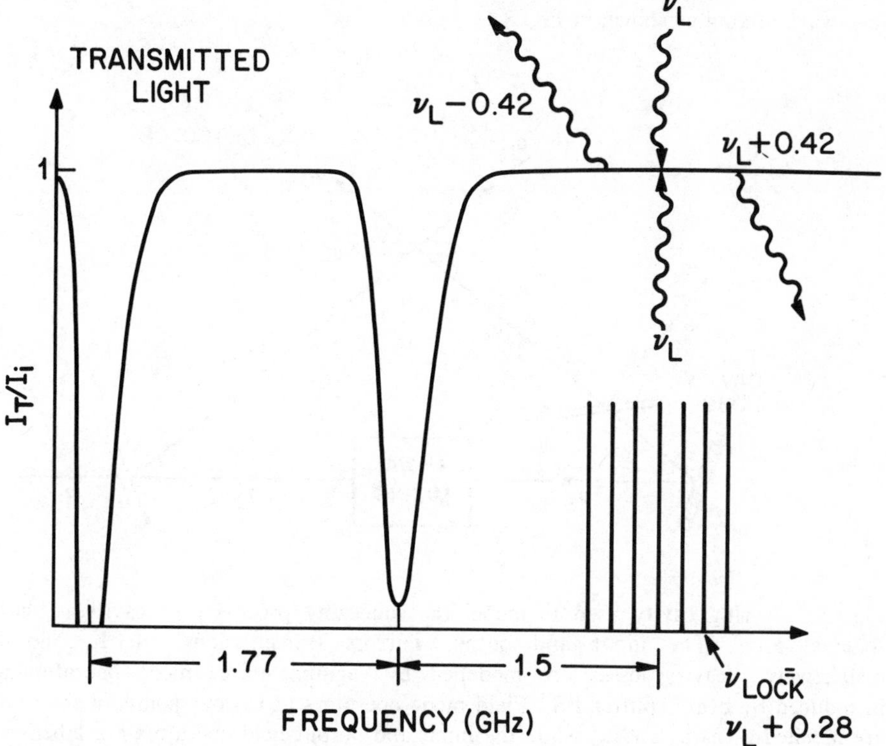

Fig. 2 Frequencies of the Na absorption, cavity modes, pump frequency and conjugate photon pairs associated with four-wave mixing are shown for the present experimental parameters. The transmitted to incident light intensity ratio I_T/I_i is shown for small intensities near the Na hyperfine pair and near the dye laser pump frequency ν_L located approximately 1.5 GHz to the high frequency side of the weaker Na resonance. Four-wave mixing generates conjugate photon pairs at frequencies near $\nu_L \pm 0.42$ coincident with a pair of cavity resonances shown as the vertical lines at the right of the figure.

The Doppler width of the Na atomic resonance for the beam geometry used in these experiments is 100 MHz and the natural linewidth is near 5 MHz. The pump frequency is fixed at a squeeze cavity resonant frequency located between 1 and 2 GHz to the high frequency side of the weaker hyperfine doublet component as shown in Fig. 2. The photon pairs generated in the cavity at $\nu_{1,2}$ are located symmetrically about this frequency. Of particular interest are the modes at $\nu_L \pm 3\nu_{sc}$ where excess noise from spontaneous emission is small enough to allow observation of squeezed state effects.

A simple phenomenological model can be used to describe the squeezing process for the configuration shown in Fig. 1. It is easier to analyze a ring cavity with forward four-wave mixing as shown in Fig. 3.

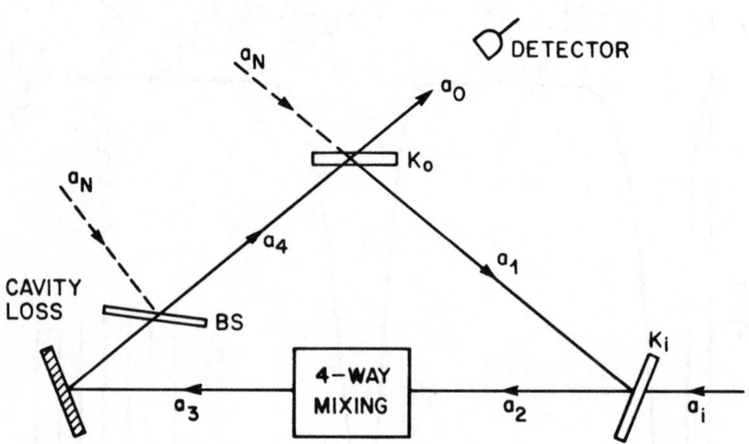

Fig. 3 A ring cavity used to model the squeezing process in a cavity is shown schematically. The input and output mirrors transmissions are K_i and K_o respectively. Cavity losses are modelled by vacuum noise (mode operator a_N) introduced by beam splitter BS. Field mode operators at various points in the cavity are labelled a_1, a_2, a_3 and a_4 while the input and output field operators are labelled a_i and a_o respectively.

It can be shown that the forward four-wave mixing ring cavity is equivalent to backward four-wave mixing in a linear confocal cavity if it is assumed that the interaction region in the ring cavity is twice as long as in the confocal cavity (i.e. there is a double pass through the interaction region for every round trip in the linear confocal cavity). A further simplifying assumption is to consider degenerate four-wave mixing instead of the wideband nondegenerate mixing analyzed by Yurke (1985a and 1985b). The results of the degenerate (all photons at the same frequency) calculations are identical to the wideband calculations at the center of the cavity resonance frequency profile. It is also assumed that the nondegenerate frequency splitting ($3\,\nu_{sc} \sim 420$ MHz for these experiments) is much less than the inverse transient time of light through the mixing medium (c/1 cm $\sim$ 10 GHz for the Na atomic beam). The wavevector mismatch due to the nondegenerate pair of photons must also correspond to a coherence length (6 cm in this case) which is long compared to the interaction region length. Finally, it is assumed that the cavity gain is below the value required for four-wave mixing oscillation.

These simplifying assumptions allow a simple set of equations to describe the four-wave mixing process in a cavity. This approach can be generalized to include nondegenerate, backward four-wave mixing as well as a full quantum treatment of the process as described by Klauder, McCall and Yurke (1986).

The input mirror is chosen to have a small transmission $K_i \simeq 0.003$ and the output mirror transmission is $K_o \simeq 0.02$. It is convenient to define

$$\cos\theta_i = K_i^{1/2} \qquad [2.2]$$

and

$$\cos\theta_o = K_o^{1/2}\,. \qquad [2.3]$$

The destruction and creation field mode operators for the resonant cavity mode are labeled respectively a_j and $a_j^\dagger$, where j = o, 1, 2, 3, 4 and N designating the fields at various points in the cavity as shown in Fig. 3. The subscript N denotes vacuum fluctuation noise which enters through cavity losses and at the input and output cavity mirrors. Cavity losses, e.g. the absorptive component of the Na atomic resonance, are modeled as a noise operator entering the cavity via the beam splitter shown in Fig. 3 which has a transmission T corresponding to a cavity loss $L = 1 - T \simeq 0.01$. The four-wave mixing process is modeled by a parameter γ which is related to the third order nonlinear susceptibility as described by Abrams and Lind (1978),

$$\gamma = 2K\ell \qquad [2.4]$$

where ℓ is the length of the interaction region and for a two level atomic model

$$K \cong \frac{2\alpha_o}{\delta} \frac{I/I_s}{(1 + 4I/I_s)^{3/2}} \qquad [2.5]$$

is the nonlinear coupling coefficient where

$$\alpha_o = \frac{p^2 \mathcal{N} T_2 k}{2 \epsilon_o \hbar} \qquad [2.6]$$

is the line-center small-signal-field attenuation coefficient

$$\delta = T_2 2\pi(\nu_L - \nu_a) \qquad [2.7]$$

is the normalized detuning parameter, T_2 is the transverse relaxation time of the Na atom (~ 32 ns), p is the effective dipole moment of the atomic resonance, ν_a is the atomic resonance frequency, k is the resonant wave vector, $\mathcal{N}$ is the atomic density, I is the pump intensity and I_s is the saturation intensity at detuning δ which is related to the line center saturation intensity I_{so} (10 mW/cm^2) by

$$I_s = (1+\delta^2)\, I_{so}\,. \qquad [2.8]$$

The factor of two in eq. 2.4 is included so that the ring calculations apply to the experimental linear confocal cavity. A high intensity, classical coherent light beam injected into the input mirror of the cavity will be used to obtain γ by measuring the phase dependent cavity power gain due to four-wave mixing.

The set of equations which describe four-wave mixing in the cavity for both the classical and quantum limits can now be written as

$$a_2 = a_1 \sin\theta_i + a_i \cos\theta_i \qquad [2.9]$$

$$a_3 = a_2 \cosh\gamma + a_2^\dagger \sinh\gamma \qquad [2.10]$$

$$a_4 = T^{1/2} a_3 + (1-T)^{1/2}\, a_N \qquad [2.11]$$

$$a_o = a_4 \cos\theta_o - a_N \sin\theta_o \qquad [2.12]$$

$$a_1 = a_4 \sin\theta_o + a_N \cos\theta_o \qquad [2.13]$$

Simultaneous solutions of these equations yields an expression for the output field operator a_o as a function of the input field operator a_i. The noise operators a_N entering at various points in the ring are assumed to be uncorrelated. In order to treat the nondegenerate problem a second set of equations similar to eqs. 2.9-2.13 is required with field operators corresponding to the two frequencies of the nondegenerate photon pair. These two sets of equations are coupled by an equation similar to eq. 2.10 except that the a and $a^\dagger$ operators correspond to the two separate frequencies of the photon pair. For the calculations analyzed here only the degenerate case will be considered.

The homodyne detection process described in Section 3 measures the mean square of the two quadrature components of the output field,

$$\langle X_1^2 \rangle = \langle i|(a_o + a_o^\dagger)^2|i\rangle \qquad [2.14]$$

and

$$\langle X_2^2 \rangle = -\langle i|(a_o - a_o^\dagger)^2|i\rangle \qquad [2.15]$$

Since $\langle X_{1,2}\rangle^2$ is zero for a vacuum state input field $|i\rangle = |0\rangle$, these mean square values are equal to the variances ΔX_1^2 and ΔX_2^2 for $|i\rangle = |0\rangle$. The variances can now be calculated using eqs. 2.9-2.13. With no four-wave mixing gain $(\gamma=0)$ and the input state as the vacuum, $|i\rangle = |0\rangle$

$$\Delta X_1^2 = \Delta X_2^2 = 1 \; ; \gamma = 0; \; |i\rangle = |0\rangle \, . \qquad [2.16]$$

In order to evaluate the parameter γ, the cavity power gain A is measured in the classical limit. In terms of the ratio of the mean square quadrature component X_1 with and without four-wave mixing,

$$A = \langle X_1^2(\gamma)\rangle / \langle X_1^2(0)\rangle \big|_{|i\rangle = |\beta\rangle} \qquad [2.17]$$

where $|\beta\rangle$ is a coherent state (experimentally a single mode laser beam at the cavity resonance frequency) and $\beta >> 1$, i.e. a high intensity, classical wave. In this classical limit eqs. 2.9-2.13 and 2.17 yield

$$A = D^2 \, [T^{1/2}\sin\theta_o\cos\theta_i \exp(\gamma) - T\sin^2\theta_o\sin\theta_i\cos\theta_i]^2$$

$$\times \left[\frac{T^{1/2}\sin\theta_o\cos\theta_i - T\sin^2\theta_o\sin\theta_i\cos\theta_i}{1 - 2T^{1/2}\sin\theta_o\sin\theta_i + T\sin^2\theta_o\sin^2\theta_i} \right]^2 \qquad [2.18]$$

where

$$D = \frac{1}{(1-2T^{1/2}\sin\theta_o\sin\theta_i\cosh\gamma + T\sin^2\theta_o\sin^2\theta_i)} \, . \qquad [2.19]$$

Using the γ value from a measurement of A, the vacuum noise squeezing can be calculated using eqs. 2.9-2.13 with $|i\rangle = |0\rangle$ to obtain,

$$\Delta X^2_{\substack{1\\2}} = D^2 \Big\{ [T^{1/2}\cos\theta_o\cos\theta_i\exp(\pm\gamma) - T\sin\theta_o\cos\theta_o\sin\theta_i\cos\theta_i]^2$$

$$+ [T^{1/2}\sin\theta_i\cosh\gamma(1 + \sin^2\theta_o) - \sin\theta_o(1 + T\sin^2\theta_i) \pm T^{1/2}\sin\theta_i\cos^2\theta_o\sinh\gamma]^2$$

$$+ [(1-T)^{1/2}\cos\theta_o - T^{1/2}(1-T)^{1/2}\sin\theta_o\cos\theta_o\sin\theta_i\exp(\mp\gamma)]^2 \Big\} \qquad [2.20]$$

The three terms in square brackets arise respectively from the vacuum noise entering the input mirror, the vacuum noise entering the output mirror and the vacuum noise entering the cavity from the beam splitter which models cavity losses.

Noise amplitudes for the two quadratures of the output field obtained from eq. 2.20 are shown in Fig. 4 as a function of classical gain.

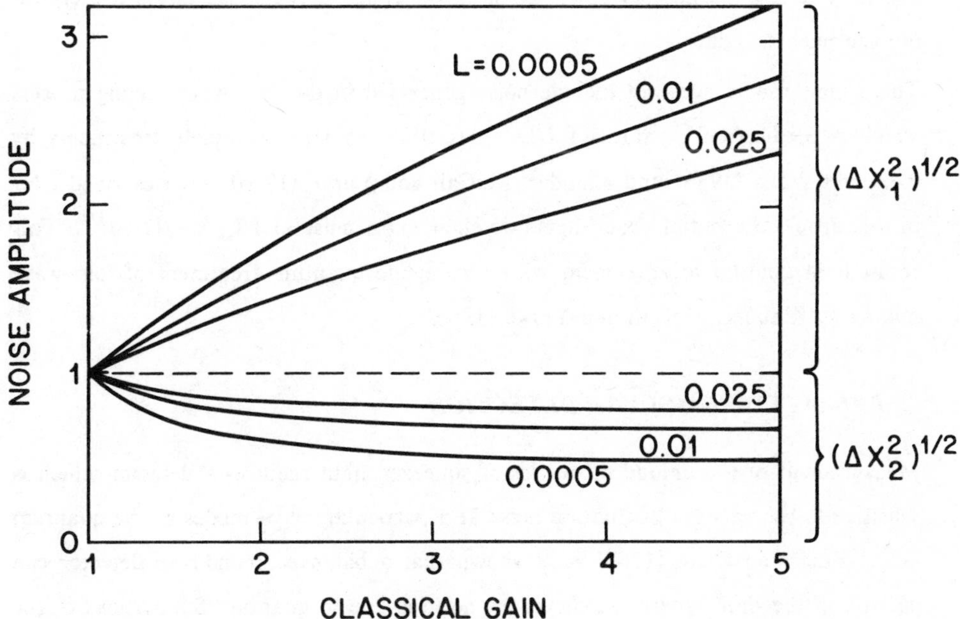

Fig. 4 Calculated values of noise amplitude squeezing are shown as a function of classical four-wave mixing power gain for values of the cavity loss L = 0.0005, 0.01 and 0.025. The output mirror transmission is the experimental value of $K_o = 0.02$ and the input mirror transmission is $K_0 = 0.003$. The squeezed noise rms levels are the group of curves labelled $(\Delta X_1^2)^{1/2}$ and the anti-squeezed values are labelled $(\Delta X_2^2)^{1/2}$. The unsqueezed vacuum fluctuation noise level is normalized to be one.

The dashed line represents the vacuum fluctuation level with no four-wave mixing gain, i.e. A = 1. For classical gains greater than one, the variance ΔX_2 drops below the vacuum noise level while ΔX_1 rises above to maintain the uncertainty relation requirement expressed in eq. 1.3. This is the squeezed state effect that is observed in the experiment and can only be obtained from a quantum model of the fields. As the classical gain is increased beyond a value of 3 in Fig. 4 the noise reduction due to squeezing approaches a limiting value near $(\Delta X_2^2)^{1/2} = 0.45$. For very small cavity losses (L >> 0.02) this limit is due to the finite transmission of the input mirror. As cavity losses increase to dominate the input mirror transmission the noise reduction is limited to smaller values as shown in Fig. 4 for L = 0.01 and L = 0.025. Typical values of L in the experiments are of the order of 0.01 due the absorptive component of the Na atomic resonance which has a Lorentzian frequency profile that decreases as $1/\delta^2$. Classical power gains in the range between 1.5 and 3 are typical of experiments to date.

This simple model does not include noise generated in the four-wave mixing process which is predicted for values of $I/I_s > 1/(10\delta)^{1/2}$ in more complete treatments by Reid and Walls (1985) and Klauder, McCall and Yurke (1986). As described later in this article the initial experiments do show extra noise at $I/I_s > 1/(10\delta)^{1/2}$. This result is in qualitative agreement with a complete quantum treatment of four-wave mixing by Klauder, McCall and Yurke (1986).

3. BALANCED HOMODYNE DETECTION

Measurement of the unique properties of squeezed light requires a detector which is sensitive to the vacuum fluctuation noise at a particular set of modes of the quantum field. Yuen and Chan (1983) have shown that a balanced homodyne detector can operate in the limit where the dominant noise is due to quantum fluctuations of the signal. A schematic diagram of this detector is shown in Fig. 5.

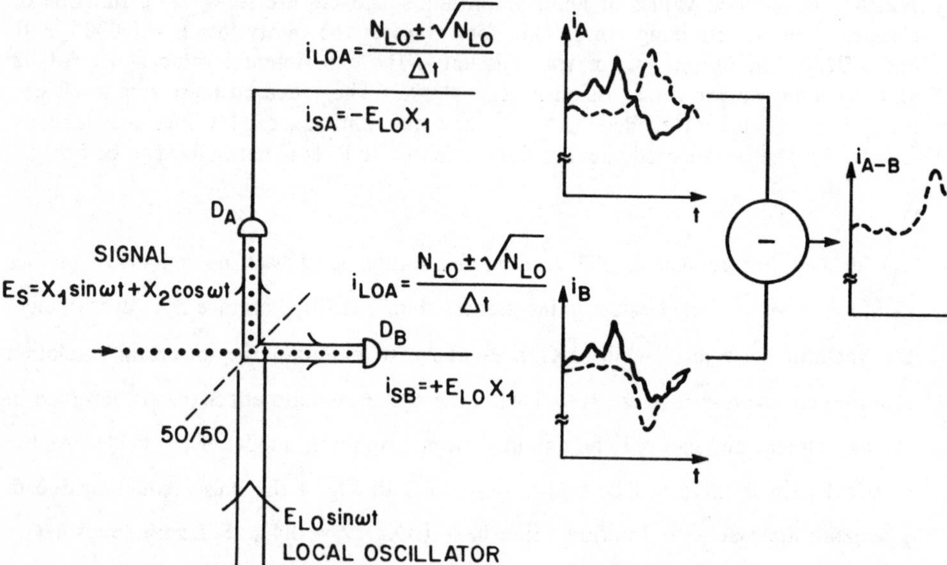

Fig. 5 A schematic diagram of the light beams and detector currents is shown for a balanced homodyne detector. Detectors D_A and D_B are aligned to respond to the split light beams from the signal and local oscillator. Photocurrents i_A and i_B from the two detectors are dominated by local oscillator noise (solid curves) and signal noise (dashed curves). The two currents are subtracted to yield a current i_{A-B} which is dominated by vacuum fluctuations in the signal channel.

Two photon detectors D_A and D_B are aligned to detect the local oscillator and signal beams split by a 50/50 beam splitter. The currents from these two detectors are dominated by the high intensity local oscillator beam but also contain a small component due to the mixing of the signal and local oscillator beams. These mixed signal currents are 180° out of phase in the two detectors. If the detector currents are subtracted, the signal currents add to form the output current. As shown below the noise in the local oscillator currents cancel in the subtraction process leaving only the signal currents. Even when there is no injected light signal input, there is still signal noise due entirely to the vacuum noise fluctuations. Thus the output current from this balanced homodyne detector is a measure of the squeezed noise phenomena.

This balanced homodyne detector is based on a beautiful measurement process which is often not familiar to those outside of this research field. It deserves a detailed treatment here which will also allow exploring the experimental parameters required to obtain quantum noise limited detector performance. Let the signal and local oscillator fields be represented by the following field mode operators. The signal field is

$$E_s = X_1 \sin \omega t + X_2 \cos \omega t \qquad [3.1]$$

where

$$X_1 = a + a^\dagger$$

and

$$X_2 = -i(a - a^\dagger)$$

and the local oscillator field is

$$E_{LO} = (b + b^\dagger) \sin \omega t - i(b - b^\dagger) \cos \omega t \, .$$

The photon state of the local oscillator is taken to be a coherent state $|\beta>$ where β^2 is the mean photon number. The signal state is taken to be the vacuum state $|0>$. At detectors A and B the fields are

$$E_A = \frac{1}{\sqrt{2}}(E_s - E_{LO}) \qquad [3.2]$$

$$= (a_A^\dagger + a_A)\sin\omega t + i(a_A^\dagger - a_A)\cos\omega t \qquad [3.3]$$

and

$$E_B = \frac{1}{\sqrt{2}}(E_s + E_{LO}) \qquad [3.4]$$

$$= (a_B^\dagger + a_B)\sin\omega t + i(a_B^\dagger - a_B)\cos\omega t \qquad [3.5]$$

Using normally ordered operators as described by Loudon (1984) the current operators for each detector can be obtained,

$$\hat{I}_A = E_A^{(-)} E_A^{(+)} = a_A^\dagger a_A \qquad [3.6]$$

where

$$E_A^{(-)} = a_A^\dagger \exp(i\omega t)\,;\; E_A^{(+)} = a_A \exp(-i\omega t) \qquad [3.7]$$

and similarly

$$\hat{I}_B = a_B^\dagger a_B \qquad [3.8]$$

The subtracted current is then

$$\hat{I}_A - \hat{I}_B = a_A^\dagger a_A - a_B^\dagger a_B\,. \qquad [3.9]$$

Of course the quantum nature of the detection process, i.e. the one-to-one correspondence between the destruction of a photon and the creation of an electron, means that the current statistics are an exact replica of the photon statistics (assuming no other noise sources and unit quantum efficiency). From the equations above the operators can be rewritten in terms of the input field operators,

$$\hat{I}_A = (a^\dagger a - a^\dagger b - b^\dagger a + b^\dagger b)/2\,, \qquad [3.10]$$

$$\hat{I}_B = (a^\dagger a + a^\dagger b + b^\dagger a + b^\dagger b)/2 \qquad [3.11]$$

and

$$\hat{I}_A - \hat{I}_B = -\,(a^\dagger b + b^\dagger a) \qquad [3.12]$$

It is now straightforward to calculate the mean values and variances of the detector currents,

$$I_A = \langle 0|\langle\beta|\hat{I}_A|\beta\rangle|0\rangle \ , \tag{3.13}$$

$$\simeq \frac{\beta^2 - \beta\langle X_1\rangle}{2} , \tag{3.14}$$

$$I_B = \langle\hat{I}_B\rangle \ , \tag{3.15}$$

$$\simeq \frac{\beta^2 + \beta\langle X_1\rangle}{2} , \tag{3.16}$$

and

$$I_A - I_B = \langle(\hat{I}_A - \hat{I}_B)\rangle \tag{3.17}$$

$$\simeq \beta\langle X_1\rangle \tag{3.18}$$

where the $\langle \quad \rangle$ is shorthand for the explicit input states written in eq. 3.13. Note that a single quadrature of the input signal (in this case X_1) is selected by the phase of the local oscillator field.

Only the leading terms in β have been retained in the results given above. Experimentally this requires that, during a time of the order of the inverse of the noise frequencies to be measured, there must be β^2 photons detected on the average so that $\beta >> 1$. For example, for the experiments described here, the local oscillator power is 4 mW and noise frequencies near 400 MHz are of interest. In a time period of the 400 MHz noise each detector responds to $\beta^2 \sim 10^7$ photons and $\beta \sim 4 \times 10^3$. It is clear that terms in the current of order β^2 and β dominate the remaining terms of order 1.

Next consider the variances in the mean current,

$$\Delta I_A^2 = \langle\hat{I}_A^2\rangle - \langle\hat{I}_A\rangle^2 \tag{3.19}$$

$$\simeq (\beta^2 + \beta^2\Delta X_1^2)/4 \tag{3.20}$$

$$\Delta I_B^2 = <\hat{I}_B^2> - <\hat{I}_B>^2 \qquad [3.21]$$

$$\simeq (\beta^2 + \beta^2 \Delta X_1^2)/4 \qquad [3.22]$$

and finally,

$$\Delta(I_A - I_B)^2 = <(\hat{I}_A - \hat{I}_B)^2> - <(\hat{I}_A - \hat{I}_B)>^2 \qquad [3.23]$$

$$\simeq \beta^2 \Delta X_1^2 \qquad [3.24]$$

Again only the dominant terms in β have been retained as described above. By taking the limit as $\Delta X_1^2 \rightarrow 0$, the noise variation in I_A relative to I_B and in the subtracted current $<I_A - I_B>$ can be deduced. Since the variance $\Delta(I_A - I_B)^2$ goes to zero with ΔX_1^2 and the variance in I_A and I_B remain at $\beta^2/4$, it is clear that the noise in the individual detector currents due to the local oscillator (the origin of the $\beta^2/4$ terms) must be exactly in phase. In the current subtraction process the noise from the local oscillator exactly cancels leaving *only* the vacuum fluctuation noise from the signal channel. This is the key result of the calculation. It is also clear that the amplitudes of the signal noise are exactly 180° out of phase in the two detectors and *add* in the current subtraction process so that the variances in the current I_A and $I_B(\sim \beta^2 \Delta X_1^2/4)$ add to given a variance of $\beta^2 \Delta X_1^2$ at the output. These currents are shown schematically in Fig. 5 for a time period of the order of the noise frequency.

Two important assumptions have been made in the discussion above which are difficult to obtain in practice. First it is assumed that a single field-mode is present in both signal and local oscillator and that the matching between the modes is perfect. In practice the cavity mode is well defined by the confocal cavity but it is difficult to align the cavity for complete degeneracy and pure fundamental mode. The curvature of the wavefronts and cross section of the fundamental cavity mode must also be matched to the local oscillator. In practice a mode matching efficiency of approximately 0.8 is typically obtained.

A second important nonideal feature of the detector is the quantum efficiency of the detectors. For the Si pin detectors used in the present experiment at a wavelength of

589 nm the quantum efficiencies are near 0.7. There are also small losses in the mirrors and lenses used between the cavity and the balanced detector.

Finally, the detector currents are amplified for recording and analysis on a spectrum analyzer. The amplifiers add approximately 10% to the total noise power.

These non-ideal detection features can be modeled as unsqueezable additions to the noise measured at the output of the detector. In terms of the rms voltage measured on the spectrum analyzer V_{rms},

$$V_{rms} = \{[\eta \Delta X_{1,2}^2 + (1-\eta)]\, V_s^2 + V_a^2\}^{1/2} \qquad [3.25]$$

where $\Delta X_{1,2}^2$ are predicted by eq. 2.20, V_s is the noise voltage due to the signal vacuum fluctuations, V_a is the noise voltage due to the amplifier and η is the effective detector efficiency

$$\eta = \eta_d\, \eta_h\, \eta_o \qquad [3.26]$$

where η_d is the detector efficiency, η_h is the homodyne mode matching efficiency and η_o is the efficiency of the optical components.

These effects degrade the amount of measured squeezing relative to the amount of squeezing present in the light field at the output of the cavity. A comparison of the ideal (eq. 2.20) and nonideal (eq. 3.25) squeezing is shown in Fig. 6.

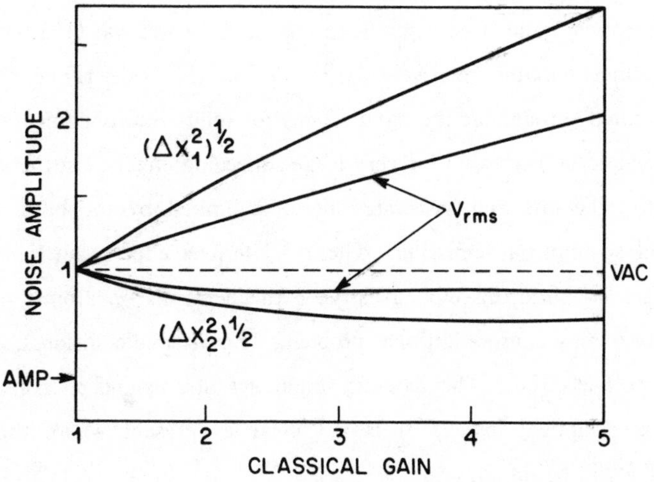

Fig. 6 Calculations of noise amplitude squeezing as a function of classical gain are shown which exhibit the effect of detection efficiency. For unit detection efficiency the curves labelled $(\Delta X_1^2)^{1/2}$ and $(\Delta X_2^2)^{1/2}$ are obtained as in Fig. 4 for a cavity loss of 0.01. A detection efficiency $\eta = 0.5$ reduces the observed squeezing to the curves labelled by the observed rms voltage V_{rms}.

Unfortunately, there are at least two remaining noise sources, described later, which will further degrade the degree of squeezing that can be measured.

4. EXPERIMENTAL APPARATUS

Observation of squeezed state effects requires careful attention to minimizing system losses and optimizing the optical phase stability. Losses in the system introduce unsqueezed vacuum fluctuations. Since the squeezed effects vary with optical phase, both the frequency and phase of all light beams must be controlled and stabilized to a small fraction of an optical cycle. For the experiments described here a Na atomic beam is used to reduce the Doppler spread of atomic absorption so that, at the observed off resonance frequencies, only the Lorentzian absorption wings fundamentally associated with the atomic resonance contribute to the system loss. Unfortunately, the Lorentzian absorption decreases slowly as $1/\delta^2$ while the nonlinear coupling coefficient K decreases as $1/\delta^3$. In principle, a glass fiber used by Levenson, et al. (1985b) or any solid allows a more favorable (smaller) ratio of loss to nonlinear coupling at a given frequency displacement from an absorption band. In a solid the fundamental absorption decreases exponentially as the frequency decreases below an absorption band (see e.g., Schreiber and Toyozawa (1982)) while the nonlinear coupling remains relatively large due to the wide range of oscillator strength distributed throughout the band. Thus for solids the fundamental limits for squeezing appear to be favorable for very large squeezing effects. In the atomic case considered here, the loss and associated noise is limited by the basic Lorentzian absorption and spontaneous emission. These deleterious effects can be reduced by tuning to larger δ. Unfortunately, at large δ the demands on atomic density and pump laser power may cause additional problems (e.g. atomic collisions). Of course, solids are not perfect either. They contain impurities, strains and phonons which can lead to increased optical losses. It is not clear at present where the optimum squeezing media will be found.

A schematic diagram of the apparatus used in the present experiments is shown in Fig. 7.

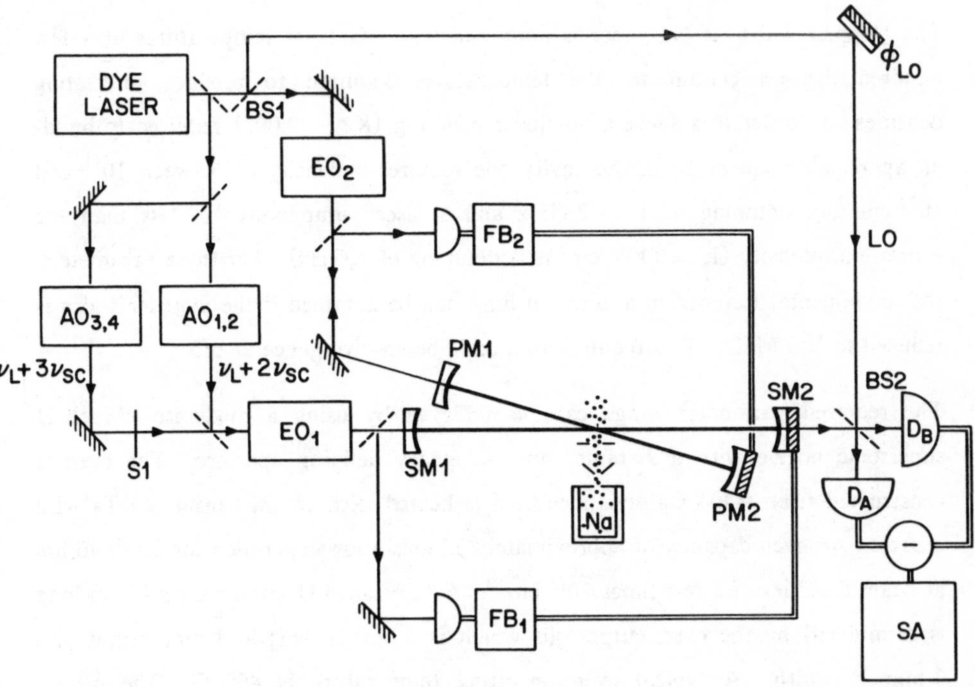

Fig. 7 A schematic diagram of the experimental apparatus for generating squeezed states using four-wave mixing. A ring dye laser pumps a Na atomic beam at the pump cavity (mirrors PM1 and PM2) resonance frequency. The pumped Na atoms generate four-wave mixing gain in the squeezing cavity (mirrors SM1 and SM2). A local oscillator beam LO is split off by beam splitter BS1. The squeezed cavity noise is detected with a balanced homodyne detector (splitter BS2, detectors D_A and D_B) and observed on a spectrum analyzer SA. The classical four-wave mixing gain is measured by opening shutter S1 and injecting a beam into the squeezing cavity which is frequency shifted from the pump by $3\nu_{sc}$. Both pump and squeezed cavities are locked to the dye laser frequency using FM sidebands generated by EO_1 and EO_2 which are detected to form feedbacks FB_1 and FB_2 to piezoelectric transducers on mirrors PM2 and SM2.

A ring dye laser is used as pump for the Na beam and a split off portion of it forms the local oscillator beam for homodyne detection. Another portion of this laser beam is split off and frequency shifted to form beams for locking both the squeezing and pump cavities. Finally a beam is split off and further frequency shifted as a probe of the squeezing cavity resonance properties. This beam is used for alignment, calibration and measurement of the four-wave mixing gain in the classical limit.

4.1 Na Atomic Beam

The Doppler width of Na vapor is approximately 1 GHz at temperatures of a few hundred degrees centigrade, the temperatures required to produce interesting densities. In order to achieve a nonlinear coupling ($K\ell \sim 0.005$) required to build-up appreciable squeezing in the cavity one requires a density of between 10^{11} and 10^{12} cm^{-3}, a detuning of 1 to 2 GHz and a laser pump power of less than the saturation intensity ($I_s \simeq 1\,kW/cm^2$ at a detuning of 1 GHz). For these parameters, the fundamental Lorentzian absorption limit can be obtained if the Doppler width is reduced to 100 MHz. This requires an atomic beam divergence of $\pm 5°$.

The required parameter range can be achieved by using a moderate (Mach 2) supersonic nozzle with a skimmer and secondary defining aperture. The oven is constructed from #304 stainless steel and is heated with ceramic insulated Ta wire heaters. An oven capacity of approximately 25 gms allows operation for 30 to 40 hrs at beam densities of a few times 10^{11} cm^{-3}. A 12.5 mm I.D. oven nozzle 10 cm long is terminated by the oven output slit which is 5 mm in length, 1 mm height and 3 mm in width. A typical oven operating temperature is 400°C. The copper skimmer assembly is positioned 2.5 cm down the beam line and is 1.5 mm in height and 4 mm in width. The skimmer tapers away from the oven for 2.5 cm and is maintained at a temperature of 110°C (10°C above the Na melting temperature) by a circulating ethylene glycol and water solution. This temperature must be uniformly controlled just above the melting point so that the Na collected by the skimmer flows freely into a collection chamber below the oven but does not lead to disruption of the Na beam due to Na vapor from the skimmer directed back into the beam. Finally, at 8 cm from the oven output a liquid nitrogen cooled copper aperture 3 mm in height and 1 cm in width confines the Na beam to the optical interaction region located 10.5 cm from the oven output slit. The entire oven assembly, interaction region and optical cavities are contained in a vacuum system which typically is operated at a pressure of 10^{-7} torr.

A typical small-intensity absorption spectrum is shown in Fig. 8.

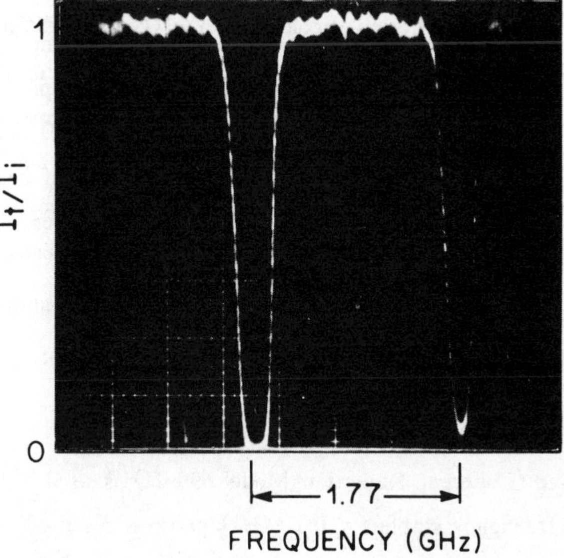

Fig. 8 A small-signal absorption spectrum due to the Na atomic beam is shown including the two hyperfine groups of the D_2 resonance. At frequency shifts from the absorption minima larger than 0.5 MHz the absorption is less than 1%.

The dye laser beam is incident at 90° to the Na beam. Individual components of the two Na D_2 hyperfine absorption lines are spaced at between 19 and 60 MHz. These are not resolved in Fig. 8. This is consistent with the measured effective Doppler width of approximately 100 MHz. The liquid nitrogen cooled aperture limits the beam to *only* the interaction region so that, at a 1.5 GHz frequency shift, the absorption over the entire 1m cavity length is less than 1% and can be limited only by the Lorentzian atomic absorption wings. This apparatus also insures that the cavity mirrors do not become coated with Na.

A computer fitting program is used to obtain the Na density and the effective line center absorption coefficient required for comparison with theoretical models. The Doppler broadened line center absorption for the weak line in Fig. 8 is 3 cm^{-1} which after deconvolution of the Doppler profile corresponds to an effective α_0 near

30 cm^{-1}. Both groups of hyperfine components (separated by 1.77 GHz in Fig. 8) contribute to the Lorentzian absorption and nonlinear coupling at the shifted laser pumping frequency. The effective saturation intensity is different for these two components due to both the different dipole moments (I_{so}) and different frequency shifts δ (see eq. 2.8). A weighted mean value for I_s and α_0 is taken for comparison with theoretical models. All of the hyperfine transitions are pumped coherently by the off resonance laser and contribute coherently to four-wave mixing. Since the Na levels are not prepumped to form an effective two level system as described by Grove, Wu and Ezekiel (1977), there is significant optical pumping, depending on δ and I. This is a difficulty for comparison with models but only decreases the effective atomic density for the four-wave mixing process.

4.2 Laser Pump

A cw ring dye laser (Coherent Radiation Model 699-21) is used to pump the Na beam. A tunable, frequency stable ($\lesssim$ 0.1 MHz) pump is required with a uniform intensity of approximately 1 kW/cm^2 over an area of 0.01 cm^2 (interaction region cross section). The Coherent Radiation laser does not meet these requirements but it is within an order of magnitude. For example, the output frequency width is 1 to 2 MHz due to dye stream fluctuations. The frequency jitters within this width at frequencies up to 50 kHz. This frequency jitter is nearly equal to the frequency width of the squeezing cavity mode and results in large ($\pm\pi$ radians) phase jitter of the pump relative to the squeezing cavity. As described in Section 5.1, this degrades the observed squeezing effects. At present a squeeze cavity stabilization loop is used to compensate for phase jitter at rates up to 10 kHz using a piezoelectric mounted output mirror as described in the next section. Further frequency stabilization of the pump laser is also planned using an acousto-optic and electro-optic modulator as described by Hall and Hansch (1984).

Power is also a limitation with the present pump laser. In single-mode, stabilized operation its output power is typically between 800 and 1000 mW. After isolating the laser from back reflection and splitting out a portion for the local oscillator, probe and locking beams, the useful power at the interaction region is typically 400 mW. This pump beam is focussed with a 50 cm focal length lens and back reflected with a spherical mirror to obtain an intensity at the interaction region between 20 and 100 W/cm^2.

Increased pump intensity can be obtained by using a confocal pump build-up cavity with a finesse of 30 as shown in Fig. 7. This cavity is locked to the pump laser frequency as described in the next section. At the higher pump intensities, the near resonant Na nonlinearities can cause instabilities in the pump build-up cavity which lead to phase jitter and decreased squeezing effects.

Fortunately, the optimum squeezing effects are found for small values of I near $I/I_s \simeq 0.02$. These pump intensities can be obtained with the present system, even without the build-up cavity. Obtaining these intensities does require focusing the pump into an area approximately equal to the area of the squeezing cavity mode ($\simeq 3 \times 10^{-3}$ cm^2). In this case the intensity over the interaction region is nonuniform. As a result both I/I_s and the effective sodium density (due to optical pumping and nonlinear effects) are nonuniform in the interaction region. Again, this limits comparison with theoretical models to within factor of the order of two in both I/I_s and effective sodium density.

4.3 Squeezing Cavity

Simulations of the squeezing process as described in Section 2 show that the optimum output mirror transmission is nearly equal to the cavity loss. Cavity losses in the present experiment are dominated by the input mirror transmission (~ 0.005 or $R_i = 0.995$) and Na absorption ($\simeq 0.005$). The output mirror transmission was chosen to be 0.02 ($R_o \simeq 0.98$) corresponding to a finesse of $\frac{\pi\sqrt{R_i R_o}}{(i - R_i R_o)} = 124$ or a frequency resolution of 1.1 MHz. The cavity mirrors are spaced at 1.07m (mirror radius) and are mounted in vacuum on mirror mounts that are separated by four invar rods. A set of bellows vibrationally isolates the mirror mounts from the main vacuum system. The squeezing cavity is at an angle of 0.86° with respect to the pump laser direction. Both pump light and squeeze cavity are at an angle of 90° to the Na beam which intersects the midpoint of the squeeze cavity.

Cavity drift and vibrations as well as pump laser frequency jitter are reduced by servo-controlling the cavity length using a piezoelectric mount for the output mirror. The error signal for the feedback signal is obtained by the method discussed by

Drever, et al. (1983) using frequency modulation of a probe beam. This avoids fast dithering of one of the cavity mirrors which can decrease the build-up of squeezed noise in the cavity. A useful criteria is that the cavity phase should change by less than a single pass nonlinear phase shift ($K\ell \sim 0.005$) in a cavity build-up time ($2d/c \times F \sim 0.85\ \mu s$). A small intensity beam ($P \sim 0.15\ \mu W$) is frequency shifted by two acousto-optic modulators ($AO_{1,2}$ in Fig. 7, where AO_1 is at a frequency of 80 MHz and AO_2 is at 201 MHz) to a frequency near 281 MHz above the pump frequency. The polarization of this beam is rotated 90° so that no four-wave mixing occurs on the locking beam and it is not detected in the homodyne detection system. The local oscillator and lock beam are at orthogonal polarizations resulting in no mixing between them. This feature was not included in the initial experiments. The locking beam passes through an electro-optic phase modulator operating at 11 MHz which produces side bands at ± 11 MHz. The beam back reflected from the squeezing cavity is detected using a photodiode and amplifier with wide-band response near 11 MHz. The resulting photocurrent is due to mixing between the totally reflected 11 MHz sidebands (well outside the cavity resonant width) and the central frequency which is near the cavity resonance and shifted either plus or minus in phase due to the resonance properties of the cavity. This mixing produces an error signal which is fed to the piezoelectric mirror mount after passing through an integrating amplifier. The feedback system responds to frequencies as high as 10 kHz. At higher frequencies the mechanical response of the piezoelectric crystal assembly and the current limits of the high voltage driver limit the effective response. Dye laser frequency jitter (± 1 MHz) at higher frequencies, between 10 and 50 kHz still contributes in a major way to phase jitter of light transmitted through the servo-controlled cavity.

The phase jitter and the classical four-wave mixing gain in the cavity are monitored by a probe beam which can be frequency tuned in the region near 420 MHz (i.e. $+3\ \nu_{sc}$ above ν_L, the pump frequency). This beam is generated by passing the beam shifted 80 MHz (using AO_1) through two additional acousto-optic modulators ($AO_{3,4}$ in Fig. 7) which operate at frequencies of 125 MHz (double pass) and 90 MHz. This beam is matched into the squeezing cavity collinearly with the lock beam using a beam splitter and a 50 cm focal length lens.

Transmission of the 420 MHz probe beam through the cavity (with the cavity locked at 281 MHz) results in a transmitted intensity which varies appreciably (maximum intensity down to 50% of maximum corresponding to phase jitter of $\pm\pi$ radians) in the frequency range between 10 and 50 kHz due primarily to the dye laser frequency jitter (± 1 MHz). The transmitted beam pattern is monitored at a distance of several meters from the cavity to optimize its Gaussian profile. Its alignment, cross-section and wavefront curvature are matched to the local oscillator beam (using optics in the local oscillator beam path). The long term stability (times between 0.1 ms and 10 minutes) of the locked squeezing cavity is quite good. The output beam from this cavity exits the vacuum system through an anti-reflection coated window. It is then incident on the 50/50 beam splitter used in the balanced homodyne detector.

4.4 Balanced Homodyne Detector

Light from the squeezing cavity output, aligned and phase matched to the local oscillator beam is split by the 50/50 beam splitter BS2 and directed with mirrors and 30 cm focal length lenses to detectors D_A and D_B as shown in Fig. 5. Both detectors are standard RCA Type 30971E Si pin photodiodes. The local oscillator power of 4 mW results in a dc photocurrent corresponding to a quantum efficiency of 0.7 at a wavelength of 589 nm. Noise due to the local oscillator photocurrent is amplified by 60dB using Trontech Model W500EF amplifiers with high pass filters at 300 MHz. The high pass filters eliminate any gain variation due to the locking signal at 281 MHz. The measured rms noise level at each detector with 4 mW of local oscillator power is equal to the "shot noise" level (or vacuum fluctuation level in the quantum picture) expected for the measured dc photocurrent and amplifier gain to within 2% after taking into account 10% of the noise power due to the amplifier.

The amplified photocurrents are subtracted using a Minicircuits Model ZFSCJ-2-1, 180°, two-way power combiner. Noise of the subtracted currents is analyzed using a Hewlett Packard Model 8568 spectrum analyzer. In addition to the uniformly distributed "shot noise" there are numerous sharp peaks in the photocurrent frequency spectrum near multiples of the longitudinal cavity-mode spacing of the Ar-ion laser used as a pump for the dye laser. These are primarily due to low level radial modes oscillating in the Ar laser which cause gain modulation of the dye laser. This results in dye laser (local oscillator) fluctuations. These peaks in the photocurrent spectrum are narrow in frequency (~ 1 MHz) and can be tuned to avoid their overlap in frequency with the squeezing cavity-mode frequencies. They can also be used to monitor the common mode subtraction process between the two detector photocurrents. Typically, this common mode subtraction is better than 20dB and it can easily be optimized to 40dB over limited ranges of the frequency spectrum.

In summary, the noise spectra displayed on the spectrum analyzer is predominately due to vacuum fluctuations in the squeezing cavity mode with a small (10%) contribution from the amplifiers. The phase of the cavity-mode noise depends on the local oscillator phase which can be scanned $\pm 1.1\ \pi$ radians using a piezoelectrically controlled mirror. This balanced detector can be stably aligned to monitor *only* the vacuum mode to be squeezed in the cavity.

5. *EXPERIMENTAL RESULTS*

This section will follow the sequence of a "typical" day in the lab. An experimental run begins with coalignment of the locking beam at $\nu_L + 2\nu_{sc}$ and probe beam at $\nu_L + 3\nu_{sc}$ in the squeezing cavity. With the cavity servo-loop optimized the beam pattern transmitted at the locked cavity resonance is checked for mode quality. This process is iterated until optimum cavity mode alignment is achieved. Next the local oscillator and squeezing cavity output mode are coaligned. The balanced homodyne detector is checked for "shot noise" level and the homodyne efficiency is measured and optimized. This efficiency depends on the alignment and wavefront matching of the local oscillator and squeezing cavity mode. A sweep of the probe beam frequency in the region of 421 MHz (by varying the 125 MHz double pass acousto-optic modulator) allows a measure of ν_{sc} and the cavity-mode frequency resolution (~ 1 MHz).

After the initial alignment and calibration the Na atomic beam is turned on. The Na introduced into the cavity changes the index of refraction n in the cavity and the associated resonant frequency

$$\nu_{sc} = c/2nd \qquad [5.1]$$

For frequencies above the Na resonance $n < 1$ and ν_{sc} increases. The index also varies with δ due to optical pumping effects proportional to the absorption at ν_L,

$$n(\delta) = 1 + \frac{K}{\nu}\left(1 - \frac{K'}{\delta^2}\right) \qquad [5.2]$$

where K and K′ are constants and ν is the optical frequency. The factor K'/δ^2 can be large, typically in the range between 0.1 and 0.8. At the Na density required for squeezing the net effect of the index variation is a downshift ν_{sc} as δ is decreased. It is also clear that ν_{sc} is not constant as a function of cavity fundamental-mode number. The shifts in ν_{sc} at the cavity resonances up and down shift by 3 mode numbers are of the order of cavity resonance width. In order to maintain phase coherence of the conjugate photon pairs generated by four-wave mixing it is expected that it is necessary to slightly detune the pump frequency from the cavity resonance. Since n changes the effective cavity length, the confocal degeneracy obtained in the initial alignment is lost to some degree. These effects have not been studied in detail.

5.1 CLASSICAL GAIN

Analysis of noise squeezing near the vacuum level relies on measured values of the four-wave mixing gain A defined in eq. 2.17. This gain includes the effects of nonuniform pumping, Na density, optical pumping, cavity build-up and phase jitter. It does not include vacuum level noise sources, e.g. spontaneous emission and detector efficiency degradation of noise squeezing. A measurement of the classical gain is shown in Fig. 9.

With the dye laser pump off the horizontal trace in Fig. 9 is the homodyne detector rms voltage near 421 MHz due to mixing of the probe beam (tuned to the cavity resonance) and the local oscillator beam. This voltage is independent of local oscillator phase ϕ_{LO} as expected. When the pump beam is turned on and the alignment of the forward and backward pump beams are optimized the modulated trace is obtained. The period of the modulation is π radians of ϕ_{LO} as expected for four-wave mixing. The probe signal has been amplified on one quadrature and deamplified on the other (phase shift of $\pi/2$). This phase dependence can also be obtained by fixing ϕ_{LO} and varying the pump standing-wave phase relative to the squeezing cavity. The value of gain in Fig. 9 is

$$A = (V_{max}/V_o)^2 = 3.5 \quad [5.3]$$

where V_{max} is the maximum voltage and V_o is the voltage with the pump off.

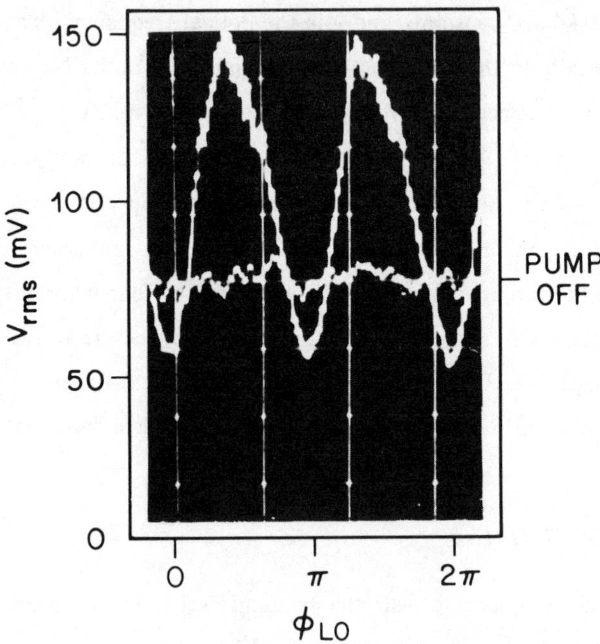

Fig. 9 A measurement of the classical four-wave mixing power gain is shown as a function of the phase of the local oscillator beam ϕ_{LO}. With the pump off the cavity output (due to an injected laser beam at frequency $\nu_L + 3\nu_{sc} = 420.6$ MHz) is constant as a function of ϕ_{LO}. With the laser pump on the light in the squeezing cavity is alternately amplified and deamplified with a period of π in ϕ_{LO}. The power gain is the ratio of the square of the maximum or minimum voltage to the voltage with the pump off. The averaging time for these data is 1 ms.

Phase jitter of the squeezing cavity relative to the pump frequency can be evaluated from the data in Fig. 9. Note that the deamplified dips correspond to a deamplification power gain

$$A_d = (V_o/V_{min})^2 = 1.9 \quad [5.4]$$

where V_{min} is the minimum deamplified voltage. This is considerably lower than

A = 3.5 obtained from the amplified peaks. It is known from the transmission characteristics of the squeezing cavity that the phase of the pump is varying rapidly (rates ~ 10 to 50 kHz) by as much as $\pm\pi/4$ radians with respect to the squeezing cavity. For the averaging times near 1 ms in Fig. 9 (and the longer averaging times near 100 ms used in observing quantum noise squeezing) the average over this phase jitter reduces the deamplification more than the amplification. This can be seen by assuming small losses and rewriting eq. 2.20 (or its classical limit) in the form

$$A \cong B + f\sqrt{(B^2-1)} \qquad [5.5]$$

and a corresponding deamplification gain

$$A_d^{-1} \simeq B - f\sqrt{(B^2-1)} \qquad [5.6]$$

where B depends on the single-pass gain γ and f is a phase jitter factor which is 1 for zero phase jitter. For f = 1, A_d^{-1} is the reciprocal of A in the limit that γ is much larger than the single pass cavity loss. At large values of B (large gain) the gain reduction effect of f on A_d^{-1} is much larger than on A. The data in Fig. 9 are consistent with f = 0.85. This value of f is typical of experimental values between 0.8 and 0.9. A value of f = 0.85 is calculated if equally weighted variation in phase is assumed over a $\pm 27°$ range of phase jitter. Since the quantum noise *deamplification* is of primary interest, the large effects of phase jitter on the deamplified quadrature are particularly important.

5.2 SPONTANEOUS EMISSION

After alignment, calibration and measurement of the classical four-wave mixing gain, the next step is to isolate the squeezing cavity inputs and study the cavity mode noise near the vacuum fluctuation level. Even with a single forward directed pump beam (no four-wave mixing), excess noise of the order of the vacuum fluctuation noise is observed at cavity resonances near the pump frequency. This excess noise has the features expected for spontaneous emission from the pumped Na atoms as describing by Slusher, Hollberg, Yurke, Mertz and Valley (1985a). The magnitude of the spontaneous emission falls rapidly with the frequency displacement from the pump laser as shown in Fig. 10. Residual Doppler broadening due to the 10° atomic beam width is responsible for the broad frequency spread out to 300 MHz. However, at

$3\nu_{sc} \simeq 421$ MHz the Gaussian Doppler profile has decreased so that the spontaneous emission noise power is less than 1/5 the vacuum fluctuation noise power. This is small enough to consider looking for noise squeezing below the vacuum fluctuation level.

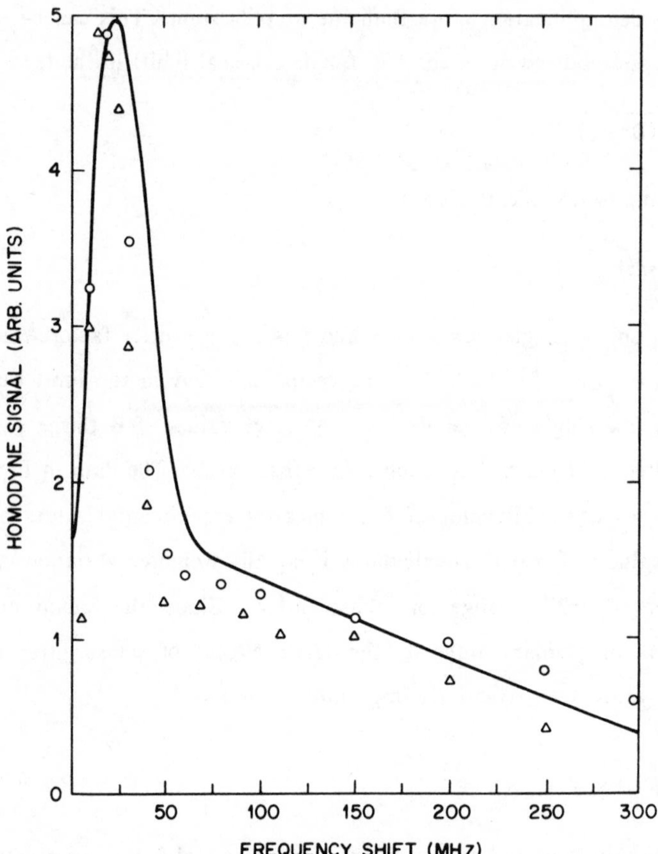

Fig. 10 Spontaneous emission as a function of frequency shift from the pump frequency for a cavity four-wave mixing configuration. The circles and triangles are two sets of experimental data obtained by deconvolving data obtained by homodyne detection. The solid line is a calculation of the spectrum assuming an effusive Na beam with an angular spread of $\pm 5°$ and a squeezing-cavity to pump-beam crossing angle of $0.7°$.

Both spontaneous emission and vacuum noise are squeezed. This has been demonstrated at ν_{sc} (see Slusher, et al. (1985b)) and at $2\nu_{sc}$ (see Slusher, et al. (1985c)). The noise at these cavity resonances has been deamplified below the total noise level (spontaneous emission plus vacuum fluctuations) but not below the vacuum fluctuation level. Clear evidence of the squeezed state effects, where the basic quantum noise is squeezed, requires a demonstration of noise reduction below the vacuum fluctuation level, i.e. the detector noise level with the squeezing cavity or pump beam completely blocked.

5.3 VACUUM NOISE SQUEEZING

Conjugate photon pairs generated by four-wave mixing with frequencies near $\nu_L \pm 3\nu_{sc}$ are detected as noise components near $3\nu_{sc} \simeq 421$ MHz. A frequency scan of the detected noise in this frequency region is shown in Fig. 11.

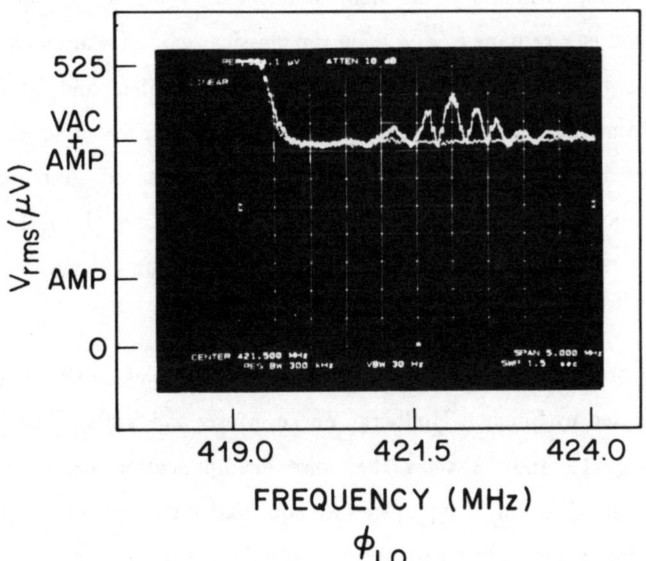

Fig. 11 Homodyne detector noise spectra due to a pair of cavity resonances spaced at $\pm\ 3\nu_{sc}$ from the pump frequency. Pump light is blocked for the nearly flat trace (dim trace labeled VAC + AMP) and the pump light is on for the brighter modulated trace. The amplifier noise level (AMP) is obtained by blocking the LO beam. Both spectrum analyzer frequency and LO phase are linearly swept in these pictures. The feature at the extreme left is an extraneous signal due to the LO and some remaining mismatch between the two detectors.

For this data the frequency of the spectrum analyzer and the phase of the local oscillator are swept slowly in time, 1.5s for the complete scan in Fig. 11. The cavity is locked to the pump laser using the 281 MHz shifted beam. The pump laser frequency is near 1.5 GHz above the frequency of the weak hyperfine component corresponding to an effective δ value near 300. The feature at the left is due to a radial mode of the Ar laser as described in Section 4. An incomplete subtraction of the common mode noise is indicated by the magnitude of this feature, however, the common mode suppression is still near 20db. The flat trace in the region of the cavity resonance (centered at 422 MHz for these data) is obtained on a separate scan with the pump light or squeezing cavity blocked. In this case the noise is predominately the vacuum fluctuation noise plus a 10% (in power) component due to the amplifier noise. The amplifier noise level is indicated by the AMP label in Fig. 11 which is obtained by blocking both the local oscillator and squeezing cavity beams. With the squeezing cavity open to the balanced detector and the forward and backward pump beams on the modulated trace in Fig. 11 is obtained. The modulation period is π radians in ϕ_{LO} as in the classical gain measurement. Vacuum fluctuation noise in the cavity mode is clearly being amplified and, at frequencies $\pm$ 0.75 MHz from the center frequency there is some evidence of squeezing below the vacuum level. Note that the four-wave mixing phased amplification and deamplification occurs over a range of frequencies nearly equal to the measured cavity resonance width as predicted in the wideband squeezing calculations by Yurke (1985a and 1985b).

The noise does drop below the vacuum fluctuation noise level as shown in Figs. 12 and 13 and reported by Slusher, Hollberg, Yurke, Mertz and Valley (1985c).
The data in Figs. 12 and 13 show the noise deamplification below the vacuum fluctuation level that is a major signature of a squeezed state. These results can only be explained using a quantum treatment of the light fields. In Figs. 12 and 13 the spectrum analyzer frequency is fixed 0.3 MHz below the resonance center frequency and ϕ_{LO} is varied. Variations in the local oscillator beam path relative to the squeezing cavity (due primarily to vibrations on the optic table) are responsible for some nonuniformity in the sweep of ϕ_{LO}. This phase is swept linearly back and forth by $\pm$ 1.1 π radians.

The vacuum fluctuation noise level in Figs. 12 and 13 is quite stable. The level in Fig. 13 was taken with a 20s sweep time. The mean level during this time is constant within the rms variation around the mean. The fluctuations in the mean

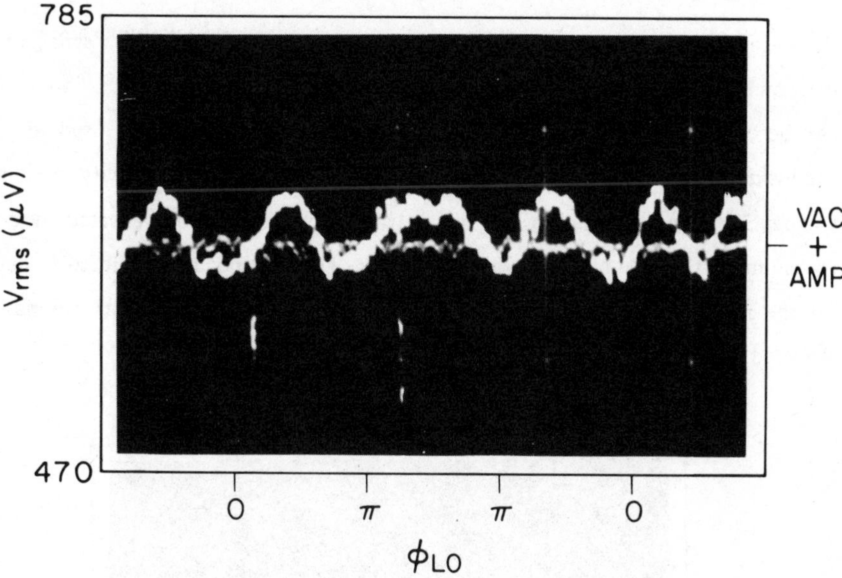

Fig. 12 Homodyne rms noise voltage at fixed rf frequency is shown as a function of the local oscillator phase with the squeeze cavity blocked (dim horizontal trace) and with the squeeze cavity open to the balanced detector and the pump light on (bright modulated trace). The spectrum analyzer frequency is fixed at 420.6 MHz, 0.3 MHz below the cavity resonance center frequency. I/I_s is near 0.01.

levels in Figs. 12 and 13 are consistent with the square root of the ratio of rf bandwidth (300 kHz) and the "video" bandwidth (30 Hz), i.e. the reciprocal of the averaging time. The squeezed noise sweep in Fig. 13 is taken with a 1s sweep time. There is no noise change or dependence on ϕ_{LO} at frequencies greater than 3 MHz from the resonance center when the cavity is open and blocked while the pump is on or off. This eliminates the possibility of effects caused by photodiode efficiency or amplifier gain modifications due to the presence of light from the cavity or pump.

A measure of the spontaneous emission level for the data in Figs.12 and 13 is obtained by misaligning the backward pump be to eliminate four-wave mixing effects. From this measurement it can be inferred that the average level of spontaneous emission with both pump beams on (nearly twice as much spontaneous emission as measured for one pump beam) is approximately 1/12 photon per mode as compared with the vacuum fluctuation level of 1/2 photon per mode. Spontaneous emission increases the level of noise that is amplified and deamplified and thus reduces the drop in squeezed noise below the vacuum fluctuation level.

Both classical and vacuum level four-wave mixing are found to be sensitive to the placement of the pump frequency relative to the cavity resonance. This placement can be controlled by tuning the 200 MHz acousto-optic modulator used to generate the locking frequency near ν_L + 281 MHz. The disparity of the cavity mode spacing frequencies above and below the pump frequency may be compensated by detuning the pump to the side of the cavity resonance. This has not been studied in detail but for the results shown here the pump frequency relative to the cavity resonance was adjusted for optimum squeezing.

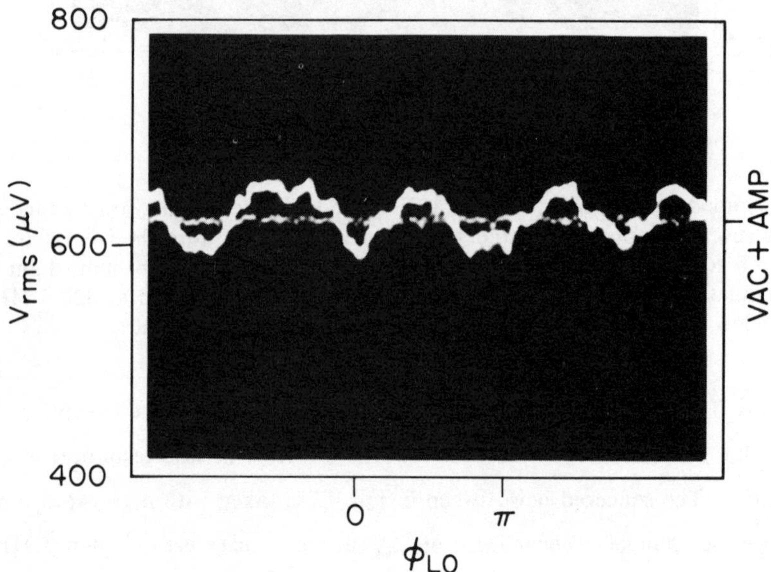

Fig. 13 Homodyne noise voltage at fixed rf frequency as a function of ϕ_{LO} with squeeze cavity blocked (dim horizontal trace) and with squeeze cavity open and pump light on (bright modulated trace). The frequency is 0.3 MHz below the cavity resonance center frequency and I/I_s is near 0.02. The cavity blocked signal is swept at a rate 20 times slower than the cavity open signal to show the stability of the vacuum fluctuation level. The effective video averaging time is increased in the 20s sweep relative to the 1s sweep (modulated trace).

Squeezing effects are observed over a range of values for δ, both above and below the Na lines. For $\delta < 100$ the squeezing effects decrease, probably due to increased atomic absorption. At larger shifts, $\delta > 400$ the atomic beam density and pump power must be increased in order to maintain large squeezing effects. In this regime the spontaneous emission may be a problem for observing squeezing below the vacuum level since the nonlinear mixing decreases as δ^{-3} and the absorption (directly related to spontaneous emission) only as δ^{-2}.

The Na density has also been varied over an order of magnitude. At low densities, the squeezing effects decrease uniformly with density. At higher densities a number of stimulated processes, including four-wave mixing oscillation, are observed for smaller values of δ.

Simultaneous comparison of noise amplification and deamplification at adjacent cavity resonances shows that the phase dependence of the process changes by π from one cavity resonance to the next. This antiphasing at adjacent cavity resonances is predicted in the analysis of Yurke (1985a and 1985b).

6. *DISCUSSION OF THE RESULTS*

The results can now be compared directly with theoretical predictions. As an example, the results in Fig. 12 are analyzed here. Using a classical gain of 1.5 in the expression from eq. 2.18, the values $\alpha_o = 24\ \text{cm}^{-1}$, $2Kl = 0.003$ and an absorption at the laser frequency of $2.6 \times 10^{-4}\ \text{cm}^{-1}$ are obtained at a I/I_s value of 0.01. There is some uncertainty (± 30%) in the classical gain since the data in Fig. 12 were taken off the cavity resonance center frequency where the gain is reduced from the center frequency value. The magnitude of α_o is consistent with the measured small-signal absorption, the Doppler width and optical pumping of the order of a factor of two.

Spontaneous emission can be included in the squeezing predictions by multiplying the variance $\Delta X_{1,2}^2$ by a factor f_{SE} so that eq. 3.25 becomes

$$V_{rms} = \left\{ \left[\eta\, f_{SE} \Delta X_{1,2}^2 + (1-\eta) \right] V_s^2 + V_a^2 \right\}^{1/2} . \qquad [5.7]$$

For the data in Fig. 12 f_{SE} is estimated to be 1.12. Spontaneous emission will vary with I/I_s, the amount of optical pumping and the detuning parameter δ. A more complete study of spontaneous emission under various conditions is now in progress.

Noise squeezing near the vacuum fluctuation level can now be calculated using eqs. 2.20 and 5.7. The predicted maximum and minimum V_{rms} are $V_{min} = 589\ \mu\text{V}$

and V_{max} = 751 μV relative to the measured vacuum fluctuation (plus amplifier noise) level of 627 μV. The observed values are V_{min} = 607 μV and V_{max} = 660 μV. At least, the comparison between observed and predicted values leaves one in the right ball park. However, the amount of observed squeezing is not as great as expected. The observed squeezing corresponds to a noise reduction below the "shot noise" level of 7% in power (after accounting for the amplifier noise).

A large discrepancy with the model predictions is observed at larger values of I/I_s. As shown in Fig. 14, the model in Section 2 predicts a continued increase in the degree of squeezing at higher values of I/I_s, at least to $I/I_s$of the order of 0.2 where saturation of the absorption becomes important. The experimental results at larger I/I_s show large increases in V_{max}, but V_{min} does not decrease but instead increases to levels slightly above the vacuum fluctuation level. This is clearly in disagreement with the predictions in Fig. 14.

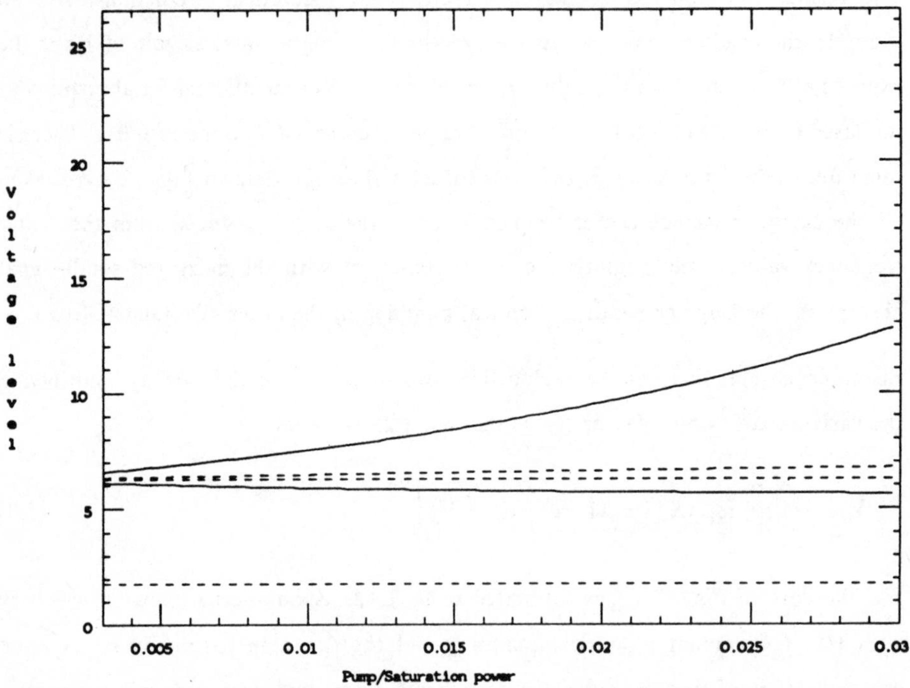

Fig. 14 A phenomenological calculation of the minimum and maximum rms squeezed noise levels (solid lines) is shown as a function of I/I_s. The dashed lines, in order of their level, are the amplifier noise, the vacuum fluctuation plus amplifier noise and this latter noise plus the spontaneous emission noise. The effective α_o is 24 cm^{-1}, $\eta = 0.55$, cavity loss is 0.01, the input mirror transmission is 0.003, the output mirror transmission is 0.02, $\delta = 300$ and the classical power gain is 1.5.

The results at $I/I_s \gtrsim 0.02$ are in better agreement with a full quantum treatment as described by Ried and Walls (1985) and Klauder, McCall and Yurke (1985). In the full quantum treatment an important noise term arises due to fluctuations in the polarization of the Na atoms caused by spontaneous emission in *all* directions. This noise term becomes important for values of I/I_s above the "ideal noise limit" $I/I_s = (10\delta)^{-1/2}$ (~ 0.018 for $\delta = 300$) defined by Reid and Walls (1985). This extra noise increases rapidly as $(I/I_s)^3$ for $I/I_s > (10\delta)^{-1/2}$ as shown in Fig. 15.

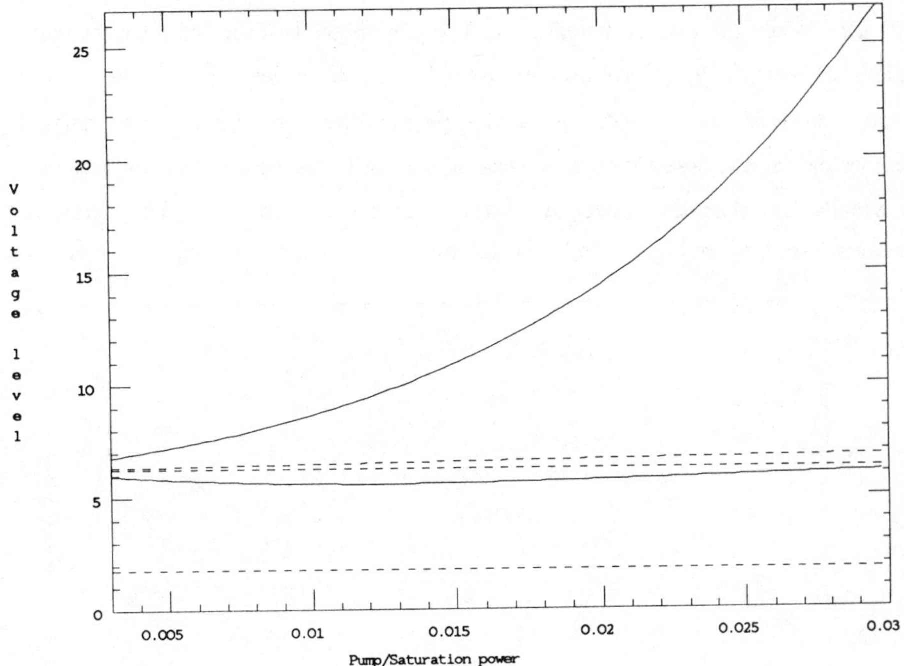

Fig. 15 A full quantum calculation of the minimum and maximum rms squeezed noise levels (solid lines) is shown as a function of I/I_s. The dashed lines are the same as in Fig. 14. Other parameters are also the same as in Fig. 15 except the input mirror is assumed to be totally reflecting and the cavity loss is 0.0002, due only to atomic absorption. The phase jitter factor is 1.0 (no phase jitter).

Comparison with Fig. 14 clearly shows how important the extra noise term becomes at higher I/I_s. Calculations in Fig. 15 are for a cavity similar to that used in the calculations for Fig. 14 except that the input mirror is assumed to be totally reflecting and the only cavity losses are due to the Na Lorentzian absorption. The full quantum model again predicts more noise squeezing than is observed for $I/I_s \sim 0.01$.

To this point, the effects of phase jitter described earlier in Section 5.1 using eqs. 5.5 and 5.6 have not been included in the model. Phase jitter effects for f = 0.85 are shown by a comparison of Fig. 15 with Fig. 16. Jitter is clearly to be avoided in measurements of squeezed noise. At $I/I_s = 0.01$ and A = 1.5 corresponding to the data in Fig. 12, this full quantum model predicts $V_{min} = 560\ \mu V$. When phase jitter f = 0.84 is included, V_{min} is increased to 589 μV, closer to the observed value of 607 μV. When the gain is increased to A =3, as might be expected near the cavity resonance center, V_{min} *increases* to 632 μV, i.e. to a level above the vacuum fluctuation level at 627 μV. Thus the phase jitter may be at least partially responsible for the observation that noise squeezing at the center resonant frequency is usually less than the squeezing at the sides of the resonance. The agreement between the data in Fig. 12 with the calculations is still not perfect but it is close

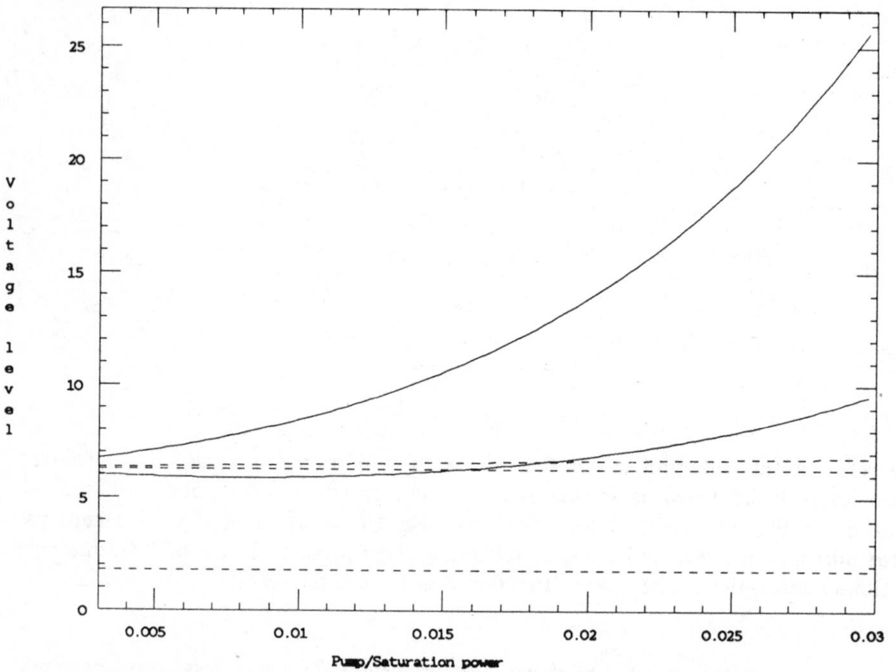

Fig. 16 Phase jitter effects on the calculated noise squeezing are shown for parameters the same as in Fig. 15 except for a phase jitter factor of 0.85.

enough, considering the estimates and assumptions involved, to inspire some confidence in the present understanding of the experimental results at $I/I_s \lesssim 0.01$. The phase jitter effects clearly make direct comparisons with the theory very difficult at $I/I_s > 0.01$.

7. *SUMMARY*

Initial experimental results and model calculations have been described for noise squeezing in an optical cavity. Resonantly enhanced nonlinear atomic polarization in a pump laser field is used to amplify and deamplify vacuum fluctuations in a cavity mode. The noise properties of the light transmitted out of the cavity are observed using a balanced homodyne detection. Noise in one quadrature of the output field is reduced 7% in power below the vacuum fluctuation level while the other quadrature component is amplified. This is a clear signature of squeezed states of light and can be explained only by a quantum description of the electromagnetic field. When the losses, phase jitter and inefficiencies in the detection process are taken into account, the noise reduction below the vacuum fluctuation level at the output of the cavity is near 25% in power. The agreement between these results and theoretical predictions is good at pump intensities well below the saturation intensity. Squeezing at higher pump intensities near the saturation intensity is not well understood at present.

Squeezing effects described here are not large enough to be useful in interferometric or other applications. Extrapolation of the model parameters to larger detuning values δ, lower cavity losses and increased output mirror reflectivities indicate that measured noise reduction as large as a factor of 50% in power may be possible near an atomic resonance. Lorentzian absorption near the atomic resonance is a limit on the degree of squeezing that can be achieved. Nonlinear processes in solids may circumvent this limitation. At present, the prospects for large squeezed noise reduction and many interesting applications seem very bright.

We especially thank L. W. Hollberg and J. C. Mertz for strong collaboration on the experiments as well as J. R. Klauder and S. L. McCall for collaboration on the theoretical models. We also thank P. Kumar, J. H. Shapiro, D. F. Walls, D. Stoler, J. Denker and C. M. Caves for many helpful discussions.

REFERENCES

Abrams, R. L. and Lind, R. C. 1978, Opt. Lett. *2*, 94-96 and errata Opt. Lett. *3*, 205.

Bondourant, R. S. and Shapiro, J. H. 1984, Phys. Rev. D*30*, 2548.

Caves, C. M. 1981, Phys. Rev. D*23*, 1693.

Caves, C. M. and Schumaker, B. L. 1985, Phys. Rev. A*31*, 3068-3092.

Davis, J. L. and Ezekiel, S. 1981, Opt. Lett. *6*, 505-507.

Drever, R. W. P., Hall, J. L., Kowalski, F. W., Hough, J., Ford, G. M., Munley, A. J. and Ward, H. 1983, Applied Phys. B*31*, 97.

Grove, R. E., Wu, F. Y. and Ezekiel, S. 1977, Phys. Rev. A*15*, 227-233.

Hall, J. L. and Hänsch, T. W. 1984, Opt. Lett. *9*, 502-504.

Klauder, J. R., McCall, S. L. and Yurke, B. 1986, submitted to Phys. Rev. A.

Levenson, M. D., Shelby, R. M. and Perlmutter, S. H. 1985a, Conference on Lasers and Electro-Optics, Baltimore, Maryland and Conference Proceeding SEICOLS 1985, Maui, Hawaii.

Levenson, M. D., Shelby, R. M., Aspect, A., Reid, M. and Walls, D. F. 1985b, Phys. Rev. A*32*, 1550-1562.

Loudon, R. 1983, The Quantum Theory of Light, Clarendon Press, Oxford, pp 162-206.

Maeda, M. W. Kumar, P. and Shapiro, J. H. 1985, Optical Instabilities, Ed. R. W. Boyd, M. G. Raymer and L. M. Narducci, Cambridge University Press, Cambridge, pp. 370-372.

Reid, M. D. and Walls, D. F. 1985, Phys. Rev. A*31*, 1622.

Schreiber, M. and Toyozawa, Y. 1982, Jour. Phys. Soc. Jap. *51*, 1528-1550.

Schumaker, B. L. 1984, Opt. Lett. *9*, 189.

Schumaker, B. L. and Caves, C. M. 1985, Phys. Rev. A*32*, 3093.

Shapiro, J. H. 1985, IEEE J. Quantum Elec. QE-21, 237.

Slusher, R. E., Hollberg, L. W., Yurke, B. Mertz, J. C. and Valley, J. F. 1985a, Phys. Rev. A*31*, 3512-3515.

Slusher, R. E., Hollberg, L. W., Yurke, B. and Mertz, J. C. 1985b, Optical Instabilities, Ed. R. W. Boyd, M. G. Raymer and L. M. Narducci, Cambridge University Press, Cambridge, pp. 380-382.

Slusher, R. E., Yurke, B., Hollberg, L. W. and Mertz, J. C. 1985c, Conference Proc. SEICOLS' 85, Maui, Hawaii.

Slusher, R. E., Hollberg, L. W., Yurke, B., Mertz, J. C. and Valley, J. F. 1985d, Phys. Rev. Lett. *55*, 2409-2412.

Walls, D. F. 1983, Nature *306*, 141-146.

Yuen, H. P., and Shapiro, J. H. 1979, Opt. Lett. *4*, 334-336.

Yuen, H. P., and Chan, V. W. S. 1983, Opt. Lett. *8*, 177-179 and errata Opt. Lett. *8*, 345.

Yurke, B. 1984, Phys. Rev. A*29*, 408-411.

Yurke, B. 1985a, Phys. Rev. A*32*, 300-310.

Yurke, B. 1985b, Phys. Rev. A*32*, 311-322.

Yurke, B., McCall, S. L. and Klauder, J. R. 1986, submitted to Phys. Rev. A.

EFFECTS OF OPTICAL PROCESSING ON NONCLASSICAL PROPERTIES OF LIGHT

R LOUDON

1. INTRODUCTION

The main properties of squeezed light are briefly reviewed, with particular emphasis on its nonclassical field and photon number fluctuations. The influences of various optical processing elements on the low-noise properties of squeezed light are determined. The elements considered include beam splitters, with application to a model for optical fibre attenuation, linear optical amplifiers, with application to a model for a long transmission link formed from alternating attenuators and amplifiers, direct detection, and homodyne detection using four-port and eight-port arrangements. Similarities between the effects of different processing elements are emphasized.

2. FLUCTUATION PROPERTIES OF SQUEEZED LIGHT

The input signal to each optical processing element is assumed to excite a single mode denoted by a subscript S. It is convenient to express the signal mode creation and destruction operators in terms of Hermitian field operators

$$\hat{X}_S = \tfrac{1}{2}(\hat{a}_S^\dagger + \hat{a}_S) \text{ and } \hat{Y}_S = \tfrac{1}{2}i(\hat{a}_S^\dagger - \hat{a}_S). \tag{1}$$

The homodyne detection considered below corresponds to measurements represented by a rotated pair of quadrature field operators

$$\begin{aligned}\hat{x} &= \tfrac{1}{2}\{\exp(i\arg\alpha_L)\hat{a}_S^\dagger + \exp(-i\arg\alpha_L)\hat{a}_S\}\\ \hat{y} &= \tfrac{1}{2}i\{\exp(i\arg\alpha_L)\hat{a}_S^\dagger - \exp(-i\arg\alpha_L)\hat{a}_S\},\end{aligned} \tag{2}$$

where the phase angle is that of the coherent local oscillator.

The operators for all such pairs of perpendicular axes have the commutation relation

$$[\hat{x},\hat{y}] = \tfrac{1}{2}i, \tag{3}$$

and the uncertainties parallel to the homodyne axes therefore satisfy

$$(\Delta x)^2_S(\Delta y)^2_S \geq 1/16. \tag{4}$$

The S subscripts signify that these are the uncertainties in the input signal states.

Minimum-uncertainty homodyne measurements are realized for squeezed-state input signals (Yuen 1976, Caves 1981, Walls 1983). The ideal squeezed state $|\alpha_S;s,\theta\rangle$ satisfies the eigen-value equation

$$(\hat{a}_S\cosh s + \hat{a}_S^\dagger\exp(i\theta)\sinh s)|\alpha_S;s,\theta\rangle$$
$$= (\alpha_S\cosh s + \alpha_S^*\exp(i\theta)\sinh s)|\alpha_S;s,\theta\rangle. \tag{5}$$

It has a complex coherent amplitude α_S and uncertainty variances that take minimum and maximum values

$$(\Delta X_S')^2 = \tfrac{1}{4}\exp(-2s) \quad \text{and} \quad (\Delta Y_S')^2 = \tfrac{1}{4}\exp(2s) \tag{6}$$

parallel to another pair of field axes inclined at angles $\tfrac{1}{2}\theta$ to the axes defined in (1). The squeeze parameter s is real and positive, and the angle θ runs from 0 to 2π. The amplitude and the uncertainty ellipse of the squeezed state are represented in Fig. 1. The nonclassical regime corresponds to field variances smaller than $\tfrac{1}{4}$.

The homodyne observables represented by the operators (2) have expectation values

$$x_S \equiv \langle\hat{x}\rangle = |\alpha_S|\cos(\arg\alpha_S - \arg\alpha_L)$$
$$y_S \equiv \langle\hat{y}\rangle = |\alpha_S|\sin(\arg\alpha_S - \arg\alpha_L) \tag{7}$$

for the ideal squeezed state. The corresponding variances are

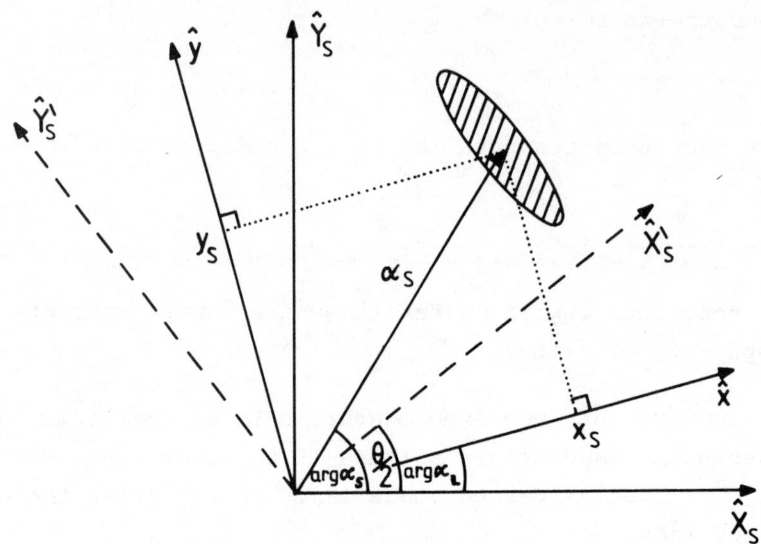

Fig. 1 Squeezed-state uncertainty ellipse and co-ordinate axes

$$(\Delta x)_S^2 = \tfrac{1}{4}\{\exp(-2s)\cos^2(\tfrac{1}{2}\theta-\arg\alpha_L)+\exp(2s)\sin^2(\tfrac{1}{2}\theta-\arg\alpha_L)\}$$
$$(\Delta y)_S^2 = \tfrac{1}{4}\{\exp(-2s)\sin^2(\tfrac{1}{2}\theta-\arg\alpha_L)+\exp(2s)\cos^2(\tfrac{1}{2}\theta-\arg\alpha_L)\}. \quad (8)$$

These expressions have obvious geometrical interpretations as projections of the uncertainty ellipse in Fig. 1 on to the homodyne axes. The minimum uncertainty product is realized when the local oscillator phase equals $\frac{1}{2}\theta$. It is shown below that the uncertainties in the corresponding measured quantities generally exceed the limit provided by (4).

The homodyne signal-to-noise ratio, defined by

$$\mathrm{SNR} = x_S^2/(\Delta x)_S^2 \quad (9)$$

and an analogous expression for the $\hat{y}$ operator, takes the form

$$\mathrm{SNR} = \frac{4|\alpha_S|^2\cos^2(\arg\alpha_S-\arg\alpha_L)}{\exp(-2s)\cos^2(\tfrac{1}{2}\theta-\arg\alpha_L) + \exp(2s)\sin^2(\tfrac{1}{2}\theta-\arg\alpha_L)} \quad (10)$$

for the ideal squeezed state. The maximum ratio, obtained when all the angles are equal, is

$$SNR_{max} = 4|\alpha_S|^2 \exp(2s) \qquad (\arg\alpha_L = \arg\alpha_S = \tfrac{1}{2}\theta). \tag{11}$$

Fig. 2 shows the dependence of the homodyne x field and its variance on the relative phase of the signal and local oscillator for the case $\arg\alpha_S = \frac{1}{2}\theta$.

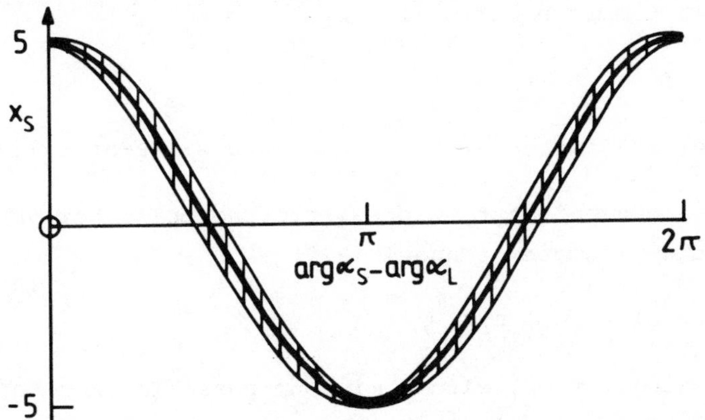

Fig. 2 Mean and variance of homodyne field as functions of local oscillator phase for $\arg\alpha_S = \frac{1}{2}\theta$, $|\alpha_S| = 5$ and $s = 1$.

The mean photon number for the ideal squeezed state is

$$n_S = \langle \hat{a}_S^{\dagger}\hat{a}_S \rangle = |\alpha_S|^2 + \sinh^2 s. \tag{12}$$

For the photon number variance $(\Delta n)_S^2$ we consider only the limit where

$$|\alpha_S| >> \exp(s) \quad \text{and} \quad n_S \approx |\alpha_S|^2. \tag{13}$$

The amplitude uncertainty $\Delta|\alpha_S|$ is then determined by the projection of the uncertainty ellipse on to the amplitude vector α_S. It is thus seen by reference to Fig. 1 that

$$(\Delta|\alpha_S|)^2 = \tfrac{1}{4}\{\exp(-2s)\cos^2(\tfrac{1}{2}\theta - \arg\alpha_S) + \exp(2s)\sin^2(\tfrac{1}{2}\theta - \arg\alpha_S)\}, \tag{14}$$

and taking account of the inequality in (13), the photon number uncertainty is related to the amplitude uncertainty by

$$(|\alpha_S| \pm \tfrac{1}{2}\Delta|\alpha_S|)^2 \approx |\alpha_S|^2 \pm |\alpha_S|\Delta|\alpha_S| = n_S \pm \tfrac{1}{2}(\Delta n)_S. \tag{15}$$

Hence

$$(\Delta n)^2_S = 4|\alpha_S|^2(\Delta|\alpha_S|)^2$$
$$= |\alpha_S|^2\{\exp(-2s)\cos^2(\tfrac{1}{2}\theta-\arg\alpha_S)+\exp(2s)\sin^2(\tfrac{1}{2}\theta-\arg\alpha_S)\}. \tag{16}$$

The signal phase angle uncertainty, determined by the angle subtended at the origin in Fig. 1 by the uncertainty ellipse, is given in the same limit (13) by

$$(\Delta\arg\alpha_S)^2 =$$
$$= \{\exp(-2s)\sin^2(\tfrac{1}{2}\theta-\arg\alpha_S) + \exp(2s)\cos^2(\tfrac{1}{2}\theta-\arg\alpha_S)\}/4|\alpha_S|^2. \tag{17}$$

It is often convenient to express the photon number variance in terms of the parameter (Mandel 1979)

$$Q_S = \{(\Delta n)^2_S - n_S\}/n_S. \tag{18}$$

Negative values of Q_S, with a minimum possible value of -1, correspond to nonclassical photon statistics. It is seen from (13) and (16) that the ideal squeezed state has a minimum parameter value of

$$Q_S = \exp(-2s) - 1 \quad \text{for} \quad \arg\alpha_S = \tfrac{1}{2}\theta. \tag{19}$$

Two special cases of the ideal squeezed state are important. For zero squeeze parameter it becomes the Glauber coherent state

$$|\alpha_S\rangle = |\alpha_S;0,\theta\rangle, \tag{20}$$

and the uncertainty ellipse becomes a circle with variances (8) equal to ¼. The photon number distribution has a Poisson form and Q_S vanishes. For the opposite limit of very large s, the ellipse tends to a straight line of zero width. Such line states have been considered by Yuen (1976). The states $|x\rangle$ and $|y\rangle$ relevant to homodyne detection correspond to angles θ equal respectively to $2\arg\alpha_L$ and $2\arg\alpha_L+\pi$, with associated eigenvalue equations

$$\hat{x}|x\rangle = x|x\rangle \quad \text{and} \quad \hat{y}|y\rangle = y|y\rangle. \tag{21}$$

The parameter Q_S takes its minimum value of -1 for these states.

3. BEAM SPLITTING: OPTICAL FIBRE ATTENUATION

Consider the configuration shown in Fig. 3 where a lossless symmetric beam splitter couples the signal and an unexcited

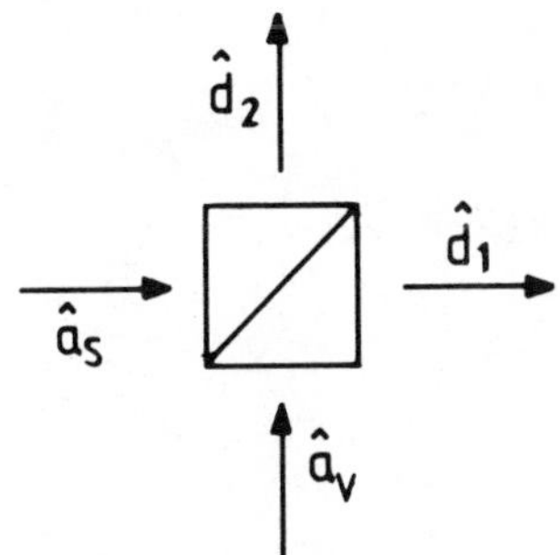

Fig. 3 Beam splitter with notation for input and output mode operators

vacuum mode to two output modes. The relation between input and output mode destruction operators is

$$\begin{bmatrix} \hat{d}_1 \\ \hat{d}_2 \end{bmatrix} = \begin{bmatrix} t & r \\ r & t \end{bmatrix} \begin{bmatrix} \hat{a}_S \\ \hat{a}_V \end{bmatrix} \tag{22}$$

where the conditions for a unitary coupling matrix are

$$|r|^2 + |t|^2 = 1 \quad \text{and} \quad tr^* + rt^* = 0. \tag{23}$$

Expectation values for the output in mode $\hat{d}_1$ are easily expressed in terms of the corresponding input quantities, for example

$$x_1 = |t|x_S \quad \text{and} \quad y_1 = |t|y_S, \tag{24}$$

and

$$(\Delta x)_1^2 = |t|^2(\Delta x)_S^2 + \tfrac{1}{4}|r|^2 \quad \text{and} \quad (\Delta y)_1^2 = |t|^2(\Delta y)_S^2 + \tfrac{1}{4}|r|^2, \tag{25}$$

where the signal means and variances are given by (7) and (8) but with every composite angle increased by argt.

Now consider a series of N identical beam splitters, where N is a very large number, the amplitude transmission coefficient $|t|$ being close to unity so that $N|r|^2$ is finite. The expectation values at the output of the Nth beam splitter, obtained by iteration of (24) and (25), are

$$x = K^{\frac{1}{2}}x_S \quad \text{and} \quad y = K^{\frac{1}{2}}y_S, \tag{26}$$

and the output variances are

$$\begin{aligned}(\Delta x)^2 &= K(\Delta x)_S^2 + \tfrac{1}{4}(1-K) \\ (\Delta y)^2 &= K(\Delta y)_S^2 + \tfrac{1}{4}(1-K),\end{aligned} \tag{27}$$

where

$$K = \exp(-N|r|^2) \tag{28}$$

and the input means and variances are given by (7) and (8) but with each angle increased by Nargt.

The series of beam splitters models the cumulative effect of Rayleigh scattering in an optical fibre, where each scatterer not only removes a small fraction of the signal amplitude but also couples in a vacuum mode. The number N of scatterers is proportional to the fibre length ℓ, and with an appropriate definition of the fibre attenuation coefficient κ, (28) can be replaced by

$$K = \exp(-\kappa\ell). \tag{29}$$

The output variances (27) contain contributions from the attenuated input fluctuations together with additional contributions that are independent of the input signal. Figs. 4 and 5 illustrate the effects of attenuation on the amplitude vectors and the uncertainties for input signals that are respectively in coherent and in ideal squeezed states. It is seen from (27), in agreement with Yuen and Shapiro (1978) and Milburn and Walls (1983), that heavy attenuation ($K \ll 1$) produces output variances close to the coherent state value of $\frac{1}{4}$, and any input squeezing is lost.

The output signal-to-noise ratio is defined by an expression analogous to (9). For the conditions of maximum input signal-to-noise ratio (11), it takes the form

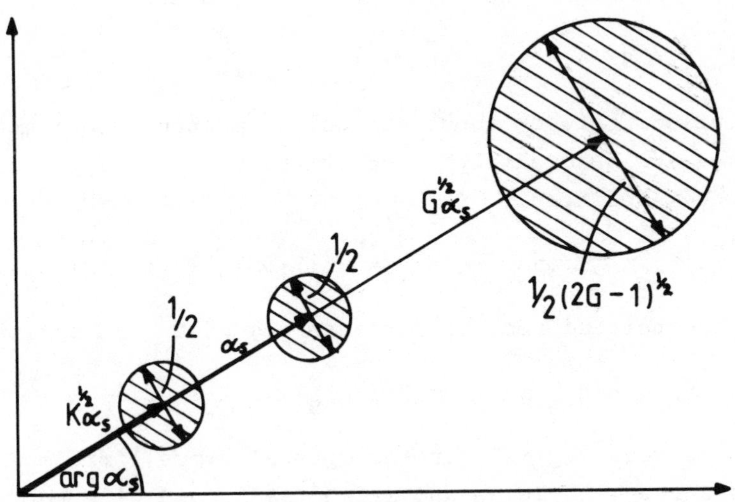

Fig. 4 Effects of attenuation with K = ¼ and phase insensitive amplification with G = 4 and P = 1 on a coherent input signal

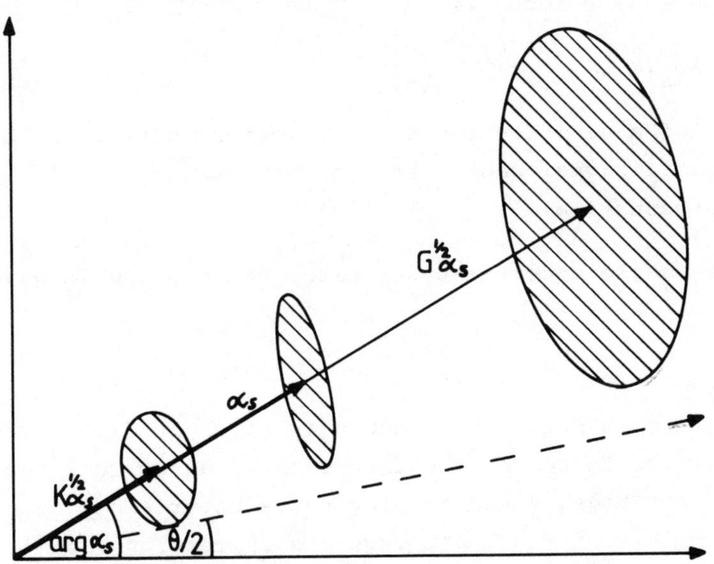

Fig. 5 Same as Fig. 4 but for a squeezed input signal

$$\mathrm{SNR}_{out} = \frac{4K|\alpha_S|^2\exp(2s)}{K + (1-K)\exp(2s)}. \tag{30}$$

This quantity falls in value with diminishing K and the enhancement of the input signal-to-noise ratio gained by squeezing the signal is lost for $K << 1$, when

$$SNR_{out} = 4K|\alpha_S|^2 \qquad (K << 1). \tag{31}$$

The relations between input and output photon-number means and variances are also easily found by the use of (22). For a single beam splitter

$$n_1 = |t|^2 n_S \quad \text{and} \quad (\Delta n)_1^2 = |t|^4 (\Delta n)_S^2 + |t|^2 |r|^2 n_S, \tag{32}$$

while the iterated results for a series of N beam splitters are

$$n = Kn_S \quad \text{and} \quad (\Delta n)^2 = K^2 (\Delta n)_S^2 + K(1-K) n_S, \tag{33}$$

where K is given by (29) in the optical fibre limit. These results agree with Teich and Saleh (1982) and with Peřina et al. (1983), and they have been discussed by Loudon and Shepherd (1984). The output variance takes a simple form when it is reexpressed in terms of a parameter Q defined by analogy with (18), and (33) reduces to

$$Q = KQ_S. \tag{34}$$

The size of a negative input Q_S parameter, corresponding to nonclassical signal photon statistics, is accordingly reduced by the attenuation.

The phase angle uncertainty $\Delta\phi$ after attenuation is given by

$$(\Delta\phi)^2 = (\Delta \arg\alpha_S)^2 + \frac{1 - K}{4K|\alpha_S|^2}, \tag{35}$$

where the input phase uncertainty is given by (17). The attenuation always increases the uncertainty, as is qualitatively evident from Figs. 4 and 5. The phase uncertainty becomes very large for $4K|\alpha_S|^2 << 1$, although the expression (35) is accurate only for small $\Delta\phi$.

4. LINEAR AMPLIFICATION

4.1 Phase-insensitive amplification

Linear optical amplifier theory has been considered in detail by Caves (1982), Friberg and Mandel (1983) and Loudon and

Shepherd (1984), and the main results are here quoted without proof. A phase-insensitive amplifier is characterized by two parameters, a gain G and a population factor P. The population factor has a range of values

$$1 \leq P < \infty. \tag{36}$$

For example, an inverted population amplifier with N_g and N_e atoms respectively in the ground and excited states has

$$P = N_e/(N_e - N_g), \tag{37}$$

while a Raman amplifier with n_P phonons excited in the scattering mode has

$$P = n_P + 1. \tag{38}$$

Both quadratures have the same gain, and denoting the output mean fields by x and y, and the corresponding inputs by x_S and y_S as before,

$$x = G^{\frac{1}{2}} x_S \quad \text{and} \quad y = G^{\frac{1}{2}} y_S. \tag{39}$$

The output variances are

$$(\Delta x)^2 = G(\Delta x)^2_S + \tfrac{1}{4}(2P-1)(G-1), \tag{40}$$

with a similar expression for Δy. These results are analogous to (26) and (27) for attenuation. The amplifier output variance likewise contains a contribution from the scaled input variance together with an additional contribution that is independent of the input. We consider particularly the case of minimum additive noise allowed by (36), where (40) reduces to

$$(\Delta x)^2 = G(\Delta x)^2_S + \tfrac{1}{4}(G-1) \qquad (P=1). \tag{41}$$

It is seen that even in this most favourable case, for a highly-squeezed input of negligible x variance, the output variance exceeds the coherent state value of ¼ for gains G greater than 2 (Loudon and Shepherd 1984). Figs. 4 and 5 illustrate the effects of amplification on the amplitude vectors and the uncertainties for input signals that are respectively in coherent and in ideal squeezed states. For large amplifications, $G \gg 1$, the additive contribution to the noise in (41) is equal to that produced by amplification of a coherent state input.

The output signal-to-noise ratio obtained from (39) and (41) in the limit of a high-gain amplifier is

$$\mathrm{SNR}_{out} = x_S^2/\{(\Delta x)_S^2 + \tfrac{1}{4}\} \qquad (P = 1,\ G \gg 1). \tag{42}$$

For the conditions of maximum input signal-to-noise ratio (11) this becomes

$$\mathrm{SNR}_{out} = 4|\alpha_S|^2/\{\exp(-2s) + 1\}. \tag{43}$$

The initial enhancement of signal-to-noise ratio by a factor exp(2s) produced by the use of a squeezed state instead of a coherent state input is thus reduced to a factor of 2 enhancement at best for the output light.

The corresponding relations between the input and output photon number means and variances for $P = 1$ are

$$n = Gn_S + G - 1 \tag{44}$$

and

$$(\Delta n)^2 = G^2(\Delta n)_S^2 + G(G-1)(n_S+1). \tag{45}$$

The greater complexity of these expressions compared to the analogous results (33) for the attenuator reflect the unavoidable habit of the amplifier of adding a spontaneous emission chaotic component to the amplified input light. It is not difficult to show that an input with negative Q_S, according to the definition (18), can at best sustain a maximum gain of $G = 2$ before the output Q parameter becomes positive.

The phase angle uncertainty after amplification is given by

$$(\Delta\phi)^2 = (\Delta\arg\alpha_S)^2 + \frac{G - 1}{4G|\alpha_S|^2} \qquad (P = 1), \tag{46}$$

consistent with Loudon and Shepherd (1984). The phase angle uncertainty is always increased by amplification, as is qualitatively evident from Figs. 4 and 5, the maximum increase for high gain being equal to $1/4|\alpha_S|^2$. Thus a coherent state input has its phase uncertainty doubled by a high-gain amplifier.

4.2 Phase-sensitive amplification

The general properties of phase-sensitive optical amplifiers have been treated in detail by Caves (1982). We consider here only the ideal degenerate parametric amplifier, where the output operators are related to the input operators by

$$\begin{aligned} \hat{a} &= J^{\frac{1}{2}}\hat{a}_S - (J-1)^{\frac{1}{2}}\exp(i\psi)\hat{a}_S^{\dagger} \\ \hat{a}^{\dagger} &= J^{\frac{1}{2}}\hat{a}_S^{\dagger} - (J-1)^{\frac{1}{2}}\exp(-i\psi)\hat{a}_S, \end{aligned} \tag{47}$$

with the real parameters J and ψ determined by the amplifier geometry and nonlinear susceptibility.

Consider the simple special case where

$$\arg\alpha_S = \tfrac{1}{2}\theta = \arg\alpha_L = \tfrac{1}{2}\psi + \tfrac{1}{2}\pi, \tag{48}$$

so that

$$x_S = |\alpha_S| \quad \text{and} \quad y_S = 0. \tag{49}$$

It is not difficult to show from (47) that the output mean fields are

$$x = G_{max}^{\frac{1}{2}} x_S = G_{max}^{\frac{1}{2}}|\alpha_S| \quad \text{and} \quad y = G_{min}^{\frac{1}{2}} y_S = 0 \tag{50}$$

with variances

$$(\Delta x)^2 = G_{max}(\Delta x)_S^2 \quad \text{and} \quad (\Delta y)^2 = G_{min}(\Delta y)_S^2, \tag{51}$$

where

$$G_{max}^{\frac{1}{2}} = J^{\frac{1}{2}} + (J-1)^{\frac{1}{2}} \quad \text{and} \quad G_{min}^{\frac{1}{2}} = J^{\frac{1}{2}} - (J-1)^{\frac{1}{2}}. \tag{52}$$

The two quadratures now suffer different gains, and indeed one of them is attenuated since

$$G_{max}\, G_{min} = 1. \tag{53}$$

There are no additive contributions to the noises given in (51), and these relations resemble (40) with a population parameter $P = \frac{1}{2}$, a value not allowed by the range (36) for a phase-insensitive amplifier.

The degenerate parametric amplifier converts an input coherent state, with its uncertainty circle, to an output squeezed state, with an uncertainty ellipse, and indeed the device was amongst the first to be considered as a potential source of squeezed light (see Milburn and Walls 1981 and earlier references therein). It is seen from (50) and (51) that the gains are related to the resulting squeeze parameter s by

$$G_{max} = \exp(2s) \quad \text{and} \quad G_{min} = \exp(-2s) \tag{54}$$

and that the squeezing is produced in the quadrature that suffers attenuation. Because of the absence of any additive noise in either quadrature, the signal-to-noise ratios are unchanged by the amplification or attenuation. Fig. 6 shows

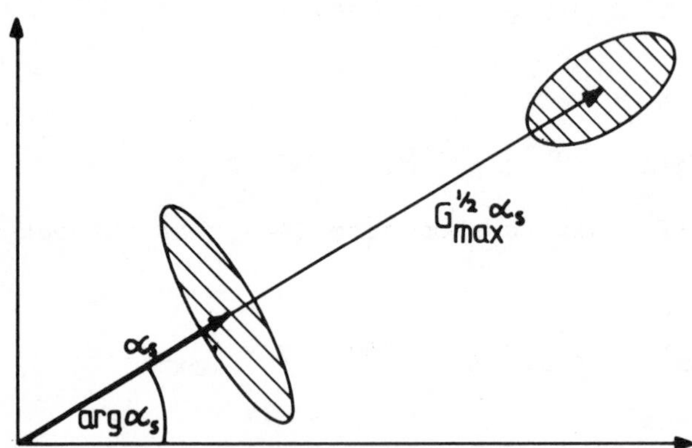

Fig. 6 Parametric amplification of a squeezed input for J = 2.5 and the angles chosen as in (48)

the effect of amplification on a squeezed input signal. Although the orientation of the uncertainty ellipse major axis is changed, the x signal-to-noise ratio is preserved. A simple consideration of the geometry of the amplitude vectors and uncertainty ellipses shows that, when the inequality in (13) is satisfied by the input signal, the output photon number and phase uncertainties are given by

$$(\Delta n)^2 = G_{max}^2(\Delta n)_S^2 \quad \text{and} \quad (\Delta\phi)^2 = G_{min}^2(\Delta \arg\alpha_S)^2. \tag{55}$$

The reduction in phase uncertainty after amplification is clearly evident in the example of Fig. 6.

The above results are valid for the choices of phase angles given in (48). We do not consider here the more varied kinds of behaviour that occur for less restrictive choices.

5. ATTENUATOR/AMPLIFIER CHAINS

A long optical transmission line is usually broken up into shorter sections separated by electrical regenerators to maintain the signal at a sufficiently high level. We consider here the noise properties of an all-optical system in which linear optical amplifiers alternate with attenuating fibre sections down a long transmission line (Mukai et al. 1982, Loudon 1985). The scheme is illustrated in Fig. 7, with a line of total

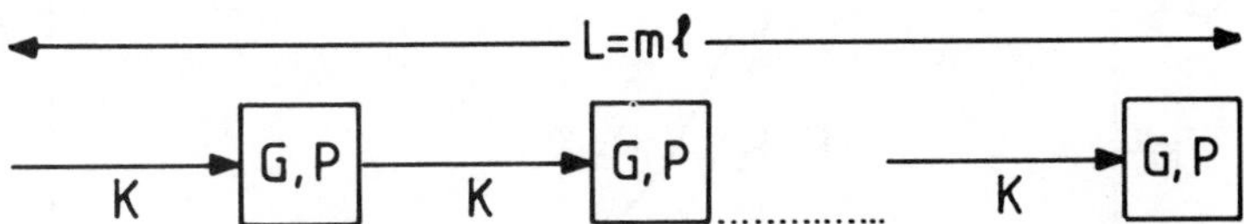

Fig. 7 Chain of alternating attenuators and linear amplifiers showing the component parameters

length L broken up into m identical sections each of length ℓ. The fibre attenuation K from (29) can be written

$$K = \exp(-\kappa L/m). \tag{56}$$

It is convenient to choose an amplifier gain that exactly compensates the fibre loss,

$$KG = 1, \tag{57}$$

so that the mean output field x equals the mean input signal field

$$x = x_S. \tag{58}$$

The relation between output and input noise is obtained by iteration of the attenuator and amplifier noise expressions given in sections 3 and 4. The phase angles are assumed to be chosen as in (11) so that the input has the maximum signal-to-noise ratio. The output variance is then easily shown to be

$$(\Delta x)^2 = \tfrac{1}{4}\exp(-2s) + \tfrac{1}{2}mP\{\exp(\kappa L/m) - 1\}, \tag{59}$$

where (56) and (57) have been used. This expression can be used both for the phase-insensitive amplifier, where minimum noise corresponds to P = 1, and for the phase-sensitive degenerate parametric amplifier, where P = ½. Fig. 8 shows how the

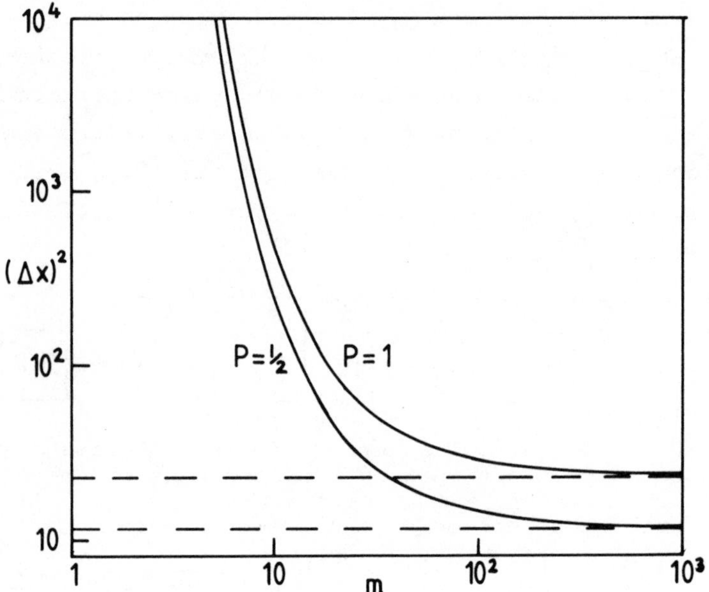

Fig. 8 Variations of output noise with the number m of attenuator/amplifier sections for parameter values given in (60)

output noise varies with the number of sections for both cases when the fixed parameter values are

$$\kappa = 0.046\ \text{km}^{-1} (0.2\ \text{dB km}^{-1}) \qquad L = 1000\ \text{km}. \tag{60}$$

It is seen that the output noise diminishes with increasing number of sections, and achieves its minimum value when m becomes very large such that (59) reduces to

$$(\Delta x)^2 = \tfrac{1}{4}\exp(-2s) + \tfrac{1}{2}P\kappa L \qquad (m >> \kappa L). \tag{61}$$

This limit corresponds physically to a system in which the fibre loss is continuously compensated by gain, as can be achieved in principle at least by Raman amplification in the optical fibre itself. The continuous limits given by (61) are indicated by the horizontal dashed lines in Fig. 8.

Although Fig. 8 is a plot of the function defined in (59), the sections of curve shown are insensitive to the first term on the right-hand side. It is thus impossible to distinguish whether the input signal is coherent (s = 0) or highly squeezed ($s \to \infty$), the low-noise advantages of the latter having been lost during the first few sections of the chain. This behaviour is shown in more detail in the upper part of Fig. 9 where

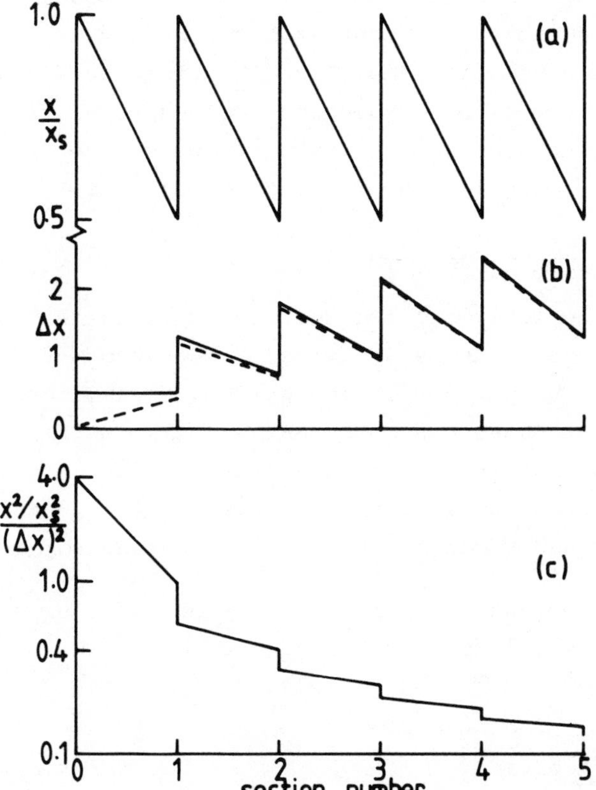

Fig. 9 (a) Variation of signal, (b) variation of noise, coherent input (——) and highly-squeezed input (---), and (c) variation of signal-to-noise ratio for coherent input. Parameter values, K = ¼, G = 4, P = 1.

the variations of signal and noise over the first five attenuators and amplifiers are shown, the distinction between coherent and squeezed inputs becoming very small after the first four sections. The decays over the attenuating sections are represented for simplicity by straight lines and are not intended to show the detailed behaviours as functions of distance along the fibre. The lower part of Fig. 9 shows the variation of signal-to-noise ratio for coherent input light. The linear optical elements considered here can never increase the signal-to-noise ratio, and indeed the ratio always decreases for the phase-insensitive amplification assumed in Fig. 9. The steep increases shown by the curves in Fig. 8 for small values of m result from the inability of the system to recover the severe reduction in signal-to-noise ratio produced in the first long length of fibre. By contrast, the modest increase in noise for the continuous amplification limit of (61), corresponding to a mere linear dependence on the total fibre length, results from the avoidance throughout the system of any disproportionate reduction in signal-to-noise ratio.

6. NOISE IN DIRECT DETECTION

In this and the following two sections we consider various optical detection schemes in terms of the information they can provide on the properties of an incident signal beam. The detection in all cases is modelled as a simple photon-counting process via a phototube with quantum efficiency η. The derivations ignore such complicating features as finite integration times and frequency-dependent detector responses.

The measured response in direct detection is determined by the photon-number statistics of the signal beam. The mean response can be written

$$M_1 = \eta\langle \hat{a}_S^\dagger \hat{a}_S \rangle = \eta n_S . \tag{62}$$

Higher-order moments are best derived via a generating function

$$J(w) = \langle \exp(-w\hat{a}_S^\dagger \hat{a}_S) \rangle . \tag{63}$$

It is convenient to put the expectation value into a normally-ordered form by use of the theorem (Louisell 1964)

$$\exp(-w\hat{a}_S^{\dagger}\hat{a}_S) = :\exp\{-[1-\exp(-w)]\hat{a}_S^{\dagger}\hat{a}_S\}:, \qquad (64)$$

where the colons denote normal ordering. The quantum efficiency can be introduced into normally-ordered expectation values simply as a factor η that multiplies the number operator. It is also useful to change the origin for the higher-order moments to the mean response (62). The generating function is transformed by these developments to

$$J'(w) = \langle :\exp\{-[1-\exp(-w)]\eta\hat{a}_S^{\dagger}\hat{a}_S + w\eta\langle\hat{a}_S^{\dagger}\hat{a}_S\rangle\}:\rangle. \qquad (65)$$

The moments are obtained from

$$M'_m = (-d/dw)^m \, J'(w)\big|_{w=0}. \qquad (66)$$

A simple special case is that of a coherent signal (20) where (65) reduces to

$$J'(w) = \exp\{-[1-w-\exp(-w)]\eta|\alpha_S|^2\}. \qquad (67)$$

This is the generating function for a Poisson distribution of mean $\eta|\alpha_S|^2$. For more general signals, it is often sufficient to evaluate only the second moment about the mean, obtained from (65) and (66) as

$$M'_2 = \eta^2(\Delta n)_S^2 + \eta(1-\eta)n_S. \qquad (68)$$

This reproduces a standard result (see for example Lee and Van der Ziel 1969). In the case of an ideal squeezed signal in the limit (13) where the coherent amplitude is much larger than the squeezing exponential, use of (16) in (68) gives

$$M'_2 = \eta^2|\alpha_S|^2\{\exp(-2s)\cos^2(\tfrac{1}{2}\theta-\arg\alpha_S)+\exp(2s)\sin^2(\tfrac{1}{2}\theta-\arg\alpha_S)\}$$
$$+ \eta(1-\eta)|\alpha_S|^2. \qquad (69)$$

It is sometimes useful to express the detected noise in terms of a parameter Q_D defined by analogy with (18) as

$$Q_D = (M'_2-M_1)/M_1. \qquad (70)$$

Then (68) may be reexpressed in the compact form

$$Q_D = \eta Q_S. \qquad (71)$$

Note that (68) and (71) have identical forms to the expressions (33) and (34) that relate the photon number variances before and after attenuation. A negative Q_S, corresponding to non-classical or sub-Poisson signal statistics, thus produces detected statistics that also have a negative Q_D parameter but whose magnitude is scaled downwards by the quantum efficiency.

7. FOUR-PORT HOMODYNE DETECTION

The quantum theory of homodyne detection has been developed in great detail and generality by Yuen and Shapiro (1978, 1980) and Shapiro et al. (1979). The outline derivation given below is intended merely to illustrate the main quantum features relevant to homodyne measurements and it omits the more technical details required for the treatment of practical systems. Fig. 10 illustrates the arrangement of signal and local oscillator input modes and the two output modes for a four-port

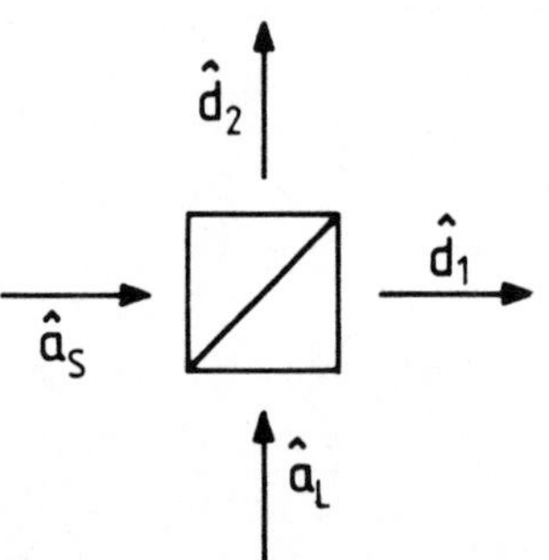

Fig. 10 Four-port homodyne detector

homodyne detector. It is assumed that the detector readings are processed to provide the difference between the photon counts in the two output arms. Such balanced homodyne detection has been considered by Oliver (1961), Carleton and Maloney (1968) and Yuen and Chan (1983). It has the advantage over measurement of a single output mode that excess noise from the local oscillator can be cancelled, as demonstrated experimentally by Carleton and Maloney (1968) and by Abbas et al. (1983).

The beam splitter included in the homodyne detector of Fig. 10 is assumed to be identical to that treated in section 3. The relation between output and input mode operators is given by (22) except that the vacuum mode operator $\hat{a}_V$ is now replaced by the local oscillator mode operator $\hat{a}_L$. The sum and difference of the output photon number operators are therefore

$$\hat{d}_1^\dagger\hat{d}_1 + \hat{d}_2^\dagger\hat{d}_2 = \hat{a}_S^\dagger\hat{a}_S + \hat{a}_L^\dagger\hat{a}_L$$

$$\hat{d}_1^\dagger\hat{d}_1 - \hat{d}_2^\dagger\hat{d}_2 = (|t|^2-|r|^2)(\hat{a}_S^\dagger\hat{a}_S-\hat{a}_L^\dagger\hat{a}_L)+(rt^*-tr^*)(\hat{a}_S^\dagger\hat{a}_L-\hat{a}_L^\dagger\hat{a}_S). \quad (72)$$

The measured response in balanced detection is represented by the latter operator, and with insertion of a quantum efficiency η, the mean difference count is

$$M_1 = \eta\langle\hat{d}_1^\dagger\hat{d}_1 - \hat{d}_2^\dagger\hat{d}_2\rangle. \quad (73)$$

The mean takes both positive and negative values. The higher-order moments and the probability distribution for the observable represented by the operator in (73) are obtained from a generating function. This is derived by the same procedure as led from (62) to (65) in the case of direct detection and the result is

$$J'(w) = \langle :\exp\{(\cosh w-1)\eta(\hat{d}_1^\dagger\hat{d}_1+\hat{d}_2^\dagger\hat{d}_2) - \eta\sinh w(\hat{d}_1^\dagger\hat{d}_1-\hat{d}_2^\dagger\hat{d}_2) + w\eta\langle\hat{d}_1^\dagger\hat{d}_1-\hat{d}_2^\dagger\hat{d}_2\rangle\}:\rangle. \quad (74)$$

Moments with respect to the mean (73) are obtained from (66). Note that the first term in the exponent represents a noise contribution determined by the total input photon number.

In the special case where the inputs are coherent states $|\alpha_S\rangle$ and $|\alpha_L\rangle$, the output modes are also coherent states $|\delta_1\rangle$ and $|\delta_2\rangle$ where

$$\delta_1 = t\alpha_S + r\alpha_L \quad \text{and} \quad \delta_2 = r\alpha_S + t\alpha_L. \quad (75)$$

The generating function can be written

$$J'(w) = \exp\{(\cosh w-1)\mu_1 - (\sinh w-w)M_1\}, \quad (76)$$

where

$$\mu_1 = \eta(|\delta_1|^2 + |\delta_2|^2) \text{ and } M_1 = \eta(|\delta_1|^2 - |\delta_2|^2). \tag{77}$$

This is the generating function for the probability distribution (Abramowitz and Stegun 1965)

$$P(\nu) = \left\{\frac{\mu_1+M_1}{\mu_1-M_1}\right\}^{\frac{1}{2}\nu} \exp(-\mu_1) I_\nu\{(\mu_1^2-M_1^2)^{\frac{1}{2}}\}, \tag{78}$$

where I_ν is a modified Bessel function. It is the probability for a photon count difference ν between the Poisson distributions of the two output modes. Except for small values of the total mean photon count μ_1, of the order unity or less, the distribution (78) is close to the Gaussian

$$P(\nu) = (2\pi\mu_1)^{-\frac{1}{2}} \exp\{-(\nu-M_1)^2/2\mu_1\}, \tag{79}$$

with mean M_1 and variance μ_1.

The remainder of the section considers the less special case where the local oscillator is excited to a coherent state but the signal state is unspecified. The beam splitter is assumed to have the properties

$$|r| = |t| = 2^{-\frac{1}{2}} \quad \text{and} \quad \arg r - \arg t = \tfrac{1}{2}\pi. \tag{80}$$

Then with the use of (72), the generating function (74) becomes

$$J'(w) = \langle :\exp\{(\cosh w-1)\eta(\hat{a}_S^\dagger \hat{a}_S + |\alpha_L|^2)$$

$$- i\eta \sinh w(\hat{a}_S^\dagger \alpha_L - \alpha_L^* \hat{a}_S) + iw\eta\langle \hat{a}_S^\dagger \alpha_L - \alpha_L^* \hat{a}_S\rangle\}:\rangle, \tag{81}$$

where this and subsequent expectation values are evaluated with respect to the signal density operator $\hat{\rho}_S$. The local oscillator is now assumed to be much stronger than the signal

$$|\alpha_L|^2 >> \langle \hat{a}_S^\dagger \hat{a}_S\rangle, \tag{82}$$

and only the terms of highest order in $|\alpha_L|$ are retained in the moments of the distribution. This is accomplished by keeping only those terms in which the power of w is no higher than the multiplying power of $|\alpha_L|$ in the exponent of (81). Thus expansion of the hyperbolic functions gives

$$J'(w) = \exp(\tfrac{1}{2}w^2\eta|\alpha_L|^2)\langle:\exp\{-2w\eta|\alpha_L|(\hat{y}-y_S)\}:\rangle, \tag{83}$$

where $\hat{y}$ is defined in (2) and y_S in (7). Use of the Baker-Hausdorff relation allows the generating function to be written equivalently in terms of a non-normally-ordered expectation value as

$$J'(w) = \exp\{\tfrac{1}{2}w^2\eta(1-\eta)|\alpha_L|^2\}\langle\exp\{-2w\eta|\alpha_L|(\hat{y}-y_S)\}\rangle. \tag{84}$$

The detected homodyne signal S, given by (73), is now

$$S \equiv M_1 = 2\eta|\alpha_L|y_S. \tag{85}$$

The noise N, given by the second moment, is determined by (66), where the forms (83) and (84) of the generating function give

$$\begin{aligned} N \equiv M_2' &= 4\eta^2|\alpha_L|^2\langle:(\hat{y}-y_S)^2:\rangle + \eta|\alpha_L|^2 \\ &= 4\eta^2|\alpha_L|^2(\Delta y)_S^2 + \eta(1-\eta)|\alpha_L|^2. \end{aligned} \tag{86}$$

The detected signal and noise are thus related via scaling factors of $2\eta|\alpha_L|$ to the mean homodyne field and its variance. In particular, for unit quantum efficiency, the detected noise can be expressed entirely in terms of the scaled signal fluctuation, as in the second line of (86) (Yuen and Shapiro 1980). For an ideal squeezed signal, where the variance is given by (8), the noise (86) in four-port homodyne detection is very similar in form to the noise (69) in direct detection, the coherent signal amplitude and phase being now replaced by the local oscillator amplitude and phase. Of course the directly detected signal given by (62) lacks the phase sensitivity of the homodyne signal given by (85) and (7).

Analogies can be drawn between homodyne detection and optical amplification. Thus the second moment (86) can be written

$$M_2' = G\{(\Delta y)_S^2 + \tfrac{1}{4}(2P-1)\} \tag{87}$$

where

$$G = 4\eta^2|\alpha_L|^2 \quad \text{and} \quad P = 1/2\eta. \tag{88}$$

This is similar to the amplifier output variance (40) in the limit of high gain, $G \gg 1$. A measurement of the photon-count second moment can thus be regarded as equivalent to a measurement of the amplified signal field variance. The population factor approaches ½ in the limit of unit quantum efficiency when the vanishing of the additive homodyne noise in (86) parallels the analogous property of a degenerate parametric amplifier.

The generating function (84) can be put in a more explicit form by insertion of the line state closure relation

$$\int dy\,|y\rangle\langle y| = 1 \tag{89}$$

into the expectation value to give

$$J'(w) = \exp\{\tfrac{1}{2}w^2\eta(1-\eta)|\alpha_L|^2\}\int dy\ P(y)\exp\{-2w\eta|\alpha_L|(y-y_S)\}, \tag{90}$$

where

$$P(y) = \langle y|\hat{\rho}_S|y\rangle. \tag{91}$$

The probability distribution for the photon count difference is thus a convolution of a Gaussian distribution that is independent of the signal state, with a probability distribution $P(y)$ for the homodyne field y that is itself independent of the detector quantum efficiency but is scaled by a factor $2\eta|\alpha_L|$ in the convolution (Yuen and Shapiro 1978, 1980; Shapiro et al. 1979).

The probability distribution is readily evaluated for a pure ideal squeezed state signal, where use of an expression for the line state overlap derived by Yuen (1976) gives a Gaussian distribution,

$$P(y) = \exp\{-(y-y_S)^2/2(\Delta y)_S^2\}/\{2\pi(\Delta y)_S^2\}^{\frac{1}{2}}, \tag{92}$$

with mean and variance given by (7) and (8) respectively. Substitution of this expression into (90) produces a generating function

$$J'(w) = \exp(\tfrac{1}{2}w^2N), \tag{93}$$

where N is given by (86), and the photon-count difference distribution is

$$P(\nu) = \exp\{-(\nu-S)^2/2N\}/(2\pi N)^{\frac{1}{2}}. \qquad (94)$$

This Gaussian distribution agrees with (79) in the case of a coherent signal and a much stronger coherent local oscillator.

8. EIGHT-PORT HOMODYNE DETECTION

Fig. 11 illustrates a more complicated form of homodyne detector with 4 input ports and 4 output ports. Two of the input

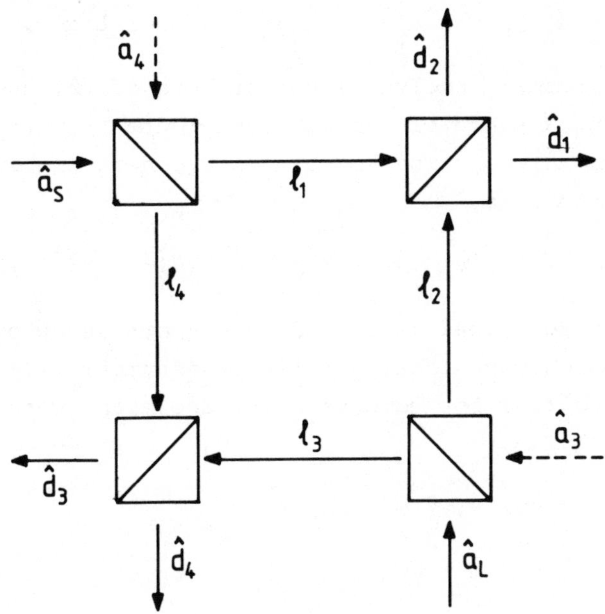

Fig. 11 Eight-port homodyne detector

ports are used for the signal and local oscillator while the remaining two inputs are unexcited. All four output ports are used, the detector readings being processed to provide the differences between the photon counts in adjacent pairs of output arms. The eight-port arrangement has been considered by Oliver (1962) and in greater detail by Walker and Carroll (1984a) and Walker (1986). It has the advantage over the four-port detector that both the signal phase and amplitude can be measured, as has been demonstrated experimentally by Walker and Carroll (1984b).

The four beam splitters included in the detector of Fig. 11 are all assumed to have the properties (80). With the path lengths chosen so that

$$k(\ell_1-\ell_2) = \tfrac{1}{2}\pi \quad \text{and} \quad k(\ell_4-\ell_3) = \pi, \tag{95}$$

where k is the optical wavevector, the operators for the photon number differences between pairs of outputs take the forms

$$\begin{aligned} \hat{d}_1^\dagger\hat{d}_1 - \hat{d}_2^\dagger\hat{d}_2 &= \tfrac{1}{2}(\hat{a}_S^\dagger\hat{a}_L + \hat{a}_L^\dagger\hat{a}_S) + \text{other terms} \\ \hat{d}_3^\dagger\hat{d}_3 - \hat{d}_4^\dagger\hat{d}_4 &= \tfrac{1}{2}i(\hat{a}_S^\dagger\hat{a}_L - \hat{a}_L^\dagger\hat{a}_S) + \text{other terms.} \end{aligned} \tag{96}$$

The 'other terms' involve either or both of the unexcited input modes and they therefore do not contribute to normally-ordered expectation values. The joint moment-generating function for these operators is

$$J(v,w) = \langle \exp\{-v(\hat{d}_1^\dagger\hat{d}_1 - \hat{d}_2^\dagger\hat{d}_2) - w(\hat{d}_3^\dagger\hat{d}_3 - \hat{d}_4^\dagger\hat{d}_4)\}\rangle . \tag{97}$$

This can be developed as for the four-port detector, and with the strong coherent local oscillator assumption (82), the form analogous to (83) for moments about the means with a detection efficiency η is

$$\begin{aligned} J'(v,w) &= \exp\{\tfrac{1}{4}(v^2+w^2)\eta|\alpha_L|^2\} \\ &\quad \times \langle :\exp\{-v\eta|\alpha_L|(\hat{x}-x_S) - w\eta|\alpha_L|(\hat{y}-y_S)\}:\rangle , \end{aligned} \tag{98}$$

and the non-normally-ordered form analogous to (84) is

$$\begin{aligned} J'(v,w) &= \exp\{\tfrac{1}{4}(v^2+w^2)\eta(1-\tfrac{1}{2}\eta)|\alpha_L|^2\} \\ &\quad \times \langle \exp\{-v\eta|\alpha_L|(\hat{x}-\hat{x}_S) - w\eta|\alpha_L|(\hat{y}-y_S)\}\rangle . \end{aligned} \tag{99}$$

The signals for the measurements of observables represented by the operators in (96) are

$$S_{12} = \eta|\alpha_L|x_S \quad \text{and} \quad S_{34} = \eta|\alpha_L|y_S . \tag{100}$$

The corresponding noises obtained by appropriate second

differentiations of (99) are

$$N_{12} = \eta^2|\alpha_L|^2(\Delta x)_S^2 + \tfrac{1}{2}\eta(1-\tfrac{1}{2}\eta)|\alpha_L|^2$$
$$N_{34} = \eta^2|\alpha_L|^2(\Delta y)_S^2 + \tfrac{1}{2}\eta(1-\tfrac{1}{2}\eta)|\alpha_L|^2. \qquad (101)$$

The detected signals and noises are thus related via scaling factors of $\eta|\alpha_L|$ to the means and variances of the two quadrature homodyne fields. However, the additional measurement capability of the eight-port detector compared to the four-port detector is gained at the expense of additional noise, since, in contrast to the four-port result (86), the additive noise contributions in (101) no longer vanish for unit quantum efficiency.

Division of the noise expressions (101) by the scaling factor $\eta^2|\alpha_L|^2$ provides results for the homodyne field fluctuations as measured by the eight-port detector. These measured variances are denoted

$$(\Delta x)^2 = (\Delta x)_S^2 + \{(2-\eta)/4\eta\}$$
$$(\Delta y)^2 = (\Delta y)_S^2 + \{(2-\eta)/4\eta\}. \qquad (102)$$

Thus taking account of (4) and the range of values allowed for η, it follows that

$$(\Delta x)^2(\Delta y)^2 \geq \tfrac{1}{4}. \qquad (103)$$

The minimum uncertainty product of measured variances therefore exceeds by a factor of 4 the corresponding product (4) of input signal variances. This agrees with a general uncertainty principle proved by Arthurs and Kelly (1965) for the simultaneous measurement of any pair of conjugate observables. It also agrees with the discussions of Haus and Townes (1962) and Personick (1970, 1971) concerning the minimum noise limits on measurements of one or two signal quadrature components. It is a feature of the more complete measurements possible with the eight-port device that extra divisions of the input beams must be made, and each beam splitter inevitably allows additional input modes to be coupled to the detectors. The doubling of the minimum noise in both measured quadratures for a coherent input signal compared to that in four-port detection can be blamed on the beating of the strong local oscillator with the doubled number of input-mode vacuum fluctuations.

The noises (101) can be written in forms similar to (87), where the effective amplifier parameters are

$$G = \eta^2|\alpha_L|^2 \quad \text{and} \quad P = 1/\eta. \tag{104}$$

The population factor can never be smaller than unity, and the addition of noise to both quadratures in an eight-port homodyne detector is analogous to the unavoidable additive noise of the phase-insensitive amplifier discussed in section 4.1. The relations between input and detected signals and noise can be represented by diagrams similar to Figs. 4 and 5, and the various expressions in section 4.1 apply in the limit of high gain after transcription according to (104). For example, the detected phase angle uncertainty is related to the input phase uncertainty by

$$(\Delta\phi)^2 = (\Delta\arg\alpha_S)^2 + \frac{2-\eta}{4\eta|\alpha_S|^2}\,. \tag{105}$$

The generating function (99) can be cast in a more explicit form if it is first put into antinormal order by further use of the Baker-Hausdorff relation. Insertion of the coherent state closure relation

$$(1/\pi)\int dx\, dy|\alpha\rangle\langle\alpha| = 1 \tag{106}$$

between the destruction and creation operator exponentials then converts the generating function to

$$J'(v,w) = \exp\{\tfrac{1}{4}(v^2+w^2)\eta(1-\eta)|\alpha_L|^2\} \times \int dx\, dy\, P(x,y)\exp\{-v\eta|\alpha_L|(x-x_S)-w\eta|\alpha_L|(y-y_S)\}, \tag{107}$$

where

$$x + iy = \exp(-i\arg\alpha_L)\alpha \tag{108}$$

and

$$P(x,y) = \langle\alpha|\hat{\rho}_S|\alpha\rangle/\pi. \qquad (109)$$

The probability distribution for the difference photon counts in the four detector arms is thus a convolution of a Gaussian distribution independent of the input with a joint distribution P(x,y) for the homodyne fields scaled by factors of $\eta|\alpha_L|$ (Walker and Carroll 1984a). The latter distribution is identical to the Q-representation of the signal field.

The probability distribution is readily evaluated for a pure ideal squeezed state signal with the use of an expression for the coherent state overlap derived by Yuen (1976). We give results only for the simple case of a local oscillator phase equal to the squeezing angle $\frac{1}{2}\theta$, when

$$P(x,y) = \frac{1}{\pi\cosh s}\exp\left\{-\frac{\exp(s)(x-x_S)^2 + \exp(-s)(y-y_S)^2}{\cosh s}\right\}. \qquad (110)$$

Insertion of this expression into (107) produces a generating function

$$J'(v,w) = \exp(\tfrac{1}{2}v^2N_{12} + \tfrac{1}{2}w^2N_{34}), \qquad (111)$$

where the noises are given by (101). The joint photon-count difference distribution is therefore a product of two Gaussian distributions similar to (94), one for each pair of detector arms.

9. CONCLUSIONS

The recent success of Slusher et al. (1985) in generating and detecting squeezed light brings closer the possibility of using such light in various applications that could take advantage of its low noise properties. It has been shown above that the squeezing effect is not very robust and it can easily be lost by maladroit optical processing. Attenuation, amplification and detection all have similar effects in scaling the input noise, down or up, and adding new noise contributions independent of the input. These additive contributions are mainly responsible for the loss of the exceptionally small field fluctuations associated with the squeezing. Of the processes considered here only ideal degenerate parametric amplification and ideal four-port homodyne detection can avoid the additive noise contribution and thus maintain the signal-to-noise ratio at its initial value.

The two homodyne schemes considered, using the four-port and eight-port arrangements of Figs. 10 and 11, are special cases of more general homodyne detectors in which other numbers of ports are used or beam splitters with dissimilar characteristics are combined. It can be shown (Walker 1986) that the general detector has a response determined by a homodyne field distribution similar to (91) or (109) but with the signal density matrix element taken with respect to an ideal squeezed state characteristic of the beam splitter and phototube configurations. The line states and coherent states employed respectively for the four-port and eight-port schemes reflect the high symmetries of these arrangements. Homodyne detection can be tailored to some extent to achieve the balance between signal and noise in the two quadratures required for a particular application.

REFERENCES

Abbas, G.L., Chan, V.W.S. and [illegible], T.K. 1983, Optics Lett. 8 419-421

Abramowitz, M. and Stegun, I.A. 1965, Handbook of Mathematical Functions, Dover, New York

Arthurs, E. and Kelly, J.L. 1965, Bell Syst. Tech. J. 44 725-729

Carleton, H.R. and Maloney, W.T. 1968, Appl. Optics 7 1241-1243

Caves, C.M. 1981, Phys. Rev. D 23 1693-1708

Caves, C.M. 1982, Phys. Rev. D 26 1817-1839

Friberg, S. and Mandel, L. 1983, Optics Commun. 46 141-148

Haus, H.A. and Townes, C.H. 1962, Proc. IRE 50 1544-1545

Lee, S.J. and Van der Ziel, A. 1969, Physica 45 379-385

Loudon, R. 1985, IEEE J. Quantum Electron. QE-21 766-773

Loudon, R. and Shepherd, T.J. 1984, Optica Acta 31 1243-1269

Louisell, W.H. 1964, Radiation and Noise in Quantum Electronics McGraw-Hill, New York

Mandel, L. 1979, Optics Lett. 4 205-207

Milburn, G.J. and Walls, D.F. 1981, Optics Commun. 39 401-404

Milburn, G.J. and Walls, D.F. 1983, Am. J. Phys. 51 1134-1136

Mukai, T., Yamamoto, Y. and Kimura, T. 1982, IEEE J. Quantum Electron. QE-18 1560-1568

Oliver, B.M. 1961, Proc. IRE 49 1960-1961

Oliver, B.M. 1962, Proc. IRE 50 1545-1546

Peřina, M.C., Saleh, B.E.A. and Teich, M.C. 1983, Optics Commun. 48 212-214
Personick, S.D. 1970, Efficient Analog Communication over Quantum Channels, Res. Lab. Electron. M.I.T. Tech. Rep. 477
Personick, S.D. 1971, Bell Syst. Tech. J. 50 213-216
Shapiro, J.H., Yuen, H.P. and Machado Mata, M.A. 1979, IEEE Trans. Inf. Theory IT-25 179-192
Slusher, R.E., Hollberg, L.W., Yurke, B., Mertz, J.C. and Valley, J.F. 1985, Phys. Rev. Lett. 55 2409-2412
Teich, M.C. and Saleh, B.E.A. 1982, Optics Lett. 7 365-367
Walker, N.G. 1986, to be published
Walker, N.G. and Carroll, J.E. 1984a, AGARD Conf. Proc. 362, Digital Optical Circuit Technology 12.1-12.12
Walker, N.G. and Carroll, J.E. 1984b, Electron. Lett. 20 981-983
Walls, D.F. 1983, Nature 306 141-146
Yuen, H.P. 1976, Phys. Rev. A 13 2226-2243
Yuen, H.P. and Chan, V.W.S. 1983, Optics Lett. 8 177-179 (erratum ibid. p. 345)
Yuen, H.P. and Shapiro, J.H. 1978, IEEE Trans. Inf. Theory IT-24 657-668
Yuen, H.P. and Shapiro, J.H. 1980, IEEE Trans. Inf. Theory IT-26 78-92

GENERATION OF SQUEEZED STATES OF LIGHT

D F WALLS AND M D REID

1. INTRODUCTION

Over the past few years there has been a growing effort to generate squeezed states of light. These are quantum states of light which have less fluctuations in one quadrature than a coherent state[1-3]. These efforts are motivated by the potential applications of squeezed states in optical communications[4] and ultrasensitive detection systems[5] which are presently quantum noise limited.

The large amount of theoretical work on this topic has now been joined by some substantial experimental efforts.

The observation of a small amount of quantum noise squeezing (~ 7%) has been reported by Slusher et al[8,50] in nondegenerate four wave mixing in atomic sodium. Squeezing of classical noise has been observed by Levenson et al[6,37] in four wave mixing in optical fibers and by Shapiro et al[7] in four wave mixing in atomic sodium.

In this paper we shall review schemes which have been predicted theoretically to generate squeezed states. We shall attempt to relate the theoretical calculations to experiments currently in progress. Some of the problems involved in generating squeezed states have recently been discussed by Leuchs[9].

2. PROPERTIES OF SQUEEZED STATES

The electric field operator for a single mode field may be written in terms of its quadrature phases as

$$E(t) = \lambda\left(X_\theta \cos(\omega t + \theta) + X_{\theta+\frac{\pi}{2}} \sin(\omega t + \theta)\right) \tag{1}$$

where λ contains spatial factors and X_θ and $X_{\theta+\frac{\pi}{2}}$ are the amplitudes of the quadrature phases

$$X_\theta = ae^{-i\theta} + a^\dagger e^{i\theta} \tag{2}$$

The fluctuations in the quadrature phases may be expressed by the variances

$$V(X_\theta) = \langle X_\theta^2\rangle - \langle X_\theta\rangle^2 \tag{3}$$

A coherent state is a minimum uncertainty state $V(X_\theta)V\left(X_{\theta+\frac{\pi}{2}}\right) = 1$ with the quantum fluctuations randomly distributed in phase $\left(V(X_\theta)=V(X_{\theta+\frac{\pi}{2}}) = 1\right)$. A reduction of quantum fluctuations in one quadrature phase to a value less than that of a coherent state is called squeezing $(V(X_\theta) < 1)$.

We may formally define a squeezed state as follows

$$|\alpha,\eta\rangle = D(\alpha)\ S(\eta)|0\rangle \tag{4}$$

where $S(\eta)$ is the squeeze operator

$$S(\eta) = \exp\left(\frac{1}{2}\eta^* a^2 - \frac{1}{2}\eta a^{\dagger 2}\right) \tag{5}$$

and

$$\eta = re^{2i\theta}$$

and $D(\alpha)$ is the displacement operator

$$D(\alpha) = e^{-\frac{1}{2}|\alpha|^2} e^{\alpha a^\dagger} e^{-\alpha^* a} \tag{6}$$

The variances in the quadrature phases of the squeezed state $|\alpha,\eta\rangle$ are given by

$$V(X_\theta) = e^{-2r}$$

$$V\left(X_{\theta+\frac{\pi}{2}}\right) = e^{2r} \tag{7}$$

An alternative but equivalent characterization of squeezed states has been given by Yuen[(2)]. If one defines a mode b by

$$b = \mu a + \nu a^\dagger \tag{8}$$

and $\mu^2 - \nu^2 = 1$, then if mode a is in a coherent state, mode b will be in a squeezed state. The squeezed state $|\alpha,\eta\rangle$ is a minimum uncertainty state. In general a sufficient condition for squeezing is $V(X_\theta) < 1$.

For a multimode field we may define a spectrum of squeezing as the Fourier transform of the two time correlation function. The squeezing spectrum is

$$V\left[X_{\theta,}\omega\right] = \int d\tau e^{-i\omega\tau}\{\langle X_\theta(\tau)\ X_\theta(o)\rangle - \langle X_\theta(\tau)\rangle\langle X_\theta(o)\rangle\} \tag{9}$$

For a coherent state $V(X_{\theta,}\omega) = 1$. Squeezing at frequency ω is characterized by $V(X_{\theta,}\omega) < 1$.

3. MEASUREMENT OF THE SQUEEZING SPECTRUM BY HOMODYNE DETECTION

The squeezing spectrum may be measured by a homodyne detection scheme[10-16] The signal field $E(t)$ is mixed on a beam splitter with a local oscillator E_{LO}. The combined field then beats on the surface of a photodetector. The photocurrent emerging from the photodetector is proportional to

$$\langle i(t)\rangle = \langle (E^{(-)}(t) + E^*_{LO})(E^{(+)}(t) + E_{LO})\rangle \tag{10}$$

The local oscillator is treated classically ($E_{LO} = |E_{LO}|e^{i\theta}$) and its amplitude is taken to be much greater than the signal field. Thus the a.c. component of the photocurrent is proportional to

$$\langle i(t)\rangle = |E_{LO}|\ \langle X_\theta(t)\rangle \tag{11}$$

where

$$X_\theta(t) = e^{i\theta}E^{(-)}(t) + e^{-i\theta}E^{(+)}(t) \tag{12}$$

The fluctuations in the photocurrent may be characterized by the two time correlation function

$$\langle i(t)\ i(t+\tau)\rangle = |E_{LO}|^2\ \langle X_\theta(t)\ X_\theta(t+\tau)\rangle \tag{13}$$

and the spectrum of fluctuations is given by

$$V(X_{\theta,}\omega) = \int \langle i(t)\ i(t+\tau)\rangle\ e^{-i\omega\tau}d\tau$$

$$= |E_{LO}|^2\int \langle X_\theta(t)X_\theta(t+\tau)\ e^{-i\omega\tau}d\tau \tag{14}$$

Thus by varying the phase θ of the local oscillator we may measure the spectrum of the quadrature phase with maximum squeezing.
In order to avoid excess noise from the local oscillator we may use the balanced homodyne detector proposed by Yuen and Chan[13] and shown in Fig.(1

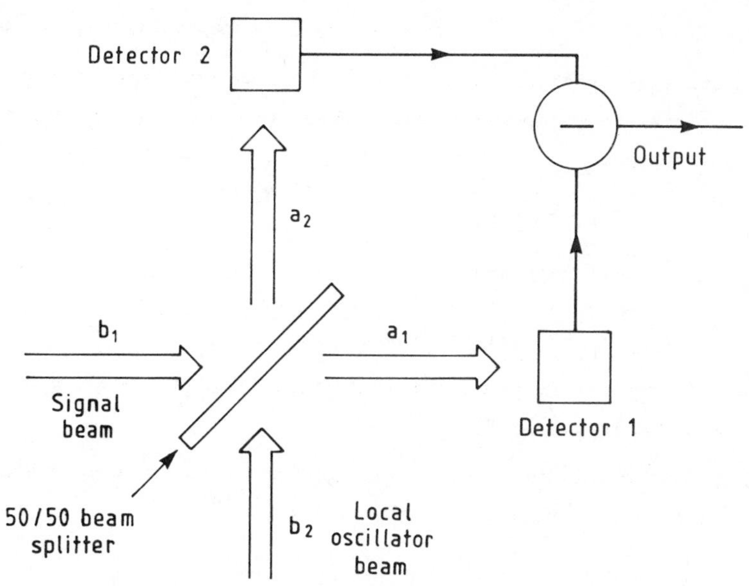

Fig.1 A balanced homodyne detector

This technique has been experimentally demonstrated by Abbas, Chan and Yee.[14] In the balanced homodyne detector the fields emerging from both ports of the beam splitter are detected on photodetectors. The two photocurrents are then subtracted and the spectrum of fluctuations in the difference current $i_D = i_1 - i_2$ analysed.

One finds

$$V(i_D,\omega) = |E_{LO}|^2\, V(X_\theta,\omega) \tag{15}$$

The above result assumes unit efficiency photodetectors. For photodetectors with photoefficiency η the result is modified to

$$V(i_D,\omega) = \eta^2\, |E_{LO}|^2\, V(X_\theta,\omega) + \eta(1-\eta) \tag{16}$$

The second term on the right hand side being noise arising from the non unit quantum efficiency of the detector.

4. IDEALIZED MODELS

Initial theoretical calculations of squeezing in nonlinear optical interactions were based on a classical treatment of the medium. This followed the formulation of nonlinear optics developed by Bloembergen[17] who expanded the polarization of the medium in terms of E as

$$P = \chi^{(1)}E + \chi^{(2)}E^2 + \chi^{(3)}E^3 \tag{17}$$

Effective Hamiltonians based on these classical nonlinear susceptibilities may be deduced. A second order nonlinear susceptibility gives rise to parametric amplification whereas a third order nonlinear susceptibility may give rise to four wave mixing. These processes may be described by the Hamiltonians

$$H = i\chi^{(n)} E_p^{(n-1)} (a^2 - a^{\dagger 2}) \tag{18}$$

in the degenerate case and

$$H = i\chi^{(n)} E_p^{(n-1)} (a_1 a_2 - a_1^{\dagger} a_2^{\dagger}) \tag{19}$$

in the non degenerate case, where the pump field has been described by the classical amplitude E_p.

These Hamiltonians give rise to minimum uncertainty squeezed states with the variance in one quadrature given by[18]

$$V(X_\theta) = e^{-2r} \tag{20}$$

(Note in the nondegenerate case the squeezing occurs in the combined modes $a_{\pm} = a_1 \pm a_2$). The squeeze parameter r is given by $r = \chi^{(n)} E_p^{(n-1)} t$ and $t = \frac{L}{c}$ where L is the interaction length. These models indicate the general class of interactions where squeezing may occur but because they neglect several sources of quantum noise they are excessively optimistic. A careful consideration of the various effects which add quantum noise random in phase and hence detrimental to the squeezing is required. It is assumed that one is able to eliminate all sources of excess classical noise and that we are working at the quantum noise limit.

Any loss mechanisms will introduce fluctuations which tend to degrade the squeezing. The nonlinear medium should itself be quantised. This will introduce quantum noise due to spontaneous emission which again degrades the squeezing. We have considered three models for the medium: (a) an ensemble of anharmonic oscillators, (b) an ensemble of two level atoms, (c) an ensemble of three level atoms. These will be discussed in detail in §7.

In order to increase the effective interaction length one may place the nonlinear medium inside an optical cavity. General considerations related to intracavity configurations are considered in section 6.

The effect of quantising the pump field is considered in the following section.

5. QUANTISATION OF THE PUMP

The Hamiltonians (18) and (19) assumed the pump field to be classical and undepleted. It is necessary to investigate the effect of quantizing the pump and allowing for pump depletion on the squeezing. We shall consider a quantised pump for the specific example of parametric amplification, where the interaction Hamiltonian (19) becomes

$$H = i\hbar\chi^{(2)} \; (a_p a_1^\dagger a_2^\dagger - a_p^\dagger a_1 a_2) \tag{21}$$

where a_p is the pump annihilation operator.

An analysis of this system by Scharf and Walls[19] shows that the variance of the maximally squeezed quadrature X_θ is given by

$$V(X_\theta) \approx \frac{1}{4}e^{-2(N_p)^{\frac{1}{2}}\chi t} + \frac{\exp\,(6(N_p)^{\frac{1}{2}}\chi t)}{1920\ N_p} \tag{22}$$

where N_p is the mean number of pump photons.

The first term is the usual term arising in the classical pump approximation. The second term is a correction arising from the pump quantisation. The variance described in eq. (22) has a minimum given by

$$V_{min}\ (X_\theta) = \frac{1}{16(10N_p)^{\frac{1}{4}}} \tag{23}$$

Thus for sufficiently intense pump fields a value of $V(X_\theta)$ close to zero is possible. $V(X_\theta)$ is plotted in Fig. (2) for $N_p = 10^4$. $N_p(\tau)$ is also plotted showing that maximum squeezing occurs in the region of minimal pump depletion.

We conclude from these results that treating the pump field classically is reasonable except at very low pump intensities.

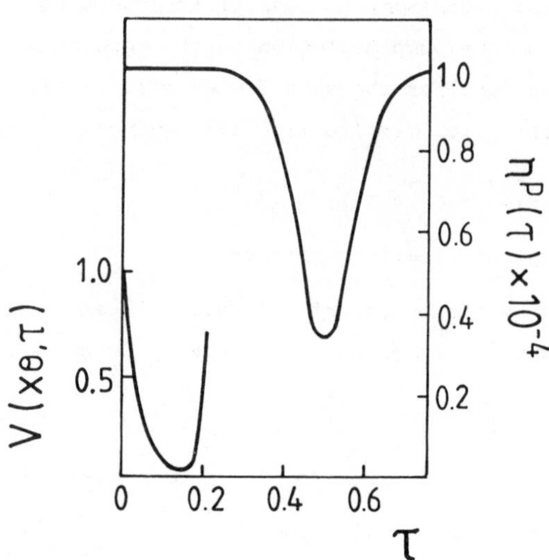

Fig.2 $V(X_\theta,\tau)$ and $N_p(\tau)$ for a parametric amplifier with $N_p(0) = 10^4$

6. INTRACAVITY CONFIGURATIONS

Calculations on intracavity nonlinear optical devices initially focussed on the photon statistics of the intracavity mode. The parametric oscillator was found to have maximum squeezing close to threshold with $V(X_\theta) = \frac{1}{2}$[20,21]. This corresponds to a reduction in fluctuations by a factor of only two. This limitation on the squeezing was generic to a number of nonlinear interactions.

However it is more relevant to calculate the squeezing in the output field of the cavity. The output field is a broadband continium and may have squeezing in certain frequency bands which exceeds that in the intracavity mode. A single mode analysis by Yurke[22] showed that the optimum cavity configuration to obtain maximum squeezing was a cavity with a single input/output port. This prevented unwanted vacuum fluctuations from entering the cavity. This theory was generalized to a multimode analysis by Collett and Gardiner[23] and others[24-26]. We shall first give a general prescription for calculating the squeezing spectrum for the output field from a cavity following the procedure of Collett and Gardiner[23].

6.1 Squeezing spectrum of the output field

Our objective is to calculate the squeezing spectrum of the output field of an optical cavity containing a nonlinear medium. In order to do this, we make use of the following expressions relating the two time correlation functions of the output field to the internal field. For an optical cavity for which the input field is entirely coherent or vacuum, it is possible to express the moments of the output field quite simply in terms of time-ordered moments of the internal field[(23)]. In particular

$$\langle a^{\dagger}_{out}(t), a_{out}(t')\rangle = 2\gamma\langle a^{\dagger}(t), a(t')\rangle \quad (24)$$

$$\langle a_{out}(t), a_{out}(t')\rangle = 2\gamma\langle a\left[\max(t,t')\right], a\left[\min(t,t')\right]\rangle \quad (25)$$

where γ is the damping constant of the internal field, and we have used the notation $\langle a,b\rangle = \langle ab\rangle - \langle a\rangle\langle b\rangle$. Now the behavior of quantum optical systems is frequently described using c number representations. Implicit in such a representation is a favored ordering of the system operators. For example, in the P function representation of the density operator[27,28]

$$\rho = \int P(\alpha)\,|\alpha\rangle\langle\alpha|\,d^2\alpha \quad (26)$$

equal time moments of the c number variables correspond to normally ordered moments of the operators. Two time moments imply precisely the time ordering of the internal operators that we need to evaluate the corresponding output moments. This can be demonstrated by noting that the evolution of the system will, in general mix $a^{\dagger}$ and a. Hence, $a(t + \tau)$ contains elements of both $a(t)$ and $a^{\dagger}(t)$. In a normally ordered two time product $a(t + \tau)$ must, therefore, stand to the left of $a(t)$; similarly, $a^{\dagger}(t + \tau)$ must stand to the right of $a^{\dagger}(t)$. That is,

$$\langle\alpha(t + \tau)\alpha(t)\rangle = \langle a(t + \tau)a(t)\rangle \quad (27)$$

$$\langle\alpha^*(t + \tau)\alpha^*(t)\rangle = \langle a^{\dagger}(t)a^{\dagger}(t + \tau)\rangle \quad (28)$$

where the left-hand side of these equations represent the average of c number variables using the P representation.

Including the factor of 2γ going from inside to outside the cavity, we can, therefore, obtain the normally ordered output correlation matrix using

$$:C_{out}(a(t+\tau),a(t)): = \begin{bmatrix} \langle a_{out}(t+\tau),a_{out}(t)\rangle & \langle a^{\dagger}_{out}(t),a_{out}(t+\tau)\rangle \\ \langle a^{\dagger}_{out}(t+\tau),a_{out}(t)\rangle & \langle a^{\dagger}_{out}(t+\tau),a^{\dagger}_{out}(t)\rangle \end{bmatrix}$$

$$=2\gamma \begin{bmatrix} \langle \alpha(t+\tau),\alpha(t)\rangle & \langle \alpha(t+\tau),\alpha^*(t)\rangle \\ \langle \alpha^*(t+\tau),\alpha(t)\rangle & \langle \alpha^*(t+\tau),\alpha^*(t)\rangle \end{bmatrix}$$

$$=2\gamma C_p(\alpha(t+\tau),\alpha(t)) \tag{29}$$

That is, the two time correlation functions for the output field may be calculated directly from the correlation functions of the stochastic variables describing the internal field using the P representation.

Note that all different time output operators commute. From Eq. (29) it follows immediately that the corresponding relation for the spectra also holds,

$$\begin{bmatrix} \langle \tilde{a}_{out}(\omega),a_{out}\rangle & \langle a^{\dagger}_{out},\tilde{a}_{out}(\omega)\rangle \\ \langle \tilde{a}^{\dagger}_{out}(\omega),a_{out}\rangle & \langle a^{\dagger}_{out}(\omega),\tilde{a}^{\dagger}_{out}\rangle \end{bmatrix} = 2\gamma \begin{bmatrix} \langle \tilde{\alpha}(\omega),\alpha\rangle & \langle \tilde{\alpha}(\omega),\alpha^*\rangle \\ \langle \tilde{\alpha}^*(\omega),\alpha\rangle & \langle \tilde{\alpha}^*(\omega),\alpha^*\rangle \end{bmatrix} \tag{30}$$

where the tilde indicates a frequency space variable.

For a linearized system $P(\alpha)$ obeys the Fokker-Planck equation

$$\frac{\partial P(\alpha)}{\partial t} = \left\{\frac{\partial}{\partial \alpha_i} A\alpha_i + \frac{1}{2} D \frac{\partial^2}{\partial \alpha_i \alpha_j}\right\} P(\alpha) \tag{31}$$

where A is the drift and D the diffusion matrix. This is equivalent to the stochastic differential equation

$$\frac{\partial}{\partial t}\alpha(t) = -A\alpha(t) + D^{\frac{1}{2}}\varepsilon(t) \tag{32}$$

where $\varepsilon(t)$ is a fluctuating force with

$$\langle \varepsilon_i(t)\rangle = 0$$

$$\langle \varepsilon_i(t)\varepsilon_j(t')\rangle = \delta_{ij}\delta(t-t'). \tag{33}$$

The spectral matrix may be obtained directly from Eq. (32) as (29)

$$S(\omega) = (A+i\omega)^{-1}D(A^T - i\omega)^{-1}. \tag{34}$$

To study squeezing we need to look at the variance of the quadrature phase X_θ where

$$X_\theta = ae^{-i\theta} + a^\dagger e^{i\theta} \tag{35}$$

The phase $e^{i\theta}$ will be chosen to maximize the squeezing. The squeezing spectrum of the output field is

$$\begin{aligned} V(X_{\theta,\omega}) &= 1 + \langle :\tilde{X}_\theta(\omega),\, X_\theta: \rangle \\ &= 1 + 2\gamma(e^{-2i\theta}S_{11}(\omega) + S_{12}(\omega) + S_{21}(\omega) + e^{2i\theta}S_{22}(\omega).) \end{aligned} \tag{36}$$

This result holds for the most favorable configuration, that of a "single-ended cavity." What is really meant by this is that the only losses from the cavity are through the front mirror which acts as an input and output port.

Thus, we now have a general presciption to calculate the squeezing in the output field from an optical cavity, provided the internal field may be described by a Fokker-Planck equation of the form of Eq.(31). In the following sections we shall apply this formalism to some particular systems.

6.2 Parametric oscillation and four wave mixing

(a) Degenerate Case. A model for degenerate parametric oscillation or intracavity four wave mixing may be given by the Hamiltonian

$$H = \frac{i\hbar}{2}(\varepsilon^* a^2 - \varepsilon a^{\dagger 2}) + a\Gamma_c^\dagger + a^\dagger\Gamma_c \tag{37}$$

where Γ_c, $\Gamma_c^\dagger$ are the reservoir operators describing the cavity damping. Following standard techniques the P function for the cavity mode obeys a Fokker Planck equation of the form (31) with

$$A = \begin{pmatrix} \frac{\gamma}{2} & \varepsilon^* \\ \varepsilon & \frac{\gamma}{2} \end{pmatrix} \qquad D = \begin{pmatrix} -\varepsilon & 0 \\ 0 & -\varepsilon^* \end{pmatrix} \tag{38}$$

The squeezing spectrum of the output light may then be calculated using eq. (36). The squeezing spectra near the oscillation threshold are plotted in Fig. (3) for a single ended cavity and a double ended cavity with two equally transmitting mirrors. The advantage of the single ended cavity is clearly apparent. A narrow spectral band around the

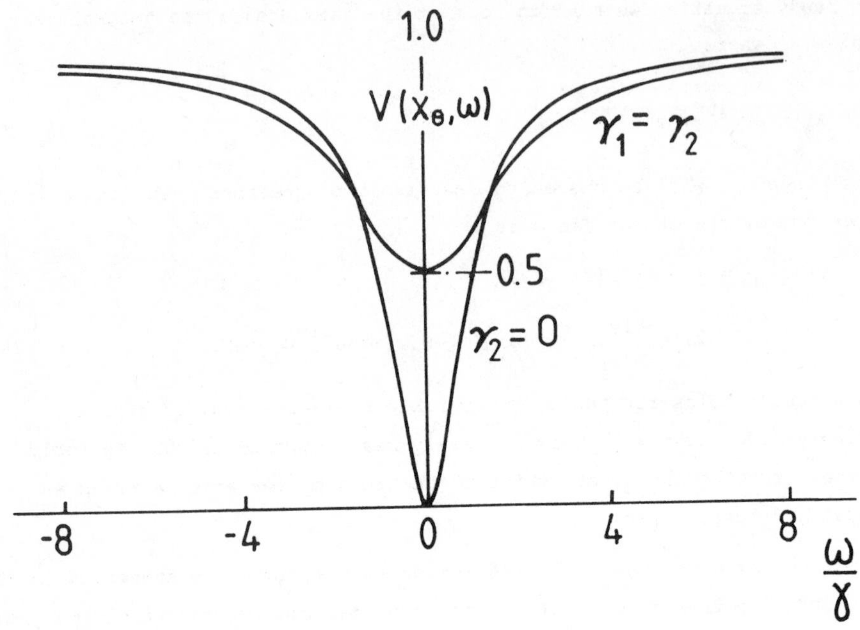

Fig.3 Squeezing spectrum for the parametric oscillator at threshold

cavity resonance exhibits a large squeezing $V(X_\theta) \to 0$, for the single ended cavity. The squeezing in this spectral band may be measured using the homodyne detection scheme described in §3.

It is clear that in order to achieve maximum squeezing in one quadrature ($V(X_\theta) \to 0$) the fluctuations in the other quadrature must tend to infinity $V(X_{\theta+\frac{\pi}{2}}) \to \infty$. Such behaviour may occur near critical points with phase sensitive critical fluctuations. In the above example the critical fluctuations occur at the threshold for parametric oscillation.

<u>(b) Non degenerate case.</u> It may be advantageous to observe the squeezing in frequencies away from the pump in order to avoid noise sources inherent in the pump. The squeezing spectrum for non degenerate four wave mixing in a cavity has been calculated by Yurke[25]. He calculated the squeezing in pairs of cavity modes symmetrically displaced from the pump field. Similar results are obtained as for the degenerate case except that the phase of the squeezed quadrature changes by $\frac{\pi}{2}$ for adjacent cavity modes.

(c) Optical bistability. We shall describe a model of dispersive optical bistability based on the phenomenological Hamiltonian[30].

$$H = \hbar\omega a^{\dagger}a + \varepsilon_p (a+a^{\dagger}) + \chi(a^{\dagger}a)^2 + a^{\dagger}\Gamma_c + a\Gamma_c^{\dagger} \tag{39}$$

where ε_p is the amplitude of the pump field, χ is the third order non-linear susceptibility and Γ_c are reservoir operators representing cavity damping. The Fokker Planck equation for the P function of the cavity mode has been derived by Drummond and Walls[30]. The drift and diffusion matrices are

$$A = \begin{pmatrix} \gamma_c + 2\varepsilon + i\phi & \varepsilon e^{i\theta} \\ -\varepsilon^* e^{-i\theta} & \gamma_c + 2\varepsilon^* - i\phi \end{pmatrix}$$

$$D = \begin{pmatrix} -\varepsilon e^{i\theta} & 0 \\ 0 & \varepsilon^* e^{-i\theta} \end{pmatrix} \tag{40}$$

where γ_c is the cavity damping rate, ϕ is the detuning of the pump field from the cavity and $\varepsilon = -2i\ \chi|\alpha_0|^2$ where α_0 is the steady state solution for cavity amplitude, $e^{i\theta}$ is the phase of the steady state solution. A calculation of the squeezing spectrum for the output field from a single ended cavity has been given by Collett and Walls[26]. At the turning points for optical bistability there are phase dependent critical fluctuations. The fluctuations of one quadrature in a small bandwidth about the cavity resonance tend to zero $V(X_\theta, 0) \to 0$.

This particular model for optical bistability is based on a classical susceptibility χ. In section 7 a quantised model for the medium is considered which places certain restrictions on the attainment of good squeezing.

To avoid noise due to pump fluctuations it is advantageous to look at the fluctuations in the cavity modes adjacent to the pump. Calculations reveal that squeezing is obtained in the cavity modes adjacent to the pump at the threshold for the pump mode to become bistable[31].

(d) Second harmonic generation. Predictions of squeezing in second harmonic generation have been made. These have either been calculations on the intracavity mode[32] or a travelling wave configuration[33]. The optimum squeezing however will occur near the threshold for self oscillation in a cavity[26]. The Hamiltonian for this system is

$$H = \hbar\omega a_1^{\dagger} a_1 + 2\hbar\omega a_2^{\dagger} a_2 + \varepsilon_p (a_1^{\dagger} + a_1) - i\hbar\chi(a_1^2 a_2^{\dagger} - a_1^{\dagger 2} a_2) + (a_1\Gamma_1^{\dagger} + a_2\Gamma_2^{\dagger} + h.c.) \tag{41}$$

where a_1 and a_2 are the boson annihilation operators for the fundamental and second harmonic modes respectively. Γ_i are the reservoir operators for each mode, and ε_p is the amplitude of the pump at the fundamental frequency.

The Fokker Planck equation for the P function of the cavity modes has been derived by Drummond et al.[34]. The drift and diffusion matrices are

$$A = \begin{pmatrix} \gamma_1 & -\varepsilon_2 & -\varepsilon_1^* & 0 \\ -\varepsilon_2^* & \gamma_1 & 0 & -\varepsilon_1 \\ \varepsilon_1 & 0 & \gamma_2 & 0 \\ 0 & \varepsilon_1^* & 0 & \gamma_2 \end{pmatrix} \tag{42}$$

$$D = \begin{pmatrix} \varepsilon_2 & 0 & 0 & 0 \\ 0 & \varepsilon_2^* & 0 & 0 \\ 0 & 0 & 0 & 0 \\ 0 & 0 & 0 & 0 \end{pmatrix} \tag{43}$$

$\varepsilon_2 = \kappa\alpha_2^0$, $\varepsilon_1 = \kappa\alpha_1^0$ where α_1^0, α_2^0 are the steady state cavity field amplitudes, and γ_1 and γ_2 are the cavity damping rates at the fundamental and second harmonic frequencies.

For values of $\varepsilon_p > \varepsilon_p^c = \gamma_1 + \gamma_2$ the output from the cavity exhibits self pulsing. At the instability point $\varepsilon_p = \varepsilon_p^c$ there are phase dependent critical fluctuations. The squeezing spectrum for the output field close to the instability point is displayed in Fig. (4)[26]. The maximum squeezing occurs in the fundamental mode for $\gamma_1 >> \gamma_2$, at $\omega = \pm\omega_c = \gamma_2$ where ω_c is the frequency of the oscillations at the critical point.

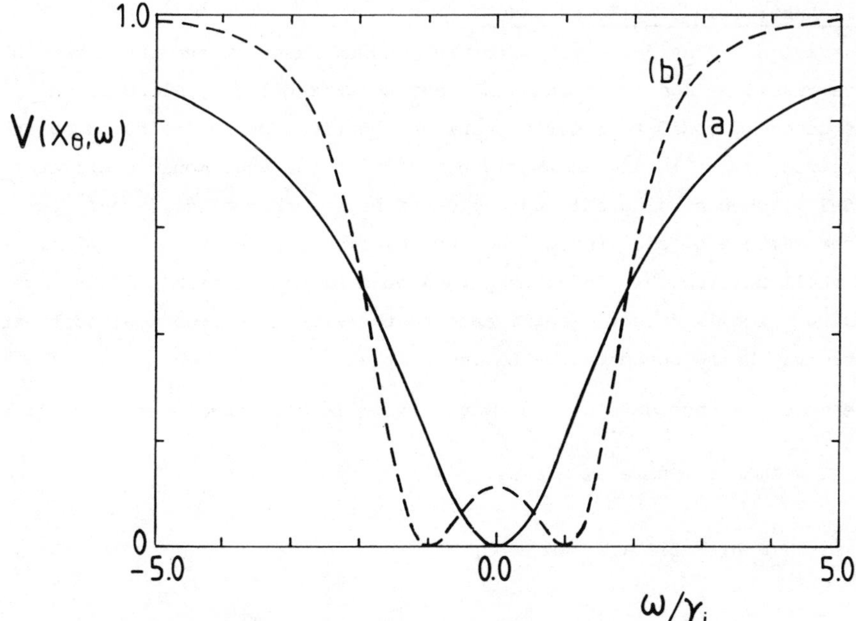

Fig.4 Squeezing spectrum for second harmonic generation
(a) Fundamental ($\gamma_1 >> \gamma_2$)
(b) Second harmonic ($\gamma_2 >> \gamma_1$)

7. QUANTISATION OF THE MEDIUM

In the previous sections the medium has been described by a classical nonlinear susceptibility χ. In order to ascertain the effects of spontaneous emission from the medium it is necessary to quantise the medium. This will enable us to determine under what conditions and how closely a real medium may be described by a classical susceptibility.

We shall discuss three models for the medium, the first an ensemble of anharmonic oscillators is a direct quantisation of a familiar classical model for nonlinear polarizability. This model is appropriate for media with a multitude of levels far from the laser frequency such as interactions in optical fibers.

In the other two cases we model the medium by an ensemble of two or three level atoms. These models are appropriate to experiments in an atomic vapour.

We shall consider four wave mixing in media described by the above models. We shall also consider optical bistability in a two level atomic medium.

7.1 Anharmonic oscillator model

We shall now consider a model for the medium based on the quantisation of a classical model for the nonlinear polarizability. Bloembergen(17) introduced a model to describe nonlinear optical phenomena based on an electron moving in an anharmonic potential. The anharmonic oscillator model is phenomenological and may describe a number of different microscopi mechanisms, e.g. electronic, phonons, excitons, polaritons, or plasmons. We shall utilize this model to give a quantum description of four wave mixing which is directly applicable to the experiments on a glass fiber currently being conducted by Levenson et al.(6)

The model we consider for four wave mixing is described by the Hamiltonian

$$\begin{aligned} H = {} & \hbar\omega_0 b^\dagger b + \hbar\omega(a_1^\dagger a_1 + a_2^\dagger a_2) \\ & + \hbar\chi(b^\dagger b)^2 + \varepsilon_p (b + b^\dagger) \\ & + \hbar g b\,(\varepsilon_p{}^* + a_1^\dagger + a_2^\dagger) + \hbar g b^\dagger(\varepsilon_p + a_1 + a_2) \\ & + b\Gamma_R{}^\dagger + b^\dagger\Gamma_R \end{aligned} \tag{44}$$

where b is the annihilation operator for the medium excitation of frequency ω_0, χ is the anharmonicity parameter. a_1, a_2 are the annihilation operators for the two signal modes of the field and ε_p is the classical amplitude of the pump mode. Damping of the medium excitation is included via coupling to a reservoir Γ_R, g is the coupling constant between the field and the medium.

In reference (35) the following stochastic differential equations for the field mode amplitudes α_1 and α_2 are derived after adiabatic elimination of the medium variables.

$$\begin{aligned} \dot{\alpha}_1 &= -\gamma\alpha_1 + \chi\alpha_2^* + \Gamma_1(t) \\ \dot{\alpha}_1 &= -\gamma\alpha_2 + \chi\alpha_1^* + \Gamma_2(t) \end{aligned} \tag{45}$$

where $\Gamma_1(t)$ and $\Gamma_2(t)$ are delta correlated fluctuating forces. The variance of fluctuations in the quadrature of the combined mode $(a_1 + a_2)$ has been calculated in ref.(35).

The anharmonic oscillator model has been used by Levenson et al.[6] to model four wave mixing in an optical fiber. In Levenson's experiment four wave mixing in a forward configuration is performed in an optical fiber. The pump beam emerging from the fiber then acts as a local oscillator beating with the sidebands directly on the photodetector. An element is introduced to give a preferential phase shift to the local oscillator in order to pick out the quadrature with maximum squeezing.

The predictions of the anharmonic oscillator model are shown in Fig.(5). where the variance is plotted against the squeeze parameter $r = |\chi|\ell$.

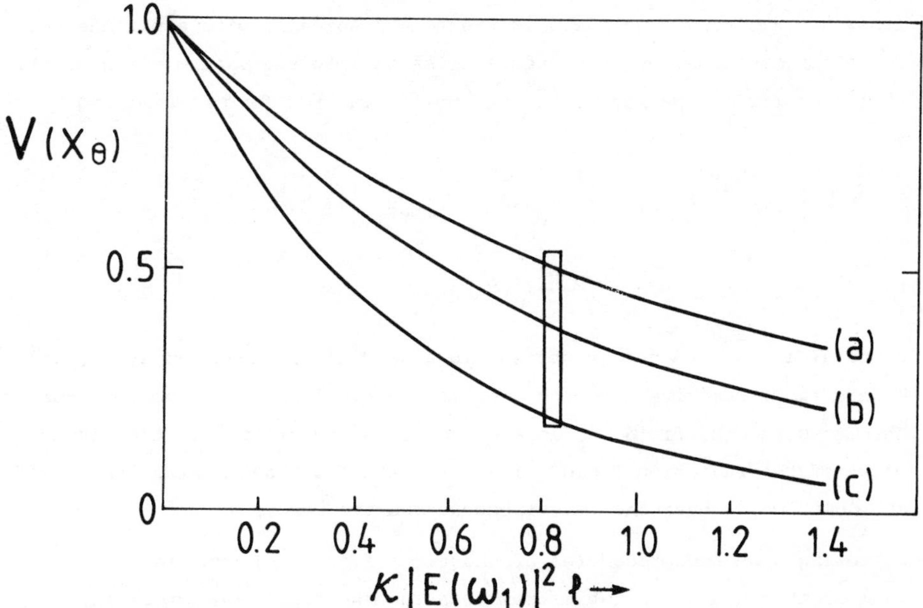

Fig.5 Squeezing via four wave mixing in an anharmonic medium (L = 100m)
(a) $\alpha = 3.10^{-3}m^{-1}$, (b) $\alpha = 1.10^{-3}m^{-1}$, (c) $\alpha = 0$

The curves shown are for an interaction length of 100m and

a) $\alpha = 3.10^{-3}m^{-1}$ b) $\alpha = 1.10^{-3}m^{-1}$

(α is the absorption per unit length). Curve (c) is the result for the ideal Hamiltonian (19) without loss. The area in the box denotes the region accessible in the glass fiber experiment (1 Watt focussed into a 4μ diameter fiber). This corresponds to a 50 % reduction in fluctuations.

These results are consistent with the predictions of Kumar and Shapiro[36] who introduced phenomenological loss terms.

Recent experiments by Levenson et al.[37] have demonstrated the squeezing of classical noise. The observation of the squeezing of quantum noise is presently obscured by guided acoustic wave Brillouin scattering occuring in the fiber.

7.2 Two level atom model.

Degenerate four wave mixing. We consider a model of degenerate four wave mixing in a medium consisting of N two level atoms, characterized by transverse and longitudinal decay rates $\gamma_\perp$ and $\gamma_{||}$. The atoms are driven off resonance with a detuning Δ scaled in units of the transverse linewidth $\gamma_\perp$. The pump field is treated classically with amplitude ε_p. The atoms may decay by spontaneous emission into the vacuum modes of the radiation field. We may write the Hamiltonian for four wave mixing as

$$H = \left[\varepsilon_p e^{i\underset{\sim}{k}_p\underset{\sim}{r}} + (a_1 e^{i\underset{\sim}{k}_1\underset{\sim}{r}} + a_2 e^{i\underset{\sim}{k}_2\underset{\sim}{r}})\right] \sum g_i \sigma_j^\dagger + \text{h.c.}$$

$$+ \sigma_j \Gamma_R^\dagger + \text{h.c.} \tag{46}$$

where a_1 and a_2 are the quantised signal and idler modes and $\underset{\sim}{k}_1$, $\underset{\sim}{k}_2$ and $\underset{\sim}{k}_p$ are the respective k vectors, g_j is the dipole matrix element coupling the atoms and the field, Γ_R is a heat bath operator for all the vacuum modes of the radiation field. In degenerate four wave mixing the signal and idler modes have the same frequency as the pump.

Following standard procedures stochastic differential equations for equivalent c number variables representing the atomic operators may be derived. After adiabatic eliminiation of the atomic variables this yields the following stochastic differential equations for the field mode variables[38]

$$\dot{\alpha}_1 = -\gamma_A(I,\Delta)\alpha_1 + \chi(I,\Delta)\alpha_2^* + \Gamma_1(t)$$

$$\dot{\alpha}_2 = -\gamma_A(I,\Delta)\alpha_2 + \chi(I,\Delta)\alpha_1^* + \Gamma_2(t) \tag{47}$$

where the complex loss and nonlinear coefficients γ ($\gamma = \gamma_A + i\gamma_I$) and χ are given for a forward configuration by

$$\gamma = \frac{\frac{g^2 N}{\gamma_\perp}}{(1 + \Delta)^2 \left(1 + \frac{I}{I_s}\right)^2}$$

$$\chi = \frac{\frac{g^2 N}{\gamma_\perp}(1 - i\Delta)}{(1 + \Delta^2)^2 \left(1 + \frac{I}{I_s}\right)^2} \tag{48}$$

$I = |\varepsilon_p|^2$, $I_s = n_o(1 + \Delta^2)$ where $n_o = \frac{\gamma_\perp \gamma_{||}}{4g^2}$ is the line centre saturation intensity and $f = \frac{\gamma_{||}}{2\gamma_\perp}$ is the collisional parameter. In the limit $f = 1$ we have pure radiative damping whereas for $f = 0$ we are in the collisional damped limit. The nonzero noise correlations are

$$\begin{aligned} &\langle \Gamma_1(t)\Gamma_2(t')\rangle = R\,\delta(t - t') \\ &\langle \Gamma_1(t)\Gamma_1^\dagger(t')\rangle = \langle \Gamma_2(t)\Gamma_2^\dagger(t')\rangle = \Lambda\delta(t - t) \end{aligned} \tag{49}$$

where $R = R_R + i\,R_I$

$$R_R = \frac{g^2 N}{\gamma_\perp}\,\frac{I}{I_s}\,\frac{1}{(1 + \Delta^2)^2 \left(1 + \frac{I}{I_s}\right)^3} \times \left\{(1 - 3\Delta^2)f + \frac{I}{n_o}(1 - f)\Delta^2 + \frac{1}{2}\left(\frac{I}{n_o}\right)^2\right\}$$

$$R_I = \frac{g^2 N}{\gamma_\perp}\,\frac{I}{I_s}\,\frac{1}{(I + \Delta^2)^2 \left(1 + \frac{I}{I_s}\right)^3}\left\{f(\Delta^3 - 3\Delta) + \frac{I}{n_o}(1 - f)\Delta\right\}$$

$$\Lambda = \frac{g^2 N}{\gamma_\perp}\,\frac{I}{I_s}\,\frac{1}{(1 + \Delta^2)^2}\,\frac{1}{\left(1 + \frac{I}{I_s}\right)^3}\left\{(1 + \Delta^2)(1 - f) + \frac{I}{n_o}\left[(2 + (1 - f)\Delta^2) + \frac{1}{2}\left(\frac{I}{n_o}\right)^2\right]\right\}$$

(a) Forward four wave mixing. Equations (47) describe four wave mixing in a phase matched forward configuration (39). Similar equations have been derived for forward four wave mixing with phenomenological loss terms.[37] In that case the noise correlations are independent of Δ and I. The dependence of the noise correlations on Δ and I arises from the quantum noise due to spontaneous emission from the atoms.

We shall consider the squeezing in the quadrature phase X_θ of the combined mode $e = a_1 + a_2$. The variance of fluctuations in the optimally squeezed quadrature is[38]

$$V(X_\theta) = 1 - \frac{A}{1 + \frac{\gamma_A}{|\chi|}}\left(1 - e^{-(\gamma_A + |\chi|)t}\right) \tag{50}$$

A is a function of the noise correlations. Provided we are in the dispersive regime ($\Delta \gg 1$) and are operating far below saturation $\left(\frac{I}{I_s} \ll 1 \text{ and } \left(\frac{I}{I_s}\right)^2 \ll \frac{1}{\Delta}\right)$ the correlations are those predicted by the ideal Hamiltonian eq. (18) and $A \to 1$. Loss is then the major barrier to squeezing and this can only be overcome at sufficiently high pump powers such that the ratio of gain to loss is enhanced, that is

$$\frac{|\chi|}{\gamma_A} \sim \frac{I}{\Delta} \gg 1$$

or (51)

$$\frac{I}{I_s} \gg \frac{1}{\Delta}$$

In Fig. (6) the steady state value of $V(X_1)$ is plotted for various detunings. For low detuning ($\Delta = 100$) as the pump intensity is increased $V(X_1)$ initially decreases as the gain exceeds the losses. However, for larger values of the pump intensity $V(X_1)$ reaches a minimum and then increases as the atoms begin to gain population in the upper state and the ensuing spontaneous emission destroys the squeezing. This increase of $V(X_1)$ at higher intensities does not occur for the constant loss mechanism considered by Kumar and Shapiro[36] and is a direct consequence of the dependence of the noise terms in eq. (47) on the pump intensity. As the detuning is increased the effect of the spontaneous emission is less. For large detunings $\Delta = 10^4$ there is a large plateau where good squeezing is attainable. In the limit of very large detuning the two level atom

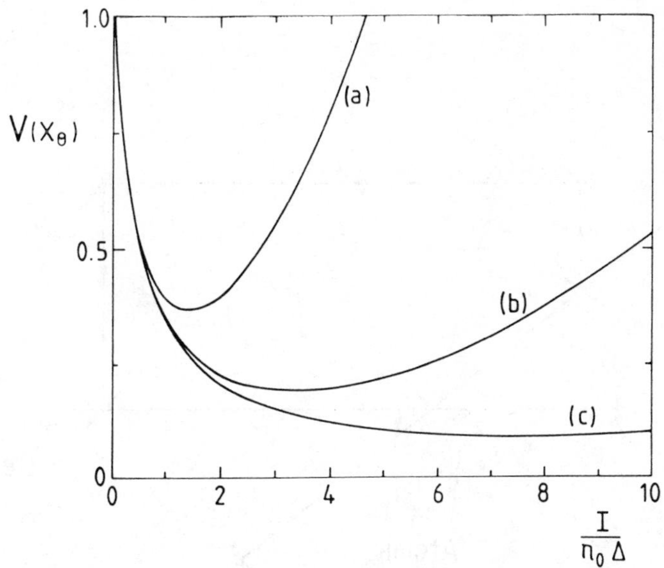

Fig.6 Squeezing via degenerate four wave mixing in a two level medium (a) $\Delta = 10^2$, (b) $\Delta = 10^3$, (c) $\Delta = 10^4$

model gives identical results to the anharmonic oscillator model.

To summarize, in order to achieve good squeezing, the gain must exceed the loss as expressed by eq. (51) $\left(\frac{I}{I_s} \gg \frac{1}{\Delta}\right)$. In addition in order to avoid the effects of spontaneous emission one must operate at large detunings ($\Delta \gg 1$) and low atomic saturation. $\left\{\frac{I}{I_s} \ll 1 \text{ and } \left(\frac{I}{I_s}\right)^2 \ll \frac{1}{\Delta}\right\}$. We note that in backward four wave mixing where a standing wave configuration is set up for the pump, spatial averaging results in the last condition being modified to $\left(\frac{I}{I_s}\right)^2 \ll \frac{1}{2.5\Delta}$. Finally in order that the gain exceeds the losses we require $\frac{I}{I_s} \gg \frac{1}{\Delta}$.

(b) Cavity configuration.

i) Degenerate four wave mixing. A four wave mixing experiment may be performed in an optical cavity as shown in Fig. (7). The scheme depicted is a ring cavity with two of the mirrors totally reflecting. The outputs of the third mirror are combined on a beam splitter and then homodyned with a local oscillator. The pumping beams are opposite in direction and external to the cavity. The equations for the field modes are given

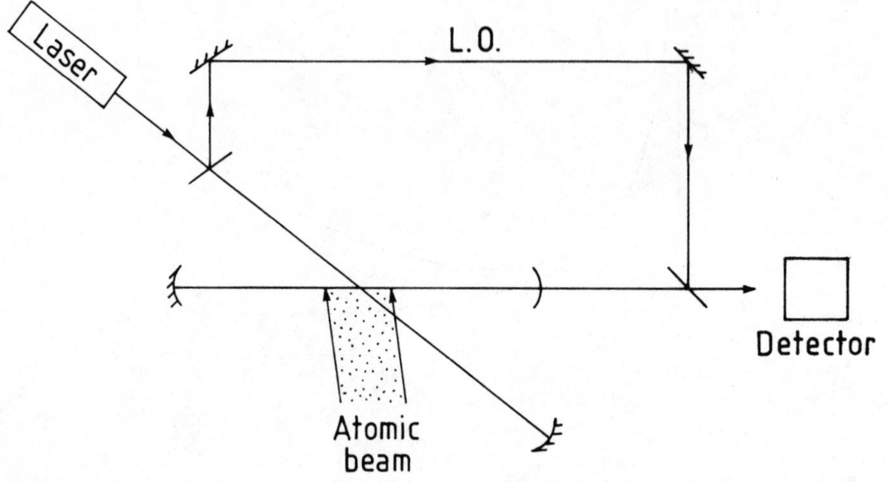

Fig.7 Experimental setup to detect squeezed states via four wave mixing

by eq. (47) where the loss term now includes the cavity loss γ_c in addition to the atomic losses. The spectrum of squeezing in the output field is calculated using the techniques described in §6. The maximum squeezing will occur in the vicinity of oscillation threshold.

In order to reach the threshold for oscillation one requires the gain to equ the losses, that is, $\gamma_c + \gamma_A = |\chi|$. For intensities satisfying $\frac{I}{I_s} << 1$ this may be written $1 + \frac{2C}{\Delta^2} = \frac{2C}{\Delta}\left(\frac{I}{I_s}\right)$ where $2C = \frac{g^2N}{\gamma_c\gamma_\perp}$. Since the atomic loss is required to be small $\left(\frac{2C}{\Delta^2} << 1\right)$ this leads to the condition $\frac{I}{I_s} >> \frac{1}{\Delta}$, which is the same as the condition for the gain to exceed the loss in the propagating case. The threshold condition implies $\frac{2C}{\Delta}\frac{I}{I_s} \sim 1$. Since we require $\frac{I}{I_s} << 1$ this implies we require a C value such that $\frac{2C}{\Delta} >> 1$. Thus there is a middle range of C values satisfying $\Delta << 2C << \Delta^2$ for which optimal squeezing is obtained. To avoid spontaneous emission due to atomic saturation we also require $\left(\frac{I}{I_s}\right)^2 << \frac{1}{\Delta}$ in a travelling wave cavity or $\left(\frac{I}{I_s}\right)^2 << \frac{1}{2.5\Delta}$ in a standing wave cavity.

The calculated results for $V(X_\theta,o)$ are shown in Fig. (8) for an atomic detuning of $\Delta = 10^3$ and different C values. C is the effective absorption, $C = \frac{\alpha_0 L}{1-R}$ where α_0 is the absorption per unit length or resonance, L is the length of the atomic sample and R is the mirror reflectivity.

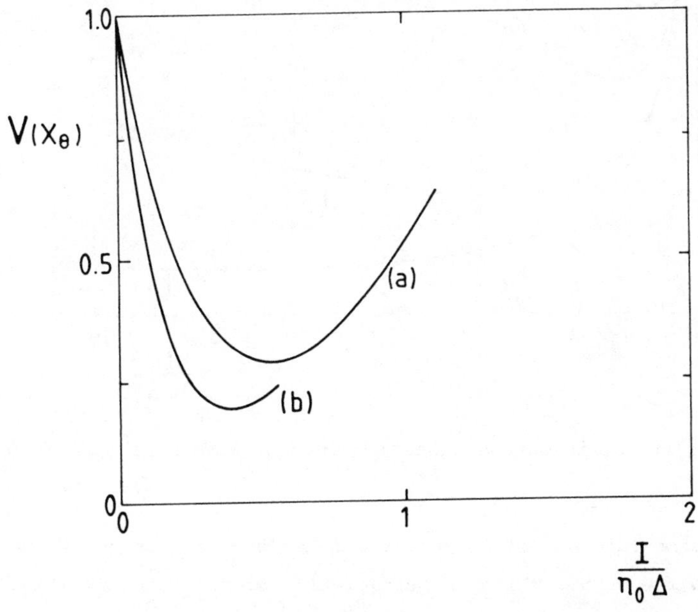

Fig.8 Squeezing via intracavity degenerate four wave mixing ($\Delta = 10^4$)

(a) $C = 2.5 \times 10^4$ (b) $C = 5 \times 10^4$

We see that good squeezing is possible with a $C \sim 5.10^4$. This C value may be obtained in a cavity with an output mirror reflectivity 0.95, and interaction length L = 10 cm and an atomic density of $\rho = 10^{11}/cm^3$. (using $\alpha_0 = \frac{3}{2\pi}\lambda^2\rho$). At densities of $10^{11}/cm^3$ the effect of collisions on the squeezing is minimal as the following argument shows. The effect of collisions on the squeezing is shown in Fig. (9) for different values of $f = \frac{\gamma_\perp}{2\gamma_\parallel}$ where $\gamma_\perp = \frac{\gamma_\parallel}{2} + \gamma_{coll}$, γ_{coll} being the collisional damping rate, showing that even for an f = 0.95 the squeezing is substantially reduced.

Resonant collisions between sodium atoms increase the collisional damping rate γ_{coll} by 1100 MHz per $10^{16}/cm^3$ pressure. A typical value for the

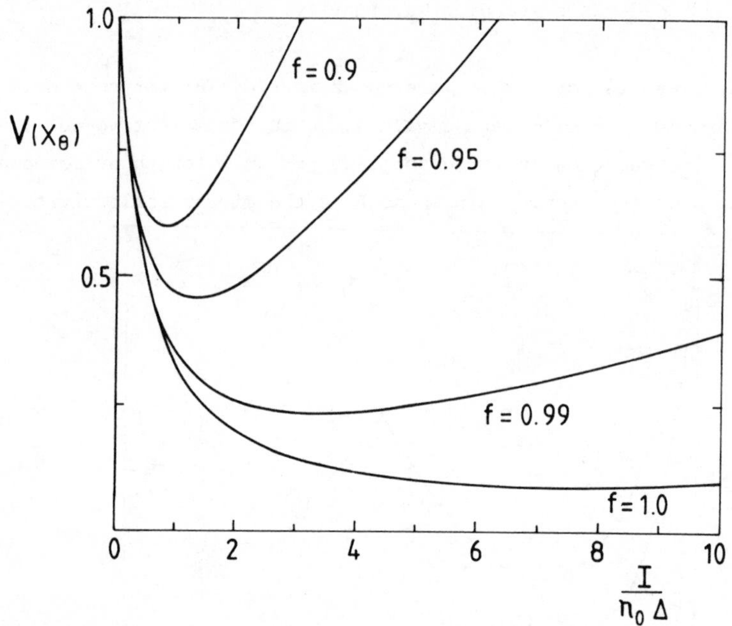

Fig.9 Effect of collisions on squeezing in four wave mixing ($\Delta = 10^4$)

radiative damping rate $\frac{\gamma_{\|}}{2}$ is 10 MHz. Hence an atomic density of $10^{11}/\text{cm}^3$ corresponds to f = 0.999, which is sufficiently close to one to allow good squeezing. We note that a detuning of $\Delta = 10^3$ is 10 times greater than the Doppler width so that the atoms are homogeneously broadened.

7.3 Optical bistability

An analysis of optical bistability based on a two level atom model for the medium may be made along similar lines to four wave mixing. In addition to the atomic detuning there is also the cavity detuning $\phi = \frac{\omega_p - \omega_c}{\gamma_c}$. Calculations for the squeezing obtained from a single ended cavity have been carried out by Reid and Walls.[40] The variance $V(X_\theta, o)$ for a small frequency band around the cavity resonance is plotted in Fig. (10) against the cavity field amplitude E_c for an atomic detuning $\Delta = 10^4$ and a cooperativity parameter $C = 2.10^6$. The maximum squeezing is obtained for a value of cavity detuning $\phi = 398$, which corresponds to the critical detuning for the onset of bistability. The value of the cavity field E_c which gives maximum squeezing corresponds to the point of inflection on the E_c versus E_p (pump field) curve.

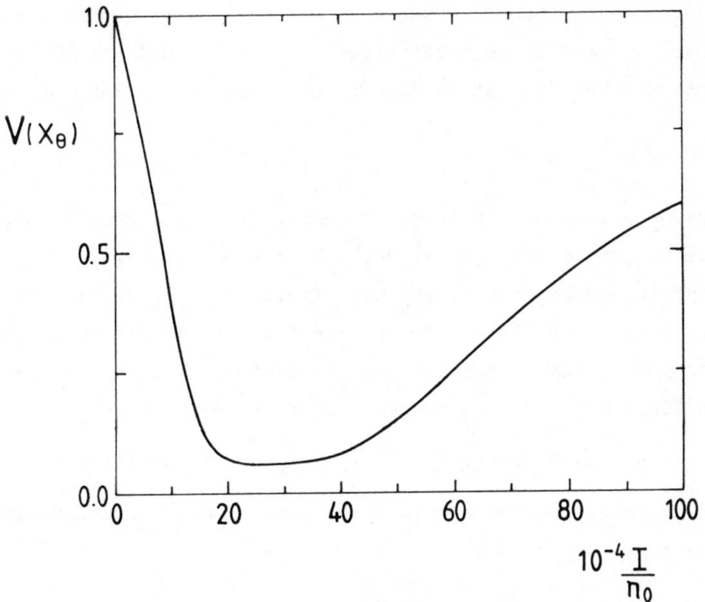

Fig.10 Squeezing via optical bistability in a two level medium
($\Delta = 10^4$, $C = 210^6$, $\phi = 398.4$)

7.4 Non-degenerate four wave mixing

Non-degenerate four wave mixing offers several advantages over degenerate four wave mixing. It allows one to look for squeezing in a spectral range where pump fluctuations and spontaneous emission noise are minimal.

Non-degenerate four wave mixing has been studied experimentally by Levenson et al.[6] in an optical fiber. Experiments on non-degenerate four wave mixing have also been performed in optical cavities, either with the pump noncollinear with the cavity as in the experiments by Slusher et al.[8] or with a collinear pump as in the experiments in a fiber ring cavity by Levenson et al.[6]

We begin with a general description of non-degenerate four wave mixing in an optical cavity.[41] The medium is modelled as N two level atoms with resonance frequency ω_0 and is interacting with three cavity modes of frequencies ω_1, ω_2 and ω_3. The cavity mode spacing is $\varepsilon = \omega_2 - \omega_1 = \omega_1 - \omega_3$. The amplitudes of the cavity modes are denoted by α_j. All three cavity modes are assumed to have the same cavity damping rate γ_c. The central cavity mode α_1 is driven by an external coherent input field E_p of frequency ω_p. The cavity detuning $\omega_1 - \omega_p$ is assumed to be much smaller than the cavity mode spacing ε. The atomic longitudianal and transverse decay rates are $\gamma_{||}$ and $\gamma_{\perp}$ respectively. The collisional parameter $f = \frac{\gamma_{||}}{2\gamma_{\perp}}$ is one for pure radiative damping and zero in the large collisional limit.

We proceed using standard techniques to derive quantum c number Langevin equations for the field mode amplitudes. We are interested in the equations describing the gain of the weak field modes α_2, α_3 in the presence of a very strong pump mode α_1. We thus treat α_1 to all orders, describing completely the saturation of the medium, while the expressions for the weak field modes α_2, α_3 are kept to first order only.

In the limit of a high Q cavity ($\kappa << \gamma_{\perp}, \gamma_{||}$) one is justified in adiabatically eliminating the atomic variables to obtain final equations for the field modes as follows.

$$\dot{\alpha}_1 = E_p - \gamma_c(1 + i\phi)\alpha_1 - \frac{2C\gamma_c\alpha_1}{(1+i\Delta_1)(1+\frac{I}{I_s})} + F_1(t) \qquad \text{(a)}$$

$$\dot{\alpha}_2 = -\gamma_c(1 + i\phi)\alpha_2 - \gamma(\delta)\alpha_2 + \chi(\delta)\alpha_3^{\dagger} + F_2(t) \qquad \text{(b)} \qquad (52)$$

$$\dot{\alpha}_3 = -\gamma_c(1 + i\phi)\alpha_3 - \gamma(-\delta)\alpha_3 + \chi(-\delta)\alpha_2^{\dagger} + F_3(t) \qquad \text{(c)}$$

and the nonzero noise correlations for the sidebands are

$$\langle F_2(t)F_3(t')\rangle = R\,\delta(t - t'), \quad R = R_R + iR_I$$

$$\langle F_2^{\dagger}(t)F_3^{\dagger}(t') = R^*\,\delta(t - t')$$

$$\langle F_2(t)F_2^{\dagger}(t')\rangle = \langle F_3(t)F_3^{\dagger}(t')\rangle$$

$$= \Lambda\;\delta(t - t').$$

The full explicit solutions are presented in ref. (40). The parameters γ, χ, R and Λ are functions of the following scaled variables: the cavity detuning, $\phi = (\omega_1 - \omega_p)/\gamma_c$; the detuning of the pump from the medium, $\Delta = (\omega_0 - \omega_p)/\gamma_{\perp}$; the detuning of the sidebands from the pump, $\delta = -\varepsilon/\gamma_{||}$; the cavity cooperativity parameter, $C = g^2N/2\gamma_{\perp}\gamma_c$; the collisional parameter f; and the scaled intracavity steady state pump intensity $I = |\alpha_1|^2$ and $I_s = n_0(1 + \Delta^2)$ where $n_0 = \gamma_{||}\gamma_{\perp}/4g^2$ is the resonant saturation intensity.

We have assumed in the derivation of equation (52) that $\Delta k = 2k_1 - k_2 - k_3$ is zero. This is always satisfied for the completely forward configuration, but not in general for alternative relative orientations of the field propagation vectors.

We have calculated the squeezing in the output light at the two sideband frequencies. Squeezing is observed in a heterodyne detection scheme where the output of the two sidebands at frequency $\omega_p \pm \varepsilon$ beat with a local oscillator $\varepsilon_{LO} = E_p e^{i\theta}$ at frequency ω_p. The local oscillator is obtained by phase shifting the external driving field. The sidebands and the local oscillator mix on the surface of a photodetector giving a photocurrent $i(\varepsilon)$. The spectrum of fluctuations in this photocurrent $\langle i^2(\varepsilon)\rangle$ are measured with a spectrum analyser. $V(X_\theta,\varepsilon)$ the spectral variance in the quadrature phase amplitude is defined by $X_\theta = ce^{-i\theta} + c^\dagger e^{i\theta}$ (where $c = \frac{1}{\sqrt{2}}(a_{2out} + a_{3out})$) is directly proportional to $\langle i^2(\varepsilon)\rangle$. Squeezing is characterized by $V(X_\theta,\varepsilon) < 1$.

The non-degenerate scheme ($\delta \neq 0$) offers advantages over the degenerate situation. The dephasing spontaneous emission noise Λ is due to the pump saturating the two-level atom and is peaked about the pump frequency. We thus look for a reduction in this noise as we increase δ. The limit of interest is that of large detuning of the pump relative to the atomic resonance ($\Delta >> 1$), low saturation ($\frac{I}{I_s} << 1$), pure radiative damping ($f = 1$), and a detuning δ of the sidebands relative to the pump, but such that $|\delta| << |\Delta|$. Then one finds a regime where γ_A and χ are essentially unchanged from the degenerate case (equation (48)), yet the desqueezing term Λ is reduced by the factor $(1 + \delta^2)$. The condition to avoid onset of spontaneous emission as one increases the pump intensity is now somewhat looser: $\left(\frac{I}{I_s}\right)^2 << \frac{(1 + \delta^2)}{\Delta}$. Thus one can afford higher pump intensities relative to the pump detuning. The implication is that squeezing becomes possible at lower pump detunings and hence, most importantly, for lower values of the cooperativity parameter C.

The dispersion term γ_I ($\mathrm{Im}\,\gamma(\delta)$) incorporates the (linear and nonlinear) change in the refractive indices due to the medium, and also any would-be phase mismatch $\Delta\underset{\sim}{k}$ that may be present in configurations where the $\underset{\sim}{k}_j$ vector have relative orientations. We analyse the simplest situation where the $\Delta\underset{\sim}{k}$ is chosen to allow perfect phase matching in the medium, that is, so that $\gamma_I = 0$ for both sidebands. Also, the pump mode α_1, distinguishable from the sidebands by a different $\underset{\sim}{k}$ direction, may or may not be in a cavity, but any cavity detuning ϕ is zero. Thus no bistability exists in this situation.

The variance of fluctuations at the sideband frequencies $V(X_\theta,\varepsilon)$ for this particular case are plotted in Figures 11 and 12 for a scaled detuning $\Delta_1 = 100$ of the pump from the atomic resonance. Solutions hold only about a stable steady state ($\alpha_2 = \alpha_3 = 0$) and the stop in the curves in Figure 11 indicate the onset of oscillation (threshold corresponds to $|\chi| = \kappa + \gamma_R$). The advantage obtainable in the nondegenerate case is immediately apparent. In the degenerate case, $\delta = 0$, the conditions $\frac{I}{I_s} >> \frac{1}{\Delta}$ and $\left(\frac{I}{I_s}\right)^2 << \frac{1}{\Delta}$ are optimised at $\frac{I}{n_o} \sim 400$ (Figure 11).

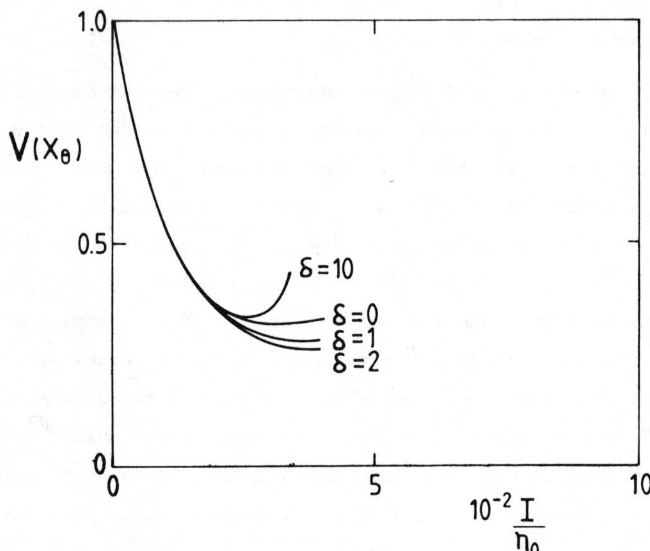

Fig.11 Squeezing in nondegenerate four wave mixing. High loss case ($\Delta = 100$, $C = 1750$)

Even at the relatively high detuning $\Delta \sim 100$ perfect optimisation has not been possible for $\delta = 0$, and the effect of atomic loss is still to limit the squeezing attainable. Thus in the degenerate situation, one would have to increase the detuning to improve squeezing ($\Delta \sim 10^4$, $2C \sim 10^6$, $I \sim 10^5$).

The conditions for the nondegenerate situation $\frac{I}{I_s} >> \frac{I}{\Delta}$ and $\left(\frac{I}{I_s}\right)^2 << \frac{1+\delta^2}{\Delta}$ mean that, for a given detuning Δ_1, one can increase intensities to a regime where loss is not important and still obtain good squeezing, as shown in Figure 12, for a C value of 250.

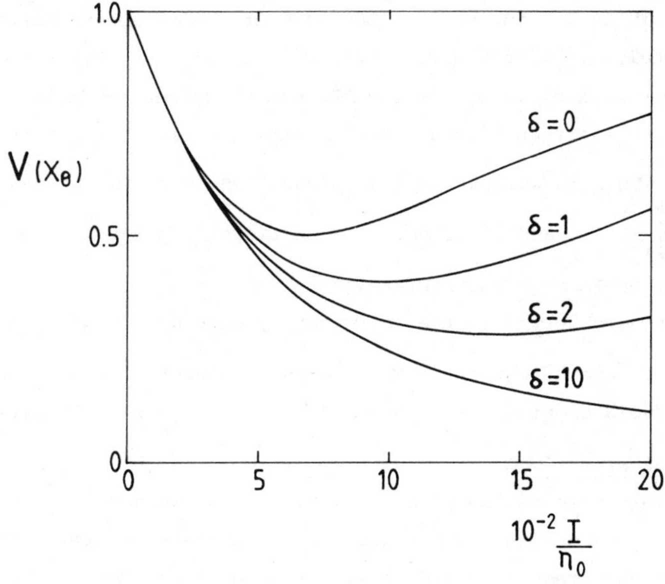

Fig.12 Squeezing in nondegenerate four wave mixing. Low loss case ($\Delta = 100$, $C = 250$)

Thus good squeezing is obtainable for lower detunings Δ and hence also for much lower C values. To avoid the regime of high atomic loss relative to the cavity loss, which still limits the squeezing possible and for which the nondegenerate situation is no advantage, we still stringently require $2C << \Delta^2$. This feature is illustrated in Figure 11, where $2C \sim \Delta^2$ and for which little improvement is seen upon increasing δ. One needs still $2C \sim \Delta$ (to avoid reaching total saturation before a reasonable order of I is obtained) but the sensitivity to this latter criterion is very much reduced and good squeezing is possible for a wider range of C values compared to the degenerate case. The advantage in terms of pump powers is not so significant, since we have to pump the medium harder to gain advantage. The lower detunings needed however do imply a net reduction in minimum powers needed to attain the same amount of good squeezing.

Comparison with Experiment

The observation of squeezing in non-degenerate four wave mixing in atomic sodium has recently been reported by Slusher et al.[50] We shall compare the foregoing theoretical conditions with the parameters achieved in this experiment. The sodium atoms are pumped at a frequency 1.5 GHz below the D_2 resonance giving a detuning $\Delta \sim 200$. The effective absorption $\alpha_0 L$ was approximately 7. With a mirror reflectivity of 0.98 this gives a C value of approximately 350, clearly the condition for low loss $2C << \Delta^2$ is satisfied, in addition $2C \gtrsim \Delta$. The intensities used were of the order of $\frac{I}{I_s} \sim 0.02$. This is well below the intensities required to reach threshold $\left(\frac{2C}{\Delta}\left(\frac{I}{I_s}\right) \sim 1\right)$ and in principle significantly greater squeezing should be achievable at higher intensities.

The squeezing was observed in the third cavity mode displaced from the pump frequency, a detuning of 400 MHz, corresponding to $\delta \sim 40$. The condition to avoid spontaneous emission $\left(\frac{I}{I_s}\right)^2 << \frac{1+\delta^2}{\Delta}$ is well satisfied.

7.5 Three level atom

From the foregoing discussion of a two level medium we have seen that the medium may become classical in the totally dispersive limit but one requires high intensities and densities to build up the nonlinearity. With an aim to reduce spontaneous emission without evoking such a high intensity we study two photon mechanisms in a three level medium.[42] A two photon detuning paramter δ is introduced. Such a mechanism unlike the two level medium, allows a nonlinearity without saturation.

The three level atoms may be in a lambda or ladder configuration see Fig. (13).

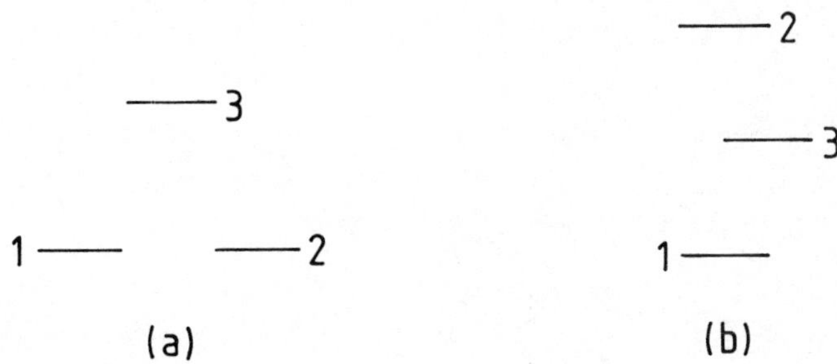

Fig.13 Three level atom
(a) lambda configuration
(b) ladder configuration

A strong inversion between levels 2 and 1 is required to enhance the two photon nonlinearity. The inversion is induced in the lambda system via an external incoherent pumping process with a rate Γ_{12} such that $\Gamma_{12} << \gamma_{||}$.

In a ladder configuration the rate is determined naturally via spontaneous emission ($\Gamma_{12} \sim \gamma_{||}$). Thus longitudinal ($\Gamma_{12}$) and transverse ($\frac{1}{\tau_2} = \frac{\Gamma_{12}}{2} + \gamma_p$, γ_p is the phase destroying collisional decay rate) decay rates are introduced between the 2 and 1 levels.

In the limits

$$\Delta >> 1$$

$$\frac{\alpha_0 L}{\Delta^2} << 1 \tag{54}$$

$$\frac{I}{I_s} << \frac{1}{\Delta}$$

$$\frac{\Gamma_{12}}{\gamma_{||}} >> \frac{I}{I_s}$$

complete elimination of the third level and the associated spontaneous emission is possible. In this limit the problem may be analysed using an effective two photon Hamiltonian [43]. The best squeezing is attained in the dispersive limit of the coherent two photon process and with low saturation.

$$|\delta| \gg \frac{\Gamma_{12}}{\gamma_{||}}, \quad \frac{1}{\tau_2 \gamma_{||}}$$
$$\frac{I}{I_s} \ll \frac{\delta}{\Delta} \tag{55}$$

In this regime the two photon susceptibility becomes

$$|\chi| L \sim \frac{\alpha_0 L}{8\ \delta} \frac{I}{I_s} \tag{56}$$

The total loss is made up of two parts - a one photon and a two photon contribution

$$\gamma L \sim \frac{\alpha_0 L}{\Delta^2} + \frac{\alpha_0 L}{8\gamma_{||}\tau_2} \frac{1}{\delta^2} \frac{I}{I_s} \tag{57}$$

The two photon loss is small compared to $|\chi|L$ in the two photon dispersive limit we are assuming. We also need the two photon loss small in absolute terms. To then enhance the $|\chi|L$ relative to the one photon loss one requires a minimum intensity such that

$$\frac{I}{I_s} \gg \frac{\delta}{\Delta^2}\ . \tag{58}$$

We rewrite (55) and (58) to summarise explicitly the minimum pump intensity required

$$\frac{I}{n_0} \gg \delta \gg \frac{\Gamma_{12}}{\gamma_{||}} \tag{59}$$

(n_0 is the resonant saturation intensity).

Importantly the pump intensity required to produce good squeezing in such three level systems is determined by the two photon detuning δ rather than the one photon detuning Δ as in the two level medium. Thus we may produce good squeezing at lower intensities in the three level system by employing a larger $|\Delta|$. There is an ultimate limit however, since $|\chi|L$ must be of significant order, implying $\alpha_0 L > \Delta$. Also we note one may reduce the intensity further in the lambda system by employing an incoherent pump rate Γ_{12} much slower than the spontaneous emission decay rate $\gamma_{||}$. The results for the squeezing for a bistable system in the limit of large one photon detuning (Δ >>1) and in the dispersive two photon limit ($\delta = 100\ \Gamma_{12}/\gamma_{||}$) is shown in Figure 14. Reid et al.[44] has suggested an alternative mechanism for squeezing based on atomic coherences in a three level atom. This will occur for low two photon detunings and because it is based on atomic coherences will be more sensitive to collisions.

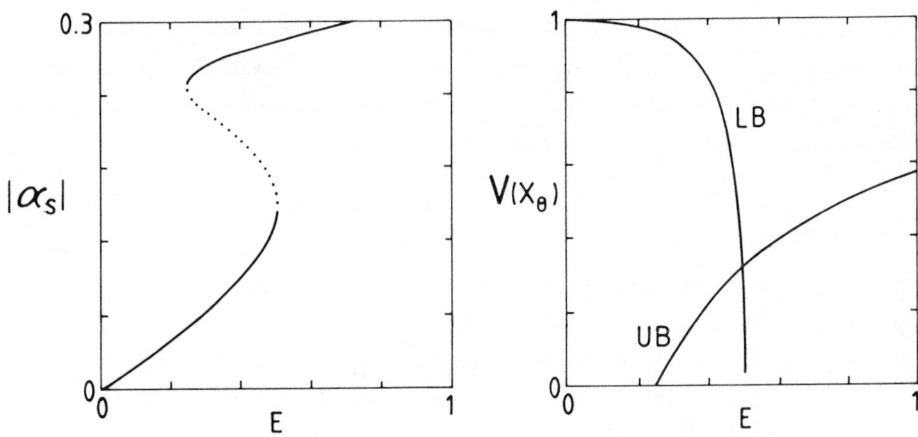

Fig.14 Squeezing via optical bistability in a three level medium $\left(\Delta >> 1,\quad \delta = 10\ \frac{\Gamma_{12}}{\gamma_{||}}\right)$

8. OTHER SCHEMES

A number of other interactions have been predicted to produce squeezing. Resonance fluorescence from a single two level atom has been predicted to produce squeezing with $V(X_1)$ min ~ 0.68[45,46]. In order to attain this minimum value all the fluorescent light must be collected before heterodyning. As a proof in principle demonstration of the existence of squeezing, resonance fluorescence from a single atom in an ion trap may be useful. However owing to the low signal strength and relatively small amount of squeezing this is hardly likely to be a practical source of squeezed state light.

Recent developments in Rydberg asers offer possibilities to produce squeezed states of microwave radiation. The squeezing in a Rydberg maser has recently been calculated by Heidmann et al.[47] There are also possibilities to produce squeezed states in Josephson parametric amplifiers working in the microwave regime[48].

Though there have been predictions of squeezing in free electron lasers, the squeezing has occurred in combined 'modes' of the radiation field and electrons. Calculations on the photon statistics of the radiation field alone[49] show no evidence of squeezing.

ACKNOWLEDGEMENTS

This paper represents a review of the work on squeezed states at the University of Waikato over the past few years. The authors gratefully acknowledge the contributions of M.J. Collett, B.J. Dalton, G.J. Milburn, C.W. Gardiner, C.M. Savage and G. Scharf. This work was supported by the New Zealand University Grants Committee.

REFERENCES

1) D. Stoler, Phys. Rev. D1 3217 (1970)

2) H.P. Yuen, Phys. Rev. A13 2226 (1976)

3) For a review see D.F. Walls Nature 306 141 (1983)

4) H.P. Yuen and J.H. Shapiro, IEEE Trans Inform Theory
IT 24 657 (1978)
IT 26 78 (1980)

5) C.M. Caves, Phys. Rev. 23D 1693 (1981)

6) M.D. Levenson, R.M. Shelby, M.D. Reid, D.F. Walls and A. Aspect Phys. Rev. A32, 1550 (1985)

7) R.S. Bondurant, P. Kumar, J.H. Shapiro and M. Maeda Phys. Rev. 30A 343 (1984)

8) R.E. Slusher, L. Hollberg, B. Yurke, J.C. Mertz and J.F. Valley Phys. Rev. 31A 3512 (1985)

9) G. Leuchs, to be published in "Non Equilibrium Quantum Statistical Physics" edited by G. Moore and M.O. Scully (Plenum Press)

10) H.P. Yuen and J.H. Shapiro, IEEE Trans Inform Theory IT-26 78 (1980)

11) J.H. Shapiro, H.P. Yuen and J.A. Machado Mata IEEE Trans. Inform Theory IT-25 179 (1979)

12) L. Mandel, Phys. Rev. Lett 49 136 (1982)

13) H.P. Yuen and V.W.S. Chan, Opt. Lett 8 177 (1983)

14) G.L. Abbas, V.W.S. Chan and T.K. Yee, Opt. Lett 8 419 (1983)

15) B.L. Schumaker, Opt Lett 9 189 (1984)

16) B. Yurke, Phys. Rev. 32A 3/1 (1985)

17) N. Bloembergen, "Non Linear Optics", Benjamin (1962)

18) B.R. Mollow and R.J. Glauber, Phys. Rev. 160 L097 (1967)

19) G. Scharf and D.F. Walls, Opt. Comm. 50 245 (1984)

20) G.J. Milburn and D.F. Walls, Opt. Comm. 39 401 (1981)

21) L.A. Lugiato and G. Strini, Opt. Comm. 41 67 (1982)

22) B. Yurke, Phys. Rev. 29A 408 (1984)

23) M.J. Collett and C.W. Gardiner, Phys Rev. 30A 1386 (1984)

24) C.W. Gardiner and C.M. Savage, Optics Comm. 50 173 (1984)

25) B. Yurke, Phys. Rev. A32 300 (1985)

26) M.J. Collett and D.F. Walls, Phys. Rev. A32 2887 (1985)

27) R.J. Glauber, Phys. Rev. 131 2766 (1963)

28) E.C.G. Sudarshan, Phys. Rev. Lett 10 277 (1963)

29) C.W. Gardiner, Handbook of Stochastic Methods, Springer (1983)

30) P.D. Drummond and D.F. Walls, J Phys. 13A 725 (1980)

31) R.M. Shelby, M.D. Levenson, D.F. Walls, A. Aspect and G.L. Milburn (to be published)

32) L.A. Lugiato, G. Strini, F. de Martini, Opt Lett 8 256 (1983)

33) L. Mandel, Opt. Comm 42 437 (1982)

34) P.D. Drummond, K.J. McNeil and D.F. Walls, Optica Acta 28 211 (1981)

35) M.D. Reid and D.F. Walls, J.O.S.A. 2B 1682 (1985)

36) P. Kumar and J.H. Shapiro, Phys. Rev. A30 1568 (1984)

37) M.D. Levenson, R. Shelby and S.H. Perlmutter, Optics Letter

38) M.D. Reid and D.F. Walls, Phys. Rev. 31A 1622 (1985)

39) Y. Prior, Appld Opt. 19 1741 (1980)

40) M.D. Reid and D.F. Walls, Phys. Rev. A32 396 (1985)

41) M.D. Reid and D.F. Walls, (to be published)

42) M.D. Reid, D.F. Walls and B.J. Dalton, (to be published)

43) C.M. Savage and D.F. Walls, (to be published)

44) M.D. Reid, D.F. Walls and B.J. Dalton, Phys. Rev. Lett 55 1288 (1985)

45) D.F. Walls and P. Zoller, Phys.Rev. Lett 47 709 (1981)

46) M.J. Collett, D.F. Walls and P. Zoller, Optics Comm. 52 145 (1984)

47) A. Heidmann, J.M. Raimond and S. Reynaud, Phys. Rev. Lett 54 326 (1985)

48) B. Yurke (private communication)

49) R. Bonifacio and F. Casagrande, Optics Comm. 50 251 (1984)

50) R.E. Slusher, L.W. Hollberg, B. Yurke, J.C. Mertz and J.F. Valley Phys. Rev. Lett. 55 2409 (1985)

POSSIBLE APPLICATIONS OF SQUEEZED STATES IN INTERFEROMETRIC TESTS OF GENERAL RELATIVITY

G LEUCHS AND M W HAMILTON

1. INTRODUCTION

There are essentially two classes of optical tests of the general theory of relativity; those performed in static and those performed in alternating gravitational fields. Tests of the first kind include the gravitational redshift, the deflection of starlight by the sun, the ranging of Venus, as well as the laser measurements of the distance between the earth and the moon (Vessot 1984). The second kind concerns the detection of gravitational radiation, a goal which is yet to be achieved. Both kinds of test are amenable to interferometric methods and it is in this context that we will discuss the possible use of squeezed states.

Most general relativistic effects for weak gravitational fields can be understood qualitatively by relying on an electromagnetic analogy (Thorne 1983). Thus one finds gravitational interactions if a test body is moving through a static gravitational field and if a rotating test body is near a spinning mass, the electromagnetic equivalents being the motional magnetic field ($\underline{v}\mathrm{x}\underline{E}$) and the magnetic moment of the spinning electron respectively. The discussion of these gravitomagnetic effects was started by Mach (1912) who postulated that not only velocity but also acceleration is a relative quantity. The quantitative theoretical treatment was given by Thirring and Lense (1918).

In the deflection of starlight by the sun one finds two equal contributions to the deflection angle, one due to the Newtonian potential of the sun acting on the mass of the photons (E/c^2) and the other one originating in general relativity. This latter is due to the curvature of space; in the gravitomagnetic picture it results from the motion of the photons through the static field of the sun. Carrying this analogy further one would also expect a contribution due to the rotation of the sun. The magnitude of this effect should be equal to the other effects if the surface of the sun rotates as fast as the photons move. However the sun revolves only once in 25 days. Consequently this last effect is only 10^{-6} of the Newtonian and the space curvature effects and is, therefore, not practicably observable (Schleich 1984). This gravitomagnetic effect of a spinning mass is also referred to as frame dragging since it "drags" the photon, increasing or decreasing the deflection angle depending on the direction of the sun's rotation. Other more promising ways of testing this gravitomagnetic effect are to use ultra-high precision gyroscopes to look at the effect of the spinning earth. The satellite based experiments using mechanical gyros, as developed by C.W.F. Everitt and co-workers at Stanford University, look promising but in this paper we will discuss the optical equivalent, gyroscopes that utilise the Sagnac effect, and their possible impact on measuring gravitomagnetic fields (Chow et al. 1985).

The other topic that will be discussed here, and which also relies on the sensitivity of interferometric devices, is the detection of gravitational waves (Thorne 1980). The interest in the detection of gravitational waves is two-fold; firstly it will provide another important test of Einstein's theory of general relativity. Unfortunately gravitational waves from any man-made source are much too weak to ever be detected. Astrophysical events on the scale of stellar catastrophes such as supernovae are necessary to produce waves which one may hope to detect. The study of such stellar catastrophes is the second reason motivating gravitational wave research since it will open new possibilities in astronomy providing information about the gravitational collapse in the final stages of stellar evolution and supernovae. Such direct information cannot be obtained by only observing electromagnetic waves.

Again we can use the analogy between gravitational and electromagnetic fields. Oscillating electric charges are the source for electromagnetic waves which are described by the relativistic generalisation of Coulomb's law. In a similar way the relativistic generalisation of Newton's laws of gravitation predicts gravitational waves. A major difference between both types of wave is that electromagnetic waves polarise matter whereas gravitational waves do not. In the electromagnetic case the strongest coupling between the oscillating charges is due to the dipole interaction. The electromagnetic power emitted by the source is proportional to the square of the second time derivative of the dipole moment ($\partial^2\underline{d}/\partial t^2$)^2. In the gravitational case no polarisation of matter is possible, therefore the interaction must be quadrupole in nature and the power emitted by an oscillating mass distribution is proportional to ($\partial^3Q/\partial t^3$)^2, where Q is the mass quadrupole moment. In addition, the gravitational coupling constant is many orders of magnitude smaller than the electromagnetic one. Hence the coupling of gravitational waves to any detector will be extremely weak, explaining why the direct observation of gravitational waves has not yet been achieved.

The effect of gravitational waves is to produce a quadrupole type of strain pattern, the wave amplitude being proportional to the strain $\Delta l/l$ (Misner et al. 1973). Pulses of gravitational waves from stellar catastrophes in the nearest cluster of galaxies, the Virgo cluster, are expected at a rate of one per month and the frequency should be in the range of 0.1–1.0 kHz. The estimate for the strain that such pulses produce on earth is $\Delta l/l < 10^{-21}$. If this strain is measured on a baseline as long as 10km the change of length due to the gravitational wave is only 0.01 of a nuclear radius. A stellar catastrophe such as a supernova occurring in our own galaxy might produce a strain amplitude as high as 10^{-18}, but only three such events are known to have occurred in the last millenium.

The first apparatus designed to detect gravitational waves was Weber's resonant bar (Weber 1979). When operated at room temperature the strain sensitivity for 1kHz pulses is about 10^{-17}. The bar operated at Stanford (Bassan et al. 1983) has been cooled to 2K and a strain sensitivity of $\Delta l/l = 10^{-18}$ has been reached. The bars are narrow band detectors by nature, although it is possible to extract information about pulsed excitation. The other scheme pursued experimentally, the laser interferometer, is a broad band detector. There are essentially three variants; the Michelson interferometer with either delay lines (Billing et al. 1983) or Fabry-Perot interferometers (Drever et al. 1983) in each arm to increase the effective arm length, and the set-up with two lasers oriented mutually orthogonally (Bagayev et al. 1981). The theoretical limit of sensitivity for the three different set-ups is the same provided that the same mirror reflectivity and linear dimension is assumed in each case. In the third variant where the beat frequency between the two lasers is measured, this sensitivity is reached if the linewidth of both lasers is limited only by spontaneous emission noise (Schawlow and Townes 1958), a requirement that is not easily met experimentally. This third variant may however be superior to the other set-ups if one can somehow correlate the spontaneous emission events in both lasers according to a proposal by Scully (1985). In the following we will however focus on the first variant, and discuss an improvement of the detection sensitivity by using squeezed states of the radiation field. A similar increase in sensitivity will be suggested for the Sagnac interferometer.

2. THE BEAMSPLITTER

Both the Michelson and Sagnac interferometers use a beamsplitter, one property of which is important to the proposals to use squeezed states. That is the half-wave phase change on reflection from the dielectric-dielectric interface when the incident beam is in the material of lower refractive index. Indeed this property is exploited in the detection of squeezed states (see the article by Richard Slusher in this volume). In the situation that we consider here i.e. that of a Michelson (Sagnac) interferometer, the beamsplitter is also a potential weakness as it couples the vacuum field fluctuations into the interferometer's mode of interest. This is illustrated by the following paradox; if we consider a light field (coherent state) to be incident on the beamsplitter from the left (Fig 1), we can ask the question; how great are the fluctuations in each of the two output beams, and how are they correlated? The incident coherent state has intensity (photon number) fluctuations which are determined by the Poisson statistics of the photon number n. To wit

$$(\Delta n_{i/p})_{rms} = \langle n\rangle^{1/2} \quad . \tag{1}$$

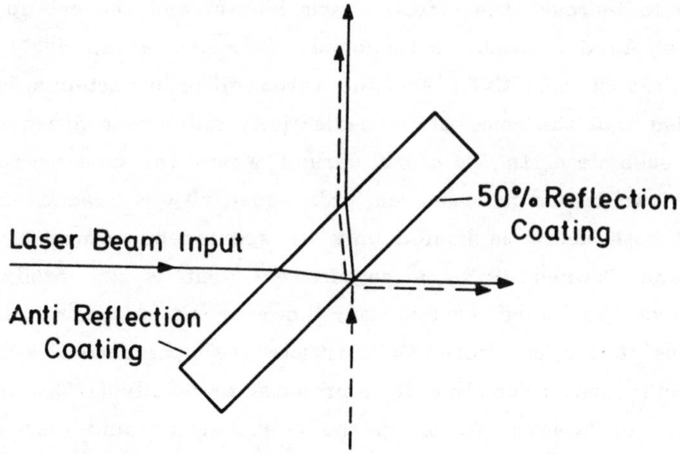

Figure 1. A typical (lossless) beamsplitter. The dashed line indicates the entry of vacuum field fluctuations.

Forgetting about Heisenberg's uncertainty principle one might expect that just as the mean intensity is divided equally between the two output beams, then so are the fluctuations. The intensity fluctuations would then be given in each output beam by

$$(\Delta n_{O/p})_{rms} = 1/2 \langle n \rangle^{1/2} \quad . \tag{2}$$

This would violate the uncertainty principle and imply that the output beams are no longer coherent states. However we can easily measure the shot noise in each of the two output beams and verify that it is still that of a coherent state. A resolution of this paradox was proffered by Caves (1980,1981) who explained the preservation of the coherent state in terms of vacuum fluctuations which are coupled into the interferometer via the unused port of the beamsplitter. Now as the two fields due to the vacuum field propagate away from the beamsplitter, one of them (the reflected) has undergone the 180° phase change. This phase change doesn't occur to the two output fields due to the main input coherent field and one then finds that the fluctuations in each arm of the interferometer are anticorrelated. This will be an important fact when we come to consider the Michelson interferometer in the next section. The full quantum-mechanical analysis shows that the vacuum field fluctuations are precisely what is required to restore the Poissonian character of the output beams from the beamsplitter.

A mode of the radiation field is conveniently described by a linear combination of a sine and a cosine wave. The operators $\hat{a}_{sin}$ and $\hat{a}_{cos}$ describing the amplitudes of the wave do not commute, leading to an uncertainty relation between the root mean square (RMS) fluctuations, $\langle \Delta a_{sin} \rangle \langle \Delta a_{cos} \rangle \geq 1/4$. Coherent states and the vacuum state are minimum uncertainty states with the same RMS fluctuations in both amplitudes. Other minimum uncertainty states that do not fulfill this relation symmetrically are called squeezed states, e.g. $\langle \Delta a_{sin} \rangle \rightarrow 0$ and $\langle \Delta a_{cos} \rangle \rightarrow \infty$ (see the contribution by D.F.Walls in this volume). If the light entering the normally unused port of the beamsplitter is in a state having fluctuations modified in this way, the output beams from the beamsplitter will also be modified, as discussed by R. Loudon in this volume. As will be seen, the possibility for modifying the fluctuations may have a direct impact on interferometers where the two light beams are recombined with a beamsplitter after having travelled through different optical paths.

3. THE MICHELSON INTERFEROMETER

A schematic of such a Michelson interferometer built to detect gravitational radiation is shown in figure 2. This design employs a delay line in each arm, rather than a Fabry-Perot, and although only two reflections on each mirror are shown, in practice more than 100 can be used in order to increase the effective length of each arm (Billing et al. 1983). Note that the delay lines do no reflect the beams back on themselves, thus avoiding the coupling of back-reflections into the laser. A change in the path-length difference between the two arms due to an impinging gravitational wave results in a phase change for the beams which one measures (hopefully) as an intensity change at the detector. If the path length difference is denoted z, then the detected intensity I_d is

$$I_d \propto \sin^2(\pi z/\lambda) \quad . \tag{3}$$

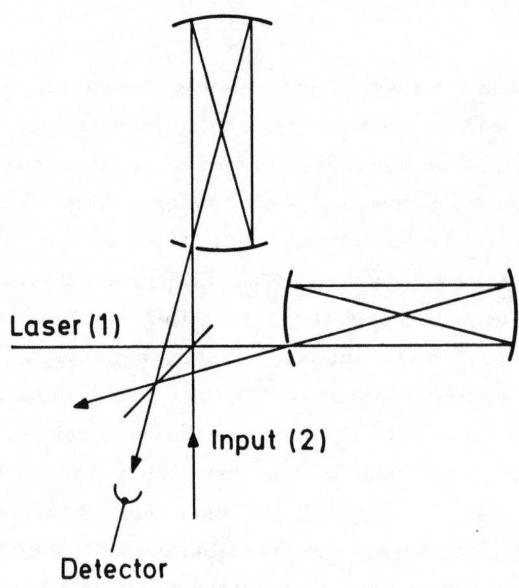

Figure 2. A Michelson Interferometer with delay lines in the arms

Ideally one would adjust z = 0 so that the intensity at the detector is at a minimum and measure changes in this value. What are the noise sources that might mask a change due to a gravitational wave? If we ignore problems due to non-fundamental noise such as laser frequency or pointing errors and vibrations, which in practice can be sufficiently reduced within a certain frequency range, we are then in a regime where photon-counting statistics limit the ultimate sensitivity (Shoemaker et al. 1985). This error, expressed as a change in the path length difference (Δz_{pc}) depends on the relative size of the photon number fluctuations to the total photon number in the main input mode, that is (assuming a coherent state)

$$\Delta z_{pc} \propto \frac{\langle \Delta n^2 \rangle^{1/2}}{\langle n \rangle} = \frac{1}{\langle n \rangle^{1/2}} . \tag{4}$$

Thus this error decreases with increasing laser power. However as the laser power increases, errors due to the anticorrelated radiation pressure fluctuations in the arms of the interferometer begin to appear. These clearly depend on the absolute size of the photon number fluctuations. For a coherent state they increase as the square root of the laser power;

$$\Delta z_{rp} \propto \langle n \rangle^{1/2} \quad . \tag{5}$$

Clearly the total error will be a minimum when these two contributions are equal (Caves 1980, Loudon 1981). The total uncertainty is then the standard quantum limit,

$$\Delta z_q = (2\hbar\tau /m)^{1/2} \quad , \tag{6}$$

which also follows directly from Heisenberg's uncertainty relation. Here m is the mass of the end mirrors and τ is the measurement time. One can then calculate the optimum power required to reach the quantum limit. One finds

$$P_{opt} \approx mc^2 \left(\omega\tau^2 b^2 \right)^{-1} \quad , \tag{7}$$

where ω is the laser angular frequency and b is the number of bounces the laser beam makes on each mirror. For the interferometer at Garching which is currently the most sensitive, having 30m long arms and upwards of 100 bounces in the delay lines (Shoemaker et al. 1985), this limit will be reached only at a power in the order of 100 kW which is not yet technically feasible. At the present sensitivity of this instrument, with 100mW of laser power, a strain $\Delta l/l$ of $6\text{x}10^{-18}$ can be measured within a bandwidth of 1kHz for 1kHz gravitational waves.

In an experiment one must also take into account the non-ideal quantum efficiency of the photodetectors (η), which up to now has been assumed to be ideal (η=1). Let us assume that the laser entering port (1) in figure 2 is a coherently excited cosine wave of power P, the field entering port (2) is described by $\hat{a}_{sin}$ and $\hat{a}_{cos}$ and τ is again the sampling time. Ignoring radiation pressure errors and operating the interferometer near an interference minimum, the minimum detectable change in the optical path length then becomes (Caves 1981)

$$\Delta z_{min} = (\lambda/2\pi)\Delta\phi \quad , \tag{8a}$$

$$\Delta\phi = \left[h\nu(1-\eta+4\eta\langle(\Delta a_{sin})^2\rangle)/(\eta P\tau_s) \right]^{1/2} \quad , \tag{8b}$$

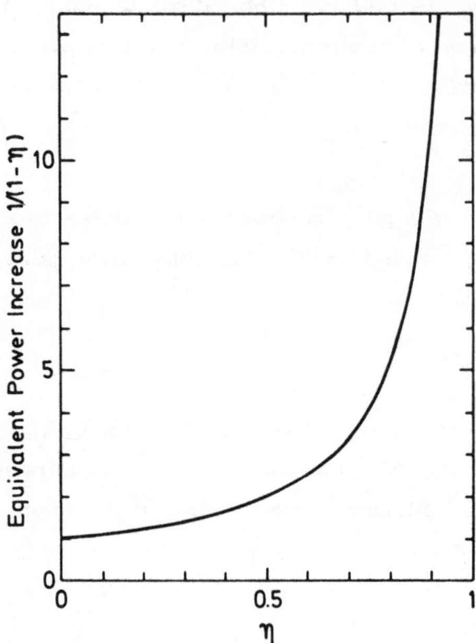

Figure 3. A plot of the equivalent power increase as determined from Eq. 8b assuming that $\langle(\Delta a_{sin})^2\rangle = 0$.

where $\Delta\phi$ is the optical phase difference. It is interesting to note that this sensitivity limit depends only on the RMS fluctuations in the amplitude of the wave that is in quadrature phase to the light entering port (1). This is precisely the reason why squeezed states could be applied here. Setting $\langle\Delta a_{sin}\rangle = 1/2$ gives the usual photon counting error (shot noise). It is clear that reducing $\langle\Delta a_{sin}\rangle$ at the expense of increasing $\langle\Delta a_{cos}\rangle$ will improve the sensitivity beyond the usual shot noise limit in much the same way as an increase in power would, when radiation pressure errors can still be neglected. It is obvious, however, that even for $\langle\Delta a_{sin}\rangle = 0$ the detector quantum efficiency (η) limits the gain in sensitivity. Figure 3 shows the factor of increase in laser power which would correspond to this maximum sensitivity as a function of η. The detector quantum efficiency has to be at least close to $\eta = 0.8$ to have a substantial impact. A critical analysis shows that there are constraints on the maximum squeezing possible and therefore on the maximum sensitivity of a Michelson (Heidmann 1984, Heidmann et al. 1984). However it is only necessary to have (Caves 1981)

$$\langle(\Delta a_{sin})^2\rangle \ll (1-\eta)/(4\eta) \quad ,$$

so that the sensitivity is limited by η (Eq. 8b). It should be stressed here, that applying squeezed states can be combined with the recycling of the laser light (Drever et al. 1983) to further increase the sensitivity.

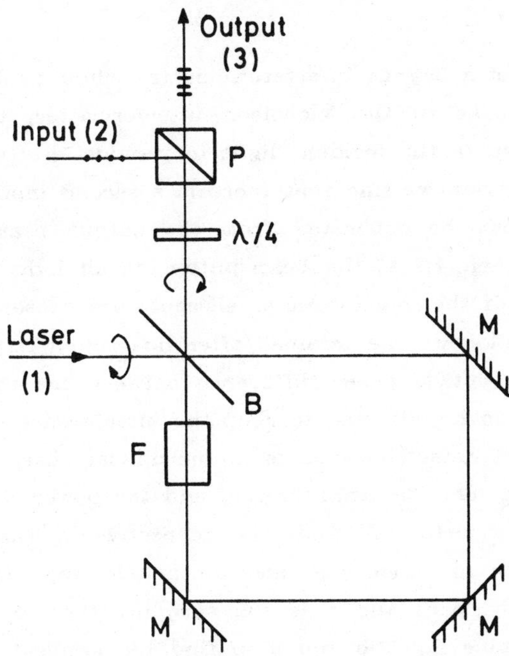

Figure 4. A Sagnac interferometer with a non-reciprocal element (Faraday rotator "F"). "B" denotes a beamsplitter, "P" a polarizing prism and "M" mirrors.

4. THE SAGNAC INTERFEROMETER

It is possible to determine the axis and frequency of a coordinate frame rotation by purely optical means (Chow et al. 1985). This idea was first introduced and demonstrated by Sagnac and it is the basis for laser gyros already used in navigation. Two light beams counter propagate along a closed path defined by mirrors or a glass fibre and interfere after one round trip. If the whole setup rotates, the round trip times for the two counter propagating beams are different due to the Doppler effect. This time difference results in an optical phase difference $\Delta\phi$ changing the interference. The phase difference depends on the wave-vector $k = 2\pi/\lambda$, the area $\underline{A}$ enclosed by the optical path, the overall rotation rate Ω and their relative orientation

$$\Delta\phi = 4\underline{A}.\underline{\Omega}\ k/c \quad . \tag{9}$$

The sensitivity of a Sagnac interferometer is limited by the photon number fluctuations. As in the Michelson interferometer, there is a beamsplitter which splits the incident light to provide the two counter propagating beams. Again we find that there is a second input port not normally used. It can be separated from the output beam using a polarising prism P (Fig. 4). At the beamsplitter, B, all light beams are circularly polarised. If the interferometer elements are chosen properly, the circular polarisation will be retained after one round trip. Using a Faraday rotator, the optical phase difference between the two counter propagating beams can be adjusted to keep the interference signal at a minimum. The smallest detectable change in the optical phase difference is again given by Eq. 8b. The amplitude, a, and the power, P, describe the fields entering ports (2) and (1) respectively. The detector efficiency is again η in which one has to include any other linear losses along the light path and τ is the sampling time interval. The final step is to combine Eqs. 8b and 9 to find the smallest detectable change in the rotation rate Ω which is measured through the feedback signal to the Faraday rotator;

$$\delta\Omega = \frac{c}{4Ak}\left[\, h\nu(1-\eta+4\eta\langle(\Delta a_{sin})^2\rangle/(\eta\tau P)\,\right]^{1/2} \quad . \tag{10}$$

In the common case when vacuum fluctuations ($\langle\Delta a_{sin}{}^2\rangle = 1/4$) enter port (2) one has

$$\delta\Omega_{coh} = c[\ 4Ak(\eta P\tau)^{1/2}\ ]^{-1} \quad . \tag{11}$$

A scheme slightly different from that in Fig. 4, is to use a ring resonator (laser gyro). Then the rotation induces a splitting in the frequency of any two counter propagating modes. In this case the difference of the mode frequencies is a measure of the rotation rate and the smallest detectable change in Ω is a factor of m smaller than for the single round trip (Eq.11), where m is the mean number of round trips of a photon in the resonator (Dorschner et al. 1980). Sanders et. al. (1981) have set up a laser gyro of 0.7m x 0.7m. For τ = 10s and P = 0.7mW they demonstrated a sensitivity of $\delta\Omega = 1.1\mathrm{x}10^{-2}$ degrees/hour or $7\mathrm{x}10^{-4}$ of the earth rotation rate Ω_e. Comparison with Eq.11 shows that this sensitivity is very close to the shot noise limit. However when increasing the sampling time interval from 10s to 90s the sensitivity increased by only a factor of two and was no longer shot noise limited.

It has been shown that the Sagnac interferometer is sensitive not only to the rotation of the interferometer but also to the gravitomagnetic effects produced by the motion of the interferometer through the gravitational field of the earth and by the rotation of the earth itself (Scully et al. 1981). These effects are of the order of only 10^{-10} Ω_e. Based on the results of Sanders et al (1981), the shot noise limited sensitivity of a 30m x 30m ring interferometer, a laser power of 4W and an averaging time of one hour will be $\delta\Omega \approx 10^{-11}\Omega_e$.

The problem with measuring the gravitomagnetic effects is that they produce a tiny change on top of a large signal. As long as the gravitomagnetic interaction cannot be switched on and off independently of the earth's rotation, one must perform a highly accurate absolute measurement on Ω. This requires a knowledge of the proportionality factor, $4Ak\cos\theta/c$ from Eq. 9, to a correspondingly high accuracy, which is not easily achieved. Here θ is the angle between $\underline{A}$ and $\underline{\Omega}$. The detection would be easier if the gravitomagnetic effects could be changed independently of the earth's rotation rate. For this purpose one could take advantage of the fact that the gravitomagnetic effects on a Sagnac interferometer depend on latitude. When moving the interferometer for example in a aeroplane from Dunedin to München the gravitomagnetic effects change whereas $\underline{\Omega}_e$ is constant. This scheme does not require an absolute measurement of Ω. Hence the high sensitivity estimated above is sufficient. The large dimension (30mx30m) may however prohibit the experiment. In this context it is interesting that one may enhance the Sagnac effect in a laser gyro by involving the nonlinear interaction in a Kerr medium (Kaplan and Meystre 1982). In addition the sensitivity of a Sagnac interferometer may be improved by employing squeezed states in the same way as for the Michelson interferometer; applying squeezed states makes up for a lack of laser power (see Eq.10). Such schemes could allow one to build a Sagnac interferometer of the required sensitivity without the necessity of going to such large dimensions, making the aforementioned experiments more viable.

ACKNOWLEDGEMENTS

We wish to thank Dana Andersen, Roland Schilling and Marlan O. Scully for helpful conversations and M.W.H. acknowledges the support of the Alexander von Humboldt Foundation.

REFERENCES

Bagayev S.N., Chebotayev V.P., Dychkov A.S. and Goldort V.G. 1981 Appl. Phys. 25, 161

Bassan M., Fairbank W.M., Mapoles E., McAshan M.S., Michelson P.F., Moskowitz B., Ralls K. and Taber R.C. 1983 Proc. of Third Marcel Grossman meeting on General Relativity, p.679 (North-Holland, Amsterdam)

Billing H., Winkler W., Schilling R., Rüdiger A., Maischberger K. and Schnupp L. 1983 in Meystre and Scully (1983)

Caves C.M. 1980 Phys.Rev.Lett. 45, 75

Caves C.M. 1981 Phys.Rev. D23, 1693

Chow W.W., Gea-Banachloche J., Pedrotti L.M., Sanders V.E., Schleich W. and Scully M.O. 1985 Rev. Mod. Phys. 57, 61

Dorschner T.A., Haus H.A., Holz M., Smith I.W., and Statz H. 1980 I.E.E.E. J. Quant. Elect. QE-16, 1376

Drever R.W.P. et al. 1983 in Meystre and Scully (1983)

Heidmann A. 1984 Thése de 3éme Cycle, L'Universitè Pierre et Marie Curie, Paris

Heidmann A., Reynaud S. and Cohen-Tannoudji C. 1984 Opt. Commun. 52, 235

Kaplan A.E. and Meystre P. 1982 Opt. Commun. 40, 229

Loudon R. 1981 Phys. Rev. Lett. 47, 815

Mach E. 1912 Die Mechanik in Ihrer Entwicklung Historisch-Kritisch Dargestellt (Brockhaus, Leipzig)

Meystre P. and Scully M.O. 1983 Quantum Optics, Experimental Gravitation and Measurement Theory (Plenum, New York)

Misner C.W., Thorne K.S. and Wheeler J.A. 1973 Gravitation (Freeman, San Francisco)

Sanders G.A., Prentiss M.G. and Ezekiel S. 1981 Opt. Lett., 569

Schawlow A.L. and Townes C.H. 1958 Phys. Rev. 112, 1940

Schleich W. 1984 Max Planck Institut für Quantenoptik, report MPQ 89

Scully M.O. 1985 Phys. Rev. Lett. 55, 2802

Scully M.O., Zubairy M.S. and Haugen M.P. 1981 Phys. Rev. A24, 2009
Shoemaker D., Winkler W., Maischberger K., Rüdiger A., Schilling R and Schnupp L. 1985 Max Planck Institut für Quantenoptik report MPQ 100
Thirring H. and Lense J. 1918 Phys. Z. 19, 156
Thorne K.S. 1980 Rev. Mod. Phys. 52, 285
Thorne K.S. 1983 in Meystre and Scully (1983)
Vessot R.F.C. 1984 Contemp. Phys. 25, 355
Walls D.F. 1983 Nature 306, 141
Weber J. 1979 in General Relativity and Gravitation A. Held (ed) (Plenum, New York)

THEORY OF QUANTUM FLUCTUATIONS IN OPTICAL BISTABILITY

H J CARMICHAEL

1. INTRODUCTION

In 1969 Szöke et al. proposed a simple scheme for a "bistable optical element" (Szöke et al., 1969). They suggested that a passive optical resonator containing a saturable absorber could exhibit two different output intensities for the same input intensity. In the same year, the same proposal was recorded independently in a patent by Seidel (1969). McCall also recognized the possibility for absorptive bistability, and published a detailed theory for a "bistable mirror" in 1974 (McCall, 1974). Two years later Gibbs et al. (1976) reported the first experimental observation of optical bistability. Their interpretation of this experiment led to the first description of bistability based on nonlinear dispersion. From these beginnings the interest in optical bistability has grown considerably. Several review works are already available (Bowden et al., 1981; Abraham and Smith, 1982; Bowden et al., 1984; Englund et al., 1984; Lugiato, 1984; Wherrett and Smith, 1984; Gibbs, 1985). Fueling much of the development is the search for a viable optical signal processing technology--the oft-quoted "optical transistor" analogy. Questions of fundamental interest, rather peripheral to this theme, have also been highlighted; optical bistability has been adopted as a paradigm for studies in quantum optics, nonequilibrium statistical mechanics, and nonlinear dynamics.

Increasingly, the field of optical bistability is identified with the development of signal processing capabilities. This is not surprising. This theme unifies a large body of work:

in the optical properties of semiconductors, new schemes for bistability, optical computing architectures, and many related areas. This review is not directly concerned with these objectives. There are, of course, questions concerning quantum fluctuations which must be addressed as the limits for miniaturized devices are approached. However, the work reviewed here draws its interest from a different direction. Our attention is focused on a system of conceptually simple design; namely, a collection of homogeneously broadened two-level atoms interacting with a single electromagnetic field mode in a cavity excited by coherent laser light. Under appropriate conditions this system exhibits both absorptive and dispersive bistability. Quantum-statistical effects directly related to bistability include noise induced transitions between bistable states (Schenzle and Brand, 1979; Farina et al., 1980; Hanggi et al., 1980; Englund et al., 1981; Farina et al., 1981; Bonifacio et al., 1981; Englund et al., 1982), and linewidth narrowing and critical slowing down (Bonifacio and Lugiato, 1976; Bonifacio and Lugiato, 1978a, Narducci et al., 1978a; Agarwal et al., 1978; Lugiato, 1979; Drummond and Walls, 1981; Carmichael et al., 1983). Other effects included in the theory of optical bistability have little to do with bistability itself: collective linewidth broadening (Bonifacio and Lugiato, 1976; Bonifacio and Lugiato, 1978a, Narducci et al., 1978; Agarwal et al., 1978; Lugiato, 1979; Carmichael et al., 1983), photon antibunching (Casagrande and Lugiato, 1980; Drummond and Walls, 1981; Carmichael et al., 1983; Carmichael, 1986a), and squeezing (Lugiato and Strini, 1982; Reid and Walls, 1985; Carmichael, 1986a). All these effects are of interest to fundamental studies in quantum optics. The unifying theme for this review is then an interest in quantum-dynamical effects in the interaction of two-level atoms and a single cavity mode. As an example of a bistable system the model outlined above is very specialized, and, indeed, idealized. It is so, however, precisely to isolate quantum-mechanical effects. The idealizations are not beyond the capabilities of quantitative experimental measurement, as demonstrated by recent experiments using optically prepumped atomic sodium beams (Rosenberger et al., 1983; Rosenberger et al., 1984; Orozco et al., 1984).

Our model occupies a central position in quantum optics. It is constructed of three parts: the interaction between a collection of two-level atoms and a single cavity mode; dissipative interactions which remove energy from the cavity field via lossy mirrors, and from the atoms via spontaneous emission; and an interaction with the incident laser which couples energy into the system. Collectively, these define the simplest possible model for the interaction of light and matter in the presence of dissipation: a driven damped harmonic oscillator coupled to a collection of damped two-level atoms. It is a simple generalization of the Tavis-Cummings model (Tavis and Cummings, 1968; Tavis and Cummings, 1969) to include dissipation. This generalization brings a new class of fundamental questions into focus. Driven far from equilibrium, macroscopic dissipative systems exhibit bifurcations, involving transitions between steady states, periodic oscillations, and chaos, all of which are contained within our model. By studying this simple quantum-mechanical system we can hope to learn something about the relationship between linear microscopic quantum theory and the nonlinearity which underlies these macroscopic phenomena. Also, much of quantum optics is contained within this one model, to be revealed for different choices of its parameters; the quantum-electromagnetic Rabi problem, spontaneous emission and superradiance, resonance fluorescence and cooperative resonance fluorescence, and, of course, optical bistability. With optical pumping added the single-mode laser and laser with injected signal are also included. Relationships between these phenomena can bring important insight, as with the relationship between photon antibunching in optical bistability and single-atom resonance fluorescence (Carmichael, 1985). Further importance is given to our subject by the current interest in the behavior of Rydberg atoms inside cavities (Kleppner, 1981; Vaidyanathan et al., 1981; Raimond et al., 1982a; Raimond et al., 1982b; Goy et al., 1983; Kaluzny et al., 1983; Sanchez-

Mondragon et al., 1983; Agarwal, 1984; Meschede et al., 1985). There are close relationships between some of the effects reported from the theory of optical bistability and cavity effects such as cavity enhanced spontaneous emission and vacuum Rabi splitting (Carmichael, 1985; Carmichael, 1986a).

This paper is intended both as a review and a tutorial introduction to the quantum theory of optical bistability and related cavity effects. As a tutorial work it attempts to offer a consistent development, rather than to merely collect results from the literature. Its function as a review is thereby slightly compromised, with emphasis being given to the author's own approach to these problems. Certainly, the references will not be comprehensive; although a thorough coverage of the central contributions, with their correct chronology, is intended. One omission should not go without mention; namely, the work concerned with bistability in mirrorless Dicke-like systems and the relationship between optical bistability with two-level atoms and cooperative resonance fluorescence (Drummond and Carmichael, 1978; Narducci et al., 1978b; Walls et al., 1978; Bowden and Sung, 1979; Puri and Lawande, 1979; Carmichael, 1980a; Hassan et al., 1980; Puri and Lawande, 1980; Puri et al., 1980; Bowden, 1981; Puri et al., 1984; Hopf and Bowden, 1985). The driven Dicke problem (cooperative resonance fluorescence) is obtained from our model as a special case. It is distinguished from conventional optical bistability by the exclusion of independent atomic decay. The relationship between this problem and optical bistability is only partially understood and is an interesting area for further study. This topic is not included in the present work.

For those unfamiliar with our subject section 2 begins with a discussion of the macroscopic theory of optical bistability, drawing only on a knowledge of the classical Maxwell's equations. The quantum-mechanical description is introduced in section 3, with a discussion of the microscopic model and the relationship between macroscopic and microscopic variables.

Section 4 deals with the Fokker-Planck equation approach to the solution of the quantum problem. Methods for obtaining a Fokker-Planck equation using the Glauber-Sudarshan-Haken representation, the Wigner representation, and the positive P-representation are discussed. The positive P-representation is highlighted and adopted for the later development. Section 5 develops the linearized theory of fluctuations, and presents results for steady-state correlations, the incoherent intensity and spectrum of the transmitted light, photon antibunching and squeezing. In section 6 the nonlinear treatment of fluctuations is considered. A discussion of the one-dimensional theory for photon statistics and transition rates is given, together with results from the numerical simulation of multidimensional stochastic differential equations derived from the Fokker-Planck equation. Section 7 draws brief conclusions regarding possible experiments and directions for future theoretical work.

2. MACROSCOPIC THEORY OF OPTICAL BISTABILITY

Before we plunge into the microscopic theory it will be useful to see how optical bistability arises in a simple macroscopic description of a nonlinear interferometer. The essential ingredients for bistability are simply nonlinearity and feedback. Recent development in this field has provided many schemes for realizing these ingredients, some, variants of the original idea for a nonlinear interferometer, and some which do not use an interferometer at all (Gibbs, 1985). Our system of two-level atoms in a cavity has become the canonical model for a general nonlinear interferometer combining both absorption and dispersion. The discussion below begins with the arguments of Szöke et al. (1969) and Gibbs et al. (1976) for purely absorptive and purely dispersive bistability, respectively. The theory which follows, combining both absorption and dispersion for a two-level medium, was developed in similar treatments by Hassan et al. (1978), Agrawal and Carmichael (1979), and Bonifacio and Lugiato (1978b).

2.1 Steady states in a nonlinear interferometer

Consider the transmission of a Fabry-Perot interferometer containing a medium of length L with absorption coefficient α and refractive index $n = 1 + \Delta n$. A standard calculation, adding repeated reflections between the mirrors, yields the transmitted intensity

$$I_t = e^{-\alpha L} TI = e^{-\alpha L} T^2 I_i / [(1 - e^{-\alpha L} R)^2 + 4e^{-\alpha L} R \sin^2(\tfrac{1}{2}\theta - \Delta n k_0 L)] \,, \tag{2.1}$$

where I_i is the incident intensity, I is the intracavity intensity at the input mirror, T is the mirror transmission coefficient, R is the mirror reflection coefficient, k_0 is the wavenumber of the incident field, and $-\theta$ is the round-trip phase detuning of the empty interferometer. The possibilities for absorptive and dispersive bistability arise when the medium is nonlinear--when α and Δn are functions of the intracavity intensity I.

The proposal for absorptive bistability (Szöke et al., 1969; Seidel, 1969; McCall, 1974) is to let α be a saturable function of intensity, such that $\exp(-\alpha L) << 1$ at low intensities, and at high intensities $\exp(\alpha L) \to 1$. For resonance ($\theta + \Delta n k_0 L = 0$), and in the absence of nonlinear dispersion, Eq.(2.1) requires

$$e^{\alpha(I)L} T^{-1} I_t / I_i = I/I_i = T/[1 - Re^{-\alpha(I)L}]^2 \,. \tag{2.2}$$

The graphical solution of this equation for I, and hence for I_t, shows the possibility for three solutions over some range of I_i, as depicted in Fig. 1(a). The plot of steady-state solutions for I_t as a function of I_i takes the characteristic S-shaped form shown in Fig. 2(a). Effectively, the interferometer has a nonlinear reflection coefficient $R'(I) = R\exp[-\alpha(I)L]$. We are "filling a leaky bucket" where the hole in the bucket shrinks as the bucket fills. In the S-shaped portion of Fig. 2(a) two stable self-consistent states are possible: a low intensity state with the medium strongly absorbing, and a high intensity state with the medium saturated.

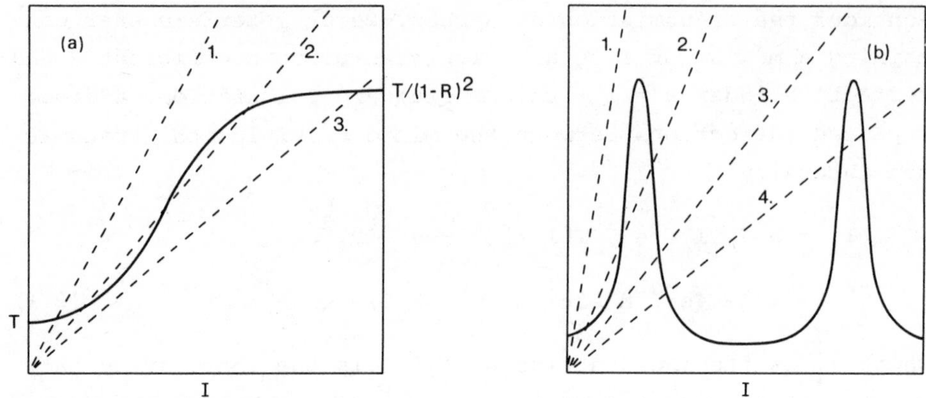

Fig. 1. Graphical solution of (a) Eq.(2.2), and (b) Eq.(2.6). The solid curves plot the functions of I on the right hand sides of these equations and the dashed lines plot I/I_i for different values of I_i.

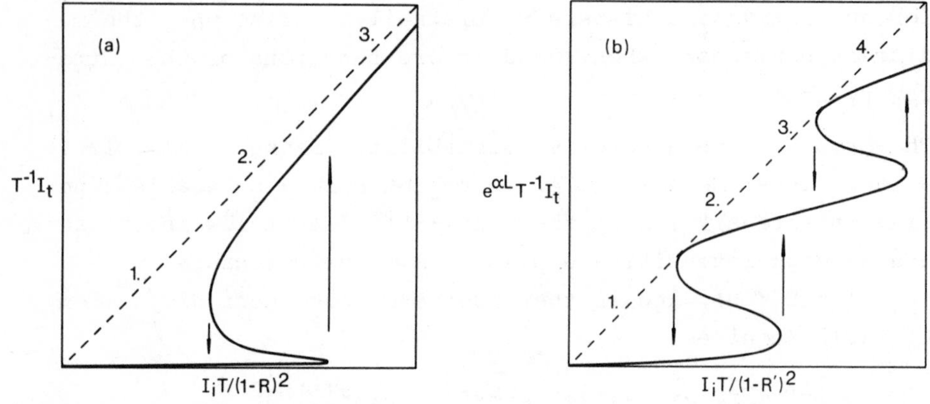

Fig. 2. Steady-state transmission curves corresponding to Fig. 1. The dashed lines have unit slope and show the empty cavity response.

Switching from low intensity to high intensity states as the input intensity is increased results from runaway around a positive feedback loop: increased intracavity intensity → decreased absorption due to saturation → further increase in the intracavity intensity. A similar feedback loop operates for switching down. The branch of solutions with negative slope ($dI_i/dI_t < 0$) is unstable, as can be appreciated from the following argument. Runaway will occur when a perturbation ΔI added to the intracavity intensity returns a field amplitude after one round trip which amplifies ΔI when superposed with the input field amplitude $(TI_i)^{\frac{1}{2}}$. The field amplitude returned after one round trip is

$$E' = (I + \Delta I)^{\frac{1}{2}} R \exp[-\alpha(I + \Delta I)L]$$

$$\simeq I^{\frac{1}{2}} R e^{-\alpha(I)L} + \Delta I \frac{d}{dI}[I^{\frac{1}{2}} R e^{-\alpha(I)L}] \ . \tag{2.3}$$

After adding this to $(TI_i)^{\frac{1}{2}}$ and using Eq.(2.2), the condition for amplification (instability) becomes

$$B \equiv 2I^{\frac{1}{2}} \frac{d}{dI}[I^{\frac{1}{2}} R e^{-\alpha(I)L}] > 1 \ , \tag{2.4}$$

or

$$T\frac{dI}{dI_i} = (TI_i/I)^{\frac{1}{2}}(1 - B) < 0 \ . \tag{2.5}$$

Absorptive bistability is difficult to achieve experimentally (Szöke et al., 1969; Austin and De Shazer, 1971; Spiller, 1971; Spiller, 1972; Gibbs et al., 1976). It relies on a strong saturable nonlinearity, and unsaturable absorption mitigates against the effect. It has now been observed, however, in a number of experiments (Gibbs et al., 1976; Sandle and Gallagher, 1980; Sandle and Gallagher, 1981; Weyer et al., 1981; Grant and Kimble, 1982; Grant and Kimble, 1983; Kimble et al., 1983; Rosenberger et al., 1983; Rosenberger et al., 1984). The work of Kimble and coworkers is of particular note (Grant and Kimble, 1982; Grant and Kimble, 1983; Kimble et al., 1983; Rosenberger et al., 1983; Rosenberger et al., 1984). It is the only work to demonstrate quantitative agreement with theory for steady-state absorptive bistability using homogeneously broadened two-level atoms.

The mechanism for dispersive bistability (Gibbs et al., 1976) is very robust. Dispersive bistability has been observed in many experiments and remains one of the prime candidates for device applications. For a nonlinear refractive index $\Delta n = \Delta n_0 + n_2 I$, and in the absence of nonlinear absorption, Eq. (2.1) reads

$$e^{\alpha L} T^{-1} I_t/I_i = I/I_i = T/[(1 - R')^2 + 4R'\sin^2(\tfrac{1}{2}\theta' - \beta I)] \ , \tag{2.6}$$

where $R' = R\exp(-\alpha L)$, $\theta' = \theta - \Delta n_0 k_0 L$, and $\beta = n_2 k_0 L$. The graphical solution of this equation is illustrated in Fig. 1(b). With an appropriate choice of signs for β and θ', as the intracavity intensity increases the nonlinear phase shift βI offsets the detuning θ' and tunes the interferometer through resonance. Three solutions for I are again possible over some range of I_i. With a large enough nonlinearity it is even feasible to tune through repeated resonances leading to repeated, and possibly overlapping, hysteresis loops, as shown in Fig. 2(b) (Felber and Marburger, 1976; Marburger and Felber, 1978). The switching dynamics are less transparent than for the resonant conditions of absorptive bistability, since we must consider the superposition of phase shifted complex field amplitudes on each round trip. However, it should be no surprise that the negative slope solutions are again unstable; we will see this explicitly further on.

2.2 Homogeneously broadened two-level atoms in a ring cavity: the mean-field single-mode limit

This simple description of optical bistability clearly illustrates the roles of nonlinearity and feedback: the latter enters via repeated reflections inside the interferometer, whereby $\alpha(I)$ and $\Delta n(I)$ effectively depend on the transmitted intensity (the relationship $I_t = TI\exp[-\alpha(I)L]$ is one-to-one). It is only approximate, however. Equation (2.1) is derived for fields propagating in a linear medium, and it is generally not correct to simply replace a constant α and Δn by $\alpha(I)$ and $\Delta n(I)$ when generalizing to nonlinear conditions. Also, Eq.(2.5) results from a rather restricted view of the dynamics. It does not generalize to dispersive bistability and it does not account for a time-dependent medium response. A more systematic development begins with Maxwell's equations to describe propagation in the medium, coupled to some time-dependent equation, or set of equations, satisfied by the medium polarization, plus boundary conditions imposed by the cavity mirrors. Such an approach, in its most general form, gives a problem which is only tractable numerically; difficulties include: the spatially modulated intensity in a standing-wave cavity (McCall, 1974; Meystre, 1978; Abraham et al.,

1980; Carmichael, 1980b; Carmichael and Hermann, 1980; Hermann, 1980; McCall and Gibbs, 1980; Roy and Zubairy, 1980), inhomogeneous broadening (Bonifacio and Lugiato, 1978b; Hassan et al., 1978; Abraham and Hassan, 1980; Agrawal and Carmichael, 1980; Carmichael and Agrawal, 1980; Gronchi and Lugiato, 1980; Carmichael and Agrawal, 1981) and the transverse spatial profile of the cavity field (Marburger and Felber, 1978; Roganov and Semenov, 1980; Drummond, 1981; Ballagh et al., 1981; Moloney and Gibbs, 1982; Moloney et al., 1982; Kimble et al., 1983). These difficulties are each manageable analytically on their own, at least under some conditions. However, together they bring considerable complexity to a problem which in essence involves rather simple physics. A simple model to describe this simple physics is found if we adopt a ring cavity geometry to eliminate standing waves, homogeneously broadened atoms for the medium, and treat field propagation in the plane-wave approximation. Even these simplifications generally leave us with a transcendental equation to solve numerically for the steady state (Bonifacio and Lugiato, 1978c; Ikeda, 1979; Carmichael and Hermann, 1980; Roy and Zubairy, 1980), and allow the possibility that nonresonant cavity modes will be unstable, whence the steady state is replaced by a complex dynamical attractor involving many degrees of freedom (Bonifacio and Lugiato, 1978d; Bonifacio et al., 1979b; Ikeda, 1979; Ikeda et al., 1980; Lugiato, 1980a; Gronchi et al., 1981; Carmichael et al., 1982; Lugiato et al., 1982a; Carmichael, 1983; Carmichael, 1984; Asquini et al., 1986). For further simplification we specialize to a high finesse cavity and a dilute medium--the mean-field limit--and assume that the medium linewidth is much smaller than the longitudinal mode spacing--the single-mode limit. These simplifications are introduced systematically in the following to arrive at the macroscopic version of the model we will quantize in the next section.

Consider a ring cavity with partially reflecting input and output mirrors, both having a reflection coefficient R and a transmission coefficient T; the mirrors which close the optical path are perfectly reflecting. The cavity field propagates from z = 0 to z = L through a medium of homogeneously broadened two-level atoms. In the slowly-varying-amplitude approximation, the complex field amplitude E satisfies Maxwell's equations in the form

$$c^{-1}\frac{\partial E}{\partial t} + \frac{\partial E}{\partial z} = i\tfrac{1}{2}(k_0/\varepsilon_0)P \ , \tag{2.7}$$

subject to the boundary condition

$$E(0,t) = \sqrt{T}\, e^{i\phi_T} E_i + Re^{-i\theta}E(L,t - \tau + L/c), \tag{2.8}$$

where E_i is the incident field amplitude, P is the polarization amplitude, ϕ_T is a phase change on transmission at the input and output mirrors, and τ is the cavity round-trip time. The transmitted field amplitude is

$$E_t = \sqrt{T}\, e^{i\phi_T} E(L) \ . \tag{2.9}$$

Cavity field and polarization amplitudes are defined with

$$\vec{E}(z,t) = \hat{e}E(z,t)e^{-i(\omega_0 t - k_0 z)} + \text{c.c.} \ , \tag{2.10}$$

$$\vec{P}(z,t) = \hat{e}P(z,t)e^{-i(\omega_0 t - k_0 z)} + \text{c.c.} \ , \tag{2.11}$$

and E_i and E_t are similarly defined.

For a dilute medium we integrate Eq.(2.7) with P taken constant. This integration is most easily performed in a retarded frame. Setting t' = t - z/c,

$$\frac{\partial E'}{\partial z} = i\tfrac{1}{2}(k_0/\varepsilon_0)P' \ , \tag{2.12}$$

where E'(z,t') = E(z,t) and P'(z,t') = P(z,t). Then, integrating from z = 0 to z = L, we write

$$E'(L,t') = E'(0,t') + i\tfrac{1}{2}(k_0 L/\varepsilon_0)P'(0,t') \ , \tag{2.13a}$$

or

$$E(L,t + L/c) = E(0,t) + i\tfrac{1}{2}(k_0 L/\varepsilon_0)P(0,t) \ . \tag{2.13b}$$

Substituting for E(L,t - τ + L/c) in Eq.(2.8) and expanding R exp(-iθ) ≃ 1 - (1 - R) - iθ for a high finesse cavity excited near resonance ($\theta \sim (1 - R) << 1$), we find

$$\kappa^{-1}[\overline{E}(t + \tau) - \overline{E}(t)]/\tau = -(1 + i\phi)\overline{E}(t) + i(1 - R)^{-1}\tfrac{1}{2}(k_0 L/\varepsilon_0)\overline{P}(t) + i(1 - R)^{-1}\sqrt{T}e^{i\phi_T} E_i \ , \tag{2.14}$$

where $\overline{E}(t) \equiv E(0,t)$ and $\overline{P}(t) \equiv P(0,t)$. Here

$$\kappa = (1 - R)/\tau \tag{2.15}$$

is the cavity decay time, and

$$\phi = (1 - R)^{-1}\theta = (\omega_c - \omega_0)/\kappa \tag{2.16}$$

is a dimensionless detuning from the cavity resonance frequency ω_c. For a homogeneously broadened two-level medium $\overline{P}$ satisfies optical Bloch equations:

$$\dot{\overline{P}} = -\gamma_\perp (1 + i\delta)\overline{P} - i\mu(\mu\overline{E}/\hbar)\overline{D} \ , \tag{2.17a}$$

$$\dot{\overline{D}} = 2i\mu^{-1}[(\mu\overline{E}/\hbar)\overline{P}^* - (\mu\overline{E}^*/\hbar)\overline{P}] - \gamma_\parallel (\overline{D} + \rho_0) \ , \tag{2.17b}$$

where $\cdot$ denotes differentiation with respect to time, $\gamma_\parallel$ and $\gamma_\perp$ are longitudinal and transverse atomic decay rates, $\mu \equiv \hat{e}\cdot\vec{\mu}$ is the atomic dipole moment, ρ_0 is the atomic density, $\overline{D}(t) \equiv D(0,t)$, where D is the atomic inversion density, and

$$\delta = (\omega_a - \omega_0)/\gamma_\perp \tag{2.18}$$

is a dimensionless detuning from the atomic resonance frequency ω_a.

The set of differential-difference equations given by Eqs. (2.14) and (2.17) comprise a general dynamical model for the mean-field limit without restriction to a single cavity mode. The single-mode limit brings further simplification. We assume that the round-trip time τ is much shorter than the time scale for change in $\overline{E}$. For a high finesse cavity $\tau << \kappa^{-1}$ and $\overline{E}(t + \tau) - \overline{E}(t)$ is guaranteed to be very small. We might then be tempted to replace the left-hand side of Eq. (2.14) by $\kappa^{-1}\dot{\overline{E}}$. However, it is possible for $\overline{E}(t + \tau) - \overline{E}(t)$ to be very small while $\bar{E}$ varies considerably during the interval τ. Large amplitude periodic (or near periodic) solutions with periods only slightly different from τ are possible. Such behavior corresponds to excitation of nonresonant longitudinal modes. To exclude this possibility we must assume that the temporal response of $\bar{P}$, which is a source for $\bar{E}$, does not contain these higher frequencies. The rate of change of $\bar{P}$ is characterized by $\gamma_{\parallel}^{-1}$ and $\gamma_{\perp}^{-1}$. We must assume $\tau \ll \kappa^{-1}$, $\gamma_{\parallel}^{-1}$ and $\gamma_{\perp}^{-1}$. Then we can replace Eq.(2.14) by the differential equation

$$\kappa^{-1}\dot{\bar{E}} = -(1 + i\phi)\bar{E} + i(1 - R)^{-1}\tfrac{1}{2}(k_0L/\varepsilon_0)\bar{P} + (1 - R)^{-1}\sqrt{T}e^{i\phi_T}E_i \ . \tag{2.19}$$

Equations (2.17) and (2.19) define our macroscopic model for optical bistability with homogeneously broadened two-level atoms in the mean-field single-mode limit.

2.3 Homogeneously broadened two-level atoms in a ring cavity: steady states and their stability

In the steady state Eqs.(2.17) give

$$\bar{P}_s = \varepsilon_0\chi(\bar{I}_s)\bar{E}_s \ , \tag{2.20}$$

with susceptibility

$$\chi(\bar{I}_s) = k_0^{-1}\alpha_0(i + \delta)(1 + \delta^2 + \bar{I}_s/I_{sat})^{-1} \ , \tag{2.21}$$

where the subscript s denotes the steady state, $\bar{I}_s = |\bar{E}_s|^2$,

$$\alpha_0 = \rho_0\mu^2\omega_0/\hbar\varepsilon_0 c\gamma_{\perp} \tag{2.22}$$

is the weak-field absorption coefficient, and

$$I_{sat} = \hbar^2\gamma_{\parallel}\gamma_{\perp}/4\mu^2 \tag{2.23}$$

is the saturation intensity. Then, from Eq.(2.19),

$$\bar{I}_s = (1 - R)^{-2}TI_i/\{[1 + (1 - R)^{-1}\tfrac{1}{2}\alpha(\bar{I}_s)L]^2 + [\phi - (1 - R)^{-1}\Delta n(\bar{I}_s)k_0L]^2\} \ , \tag{2.24}$$

where

$$\alpha(\bar{I}_s) = k_0 \mathrm{Im}[\chi(\bar{I}_s)] = \alpha_0(1 + \delta^2 + \bar{I}_s/I_{sat})^{-1} , \qquad (2.25a)$$

$$\Delta n(\bar{I}_s) = \tfrac{1}{2}\mathrm{Re}[\chi(\bar{I}_s)] = \tfrac{1}{2}k_0^{-1}\alpha_0\delta(1 + \delta^2 + \bar{I}_s/I_{sat})^{-1} . \qquad (2.25b)$$

Equation (2.24) is just the first-order expansion of Eq.(2.1) for $\theta \sim \Delta nk_0 L \sim \alpha L \sim (1 - R) << 1$. (We compare Eq.(2.1) for a medium of length L/2 since in a Fabry-Perot the field propagates twice through the medium on each round trip.) Equations (2.24) and (2.25) are traditionally written as a single polynomial state equation (Bonifacio and Lugiato, 1976; Hassan et al., 1978; Agrawal and Carmichael, 1979; Bonifacio et al., 1979a):

$$Y^2 = X^2\{[1 + 2C/(1 + \delta^2 + X^2)]^2 + [\phi - 2C\delta/(1 + \delta^2 + X^2]^2\} \qquad (2.26)$$

where

$$Y = (1 - R)^{-1}\sqrt{T}|E_i|\sqrt{I_{sat}} , \qquad (2.27a)$$

$$X = |\bar{E}_s|/\sqrt{I_{sat}} , \qquad (2.27b)$$

and

$$C = (1 - R)^{-1}\alpha_0 L/4 . \qquad (2.28)$$

Some examples of the solution for X as a function of Y are plotted in Fig. 3. The full behavior of solutions for X as a function of the parameters Y, C, δ and ϕ is discussed in the cited references.

In section 5 we will discuss the linearized quantum theory, where the regression of fluctuations is governed by the linearized form of Eqs.(2.17) and (2.19). It is useful to first meet these linearized equations here, since they also determine the stability of the steady states. In terms of dimensionless variables:

$$\bar{\alpha} = \exp[-i(\phi_T + \phi_i)]\bar{E}/\sqrt{I_{sat}} , \qquad (2.29a)$$

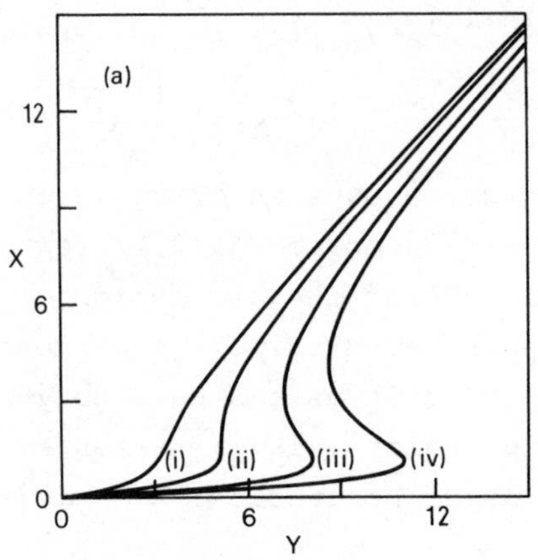

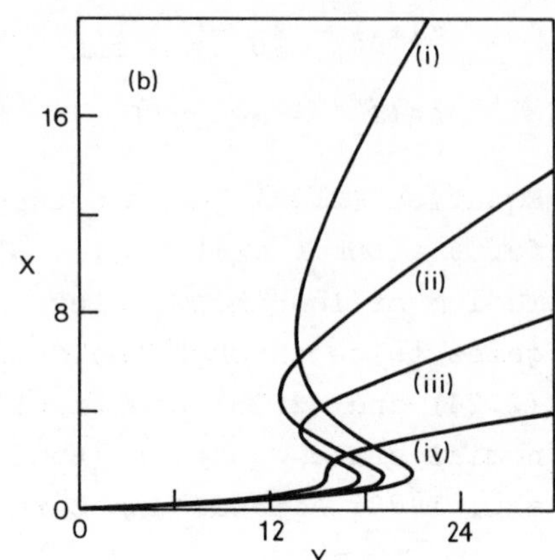

Fig. 3. Bistability curves plotted from Eq.(2.26): (a) $\delta = \phi = 0$ and (i) C = 2, (ii) C = 4, (iii) C = 7, (iv) C = 10; (b) C = 20, $\delta = 1$, and (i) $\phi = 0$, (ii) $\phi = 2$, (iii) $\phi = 4$, (iv) $\phi = 9$.

$$\bar{v} = \exp[-i(\phi_T + \phi_i]2i(\mu\rho_0)^{-1}(\gamma_\perp/\gamma_{||})^{\frac{1}{2}}\,\bar{P}\ , \tag{2.29b}$$

$$\bar{m} = D/\rho_0\ , \tag{2.29c}$$

where ϕ_i is the phase of the incident field, Eqs.(2.19), (2.17a), and (2.17b) read, respectively:

$$\kappa^{-1}\dot{\bar{\alpha}} = -(1 + i\phi)\bar{\alpha} + 2C\bar{v} + Y\ , \tag{2.30a}$$

$$\gamma_\perp^{-1}\dot{\bar{v}} = -(1 + i\delta)\bar{v} + \overline{\alpha m}\ , \tag{2.30b}$$

$$\gamma_{||}^{-1}\dot{\bar{m}} = -(\bar{m} + 1) - \tfrac{1}{2}\overline{\alpha v^*} - \tfrac{1}{2}\bar{\alpha}^*\bar{v}\ , \tag{2.30c}$$

with steady state solutions:

$$\bar{\alpha}_s = \exp(i\phi_X)X\ , \tag{2.31a}$$

$$\bar{v}_s = -\exp(i\phi_X)X(1 - i\delta)/(1 + \delta^2 + X^2)\ , \tag{2.31b}$$

$$\bar{m}_s = -(1 + \delta^2)/(1 + \delta^2 + X^2)\ , \tag{2.31c}$$

where X satisfies the state equation, Eq.(2.26), and the cavity field phase ϕ_X is given by

$$\tfrac{1}{2}\phi_X = \arg[(1 - i\phi)(1 + \delta^2 + X^2) + 2C(1 + i\delta)]\ . \tag{2.32}$$

For small deviations from the steady state, we write

$$\vec{\xi} = (\bar{\alpha},\bar{\alpha}^*,\bar{v},\bar{v}^*,\bar{m})^T = \vec{\xi}_s + \vec{\eta} \ , \tag{2.33}$$

where T denotes the transpose. Then, linearized equations for the perturbation $\vec{\eta}$ read

$$\dot{\vec{\eta}} = A\vec{\eta} \tag{2.34}$$

with

$$A = \begin{pmatrix} -\kappa(1+i\phi) & 0 & \kappa 2C & 0 & 0 \\ 0 & -\kappa(1-i\phi) & 0 & \kappa 2C & 0 \\ \gamma_\perp \bar{m}_s & 0 & -\gamma_\perp(1+i\delta) & 0 & \gamma_\perp \bar{\alpha}_s \\ 0 & \gamma_\perp \bar{m}_s & 0 & -\gamma_\perp(1-i\delta) & \gamma_\perp \bar{\alpha}_s^* \\ -\frac{1}{2}\gamma_\parallel \bar{v}_s^* & -\frac{1}{2}\gamma_\parallel \bar{v}_s & -\frac{1}{2}\gamma_\parallel \bar{\alpha}_s^* & -\frac{1}{2}\gamma_\parallel \bar{\alpha}_s & -\gamma_\parallel \end{pmatrix} \tag{2.35}$$

The stability of the steady state $\vec{\xi}_s$ is governed by the eigenvalues of A--i.e. by the solutions λ to the characteristic equation

$$\det(A - \lambda I) = 0 \ , \tag{2.36}$$

where I is the unit matrix. Stable steady states have $\mathrm{Re}(\lambda) < 0$ for all solutions to Eq.(2.36). Unstable steady states have one or more eigenvalues with $\mathrm{Re}(\lambda) > 0$. The change of stability at bifurcation points is indicated by one or more eigenvalues with $\mathrm{Re}(\lambda) = 0$.

We can show quite generally that the turning points ($dY/dX=0$) of the S-shaped bistability curve always mark a change of stability, with one eigenvalue vanishing. We write Eqs.(2.30) in the form

$$\dot{\vec{\xi}} = \vec{a}(\vec{\xi}) \ , \tag{2.37}$$

where $\vec{a}(\vec{\xi})$ is the five-component complex vector given by:

$$\kappa^{-1}a_1 = -(1 + i\phi)\xi_1 + 2C\xi_3 + Y \ , \tag{2.38a}$$

$$\kappa^{-1} a_2 = -(1 - i\phi)\xi_2 + 2C\xi_4 + Y \,, \tag{2.38b}$$

$$\gamma_\perp^{-1} a_3 = -(1 + i\delta)\xi_3 + \xi_1\xi_5 \,, \tag{2.38c}$$

$$\gamma_\perp^{-1} a_4 = -(1 - i\delta)\xi_4 + \xi_2\xi_5 \,, \tag{2.38d}$$

$$\gamma_\parallel^{-1} a_5 = -(\xi_5 + 1) - \tfrac{1}{2}\xi_1\xi_4 - \tfrac{1}{2}\xi_2\xi_3 \,. \tag{2.38e}$$

Then the matrix A is the Jacobian matrix

$$A_{ij} = \left.\frac{\partial a_i}{\partial \xi_j}\right|_s \,. \tag{2.39}$$

The steady state is defined by

$$a_i(\vec{\xi}_s) = 0 \,, \qquad i = 1,2,\ldots 5 \,. \tag{2.40}$$

It follows, after differentiating Eq.(2.40) with respect to X, that

$$A \frac{d\vec{\xi}_s}{dX} + \frac{dY}{dX}(1,1,0,0,0)^T = 0 \,. \tag{2.41}$$

Thus, since $d\vec{\xi}_s/dX$ is not zero, $dY/dX = 0 \Rightarrow \det A = 0$, and $\lambda = 0$ is then a solution to Eq.(2.36).

This vanishing eigenvalue holds a number of consequences for the theory of quantum fluctuations, which will be highlighted further on. Of course, further work is necessary to establish the stability of all positive slope solutions ($dY/dX > 0$) to Eq.(2.26), and instability of all negative slope solutions ($dY/dX < 0$). In fact, all positive slope solutions are not stable. It has been shown that for dispersive systems with similar atomic and cavity decay rates, $\gamma_\parallel \sim \kappa$, a portion of the high intensity branch of the bistability curve may be unstable (Ikeda and Akimoto, 1982; Lugiato et al., 1982b; Lugiato and Narducci, 1984; Lugiato et al., 1984). For some parameter values this instability leads to period-doubling and chaos. The parameters required for this single-mode instability tend to be rather difficult to access in a system of homogeneously broadened two-level atoms; although, periodic self-oscillation, but not period-doubling and chaos, has been observed in such a system by Orozco et al. (1984). The quantum mechanical theory developed in the following sections can in principle be applied to these nonstationary attractors. However, no explicit consideration of this interesting regime is given in the present work. A quantum theory of weak side-mode gain, applicable to the onset of instabilities in optical bistability, has recently been developed as a generalization of Scully-Lamb laser theory (Sargent et al., 1983; Sargent et al., 1985; Stenholm et al., 1985).

3. MICROSCOPIC MODEL

3.1 Master equation for optical bistability

The quantum-mechanical version of the foregoing model is now easily constructed. We focus first on the system of atoms and electromagnetic field inside the cavity. The single-mode cavity field is described by a quantized harmonic oscillator, with Hamiltonian

$$H_F = \hbar\omega_c a^\dagger a \ , \tag{3.1}$$

where $a^\dagger$ and a are creation and annihilation operators obeying boson commutation relations $[a,a^\dagger] = 1$. The electric field operator for the cavity mode is

$$\vec{E}(z,t) = i\hat{e}(\hbar\omega_0/2\varepsilon_0 V_Q)^{\frac{1}{2}} a(t) e^{ik_0 z} + \text{h.c.} \ , \tag{3.2}$$

where V_Q is the cavity quantization volume. Here, and below in Eq.(3.6), we use the frequency ω_0 rather than ω_c in anticipation of the off-resonant driving of the cavity mode. This oscillator couples to the induced polarization

$$\vec{P}(z,t) = \hat{e}\mu \sum_{j=1}^{N} \sigma_-^j \delta(z - z_j) + \text{h.c.} \tag{3.3}$$

for a collection of N two-level atoms, located at positions $z = z_j$, $j = 1,2,...N$. The atomic Hamiltonian is

$$H_A = \hbar\omega_a J_z \ , \tag{3.4}$$

and the atom-field interaction Hamiltonian in the dipole and rotating-wave approximations is

$$H_{AF} = i\hbar g(a^{\dagger}J_{-} - aJ_{+}) \ , \tag{3.5}$$

where

$$g = \mu(\omega_0/2\hbar\varepsilon_0 V_Q)^{\frac{1}{2}} \tag{3.6}$$

is the atom-field coupling constant, and J_+, J_-, and J_z are collective spin operators:

$$J_{\pm} = \sum_{j=1}^{N} \exp(\pm ik_0 z_j)\sigma_{\pm}^{j} \ , \tag{3.7a}$$

$$J_z = \sum_{j=1}^{N} \sigma_z^{j} \ . \tag{3.7b}$$

The atoms are independent, with commutation relations

$$[\sigma_{+}^{j},\sigma_{-}^{k}] = 2\sigma_{z}^{j}\delta_{jk} \ , \tag{3.8a}$$

$$[\sigma_{\pm}^{j},\sigma_{z}^{k}] = \mp\sigma_{\pm}^{j}\delta_{jk} \ , \tag{3.8b}$$

whence

$$[J_{+},J_{-}] = 2J_{z} \ , \tag{3.9a}$$

$$[J_{\pm},J_{z}] = \mp J_{\pm} \ . \tag{3.9b}$$

This coupled atom-field system is just the Tavis-Cummings problem (Tavis and Cummings, 1968; Tavis and Cummings, 1969). In the Schrödinger picture it is described by an equation of motion for the density operator ρ in the form

$$\dot{\rho} = (1/i\hbar)[H_A + H_F + H_{AF},\rho] \ . \tag{3.10}$$

The generalization to optical bistability includes interactions with the environment describing dissipation and energy input through the partially transmitting input mirror. First, dissipation. Both the cavity mode and the atoms are damped: the cavity mode by losses at the input and output mirrors, and the atoms by spontaneous emission to modes other than the privileged cavity mode (out the sides of the cavity). For generality we will also include dephasing atomic collisions, although these are absent in atomic beam experiments (Rosenberger et al., 1983; Rosenberger et al., 1984). These dissipative processes are treated by introducing reservoir interactions (Senitzky, 1960; Senitzky, 1961) and deriving a master equation in the Born-Markoff approximation. The method is identical to that used in the quantum theory of the laser (Weidlich and Haake, 1965a; Weidlich and Haake, 1965b; Haken, 1970). With the interaction H_{AF} dropped for the derivation of the irreversible terms and reintroduced a postiori (Carmichael and Walls, 1973), the generalization of Eq.(3.10) describes a damped harmonic oscillator coupled to damped two-level atoms:

$$\dot{\rho} = (1/i\hbar)[H_A + H_F + H_{AF}, \rho] + L^A_{irrev}\rho + L^F_{irrev}\rho , \tag{3.11}$$

where L^A_{irrev} and L^F_{irrev} are generalized Liouville operators:

$$L^A_{irrev}\rho = \tfrac{1}{2}\gamma_{\parallel} \sum_{j=1}^{N} (2\sigma^j_-\rho\sigma^j_+ - \sigma^j_+\sigma^j_-\rho - \rho\sigma^j_+\sigma^j_-) + (\gamma_{\parallel} - 2\gamma_{\perp}) \sum_{j=1}^{N} (\sigma^j_z\rho\sigma^j_z - \rho) , \tag{3.12}$$

$$L^F_{irrev}\rho = \kappa(2a\rho a^\dagger - a^\dagger a\rho - \rho a^\dagger a) . \tag{3.13}$$

We omit terms describing thermal fluctuations from these equations, since at optical frequencies they are negligible. Finally, coherent laser light is coupled into the cavity by driving the cavity oscillator with a classical field $\tilde{E}_i\exp(-i\omega_0 t)$; we introduce the interaction

$$H_{LF} = i\hbar(\tilde{E}_i e^{-i\omega_0 t}a^\dagger - \tilde{E}^*_i e^{i\omega_0 t}a) , \tag{3.14}$$

with

$$\kappa^{-1}\tilde{E}_i = -i(2\varepsilon_0 V_Q/\hbar\omega_0)^{\frac{1}{2}}(1 - R)^{-1}\sqrt{T}e^{i\phi_T}E_i , \tag{3.15}$$

where $\tilde{E}_i$ has been chosen to give the driving term in Eq.(2.19) from the Heisenberg equation $\dot{a} = (1/i\hbar)[a, H_{LF}]$.

Our complete quantum-mechanical model is now defined by the master equation

$$\dot{\rho} = (1/i\hbar)[H_A + H_F + H_{AF} + H_{LF},\rho] + L^A_{irrev}\rho + L^F_{irrev}\rho \ . \tag{3.16}$$

Equation (3.16) was first proposed as a model for absorptive bistability ($\omega_a = \omega_c = \omega_0$) by Bonifacio and Lugiato (1978e), and in its general form, including dispersion, by Drummond and Walls (1981). Other authors have made direct passage to master equations for either the good-cavity limit, $\kappa \ll \gamma_{\parallel}, \gamma_{\perp}$ (Willis, 1977; Bonifacio et al., 1978; Willis, 1978; Willis and Day, 1979), or the bad-cavity limit, $\kappa \gg \gamma_{\parallel}, \gamma_{\perp}$ (Narducci et al., 1978a; Agarwal et al., 1978). In fact, all authors, with the exception of recent work by Carmichael (1986a) and Carmichael et al. (1986), have specialized to one or other of these limits, either in the master equation itself or at a later stage in their analysis. Adiabatic elimination of the atomic variables (good cavity) at the master equation level has not been entirely successful (Drummond and Walls, 1981); elimination of the field variables (bad cavity) has produced identical results whether made in the master equation or at a later stage. In the present work we will perform both eliminations from stochastic differential equations in section 4. It is useful, however, to view the master equation for the bad-cavity limit before moving on. For absorptive bistability ($\omega_a = \omega_c = \omega_0$) this equation reads (Narducci et al., 1978a; Agarwal et al., 1978)

$$\dot{\rho} = (1/i\hbar)[H_A + H_{LA},\rho] + L^A_{irrev}\rho + L^{A'}_{irrev}\rho \ , \tag{3.17}$$

with

$$H_{LA} = i\hbar(g/\kappa)[\tilde{E}_i^* \exp(i\omega_a t)J_+ - \tilde{E}_i \exp(-i\omega_a t)J_-] \ , \tag{3.18}$$

$$L^{A'}_{irrev}\rho = (g^2/\kappa)(2J_-\rho J_+ - J_+J_-\rho - \rho J_+J_-) \ . \tag{3.19}$$

This clearly shows the relationship between our model and quantum-statistical models for superradiance and cooperative resonance fluorescence. Equation (3.19) describes collective spontaneous emission into the privileged cavity mode, and forms the basis for the single-mode model for superradiance (Bonifacio et al., 1971). With the addition of a coherent driving field (the interaction H_{LA}) superradiance becomes co-operative resonance fluorescence (Senitzky, 1972). Both of these phenomena only involve processes which conserve the total spin $J^2 = J_z^2 + \frac{1}{2}(J_+J_- + J_-J_+)$; they are described by master equations written entirely in terms of collective atomic operators. Our model for optical bistability includes the term $L^A_{irrev}\rho$ which involves single-atom operators. Leaving collisions aside, this term describes single-atom spontaneous emission to modes other than the privileged cavity mode. This breaks the J^2 conservation and makes the important distinction between long-time behavior in our present model and that for cooperative resonance fluorescence (Drummond and Carmichael, 1978; Narducci et al., 1978b; Walls et al., 1978; Puri and Lawande, 1979; Carmichael, 1980a; Hassan et al., 1980; Puri and Lawande, 1980; Puri et al., 1980; Puri et al., 1984).

3.2 Connection with macroscopic theory

It will be convenient from this point to adopt a frame rotating at the frequency of the incident field. It is cumbersome to introduce new notation and therefore we make this transformation with

$$\exp[i\omega_0(J_z + a^\dagger a)t]\rho\exp[-i\omega_0(J_z + a^\dagger a)t] \to \rho \ , \tag{3.20a}$$

$$e^{i\omega_0 t}(a,\sigma_-^j,J_-) \to (a,\sigma_-^j,J_-) \ , \tag{3.20b}$$

$$e^{-i\omega_0 t}(a^\dagger,\sigma_+^j,J_+) \to (a^\dagger,\sigma_+^j,J_+) \ . \tag{3.20c}$$

Then Eq.(3.16) becomes

$$\begin{aligned}\dot{\rho} = {} & (\omega_a - \omega_0)[J_z,\rho] + (\omega_c - \omega_0)[a^\dagger a,\rho] \\ & + g[a^\dagger J_- - aJ_+,\rho] + [\tilde{E}_i a^\dagger - \tilde{E}_i^* a,\rho] \\ & + \tfrac{1}{2}\gamma_{\parallel}\left(\sum_{j=1}^{N} 2\sigma_-^j\rho\sigma_+^j - J_z\rho - \rho J_z - N\rho\right) \\ & + (\gamma_{\parallel} - 2\gamma_{\perp})\left(\sum_{j=1}^{N} \sigma_z^j\rho\sigma_z^j - N\rho\right) \\ & + \kappa(2a\rho a^\dagger - a^\dagger a\rho - \rho a^\dagger a) \ ,\end{aligned} \tag{3.21}$$

where we have set $\sigma_+^j\sigma_-^j = \sigma_z^j + \frac{1}{2}$, and the field and polarization operators are now

$$\vec{E}(z,t) = i\hat{e}(\hbar\omega_0/2\varepsilon_0 V_Q)^{\frac{1}{2}} a e^{-i(\omega_0 t - k_0 z)} + \text{h.c.} , \qquad (3.22)$$

and

$$\vec{P}(z,t) = \hat{e}\mu \sum_{j=1}^{N} \sigma_-^j e^{-i\omega_0 t}\delta(z - z_j) + \text{h.c.} . \qquad (3.23)$$

Equations for expectation values follow from Eq.(3.21) in the form:

$$\kappa^{-1}\langle\dot{a}\rangle = -(1 + i\phi)\langle a\rangle + (g/\kappa)\langle J_-\rangle + \kappa^{-1}\tilde{E}_i , \qquad (3.24a)$$

$$\gamma_\perp^{-1}\langle\dot{J}_-\rangle = -(1 + i\delta)\langle J_-\rangle + \tfrac{1}{2}(g/\gamma_\perp)\langle aJ_z\rangle , \qquad (3.24b)$$

$$\gamma_\parallel^{-1}\langle\dot{J}_z\rangle = -(\langle J_z\rangle + N/2) - (g/\gamma_\parallel)(\langle aJ_+\rangle + \langle a^\dagger J_-\rangle) . \qquad (3.24c)$$

Connection with the macroscopic theory of the last section is made by factorizing the operator products--i.e. by neglecting quantum noise. We first make the connection between the variables of the two theories. From Eqs. (3.22) and (2.10), the classical field amplitude is

$$\overline{E}(t) \equiv E(0,t) = i(\hbar\omega_0/2\varepsilon_0 V_Q)^{\frac{1}{2}}\langle a(t)\rangle . \qquad (3.25)$$

Then, since the atoms differ only by a phase determined by their positions in the field, we may set

$$\langle\sigma_\pm^j\rangle = N^{-1}\langle J_\pm\rangle\exp(\mp ik_0 z_j) , \qquad (3.26)$$

whence, from Eqs.(3.23) and (2.11), the classical polarization amplitude is

$$\overline{P}(t) \equiv P(0,t) = \mu N^{-1}\langle J_-\rangle\Big[\sum_{j=1}^{N}\delta(z - z_j)\Big]_{z=0}$$
$$= \mu\rho_0 N^{-1}\langle J_-\rangle , \qquad (3.27)$$

where we have replaced the singular atomic distribution by a spatially uniform distribution with density ρ_0. The inversion density is

$$D = 2\rho_0 N^{-1}\langle J_z\rangle . \qquad (3.28)$$

The variables $\bar{\alpha}$, $\bar{v}$, and $\bar{m}$ defined by Eqs. (2.29) are now

$$\bar{\alpha} = i\,\exp[-i(\phi_T + \phi_i)]\langle a\rangle/n_{sat}^{1/2} , \tag{3.29a}$$

$$\bar{v} = i\,\exp[-i(\phi_T + \phi_i)]2N^{-1}(\gamma_\perp/\gamma_\parallel)^{1/2}\langle J_-\rangle , \tag{3.29b}$$

$$\bar{m} = 2\langle J_z\rangle/N , \tag{3.29c}$$

where

$$n_{sat} = (\hbar\omega_0/2\varepsilon_0 V_Q)^{-1} I_{sat} = \gamma_\parallel\,\gamma_\perp/4g^2 \tag{3.30}$$

is the saturation photon number. After factorization, and scaling according to Eqs.(3.29), Eqs.(3.24) reproduce Eqs. (2.30). We use Eqs.(3.15) and (2.27a) to relate $\tilde{E}_i$ and Y. The bistability parameter C enters in the alternative form

$$C = Ng^2/2\kappa\gamma_\perp . \tag{3.31}$$

To show the equivalence between this form and Eq.(2.28) we may substitute for g and κ to write

$$\begin{aligned} C &= N(\mu^2\omega_0/2\hbar\varepsilon_0 V_Q)[2\gamma_\perp(1 - R)/\tau]^{-1} \\ &= \rho_0^{-1}[N(V_Q L/L')^{-1}](1 - R)^{-1}(\rho_0\mu^2\omega_0/\hbar\varepsilon_0 c\gamma_\perp)L/4 , \end{aligned} \tag{3.32}$$

where $L' = c\tau$ is the cavity round-trip distance. Since $V_Q L/L'$ is just the cavity volume occupied by the atoms, $\rho_0 = N(V_Q L/L')^{-1}$, and Eq.(3.32) reproduces Eq.(2.28).

4. FOKKER-PLANCK EQUATION FOR OPTICAL BISTABILITY

Equation (3.21) cannot be solved in operator form. The approach taken by most authors has been to pass from this operator equation to a Fokker-Planck equation for a quasidistribution function which represents ρ in a classical phase space. An alternative Heisenberg picture approach using quantum-mechanical Langevin equations has been followed in a linearized analysis by Agarwal et al. (1978). This alternative approach will not be discussed here. An important issue is raised by the derivation of a Fokker-Planck equation for optical bistability. The standard methods based on the Glauber-Sudarshan-Haken representation (Glauber, 1963a; Glauber,

1963b; Glauber, 1963c; Sudarshan, 1963; Haken, 1970), so successfully employed to treat the laser, fail: the Fokker-Planck equation does not have positive definite diffusion (Gronchi and Lugiato, 1978; Agarwal et al., 1980). This difficulty actually makes our subject more interesting, since it is related to the nonclassical effects such as photon antibunching and squeezing which are described within our model. In this section we first review the derivation of the Fokker-Planck equation using the Glauber-Sudarshan-Haken representation. Then two approaches which have been taken to avert the difficulty with non-positive-definite diffusion will be discussed--one based on the Wigner representation (Wigner, 1932; Hillery et al., 1984) and the other based on the positive P-representation (Drummond and Gardiner, 1980).

4.1 The Glauber-Sudarshan-Haken representation

This representation is based on the normally-ordered characteristic function

$$\chi_N(\beta,\beta^*,\xi,\xi^*,\eta) = \mathrm{tr}(\rho e^{i\beta^* a^\dagger} e^{i\beta a} e^{i\xi^* J_+} e^{i\eta J_z} e^{i\xi J_-}) \ . \quad (4.1)$$

The quasidistribution function P is defined as the Fourier transform of χ_N:

$$P(\alpha,\alpha^*,v,v^*,m) = (2\pi^5)^{-1}\int d^2\beta \int d^2\xi \int d\eta \ \chi_N(\beta,\beta^*,\xi,\xi^*,\eta) \\ e^{-i\beta^*\alpha^*} e^{-i\beta\alpha} e^{-i\xi^* v^*} e^{-i\xi v} e^{-i\eta m} \ , \quad (4.2)$$

with χ_N given in terms of P by the inverse transform:

$$\chi_N(\beta,\beta^*,\xi,\xi^*,\eta) = \int d^2\alpha \int d^2 v \int dm \ P(\alpha,\alpha^*,v,v^*,m) \\ e^{i\beta^*\alpha^*} e^{i\beta\alpha} e^{i\xi^* v^*} e^{i\xi v} e^{i\eta m} \ . \quad (4.3)$$

It follows from Eq.(4.3) that normally ordered operator averages can be evaluated as "classical" averages of corresponding phase-space variables against the quasidistribution function P. We have

$$\langle a^{\dagger n} a^m J_+^p J_z^r J_-^q \rangle = \mathrm{tr}(\rho a^{\dagger n} a^m J_+^p J_z^r J_-^q)$$

$$= \frac{\partial^n}{\partial(i\beta^*)^n} \frac{\partial^m}{\partial(i\beta)^m} \frac{\partial^p}{\partial(i\xi^*)^p} \frac{\partial^r}{\partial(i\eta)^r} \frac{\partial^q}{\partial(i\xi)^q} \chi_N \Big|_{\substack{\beta=\beta^*=0 \\ \xi=\xi^*=\eta=0}}$$

$$= \int d^2\alpha \int d^2v \int dm \; \alpha^{*n} \alpha^m v^{*p} m^r v^q \; P \; . \tag{4.4}$$

Our objective is to pass from Eq.(3.21) to an equation of motion for P. The details of this calculation are rather tedious and can be found elsewhere (Haken, 1970; Carmichael, 1986b). The resulting equation for P is

$$\frac{\partial P}{\partial t} = (L_A + L_F + L_{AF} + L_{LF})P \; , \tag{4.5}$$

with

$$L_A = [\gamma_\perp + i(\omega_a - \omega_0)]\frac{\partial}{\partial v} v + \text{c.c.}$$
$$+ \gamma_\parallel (D^m_{+1} - 1)(m + N/2) + (2\gamma_\perp - \gamma_\parallel) D^m_{-1} \frac{\partial^2}{\partial v \partial v^*}(m + N/2) \; , \tag{4.6a}$$

$$L_F = [\kappa + i(\omega_c - \omega_0)]\frac{\partial}{\partial \alpha}\alpha + \text{c.c.} \; , \tag{4.6b}$$

$$L_{AF} = -g\{\frac{\partial}{\partial \alpha} v + [(D^m_{-1} - 1)v^* + 2\frac{\partial}{\partial v} m - \frac{\partial^2}{\partial v^2} v]\alpha\} + \text{c.c.} \; , \tag{4.6c}$$

$$L_{LF} = -\tilde{E}_i \frac{\partial}{\partial \alpha} + \text{c.c.} \; , \tag{4.6d}$$

where $D^m_{\pm 1} = \exp(\pm\partial/\partial m)$ are shift operators:

$$D^m_{\pm 1} f(m) = \exp(\pm\partial/\partial m) f(m) = f(m \pm 1) \; . \tag{4.7}$$

This is not yet a Fokker-Planck equation since derivatives in the inversion variable m appear to all orders. In fact, we have a set of partial differential difference equations. It can be shown quite generally that P is strictly always a singular function in m, defined on the discrete set of inversion states $2m = -N, -N+2, \ldots, N-2, N$ as a sum of weighted δ-

functions (Carmichael, 1986b). In Eq.(4.5) the dynamics is restricted to these discrete states by the shift operators $D^m_{\pm 1}$. Partial differential difference equations are certainly as difficult to solve as Eq.(3.21). Some reasonable approximation must therefore be made. If N is very large, and the distribution over inversion states varies smoothly between neighboring states, we might overlook the singular structure of P, and seek a smooth envelope function as indicated in Fig. 4. We can impose this structure on P by truncating the derivatives in $D^m_{\pm 1}$. But at what order? The rationale for the truncation we use is plausibly argued as follows: A smooth envelope function interpolates between the discrete states. Equation (4.5) couples each value of m to the two neighboring values $m \pm 1$. We therefore envisage making the interpolation by fitting a quadratic to $P(m - 1)$, $P(m)$ and $P(m + 1)$. Since the action of $\exp(\pm\partial/\partial m)$ on a quadratic terminates at the second derivative we will truncate the shift operators at the second order. If collisions are to be included ($2\gamma_\perp \neq \gamma_\parallel$), we must also drop the third and fourth order derivatives which remain in Eq.(4.6a) to arrive at a Fokker-Planck equation. This can be justified by a scaling argument similar to that used by Haken (1970) for the laser. The resulting Fokker-

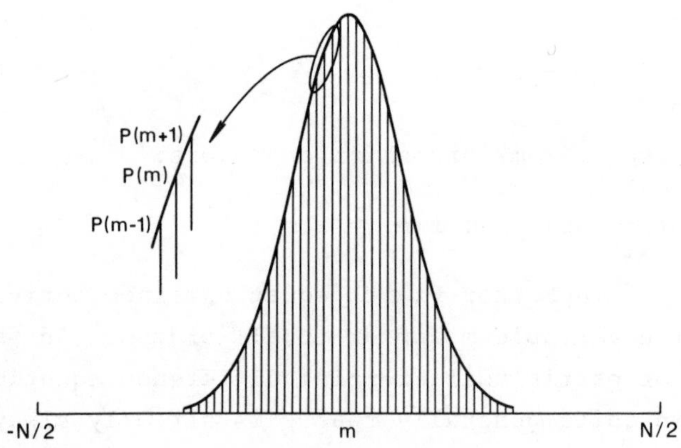

Fig. 4. Envelope construction used to truncate shift operators $D^m_{\pm 1}$ in Eqs.(4.6a) and (4.6c).

Planck equation for optical bistability then reads:

$$\frac{\partial P}{\partial t} = \{-\frac{\partial}{\partial\bar{\alpha}}\kappa[-(1+i\phi)\bar{\alpha} + 2C\bar{v} + Y] + \text{c.c.}$$

$$-\frac{\partial}{\partial\bar{v}}\gamma_\perp[-(1+i\delta)\bar{v} + \bar{\alpha}\bar{m}] + \text{c.c.}$$

$$-\frac{\partial}{\partial\bar{m}}\gamma_\parallel[-(\bar{m}+1) - \tfrac{1}{2}\bar{\alpha}\bar{v}^* - \tfrac{1}{2}\bar{\alpha}^*\bar{v}]$$

$$+\gamma_\perp N^{-1}(\frac{\partial^2}{\partial\bar{v}^2}\bar{\alpha}\bar{v} + \text{c.c.}) + 2\gamma_\perp N^{-1}(2\gamma_\perp/\gamma_\parallel - 1)\frac{\partial^2}{\partial\bar{v}\partial\bar{v}^*}(\bar{m}+1)$$

$$+\gamma_\parallel N^{-1}\frac{\partial^2}{\partial\bar{m}^2}(\bar{m} + 1 - \tfrac{1}{2}\bar{\alpha}\bar{v}^* - \tfrac{1}{2}\bar{\alpha}^*\bar{v})\}P\ , \qquad (4.8)$$

where the scaling of Eqs.(2.29) and (3.29) has been introduced with

$$\bar{\alpha} = i\exp[-i(\phi_T + \phi_i)]\alpha/n_{sat}^{1/2}\ , \qquad (4.9a)$$

$$\bar{v} = i\exp[-i(\phi_T + \phi_i)]2N^{-1}(\gamma_\perp/\gamma_\parallel)^{1/2}v\ , \qquad (4.9b)$$

$$\bar{m} = 2m/N\ . \qquad (4.9c)$$

To demonstrate the difficulty with Eq.(4.8) let us define the five-dimensional vector

$$\vec{z} = (\bar{\alpha}_x, \bar{\alpha}_y, \bar{v}_x, \bar{v}_y, \bar{m})^T\ , \qquad (4.10)$$

where $\bar{\alpha} = \bar{\alpha}_x + i\bar{\alpha}_y$ and $\bar{v} = \bar{v}_x + i\bar{v}_y$. Then we can write our Fokker-Planck equation in the general form

$$\frac{\partial P}{\partial t} = [-\sum_j \frac{\partial}{\partial z_j}\bar{a}_j(\vec{z}) + \tfrac{1}{2}\sum_{jk}\frac{\partial^2}{\partial z_j\partial z_k}\bar{d}_{jk}(\vec{z})]P\ , \qquad (4.11)$$

with

$$\bar{a}_1 = \kappa(-z_1 + \phi z_2 + 2Cz_3 + Y)\ , \qquad (4.12a)$$

$$\bar{a}_2 = \kappa(-z_2 - \phi z_1 + 2Cz_4)\ , \qquad (4.12b)$$

$$\bar{a}_3 = \gamma_\perp(-z_3 + \delta z_4 + z_1 z_5)\ , \qquad (4.12c)$$

$$\bar{a}_4 = \gamma_\perp(-z_4 - \delta z_3 + z_2 z_5) , \tag{4.12d}$$

$$\bar{a}_5 = \gamma_\parallel(-z_5 - 1 - z_1 z_3 - z_2 z_4) , \tag{4.12e}$$

and nonzero elements of the diffusion matrix given by:

$$\bar{d}_{33} = \gamma_\perp N^{-1}[\Gamma(z_5 + 1) + (z_1 z_3 - z_2 z_4)] , \tag{4.13a}$$

$$\bar{d}_{44} = \gamma_\perp N^{-1}[\Gamma(z_5 + 1) - (z_1 z_3 - z_2 z_4)] , \tag{4.13b}$$

$$\bar{d}_{34} = \bar{d}_{43} = \gamma_\perp N^{-1}(z_1 z_4 + z_2 z_3) , \tag{4.13c}$$

$$\bar{d}_{55} = 2\gamma_\parallel N^{-1}(z_5 + 1 - z_1 z_3 - z_2 z_4) , \tag{4.13d}$$

where

$$\Gamma = 2\gamma_\perp/\gamma_\parallel - 1 . \tag{4.14}$$

Now, in classical statistical physics, for $\bar{d} = (\bar{d}_{jk})$ to be an acceptable diffusion matrix the quadratic form $\vec{x}^T \bar{d} \vec{x}$ must be positive definite. Then $\bar{d}$ may be decomposed with

$$\bar{d} = \bar{b}\bar{b}^T , \tag{4.15}$$

and the Fokker-Planck description is equivalent to the set of Ito stochastic differential equations (Gardiner, 1983)

$$d\vec{z} = \vec{\bar{a}}(\vec{z})dt + \bar{b}(\vec{z})d\vec{W} , \tag{4.16}$$

where $d\vec{W}$ is a vector of independent Wiener processes. The matrix defined by Eqs.(4.13) is not positive definite, and the decomposition leading to Eq.(4.16) is only possible if $\bar{b}$ becomes complex. Specifically, for our present application $\bar{d}$ can be decomposed by a matrix $\bar{b}$ with nonzero elements:

$$\bar{b}_{33} = (\gamma_\perp/N)^{\frac{1}{2}}[\Gamma(z_5 + 1) + (z_1 z_3 - z_2 z_4)]^{\frac{1}{2}} , \tag{4.17a}$$

$$\bar{b}_{43} = (\gamma_\perp/N)^{\frac{1}{2}}(z_1 z_4 + z_2 z_3)[\Gamma(z_5 + 1) + (z_1 z_3 - z_2 z_4)]^{-\frac{1}{2}}, \tag{4.17b}$$

$$\bar{b}_{44} = (\gamma_\perp/N)^{\frac{1}{2}}\{[\Gamma(z_5 + 1) - (z_1 z_3 - z_2 z_4)] - (z_1 z_4 + z_2 z_3)^2/[\Gamma(z_5 + 1) + (z_1 z_3 - z_2 z_4)]\}^{\frac{1}{2}} , \tag{4.17c}$$

$$\bar{b}_{55} = (2\gamma_{\parallel}/N)^{\frac{1}{2}}(z_5 + 1 - z_1 z_3 - z_2 z_4)^{\frac{1}{2}} \ . \tag{4.17d}$$

Of course, this is not unique. $\bar{b}$ can be multiplied on the right by any unitary matrix. Clearly $\bar{b}$ is generally complex. But $\vec{z}$ is real by construction. There is a difficulty then with proceeding on the basis of Eq. (4.16). In particular, a linearized theory of fluctuations evaluates $\bar{b}$ at the steady state, using Eqs. (2.31). Consider the example of absorptive bistability ($\delta = \phi = 0$) with purely radiative damping ($\Gamma = 0$). The nonzero elements of $\bar{b}$ in the steady state are:

$$\bar{b}^s_{33} = i(\gamma_{\perp}/N)^{\frac{1}{2}}X/(1 + X^2)^{\frac{1}{2}} \ , \tag{4.18a}$$

$$\bar{b}^s_{44} = (\gamma_{\perp}/N)^{\frac{1}{2}}X/(1 + X^2)^{\frac{1}{2}} \ , \tag{4.18b}$$

$$\bar{b}^s_{55} = 2(\gamma_{\parallel}/N)^{\frac{1}{2}}X/(1 + X^2)^{\frac{1}{2}} \ . \tag{4.18c}$$

Equation (4.18a) gives a pure imaginary noise source driving the real part of the polarization $\bar{v}_x$. Apparently the Glauber-Sudarshan-Haken representation will be of no use for a theory of optical bistability.

In fact, it is not clear that such complex noise need worry us at all, since we are dealing with a quantum-mechanical theory where, in the final analysis, the passage to Eq. (4.16) is simply a formal construction. P is a quasidistribution which need not preserve all the properties of a classical probability--for example, it need not be positive. Actually, P need not exist at all as a nonsingular function. It is necessary to introduce singular distributions such as the derivatives of δ-functions to describe some nonclassical states of the field (Klauder and Sudarshan, 1968). Non-positive-definite diffusion indicates that the quasidistribution for optical bistability needs to exploit this potential for taking a non-classical form. However, traditionally Fokker-Planck equations have been used precisely because they lead to an acceptable "classical" statistical description, from which many things can be calculated using familiar methods. In laser theory the terms which lead to the difficulty we have encountered here are dropped on the basis of scaling arguments

(Haken, 1970; Gronchi and Lugiato, 1978). This cannot be done for optical bistability. Prior to the work of Drummond and Gardiner (1980), to which we return shortly, it seemed clear to most workers that one cannot proceed with any confidence on the basis of nonclassical equations such as Eqs.(4.8) and (4.16). A different approach must therefore be taken. The alternative taken by Lugiato and coworkers (Gronchi and Lugiato, 1978; Lugiato et al., 1982c) and Agarwal et al. (1980) is to abandon the Glauber-Haken-Sudarshan representation for the Wigner representation.

4.2 The Wigner representation

The Wigner representation is the grandfather of these phase space techniques. Its use in quantum statistical mechanics (Wigner, 1932) predates the popular rise of the Glauber-Sudarshan representation for applications in quantum optics by many years. The formal introduction of the Wigner function follows Eqs.(4.1)-(4.3) very closely. It is based on the symmetrically- or Weyl-ordered characteristic function

$$\chi_W(\beta,\beta^*,\xi,\xi^*,\eta) = \mathrm{tr}(\rho e^{i\beta^* a^\dagger + i\beta a + i\xi^* J_+ + i\eta J_z + i\xi J_-}) \;. \tag{4.19}$$

The Wigner function W is defined as the Fourier transform of χ_W:

$$W(\alpha,\alpha^*,v,v^*,m) = (2\pi^5)^{-1}\int d^2\beta \int d^2\xi \int d\eta\; \chi_W(\beta,\beta^*,\xi,\xi^*,\eta)\, e^{-i\beta^*\alpha^*} e^{-i\beta\alpha} e^{-i\xi^* v^*} e^{-i\xi v} e^{-i\eta m} \;, \tag{4.20}$$

with

$$\chi_W(\beta,\beta^*,\xi,\xi^*,\eta) = \int d^2\alpha \int d^2 v \int dm\; W(\alpha,\alpha^*,v,v^*,m)\, e^{i\beta^*\alpha^*} e^{i\beta\alpha} e^{i\xi^* v^*} e^{i\xi v} e^{i\eta m} \;. \tag{4.21}$$

Averages of phase-space variables give symmetrically-ordered operator averages; for example

$$\int d^2\alpha \int d^2 v \int dm\; \alpha^*\alpha\, W = \tfrac{1}{2}(\langle a^\dagger a\rangle + \langle a a^\dagger\rangle) = \langle a^\dagger a\rangle + \tfrac{1}{2} \;. \tag{4.22}$$

Use of the commutation relations to relate symmetrically- and normally-ordered averages, as in Eq.(4.22), indicates that W must obey an equation with quite different noise properties to the equation for P. Nonetheless, it is not guaranteed that W obeys a Fokker-Planck equation with positive definite diffusion. The Wigner function can take negative values just like P. It does always exist as an ordinary (nonsingular) function, however. Its properties are discussed at some length by Hillery et al. (1984).

The Fokker-Planck equation for optical bistability based on the Wigner representation has been derived by Gronchi and Lugiato (1978) and Agarwal et al. (1980). We will not follow the details of this derivation here. A similar truncation of derivatives must be used to obtain an equation in the Fokker-Planck form as used with the Glauber-Sudarshan-Haken representation. Actually, the derivation of a formally exact equation corresponding to Eq.(4.5) is not at all straightforward in the Wigner representation due to difficulties with disentangling the collective atomic operators appearing in Eq.(4.19). Both of the references cited above bypass this step and impose the Fokker-Planck form on the equation of motion for W a priori. The drift and diffusion coefficients for this Fokker-Planck equation are then identified from the hierarchy of moment equations which follow from the operator master equation. The resulting equation reads:

$$
\begin{aligned}
\frac{\partial W}{\partial t} = \Big\{ & -\frac{\partial}{\partial\bar{\alpha}}\kappa[-(1+i\phi)\bar{\alpha}+2C\bar{v}+Y]+\text{c.c.} \\
& -\frac{\partial}{\partial\bar{v}}\gamma_{\perp}[-(1+i\delta)\bar{v}+\overline{\alpha m}]+\text{c.c.} \\
& -\frac{\partial}{\partial\bar{m}}\gamma_{\parallel}[-(\bar{m}+1)-\tfrac{1}{2}\overline{\alpha v^*}-\tfrac{1}{2}\bar{\alpha}^*\bar{v}] \\
& +\kappa n_{sat}^{-1}\frac{\partial^2}{\partial\bar{\alpha}\partial\bar{\alpha}^*}+2\gamma_{\perp}N^{-1}(2\gamma_{\perp}/\gamma_{\parallel})\frac{\partial^2}{\partial\bar{v}\partial\bar{v}^*} \\
& +\gamma_{\parallel}N^{-1}\Big(\frac{\partial^2}{\partial\bar{v}\partial\bar{m}}\bar{v}+\text{c.c.}\Big)+\gamma_{\parallel}N^{-1}\frac{\partial^2}{\partial\bar{m}^2}(\bar{m}+1)\Big\}W\,.
\end{aligned}
\tag{4.23}
$$

This equation may be written in the form of Eq.(4.11) with

$\bar{a}_1 \ldots \bar{a}_5$ as in Eqs.(4.12), and nonzero elements of the diffusion matrix given by:

$$\bar{d}_{11} = \bar{d}_{22} = \tfrac{1}{2}\kappa n_{sat}^{-1} , \tag{4.24a}$$

$$\bar{d}_{33} = \bar{d}_{44} = \gamma_\perp N^{-1} f^{-1} , \tag{4.24b}$$

$$\bar{d}_{35} = \bar{d}_{53} = \gamma_{\|} N^{-1} z_3 , \tag{4.24c}$$

$$\bar{d}_{45} = \bar{d}_{54} = \gamma_{\|} N^{-1} z_4 , \tag{4.24d}$$

$$\bar{d}_{55} = 2\gamma_{\|} N^{-1}(z_5 + 1) , \tag{4.24e}$$

with

$$f = \gamma_{\|}/2\gamma_\perp . \tag{4.25}$$

The decomposition of $\bar{d}$ in Eq.(4.15) is made with nonzero elements of the matrix $\bar{b}$ given by:

$$\bar{b}_{11} = \bar{b}_{22} = (\kappa/2n_{sat})^{\frac{1}{2}} , \tag{4.26a}$$

$$\bar{b}_{33} = \bar{b}_{44} = (\gamma_\perp/N)^{\frac{1}{2}} f^{-\frac{1}{2}} , \tag{4.26b}$$

$$\bar{b}_{53} = (2\gamma_{\|}/N)^{\frac{1}{2}} f z_3, \tag{4.26c}$$

$$\bar{b}_{54} = (2\gamma_{\|}/N)^{\frac{1}{2}} f z_4, \tag{4.26d}$$

$$\bar{b}_{55} = (2\gamma_{\|}/N)^{\frac{1}{2}}[(z_5 + 1) - f^2(z_3^2 + z_4^2)]^{\frac{1}{2}} . \tag{4.26e}$$

This almost escapes the difficulty with non-positive-definite diffusion; however, $\bar{b}_{55}$ may still become complex. Thus, in the general nonlinear situation the Wigner representation also leads to a Fokker-Planck equation with non-positive-definite diffusion. On the other hand, when $\bar{b}_{55}$ is evaluated at the steady state, using Eqs.(2.31), we find

$$\bar{b}_{55}^s = (2\gamma_{\|}/N)^{\frac{1}{2}} X(1 + \delta^2 + X^2)^{-1}[(1 + \delta^2)(1 - f^2) + X^2]^{\frac{1}{2}} . \tag{4.27}$$

This is always real since $f = \gamma_{\|} / 2\gamma_\perp \leq 1$. A linearized theory of fluctuations can therefore be built on the basis of Eq.(4.23) without difficulty. Such a theory has been used to obtain results for the transmitted spectrum, photon antibunching, and squeezing, by Lugiato and coworkers (Lugiato, 1979; Casagrande and Lugiato, 1980; Lugiato and Strini, 1982). These results are reviewed in section 5. A Fokker-Planck equation with positive definite diffusion is obtained even in the nonlinear situation if the atomic variables are first adiabatically eliminated (Lugiato et al., 1982c). Results from the nonlinear theory are reviewed in section 6. The full nonlinear description without adiabatic elimination does still meet with difficulties, however, in Eq.(4.26e). This last obstacle is overcome with use of the positive P-representation.

4.3 The positive P-representation

The positive P-representation is a close relative of the Glauber-Sudarshan-Haken representation. To appreciate this relationship let us reconstitute Eq.(4.16) in complex form with the definition of new variables:

$$\bar{\alpha} = z_1 + iz_2 , \tag{4.28a}$$

$$\bar{\alpha}_* = z_1 - iz_2 , \tag{4.28b}$$

$$\bar{v} = z_3 + iz_4 , \tag{4.28c}$$

$$\bar{v}_* = z_3 - iz_4 , \tag{4.28d}$$

$$\bar{m} = z_5 . \tag{4.28e}$$

Since Eq.(4.16) does not preserve $\vec{z}$ as a real vector we must regard $\bar{\alpha}$, $\bar{\alpha}_*$, $\bar{v}$, $\bar{v}_*$, and $\bar{m}$ as five independent complex variables; $\bar{\alpha}$ and $\bar{\alpha}_*$, and $\bar{v}$ and $\bar{v}_*$ are not complex conjugate, and $\bar{m}$ is not real. Using Eqs.(4.12) and (4.17), Eqs.(4.16) become:

$$d\bar{\alpha} = \kappa[-(1 + i\phi)\bar{\alpha} + 2C\bar{v} + Y]dt , \tag{4.29a}$$

$$d\bar{\alpha}_* = \kappa[-(1 - i\phi)\bar{\alpha}_* + 2C\bar{v}_* + Y]dt , \tag{4.29b}$$

$$d\bar{v} = \gamma_\perp[-(1 + i\delta)\bar{v} + \bar{\alpha}\bar{m}]dt + dV , \tag{4.29c}$$

$$d\bar{v}_* = \gamma_\perp[-(1 - i\delta)\bar{v}_* + \bar{\alpha}_*\bar{m}]dt + dV_* , \tag{4.29d}$$

$$d\bar{m} = \gamma_\parallel[-(\bar{m} + 1) - \tfrac{1}{2}\bar{\alpha}\bar{v}_* - \tfrac{1}{2}\bar{\alpha}_*\bar{v}]dt + dM , \tag{4.29e}$$

where the complex noise terms dV, dV_* and dM are defined in terms of the independent Weiner processes dW_3, dW_4 and dW_5:

$$dV = (\bar{b}_{33} + i\bar{b}_{43})dW_3 + i\bar{b}_{44}dW_4 , \tag{4.30a}$$

$$dV_* = (\bar{b}_{33} - i\bar{b}_{43})dW_3 - i\bar{b}_{44}dW_4 , \tag{4.30b}$$

$$dM = \bar{b}_{55}dW_5 . \tag{4.30c}$$

It can be shown from Eqs.(4.17) and (4.28) that the nonzero correlations for these noise terms are:

$$(dVdV)_{av} = 2\gamma_\perp N^{-1}\overline{\alpha v}dt , \tag{4.31a}$$

$$(dV_*dV_*)_{av} = 2\gamma_\perp N^{-1}\overline{\alpha}_*\overline{v}_*dt , \tag{4.31b}$$

$$(dVdV_*)_{av} = 2\gamma_\perp N^{-1}(2\gamma_\perp/\gamma_\parallel - 1)(\overline{m} + 1)dt , \tag{4.31c}$$

$$(dMdM)_{av} = 2\gamma_\parallel N^{-1}(\overline{m} + 1 - \tfrac{1}{2}\overline{\alpha v_*} - \tfrac{1}{2}\overline{\alpha}_*\overline{v})dt , \tag{4.31d}$$

where the notation $(\ )_{av}$ distinguishes the stochastic ensemble average from the operator average $\langle\ \rangle$. Now Eqs.(4.29) are equivalent to Eq.(4.16). They are just the equations which follow from the Glauber-Sudarshan-Haken representation if we argue that the difficulty with non-positive-definite diffusion is really no difficulty at all--that the formal construction in quantum mechanics need not be confined by the restrictions of classical statistical physics. Note the relationship between Eqs.(4.31) and the diffusion terms of Eq.(4.8). However, perhaps acceptance of Eqs.(4.29) is entirely unjustified. There is surely an inconsistency in passing from the five-dimensional space in which the Glauber-Sudarshan-Haken representation was constructed to the ten-dimensional space required by the dynamics of Eqs.(4.29). It turns out that we can accept Eqs.(4.29). The positive P-representation provides the formal justification.

We will only look briefly at the justification of Eqs.(4.29) on the basis of the positive P-representation. The details of the calculation outlined below can be found in Drummond (1979) and Carmichael (1986b). For bosons the rigorous foundation for the positive P-representation has been given in some detail by Drummond and Gardiner (1980). Its application to atomic systems is a straightforward generalization of the methods used for bosons, much along the lines of Haken's generaliztion of the Glauber-Sudarshan representation (Haken, 1970). Rigorous results comparable to those provided by Drummond and Gardiner for the boson case have not yet been given, however, for the representation of atomic operators.

Since Eq.(4.16) wishes to explore a higher dimensional space we must provide the extra dimensions from the outset. The positive P-representation is based on the generalization of Eq.(4.1) to an analytic characteristic function

$$\chi_N(\beta,\beta_*,\xi,\xi_*,\eta) = \mathrm{tr}(\rho e^{i\beta_* a^\dagger} e^{i\beta a} e^{i\xi_* J_+} e^{i\eta J_z} e^{i\xi J_-}), \qquad (4.32)$$

where β, β_*, ξ, ξ_*, and η are independent complex variables. A distribution function P is then postulated which gives

$$\chi_N(\beta,\beta_*,\xi,\xi_*,\eta) = \int d^2\alpha \int d^2\alpha_* \int d^2v \int d^2v_* \int d^2m \; P(\alpha,\alpha_*,v,v_*,m)$$
$$e^{i\beta_*\alpha_*} e^{i\beta\alpha} e^{i\xi_* v_*} e^{i\xi v} e^{i\eta m} , \qquad (4.33)$$

where α, α_*, v, v_*, and m are also independent complex variables. Note that this is not a Fourier transform, and we cannot write the inverse relationship corresponding to Eq.(4.2). However, from Eq.(4.33) we have, as in Eq.(4.4),

$$\langle a^{\dagger n} a^m J_+^p J_z^r J_-^q \rangle = \int d^2\alpha \int d^2\alpha_* \int d^2v \int d^2v_* \int d^2m \; \alpha_*^n \alpha^m v_*^p m^r v^q \, P$$
$$= (\alpha_*^n \alpha^m v_*^p m^r v^q)_{av} . \qquad (4.34)$$

While α and α_* and v and v_* are not complex conjugate, they must be so in the mean; and while m is not real, it must be real in the mean. The distribution function P is not unique. Drummond and Gardiner show for the boson case (Drummond and Gardiner, 1980) that P may be chosen to be everywhere positive--i.e. to be a true probability distribution. Hence the name positive P-representation.

The passage from our operator master equation to a Fokker-Planck equation can now follow almost identical lines to the derivation of Eq.(4.8). Precisely the same equation is obtained with $\bar{\alpha}^*$ replaced by $\bar{\alpha}_*$ and $\bar{v}^*$ replaced by $\bar{v}_*$; also $\bar{m}$ is now complex. The central point comes with the interpretation of this equation. This makes use of the analyticity of χ_N, or, more precisely, with the analyticity of the product of exponentials appearing in Eqs.(4.33). The following can be appreciated from the detailed calculations mentioned above. Each of the operators $\partial/\partial\bar{\alpha}$, $\partial/\partial\bar{\alpha}_*$, $\partial/\partial\bar{v}$, $\partial/\partial\bar{v}_*$, and $\partial/\partial\bar{m}$ enters the equation for P by virtue of its action on this product of exponentials. The analyticity allows each operator, $\partial/\partial\nu$, to be written either as $\partial/\partial\nu_x$, or $-i\partial/\partial\nu_y$. The choice is judiciously made to force the resulting Fokker-Planck equation to have positive definite diffusion in the ten-dimensional space

$$\vec{z}' = (\bar{\alpha}^x, \bar{\alpha}_*^x, \bar{v}^x, \bar{v}_*^x, \bar{m}^x, \bar{\alpha}^y, \bar{\alpha}_*^y, \bar{v}^y, \bar{v}_*^y, \bar{m}^y)^T , \tag{4.35}$$

where $\nu = \nu^x + i\nu^y$, for $\nu \equiv \bar{\alpha}$, $\bar{\alpha}_*$, $\bar{v}$, $\bar{v}_*$, or $\bar{m}$. Thereby P is chosen to be a "classical" probability distribution in this ten-dimensional space. The Fokker-Planck equation takes the explicit form:

$$\frac{\partial P}{\partial t} = [- \sum_j \frac{\partial}{\partial z_j'} a_j'(\vec{z}') + \tfrac{1}{2} \sum_{jk} \frac{\partial^2}{\partial z_j' \partial z_k'} d_{jk}'(\vec{z}')]P , \tag{4.36}$$

with

$$\vec{a}' = (a_j') = \begin{pmatrix} \vec{a}_x \\ \vec{a}_y \end{pmatrix} , \tag{4.37}$$

$$d' = (d_{jk}') = \begin{pmatrix} b_x & 0 \\ b_y & 0 \end{pmatrix} \begin{pmatrix} b_x^T & b_y^T \\ 0 & 0 \end{pmatrix} , \tag{4.38}$$

where $\vec{a} = \vec{a}_x + i\vec{a}_y$ is the five component complex vector given by:

$$\kappa^{-1} a_1 = -(1 + i\phi)\bar{\alpha} + 2C\bar{v} + Y , \tag{4.39a}$$

$$\kappa^{-1} a_2 = -(1 - i\phi)\bar{\alpha}_* + 2C\bar{v}_* + Y , \tag{4.39b}$$

$$\gamma_\perp^{-1} a_3 = -(1 + i\delta)\bar{v} + \bar{\alpha}\bar{m} , \tag{4.39c}$$

$$\gamma_\perp^{-1} a_4 = -(1 - i\delta)\bar{v}_* + \bar{\alpha}_*\bar{m} , \tag{4.39d}$$

$$\gamma_{\|}^{-1}\, a_5 = -(\bar{m} + 1) - \tfrac{1}{2}\bar{\alpha}\bar{v}_* - \tfrac{1}{2}\bar{\alpha}_*\bar{v}\ , \tag{4.39e}$$

and $b = b_x + ib_y$ is a 5x5 complex matrix defined by

$$d = bb^T \tag{4.40}$$

where d has nonzero elements corresponding to Eqs.(4.31):

$$d_{33} = 2\gamma_\perp N^{-1}\bar{\alpha}\bar{v}\ , \tag{4.41a}$$

$$d_{44} = 2\gamma_\perp N^{-1}\bar{\alpha}_*\bar{v}_*\ , \tag{4.41b}$$

$$d_{34} = d_{43} = 2\gamma_\perp N^{-1}(2\gamma_\perp/\gamma_{\|} - 1)(\bar{m} + 1)\ , \tag{4.41c}$$

$$d_{55} = 2\gamma_{\|} N^{-1}(\bar{m} + 1 - \tfrac{1}{2}\overline{\alpha v_*} - \tfrac{1}{2}\bar{\alpha}_*\bar{v}). \tag{4.41d}$$

The stochastic differential equations derived from Eq.(4.36) are

$$d\vec{\xi}' = \vec{a}(\vec{\xi}')dt + b(\vec{\xi}')d\vec{W} \tag{4.42}$$

where $\vec{\xi}'$ is the five-component complex vector

$$\vec{\xi}' = (\bar{\alpha},\bar{\alpha}_*,\bar{v},\bar{v}_*,\bar{m})^T\ . \tag{4.43}$$

These are precisely Eqs.(4.16)--with $\vec{z}$ complex and related to $\vec{z}'$ via Eqs.(4.28) and (4.35)--or, equivalently, Eqs.(4.29). They are the equations suggested by the Glauber-Sudarshan-Haken representation by disregarding the non-positive-definite diffusion, now interpreted in terms of a Fokker-Planck equation which does have positive definite diffusion. Remember, when comparing Eqs.(4.42) and (4.16), the noise terms must only be equal in the sense of their correlations; neither b nor $\bar{b}$ is unique.

In what follows, unless otherwise stated, the development will be made on the basis of Eq.(4.42). Without the noise terms these are a complexified version of the vector form of the Maxwell-Bloch equations given in Eqs.(2.37) and (2.38). Solutions for initial conditions in the subspace defined by $\bar{\alpha}_* = \bar{\alpha}^*$, $\bar{v}_* = \bar{v}^*$, $\bar{m} = \bar{m}^*$, remain within this subspace, and are just solutions to the ordinary Maxwell-Bloch equations. Individual trajectories can leave this "physical" subspace with the addition of the quantum noise; however, as noted above, by virtue of Eq.(4.34), ensemble averages must preserve $(\bar{\alpha}_*)_{av} = (\bar{\alpha})^*_{av}$, $(\bar{v}_*)_{av} = (\bar{v})^*_{av}$, $(\bar{m})_{av} = (\bar{m})^*_{av}$.

4.4 Adiabatic elimination of the atoms and the field

Much of the work on the quantum theory of optical bistability is restricted to either the good cavity ($\kappa \ll \gamma_{\parallel}, \gamma_{\perp}$) or the bad cavity ($\kappa \gg \gamma_{\parallel}, \gamma_{\perp}$) limit. These limits simplify theoretical analysis by allowing, respectively, for the adiabatic elimination of the atoms or the field. Adiabatic elimination from Eqs.(4.29) is described by Drummond and Walls (1981) and Carmichael et al. (1983).

Elimination of the field variables in the bad cavity limit is quite straightforward. Equations (4.29a) and (4.29b) are solved in the steady state for fixed $\bar{v}$ and $\bar{v}_*$ with

$$\bar{\alpha} = (1 + i\phi)^{-1}(2C\bar{v} + Y) , \tag{4.44a}$$

$$\bar{\alpha}_* = (1 - i\phi)^{-1}(2C\bar{v}_* + Y) . \tag{4.44b}$$

Then substitution into the equations for the atomic variables gives a set of equations in the form of Eq.(4.42), where $\vec{\xi}'$ is now the three-component complex vector $(\bar{v},\bar{v}_*,\bar{m})^T$, and

$$\gamma_{\perp}^{-1}a_1 = -(1 + i\delta)\bar{v} + (1 + i\phi)^{-1}(2C\bar{v} + Y)\bar{m} , \tag{4.45a}$$

$$\gamma_{\perp}^{-1}a_2 = -(1 - i\delta)\bar{v}_* + (1 - i\phi)^{-1}(2C\bar{v}_* + Y)\bar{m} , \tag{4.45b}$$

$$\gamma_{\parallel}^{-1}a_3 = -(\bar{m} + 1) - \tfrac{1}{2}(1 + i\phi)^{-1}(2C\bar{v} + Y)\bar{v}_* - \tfrac{1}{2}(1 - i\phi)^{-1}(2C\bar{v}_* + Y)\bar{v} . \tag{4.45c}$$

The nonzero elements of the complex diffusion matrix are:

$$d_{11} = 2\gamma_{\perp} N^{-1}(1 + i\phi)^{-1}(2C\bar{v} + Y)\bar{v} , \tag{4.46a}$$

$$d_{22} = 2\gamma_{\perp} N^{-1}(1 - i\phi)^{-1}(2C\bar{v}_* + Y)\bar{v}_* , \tag{4.46b}$$

$$d_{12} = d_{21} = 2\gamma_{\perp} N^{-1}(2\gamma_{\perp}/\gamma_{\parallel} - 1)(\bar{m} + 1) , \tag{4.46c}$$

$$d_{33} = 2\gamma_{\parallel}N^{-1}[\bar{m} + 1 - \tfrac{1}{2}(1 + i\phi)^{-1}(2C\bar{v} + Y)\bar{v}_* - \tfrac{1}{2}(1 - i\phi)^{-1}(2C\bar{v}_* + Y)\bar{v}] . \tag{4.46d}$$

Elimination of the atomic variables is complicated by the appearance of noise terms in Eqs.(4.29c)-(4.29e). The method used by Drummond and Walls (1981) is valid for large N, which, as we have seen, is already assumed en route to the Fokker-Planck equation. The resulting stochastic differential equations for the good cavity limit are written in the notation of Eq.(4.42) with $\vec{\xi}'$ as a two-component complex vector $(\bar{\alpha},\bar{\alpha}_*)^T$, and

$$\kappa^{-1}a_1 = -(1 + i\phi)\bar{\alpha} + 2C(1 - i\delta)(1 + \delta^2 + \bar{\alpha}_*\bar{\alpha})^{-1}\bar{\alpha} + Y \ , \tag{4.47a}$$

$$\kappa^{-1}a_2 = -(1 - i\phi)\bar{\alpha}_* + 2C(1 + i\delta)(1 + \delta^2 + \bar{\alpha}_*\bar{\alpha})^{-1}\bar{\alpha}_* + Y \ . \tag{4.47b}$$

The elements of the complex diffusion matrix are:

$$d_{11} = -2C\kappa n_{sat}^{-1}\bar{\alpha}^2(1 + \delta^2 + \bar{\alpha}_*\bar{\alpha})^{-3}[(1 + i\delta)^3 f - i\delta\bar{\alpha}_*\bar{\alpha}(1 + i\delta)(1 - f) - \tfrac{1}{2}\bar{\alpha}_*^2\bar{\alpha}^2] \ , \tag{4.48a}$$

$$d_{22} = -2C\kappa n_{sat}^{-1}\bar{\alpha}_*^2(1 + \delta^2 + \bar{\alpha}_*\bar{\alpha})^{-3}[(1 - i\delta)^3 f + i\delta\bar{\alpha}_*\bar{\alpha}(1 - i\delta)(1 - f) - \tfrac{1}{2}\bar{\alpha}_*^2\bar{\alpha}^2] \ , \tag{4.48b}$$

$$d_{12} = d_{21} = 2C\kappa n_{sat}^{-1}\bar{\alpha}_*\bar{\alpha}(1 + \delta^2 + \bar{\alpha}_*\bar{\alpha})^{-3}[(1 + \delta^2 + \delta^2\bar{\alpha}_*\bar{\alpha})(1 - f) + \bar{\alpha}_*\bar{\alpha}(2 + \tfrac{1}{2}\bar{\alpha}_*\bar{\alpha})] \ . \tag{4.48c}$$

These equations for the good cavity limit have recently been generalized to gaussian beam and standing wave theories of optical bistability (Xiao et al., 1986).

5. LINEARIZED THEORY OF FLUCTUATIONS

Attention was first focused on the quantum theory of optical bistability by Bonifacio and Lugiato (1976) who predicted that interesting cooperative effects might be seen in the fluorescent spectrum in absorptive bistability. At the time of their work there had recently been much activity surrounding study of the Mollow spectrum and photon antibunching in resonance fluorescence (Cresser et al., 1982). Naturally, many workers were attracted to the possibility of studying the new features predicted for absorptive bistability. The original suggestions by Bonifacio and Lugiato were drawn from an analysis of the eigenvalues of the linearized equations which govern dynamics in the vicinity of the steady state (Eqs.(2.34) and (2.35)). Via the quantum regression theorem (Lax, 1963; Lax, 1967), these eigenvalues also govern the regression of fluctuations, and, hence, the linewidths and frequency splittings which appear in the spectrum of fluctuations. On this basis Bonifacio and Lugiato drew some qualitative conclusions regarding the fluorescent spectrum. To complete their work a full linearized theory of fluctuations around the steady state was called for. Such a theory supposes that the deterministic steady state remains stable for long periods on the time scale of the local fluctuations: that in the bistable region local equilibriation is very rapid compared with the time for transitions between states. This is well justified away from the turning points. Linearized theory gives a clear signature of its own breakdown close to the turning points--and at the critical point, where the bistability curve shows infinite differential gain (Fig. 3). In this section we review results for the quantum-statistical effects derived from a linearized treatment of fluctuations.

An important point, not fully appreciated in early work, is that a clear distinction must be made between the properties of the fluorescent light, observed from the sides of the cavity, and those of the transmitted light seen at the cavity output. The prediction of collective effects in the fluorescent spectrum was surprising to some, since it is only the privileged cavity mode which sees a phased interaction with all of the atoms. Certainly, it is reasonable to expect collective effects in the transmitted light. But why in the fluorescence? This would, of course, be very nice for experiments, since the fluorescent light is not dominated by the incident laser as is the light in transmission. Unfortunately, the doubts were quite justified. The early calculations of the fluorescent spectrum have been shown to be in error; they do not take correct account of the scaling of atom-atom correlations with system size (Lugiato, 1980b; Carmichael, 1981; Agarwal, 1982). From Eqs.(4.24), (4.41), (4.46) and (4.48), the number of atoms N and the saturation photon number n_{sat} play an important role in characterizing the size of the quantum noise. These are large numbers in a macroscopic system. The recent experiments of Rosenberger et al. (1983, 1984), for example, have $N \sim 10^4 - 10^5$ and $n_{sat} \sim 10^3$. When these numbers are large the statistical properties of the fluorescent light are just those derived from the theory of single-atom resonance fluorescence for independent atoms driven by the mean cavity field. Interesting new effects are to be seen only in transmission. We will only discuss the properties of the transmitted light here.

Linearized theories for absorptive bistability have been developed using quantum-mechanical Langevin equations (Narducci et al., 1978a; Agarwal et al., 1978), the Fokker-Planck equation, Eq.(4.23), in the Wigner representation (Lugiato, 1979; Casagrande and Lugiato, 1980; Lugiato and Strini, 1982), and the Fokker-Planck equation, Eqs.(4.36)-(4.41), in the positive P-representation (Drummond and Walls, 1981; Carmichael et al., 1983; Carmichael, 1986a). All of this work, with the exception of Carmichael (1986a), specializes to either the good cavity or the bad cavity limit. The results agree where the same quantities have been calculated with different methods. The theory based on the positive P-representation has also been analyzed for cases of mixed absorptive and dispersive bistability: both for the good cavity limit (Drummond and Walls, 1981; Reid and Walls, 1985), and for the bad cavity limit (Carmichael et al., 1983). In related work, Drummond and Walls (1980) have developed a theory for the dispersive limit (large atomic detuning and weak nonsaturating fields) based on a nonlinear polarizability model. Agarwal and Tewari (1980) include atomic detuning in a linearized treatment of quantum fluctuations but do not include the cavity detuning necessary for a full theory of dispersive bistability.

We will first discuss the formal expressions from which steady-state correlations and the spectrum of fluctuations can be calculated in a linearized theory based on the positive P-representation. Then, formal expressions for the incoherent transmitted intensity and spectrum, and for photon antibunching and squeezing in the transmitted light will be given. It will not be possible to include explicit results for the many situations covered in the cited references. Examples illustrating each feature will be drawn for the most part from the case of absorptive bistability with radiative damping. For this case compact analytic expressions can generally be derived. The formal expressions are quite easy to implement numerically for other parameter values.

5.1 Linearized theory in the positive P-representation

If N is very large the noise terms in Eq.(4.42) are small. Then fluctuations around the deterministic steady state can be adequately treated by the linearized set of equations:

$$d\vec{\eta}' = A\vec{\eta}'dt + Bd\vec{W} , \tag{5.1}$$

where we set

$$\vec{\xi}' = \vec{\xi}'_s + \eta' \tag{5.2}$$

with

$$\vec{\xi}'_s = \vec{\xi}_s = (\bar{\alpha}_s, \bar{\alpha}^*_s, \bar{v}_s, \bar{v}^*_s, \bar{m})^T ; \tag{5.3}$$

A is the Jacobian matrix given by Eq.(2.35),

$$A_{ij} = \frac{\partial a_i}{\partial \xi'_j}\bigg|_s , \tag{5.4}$$

and

$$B = b(\vec{\xi}'_s) , \tag{5.5}$$

with constant diffusion matrix

$$D = BB^T = d(\vec{\xi}'_s) . \tag{5.6}$$

From Eq.(4.34), and the scaling of Eqs.(4.9), moments of the stochastic variables $\eta'_1 \ldots \eta'_5$ give normally-ordered operator averages

$$\langle\Delta\bar{a}^{\dagger n}\Delta\bar{a}^{m}\Delta\bar{J}_{+}^{p}\Delta\bar{J}_{z}^{r}\Delta\bar{J}_{-}^{q}\rangle = (\eta_2'^{n}\eta_1'^{m}\eta_4'^{p}\eta_5'^{r}\eta_3'^{q})_{av} , \tag{5.7}$$

where the operators $\Delta\bar{a}$, $\Delta\bar{a}^{\dagger}$, $\Delta\bar{J}_{-}$, $\Delta\bar{J}_{+}$, and $\Delta\bar{J}_{z}$ describe perturbations from the steady state:

$$\Delta\bar{a} = i\exp[-i(\phi_T + \phi_i)]n_{sat}^{-\frac{1}{2}}(a - \langle a\rangle_s) , \tag{5.8a}$$

$$\Delta\bar{J}_{-} = i\exp[-i(\phi_T + \phi_i)]2N^{-1}(\gamma_{\perp}/\gamma_{\parallel})^{\frac{1}{2}}(J_{-} - \langle J_{-}\rangle_s) , \tag{5.8b}$$

$$\Delta\bar{J}_{z} = 2N^{-1}(J_z - \langle J_z\rangle_s) . \tag{5.8c}$$

Steady-state averages $\langle a\rangle_s$, $\langle J_{-}\rangle_s$ and $\langle J_z\rangle_s$ are given, from Eqs.(3.29) and (2.31), by

$$i\exp[-i(\phi_T + \phi_i)]n_{sat}^{-\frac{1}{2}}\langle a\rangle_s = \exp(i\phi_X)X , \tag{5.9a}$$

$$i\exp[-i(\phi_T + \phi_i)]2N^{-1}(\gamma_{\perp}/\gamma_{\parallel})^{\frac{1}{2}}\langle J_{-}\rangle_s$$
$$= -\exp(i\phi_X)X(1 - i\delta)/(1 + \delta^2 + X^2) , \tag{5.9b}$$

$$2N^{-1}\langle J_z\rangle_s = -(1 + \delta^2)/(1 + \delta^2 + X^2) . \tag{5.9c}$$

Equation (5.1) defines a five-dimensional complex Ornstein-Uhlenbeck process which evolves to a stationary state with Gaussian statistics. We are interested in the quantum-statistical properties in this stationary state. All normally-ordered operator averages can be calculated from the covariance matrix

$$G_s = (\vec{\eta}'\vec{\eta}'^{T})_{av} \tag{5.10}$$

and the rule of moments for Gaussian statistics (Gardiner, 1983, pp. 36-37). Time-dependent properties are characterized by the correlation matrix

$$C_s(t) = (\vec{\eta}'(t)\vec{\eta}'(0)^{T})_{av} , \tag{5.11}$$

and the spectrum of fluctuations

$$S(\omega) = (2\pi)^{-1}\int_{-\infty}^{\infty} dt\, e^{-i\omega t}C_s(t) . \tag{5.12}$$

In general, the relationship between multi-time operator averages and multi-time stochastic averages is not as straightforward as that for equal-time averages. A full discussion can be found elsewhere (Carmichael, 1986b). For our present purposes it is sufficient to observe that time-ordered normally-ordered averages of the field operators are given by ($t > 0$):

$$\langle\Delta\bar{a}(t)\Delta\bar{a}(0)\rangle_s = (\eta_1'(t)\eta_1'(0))_{av} , \tag{5.13a}$$

$$\langle\Delta\bar{a}^\dagger(0)\Delta\bar{a}^\dagger(t)\rangle_s = (\eta_2'(t)\eta_2'(0))_{av} , \tag{5.13b}$$

$$\langle\Delta\bar{a}^\dagger(t)\Delta\bar{a}(0)\rangle_s = (\eta_2'(t)\eta_1'(0))_{av} , \tag{5.13c}$$

$$\langle\Delta\bar{a}^\dagger(0)\Delta\bar{a}(t)\rangle_s = (\eta_1'(t)\eta_2'(0))_{av} . \tag{5.13d}$$

Thus, we can calculate correlation functions and spectral densities for the field from matrix elements of $C_s(t)$ and $S(\omega)$.

Our theory must now focus on the calculation of G_s, $C_s(t)$ and $S(\omega)$. These are each given by well-known results in terms of the matrices A and D (Gardiner, 1983, pp. 109-112). These results hold equally well for our complex equations as they do for equations which are real: The covariance matrix satisfies the equation

$$AG_s + G_sA^T = -D . \tag{5.14}$$

The correlation matrix obeys the equation of motion ($t > 0$)

$$\dot{C}_s = AC_s \tag{5.15}$$

with $C_s(0) = G_s$, and $C_s(-t) = C_s(t)^T$. The spectrum of fluctuations is given by

$$S(\omega) = (2\pi)^{-1}(A - i\omega I)^{-1}D(A^T + i\omega I)^{-1} , \tag{5.16}$$

where I is the 5x5 identity matrix. Equations (5.14)-(5.16) form the basis of our linearized theory, together with the relationship between operator averages and stochastic averages given by Eqs.(5.7) and (5.13).

It will be noticed that Eqs.(5.14)-(5.16) are matrix equations in five dimensions. In fact, A and D are precisely the drift and diffusion matrices which would be obtained from the Fokker-Planck equation in the Glauber-Sudershan-Haken representation overlooking the non-positive-definite diffusion. This is an important observation. It means that no more computational effort is required to work with the positive P-representation than with the Glauber-Sudarshan-Haken representation, or the Wigner representation, despite its doubling of the dimensions. The extra dimensions are not hidden in the complex character of A and D. This simply reflects the fact that the variables $\vec{\eta}'$ generate averages of non-hermitian operators, and Eqs. (5.14)-(5.16) can readily be transformed so that A and D are replaced by real matrices. The extra dimensions required by the positive P-representation have somehow been "slipped under the rug." However, we really do have a stochastic process in ten dimensions. It is interesting to see how Eqs.(5.14)-(5.16) arise in the context of the ten-dimensional Fokker-Planck equation, Eq.(4.36).

After linearization the Fokker-Planck equation reads:

$$\frac{\partial P}{\partial t} = [- \sum_{jk} A'_{jk} \frac{\partial}{\partial z'_j} z'_k + \tfrac{1}{2} \sum_{jk} D'_{jk} \frac{\partial^2}{\partial z'_j \partial z'_k}]P , \tag{5.17}$$

with

$$A' = (A'_{jk}) = \begin{pmatrix} A_x & -A_y \\ A_y & A_x \end{pmatrix} , \tag{5.18}$$

$$D' = (D'_{jk}) = \begin{pmatrix} B_x & 0 \\ B_y & 0 \end{pmatrix} \begin{pmatrix} B_x^T & B_y^T \\ 0 & 0 \end{pmatrix} , \tag{5.19}$$

where $A = A_x + iA_y$, and $D = BB^T$, with $B = B_x + iB_y$, are the matrices appearing in Eqs.(5.14)-(5.16). A' and D' are 10x10 matrices. In the ten-dimensional real space we may define covariance and correlation matrices

$$G'_s = (\vec{z}'\vec{z}'^T)_{av} \tag{5.20}$$

and

$$C'_s(t) = (\vec{z}'(t)\vec{z}'(0)^T)_{av} , \tag{5.21}$$

and write matrix equations in ten dimensions corresponding to Eqs.(5.14)-(5.16):

$$A'G'_s + G'_s A'^T = -D' \ , \quad (5.22a)$$

$$\dot{C}'_s = A'C'_s \ , \quad (5.22b)$$

$$S'(\omega) = (2\pi)^{-1}(A' - i\omega I')^{-1} D'(A'^T + i\omega I')^{-1} \ , \quad (5.22c)$$

where I' is the 10x10 identity matrix. The central point now is that operator averages are related in a very particular wa to moments of the ten-dimensional process--namely, via the variables

$$\vec{\eta}' = M\vec{z}' , \quad (5.23)$$

where M is the 5x10 matrix

$$M = (I, iI) \ . \quad (5.24)$$

Quantities of physical interest are contained in G_s, $C_s(t)$ an $S(\omega)$, and lie embedded within Eqs.(5.22). They may be extracted using Eq.(5.23). It is straightforward to show

$$G_s = MG'_s M^T \ , \quad (5.25a)$$

$$C_s(t) = MC'_s(t)M^T \ , \quad (5.25b)$$

$$S(\omega) = MS'(\omega)M^T \ , \quad (5.25c)$$

and

$$MA'M^T = 0 \ , \quad (5.26a)$$

$$MA'M^\dagger = 2A \ , \quad (5.26b)$$

$$MD'M^T = D \ , \quad (5.26c)$$

where † denotes the Hermitian conjugate. Also, the 10x10 identity matrix can be written

$$I' = \tfrac{1}{2}(M^\dagger M + M^T M^*) \ . \quad (5.27)$$

Then, taking (5.22a) as an example, we may write

$$\begin{aligned}
-D &= -MD'M^T \\
&= M(A'G_s' + G_s'A'^T)M^T \\
&= MA'\tfrac{1}{2}(M^\dagger M + M^T M^*)G_s'M^T \\
&+ MG_s'\tfrac{1}{2}(M^\dagger M + M^T M^*)A'^T M^T \\
&= \tfrac{1}{2}(MA'M^\dagger)(MG_s'M^T) + \tfrac{1}{2}(MG_s'M^T)(MA'M^\dagger)^T \\
&+ \tfrac{1}{2}(MA'M^T)(MG_s'M^\dagger)^T + \tfrac{1}{2}(MG_s'M^\dagger)(MA'M^T)^T \\
&= AG_s + G_sA^T ,
\end{aligned}$$

which is just Eq.(5.14). Similar calculations give Eqs.(5.15) and (5.16) from Eqs.(5.22b) and (5.22c), respectively.

The linearized theory has been presented here within the context of the full model without adiabatic elimination of the atoms or the field. The drift and diffusion matrices are to be calculated from Eqs.(4.39) and (4.41). Theories for the good and bad cavity limits follow identical lines in two- and three-dimensional complex spaces (four- and six-dimensional real spaces), respectively. In the good cavity limit A and D are calculated from Eqs.(4.47) and (4.48); in the bad cavity limit they are calculated from Eqs.(4.45) and (4.46).

5.2 Steady-state correlations for absorptive bistability

There is much to be learned about the quantum fluctuations from the solution to Eq.(5.14). To provide an explicit illustration let us consider absorptive bistability ($\delta = \phi = 0$) and purely radiative damping ($\gamma_{\|} = 2\gamma_{\perp}$). With this simplification Eq.(5.14) can be solved to give compact analytical expressions for the correlations between pairs of system operators without restriction to either the good or the bad cavity limit (Carmichael, 1986a).

With the use of Eqs.(4.39) and (4.41), and Eqs.(5.3) and (2.31), the drift and diffusion matrices calculated from Eqs. (5.4) and (5.6) take the explicit form

$$A = \begin{pmatrix} -\kappa & 0 & \kappa 2C & 0 & 0 \\ 0 & -\kappa & 0 & \kappa 2C & 0 \\ -\tfrac{1}{2}\gamma_{\|}(1+X^2)^{-1} & 0 & -\tfrac{1}{2}\gamma_{\|} & 0 & \tfrac{1}{2}\gamma_{\|}X \\ 0 & -\tfrac{1}{2}\gamma_{\|}(1+X^2)^{-1} & 0 & -\tfrac{1}{2}\gamma_{\|} & \tfrac{1}{2}\gamma_{\|}X \\ \tfrac{1}{2}\gamma_{\|}X(1+X^2)^{-1} & \tfrac{1}{2}\gamma_{\|}X(1+X^2)^{-1} & -\tfrac{1}{2}\gamma_{\|}X & -\tfrac{1}{2}\gamma_{\|}X & -\gamma_{\|} \end{pmatrix}, \tag{5.28}$$

and

$$D = \gamma_{\|}N^{-1}X^2(1 + X^2)^{-1}\mathrm{diag}(0,0,-1,-1,4) \ . \tag{5.29}$$

Note that D is not positive definite. Equation (5.14) is equivalent to a set of fifteen coupled equations, which decouple here into a set of six equations for the imaginary part of G_s, with the trivial solution $G_s^y = 0$, and a set of nine equations for the real part of G_s. The solution to the set of nine equations gives the following results (Carmichael, 1986a):

1. <u>Field-field correlations</u>

$$\langle\Delta\bar{a}^{\dagger}\Delta\bar{a}\rangle_s = N^{-1}X^2\lambda \ , \tag{5.30a}$$

$$\langle\Delta\bar{a}\Delta\bar{a}\rangle_s = N^{-1}X^2(\lambda - P) \ , \tag{5.30b}$$

where

$$P = 2C\mu(\mu + 1)^{-1}(1 - X/Y) \ , \tag{5.31}$$

$$\lambda = PX^2(1 + X^2)^{-1}(dY/dX)^{-1}\{1 + \tfrac{1}{2}(Y/X) \frac{(\mu + 3)(2 - X^2) + \mu(1 + X^2)dY/dX}{(\mu + 3)[1 + X^2 + \mu(1 + \tfrac{1}{2}Y/X)] - \mu(1 + X^2)dY/dX}\} \ , \tag{5.32}$$

with

$$\mu = 2\kappa/\gamma_{\|} \ . \tag{5.33}$$

2. Atom-field correlations

$$\langle \Delta\bar{a}\Delta\bar{J}_+\rangle_s = N^{-1}(2C)^{-1}X^2\lambda \ , \tag{5.34a}$$

$$\langle \Delta\bar{a}\Delta\bar{J}_-\rangle_s = N^{-1}(2C)^{-1}X^2(\lambda - P) \ , \tag{5.34b}$$

$$\langle \Delta\bar{a}\Delta\bar{J}_z\rangle_s = N^{-1}(2C)^{-1}X(\lambda dY/dX - Q) \ , \tag{5.34c}$$

where

$$Q = (\mu + 3)^{-1}[2(1 + X^2)\lambda dY/dX + X^2(Y/X - 2)P] \ . \tag{5.35}$$

3. Atom-atom correlations

$$\langle \Delta\bar{J}_+\Delta\bar{J}_-\rangle_s = N^{-1}(2C)^{-2}X^2(\lambda + \mu^{-1}Q) \ , \tag{5.36a}$$

$$\langle \Delta\bar{J}_-\Delta\bar{J}_-\rangle_s = N^{-1}(2C)^{-2}X^2[\lambda - P + \mu^{-1}(Q - P\cdot Y/X)] \ , \tag{5.36b}$$

$$\langle \Delta\bar{J}_z\Delta\bar{J}_-\rangle_s = N^{-1}(2C)^{-2}X(\lambda Y/X + \mu^{-1}Q) \ , \tag{5.36c}$$

$$\langle \Delta\bar{J}_z\Delta\bar{J}_z\rangle_s = N^{-1}(2C)^{-2}X^2[(Y/X - 2)\lambda Y/X - (Y/X - 1)Q + 2PY/X - \mu^{-1}(Q - 2PY/X)] \ . \tag{5.36d}$$

In these expressions Y/X and dY/dX are calculated from Eq. (2.26).

Equations (5.30)-(5.36) illustrate a number of general points. Notice that λ is inversely proportional to dY/dX. Therefore, at the turning points on bistability curves, where dY/dX = 0, and at the critical point, where $dY/dX = d^2Y/dX^2 = 0$, the fluctuations given by linearized theory diverge. This divergence is related to the change of stability discussed below Eq.(2.41). It arises mathematically because detA = 0 when dY/dX = 0; which is also the condition leading to a vanishing eigenvalue in the linearized dynamics. Clearly, this divergence reflects a failure of the linearized theory. Nonlinearities actually set a limit to the size of the fluctuations, a point we return to in section 6. On the other hand, the trend of increasing fluctuations at the turning points and the critical point is a reliable prediction. Equations (5.30)-(5.36) also illustrate the way fluctuations scale with system size.

We consider the dependence on N^{-1} and μ with C fixed. μ is chosen as a distinguished parameter because it takes extreme values in the good cavity and bad cavity limits. Here we will designate these limits by $\mu \ll 1$ and $\mu \gg 1$, respectively. A more detailed specification of the conditions under which the results obtained with adiabatic elimination are recovered is given elsewhere (Carmichael, 1986a). For $\mu \gtrsim 1$, including the bad cavity limit, fluctuations scale as N^{-1}. The situation is a little different in the good cavity limit. When $\mu \ll 1$, λ, P and Q are all proportional to μ and it is natural to replace the scaling factor μN^{-1} by

$$n_{sat}^{-1} = N^{-1}4C\mu \ . \tag{5.39}$$

This relationship follows from Eqs.(3.30), (3.31), and (5.33). Fluctuations then scale as n_{sat}^{-1} everywhere except for in Eqs. (5.36). There we find terms involving $\mu^{-1}P$ and $\mu^{-1}Q$. These still scale as N^{-1}. Thus, in the good cavity limit the atom-atom correlations have two distinct contributions: the first scaling as n_{sat}^{-1}, and the second (terms proportional to $\mu^{-1}P$ and $\mu^{-1}Q$), the dominant contribution, scaling as N^{-1}. It is instructive to look carefully at the significance of the two pieces.

For an illustration we take Eq.(5.36a). If we expand the collective atomic operators to display the individual single atom contributions, we may distinguish terms Σ and Σ' arising from correlations between operators of the same atom and between operators of different atoms, respectively:

$$\langle\Delta\bar{J}_+\Delta\bar{J}_-\rangle_s = \Sigma + \Sigma' \ , \tag{5.40}$$

with

$$\Sigma = 2N^{-1}\langle\Delta\hat{\sigma}_+^j\Delta\hat{\sigma}_-^j\rangle_s \ , \tag{5.41a}$$

$$\Sigma' = 2\langle\Delta\hat{\sigma}_+^j\Delta\hat{\sigma}_-^{k\neq j}\rangle_s \ , \tag{5.41b}$$

where the operators $\hat{\sigma}_+^j$ and $\hat{\sigma}_-^j$ absorb the phase factors appearing in Eq.(3.7a), and we use the fact that the phased atoms are then identical; Σ' is given to lowest order in N^{-1}. We may now use Eqs.(5.9) to write

$$\begin{aligned}\Sigma &= 2N^{-1}(\langle\hat{\sigma}^j_+\hat{\sigma}^j_-\rangle_s - \langle\hat{\sigma}^j_+\rangle_s\langle\hat{\sigma}^j_-\rangle_s)\\ &= 2N^{-1}(\tfrac{1}{2} + \langle\sigma^j_z\rangle_s - \langle\hat{\sigma}^j_+\rangle_s\langle\hat{\sigma}^j_-\rangle_s)\\ &= N^{-1}[1 + 2N^{-1}\langle J_z\rangle_s - (\sqrt{2}N^{-1}\langle J_+\rangle_s)(\sqrt{2}N^{-1}\langle J_-\rangle_s)]\\ &= N^{-1}X^4(1 + X^2)^{-2} ,\end{aligned}$$

and thus

$$\langle\Delta\bar{J}_+\Delta\bar{J}_-\rangle_s = N^{-1}X^4(1 + X^2)^{-2} + \Sigma' . \tag{5.42}$$

Alternatively, from Eq.(5.36a), in the good cavity limit we find

$$\langle\Delta\bar{J}_+\Delta\bar{J}_-\rangle_s = N^{-1}X^4(1 + X^2)^{-2} + n_{sat}^{-1}[(2C)^{-2}X^2\lambda/4C\mu] \tag{5.43}$$

with

$$(2C)^{-2}X^2\lambda/4C\mu = (1/8C^2)(1 - X/Y)X^4(1 + X^2)^{-1}(dY/dX)^{-1}$$
$$[1 + \tfrac{1}{2}(Y/X)(1 + X^2)^{-1}(2 - X^2)] . \tag{5.44}$$

The term scaling as N^{-1} in Eq.(5.43) is just the single-atom term Σ. The term scaling as n_{sat}^{-1} is then identified with Σ' and accounts for correlations between atoms. A similar result follows for each of Eqs.(5.36b)-(5.36d). Actually, these results are perhaps to be expected. In the good cavity limit there is a clear separation of time scales between the fluctuations which drive the atomic variables directly, in Eqs. (4.29c)-(4.29e), and those which feed into the field and back into the atomic equations via the atom-field interaction. The field mediates correlations between different atoms. It seems reasonable then that the direct fluctuations should be associated with Σ and a scaling with N^{-1}, while the indirect fluctuations should be associated with Σ' and a scaling with n_{sat}^{-1}. Since $n_{sat}^{-1}/N^{-1} \sim \mu \ll 1$ in the good cavity limit, it follows that atom-atom correlations are negligible in this limit (Carmichael, 1986a; Lugiato, 1986). This is so at least away from instability points. Notice that it is Σ', not Σ, which contains the dependence on λ, and hence the divergence for $dY/dX = 0$. This observation is in fact quite general, since Σ is calculated solely from steady-state operator averages which are independent of μ. Thus, in the bad cavity limit, where field fluctuations directly monitor the atomic fluctuations (Eqs.(4.44)), increased fluctuations in the transmitted light at instability points provide a direct measure of atom-atom correlations.

5.3 Incoherent intensity and spectrum for the transmitted light

From Eq.(3.22), and Eqs.(5.8a), (5.9a) and (3.30), the mean intensity of the transmitted light including quantum fluctuations is given by

$$I_t = T(\hbar\omega_0/2\varepsilon_0 V_Q)\langle a^\dagger a\rangle_s$$
$$= TI_{sat}(X^2 + \langle\Delta\bar{a}^\dagger\Delta\bar{a}\rangle_s) = TI_{sat}(X^2 + G_s^{21}) \ . \tag{5.45}$$

The ratio of incoherent and coherent intensities,

$$I_{inc}/I_{coh} = \langle\Delta\bar{a}^\dagger\Delta\bar{a}\rangle_s/X^2 = G_s^{21}/X^2 \ , \tag{5.46}$$

measures the size of fluctuations in the transmitted field. For absorptive bistability and radiative damping Eq.(5.30a) gives

$$I_{inc}/I_{coh} = N^{-1}2C\mu(\mu+1)^{-1}(1 - X/Y)X^2(1 + X^2)^{-1}(dY/dX)^{-1}$$
$$\{1+\tfrac{1}{2}(Y/X)\frac{(\mu+3)(2-X^2)+\mu(1+X^2)dY/dX}{(\mu+3)[1+X^2+\mu(1+\tfrac{1}{2}Y/X)]-\mu(1+X^2)dY/dX}\}. \tag{5.47}$$

This ratio scales as N^{-1} in the bad cavity limit, as n_{sat}^{-1} in the good cavity limit, and diverges at instability points.

The spectrum of the transmitted light is the sum of coherent and incoherent components:

$$I_t(\omega) = TI_{sat}X^2\delta(\omega - \omega_0) + I_{inc}(\omega) \ , \tag{5.48}$$

with

$$I_{inc}(\omega) = TI_{sat}(2\pi)^{-1}\int_{-\infty}^{\infty} dt \exp[-i(\omega-\omega_0)t]\langle\Delta\bar{a}^\dagger(t)\Delta\bar{a}(0)\rangle_s$$

$$= TI_{sat}S(\omega-\omega_0)_{21}. \tag{5.49}$$

Explicit analytic expressions have been obtained for the incoherent spectrum in absorptive bistability in both the good cavity limit (Lugiato, 1979) and the bad cavity limit (Bonifacio and Lugiato, 1978; Narducci et al., 1978a; Agarwal et al., 1978; Lugiato, 1979). Also, for weak fields in dispersive bistability in the bad cavity limit (Carmichael et al., 1983). These will not be reproduced here. The spectrum is readily calculated numerically for arbitrary parameters directly from Eq.(5.49) and the formal results given in section 5.1. We will review its main features for the bad cavity limit with reference to Figs. 5-7. The bad cavity limit is particularly interesting because in this limit atomic dynamics are revealed in the fluctuations of the transmitted light. Here, with the adiabatic elimination of the field,

$$\langle\Delta\bar{a}^\dagger(t)\Delta\bar{a}(0)\rangle_s = 4C^2\langle\Delta\bar{J}_+(t)\Delta\bar{J}_-(0)\rangle_s , \tag{5.50}$$

and then $I_{inc}(\omega)$ gives the spectrum of fluctuations in the collective polarization. This collective spectrum can be compared with the spectrum of single-atom resonance fluorescence in the spirit of Bonifacio and Lugiato's original work (Bonifacio and Lugiato, 1976). Note that the implicit assumption of the adiabatic elimination is that the entire atomic spectrum should fit comfortably under the empty cavity line. For example, the cavity line must extend beyond the Rabi sidebands on the upper branch (Figs. 5 and 7)--i.e. the cavity must be fast enough to follow the Rabi oscillations. This is a more stringent requirement than might be assumed on the basis of the rather imprecise condition $\kappa \gg \gamma_{\parallel}, \gamma_{\perp}$ often quoted (Carmichael, 1986a). Since the single-mode assumption must be preserved the cavity must be fast by being short. We cannot broaden the line indefinitely by simply spoiling the mirrors.

Figure 5 shows the incoherent transmitted spectrum for absorp-

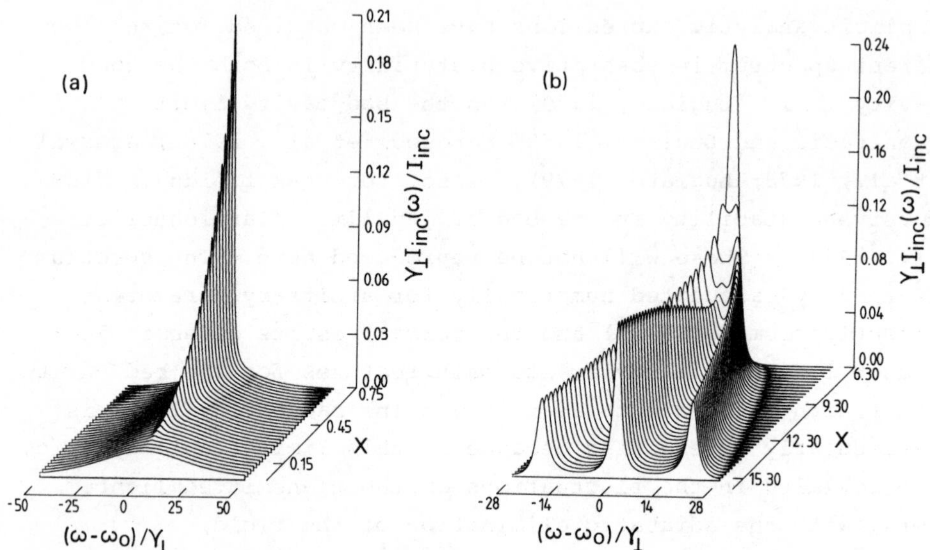

Fig. 5. The incoherent part of the transmitted spectrum for C = 20, $\delta = \phi = 0$, and $\gamma_{||} = 2\gamma_{\perp}$. Turning points in the bistability curve are at X = 1.05, Y = 21.0 and X = 6.07, Y = 12.5. (a) Lower branch, (b) upper branch.

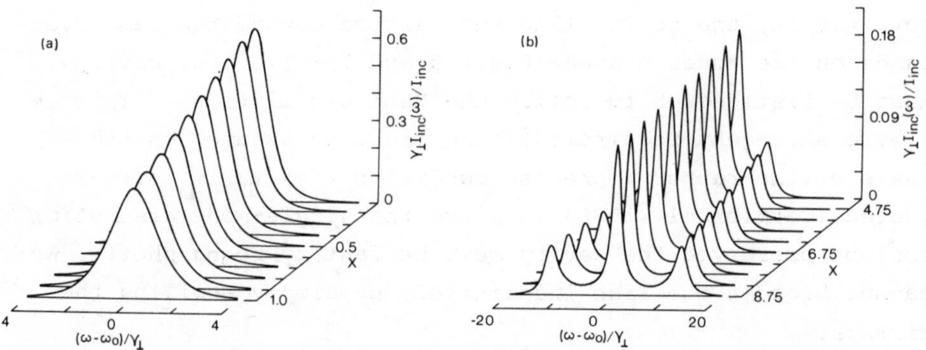

Fig. 6. The incoherent part of the spectrum for single-atom resonance fluorescence.

tive bistability and radiative damping. We focus on the collective features which distinguish this spectrum from that obtained from the theory of single-atom resonance fluorescence (Mollow, 1969) for independent atoms driven by the mean intracavity intensity $X^2 I_{sat}$. For comparison the incoherent component of the single-atom spectrum is shown in Fig. 6. First, we find a spectral component with a broad collective linewidth $\sim 2C\gamma_{\perp} = 2C\gamma_{\parallel}/2$ along the lower branch--to be compared with the width $\gamma_{\perp} = \gamma_{\parallel}/2$ for single-atom resonance fluorescence. Then, at the turning points of the bistability curve the spectral density coalesces into a single δ-function spike. This linewidth narrowing is associated with the vanishing eigenvalue discussed below Eq.(2.41). Recall from the last section that the divergence of fluctuations at these instability points is carried by the atom-atom correlations. Finally, along the upper branch, the Mollow triplet seen at high intensities follows a quite different development as the intracavity intensity decreases than in the spectrum of single-atom resonance fluorescence. A well developed triplet would still be expected at the instability point, for example, while the spectrum for bistability actually coalesces prematurely into a single peak. The ratio of peak heights can be very different from that in single-atom resonance fluorescence; the central peak actually becomes smaller than the sidebands in Fig. 5(b). Figure 7 shows that this change in relative peak heights can work in both directions as the upper branch instability point is approached. Further examples of spectra for the bad cavity limit, showing the effects of dispersion and nonradiative decay, are given by Carmichael et al. (1983). Examples of spectra for the good cavity limit are given by Lugiato (1979) and Drummond and Walls (1981).

These collective spectral features were the subject of the early interest in the quantum theory of optical bistability. As noted above, similar features have on occasion been attributed to the fluorescence seen from the sides of the cavity, but such assertions proved to be incorrect (Lugiato, 1980b; Carmichael, 1981; Agarwal, 1982). The collective effects in the transmitted spectrum have not yet been observed in the

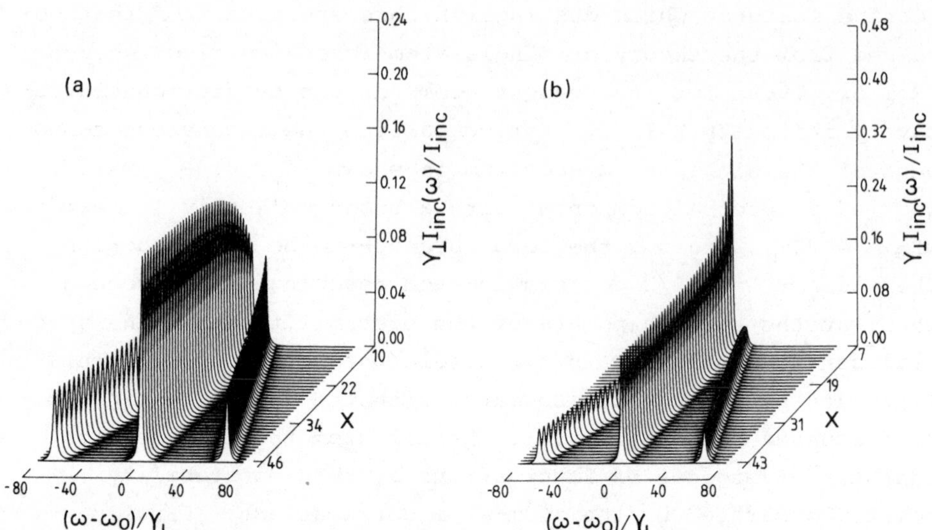

Fig. 7. The incoherent part of the transmitted spectrum along the upper branch for $C = 20$, $\gamma_{||} = 2\gamma_\perp$, and (a) $\delta = \phi = 1$, (b) $\delta = 1$, $\phi = -2$.

laboratory. One reason is that the ratio given by Eqs.(5.46) and (5.47) is very small if N is large, except near instability points. Near the instability points experimental systems are very susceptible to other external fluctuations and the clear demonstration of a quantum-statistical effect becomes rather problematic. It is hoped, however, that with the recent experience gained in atomic beam experiments (Grant and Kimble, 1983; Kimble et al., 1983; Rosenberger et al., 1983; Rosenberger et al., 1984; Orozco et al., 1984) a new generation of smaller systems might make measurements possible. Spectral features are then not the only interesting quantum-statistical effects to be observed.

5.4 Photon antibunching and squeezing

Photon antibunching was first predicted in absorptive bistability by Casagrande and Lugiato (1980). Their results were extended to include dispersion by Drummond and Walls (1981) and Carmichael et al. (1983). Squeezing in optical bistability is closely related to the existence of photon antibunching. It was first reported for absorptive bistability by

Lugiato and Strini (1982), and in the nonlinear polarizability model for dispersive bistability (Drummond and Walls, 1980) by Walls and Milburn (1983). Reid and Walls (1985) have recently studied the squeezing obtainable in the good cavity limit for mixed absorptive and dispersive bistability.

We consider the second-order correlation function for the transmitted light

$$g^{(2)}(t) = \langle a^\dagger a\rangle_s^{-2}\langle a^\dagger(0)a^\dagger(t)a(t)a(0)\rangle_s \ . \tag{5.51}$$

Using Eqs.(5.8a) and (5.9a) to write the field operators as steady-state averages plus a perturbation, Eq.(5.51) may be reexpressed in the form

$$\begin{aligned} g^{(2)}(t) = 1 + (X^2 + \langle\Delta\bar{a}^\dagger\Delta\bar{a}\rangle_s)^{-2}[&X^2 4\langle:\Delta\bar{a}_X(t)\Delta\bar{a}_X(0):\rangle_s \\ &+ X\exp(i\phi_X)\langle\Delta\bar{a}^\dagger(0)\Delta\bar{a}^\dagger(t)(\Delta\bar{a}(t) + \Delta\bar{a}(0))\rangle_s \\ &\qquad + \text{c.c.} \\ &+ \langle\Delta\bar{a}^\dagger(0)\Delta\bar{a}^\dagger(t)\Delta\bar{a}(t)\Delta\bar{a}(0)\rangle_s - \langle\Delta\bar{a}^\dagger\Delta\bar{a}\rangle_s^2] \ , \end{aligned} \tag{5.52}$$

where :: denotes normal and time ordering and

$$\Delta\bar{a}_X = \tfrac{1}{2}[\exp(-i\phi_X)\Delta\bar{a} + \exp(i\phi_X)\Delta\bar{a}^\dagger] \tag{5.53}$$

is the field quadrature operator describing fluctuations which are in phase with the mean cavity field. In this form $g^{(2)}(t)$ describes the self-homodyning of fluctuations in the transmitted light with the strong coherent transmitted field. The connection between photon antibunching and squeezing is then immediate, since homodyning techniques provide the practical method for detecting squeezed light (Mandel, 1982; Yuen and Chan, 1983; Schumaker, 1984). Now, since linearized theory gives Gaussian statistics, the third-order moments in Eq. (5.52) are zero; also the fourth-order moment is expandable in terms of products of moments of second order. Then, since second-order moments scale as N^{-1} (or n_{sat}^{-1}), to lowest order in N^{-1} we find

$$g^{(2)}(t) = 1 + X^{-2} 4 \langle :\Delta\bar{a}_x(t)\Delta\bar{a}_x(0): \rangle_s \tag{5.54a}$$

$$= 1 + X^{-2}[\exp(-2i\phi_x)C_s(t)_{11} + \exp(2i\phi_x)C_s(t)_{22} + C_s(t)_{21} + C_s(t)_{12}]. \tag{5.54b}$$

Photon antibunching is often defined by the condition (Walls, 1979; Loudon, 1980; Paul, 1982)

$$g^{(2)}(0) < 1 . \tag{5.55}$$

Alternatively, it has been defined to correspond to an initially rising correlation function (Kimble et al., 1977)--by the condition

$$\lim_{t\to 0} \frac{dg^{(2)}(t)}{dt} > 0 . \tag{5.56}$$

The distinction is often academic, since in many situations both conditions are satisfied simultaneously. Nevertheless, this need not be so; and in section 6 we will meet an example which satisfies Eq.(5.56) but not Eq.(5.55). For the present we will speak of photon antibunching in the sense of Eq. (5.55). In this section we focus on results in the limit $X^2 \to 0$ where both conditions are always satisfied.

Quantum squeezing occurs when the fluctuations in one field quadrature are reduced below the quantum limit for a coherent state at the expense of increased fluctuations in the other quadrature. An introductory discussion is given by Walls (1983). In a homodyne detection scheme photon counting may be used to measure the squeezing in terms of subpoissonian counting statistics, characterized by the parameter (Mandel, 1982)

$$Q = \langle \Delta n^2 \rangle / \langle n \rangle - 1 . \tag{5.57}$$

When the incoherent component of the transmitted field in optical bistability is homodyned with a local oscillator in phase with the mean transmitted field--for example, with the mean transmitted field itself--for short counting times the photon counting distribution measures

$$Q_x(T') = \eta\kappa T' n_{sat} 4\langle :\Delta\bar{a}_x^2 : \rangle_s \tag{5.58a}$$

$$= \eta\kappa T' n_{sat}[\exp(-2i\phi_x)G_s^{11} + \exp(2i\phi_x)G_s^{22} + G_s^{21} + G_s^{12}] , \tag{5.58b}$$

where T' is the counting time and η is the counting efficiency. For long counting times it measures

$$Q_x(\infty) = 2\eta\kappa n_{sat} \int_0^\infty dt 4\langle :\Delta\bar{a}_x(t)\Delta\bar{a}_x(0): \rangle_s \tag{5.59a}$$

$$= 2\eta\kappa n_{sat} \int_0^\infty dt[\exp(-2i\phi_x)C_s(t)_{11} + \exp(2i\phi_x)C_s(t)_{22} + C_s(t)_{21} + C_s(t)_{12}] . \tag{5.59b}$$

These results, and Eq.(5.62) below, are given for a cavity with equal photon escape rates at the input and output mirrors, consistent with the equal mirror transmissions adopted in section 2. The relationship between photon antibunching and squeezing can now be stated explicitly. From Eqs.(5.58a) and (5.59a), and Eq.(5.54a):

$$Q_x(T') = \eta\kappa T' X^2 n_{sat}[g^{(2)}(0) - 1] , \tag{5.60}$$

for short counting times, and

$$Q_x(\infty) = 2\eta\kappa X^2 n_{sat} \int_0^\infty dt[g^{(2)}(t) - 1] . \tag{5.61}$$

In particular, for short counting times photon antibunching implies the existence of squeezing, $Q_x(T') < 0$, and vice versa. Squeezing may be calculated for quadratures other than that phased with the coherent transmission by substituting a variable phase θ for ϕ_x in Eqs.(5.58) and (5.59). The recent work of Reid and Walls discusses squeezing in this more general context (Reid and Walls, 1985). These authors give their results of the spectrum of squeezing (Gardiner and Savage, 1984). For the in phase quadrature this is defined by

$$S_x(\omega) = \tfrac{1}{2}\kappa n_{sat} \int_0^\infty dt \cos\omega t \; 4\langle :\Delta\bar{a}_x(t)\Delta\bar{a}_x(0): \rangle , \tag{5.62}$$

from which the following relationship to Q_x is obtained:

$$Q_x(T') = 4\eta T'(2\pi)^{-1} \int_{-\infty}^\infty d\omega S_x(\omega) \tag{5.63}$$

for short counting times, and

$$Q_x(\infty) = 4\eta S_x(0) \ . \tag{5.64}$$

Let us now take absorptive bistability with radiative damping again as an example. From Eqs.(5.53), (5.30) and (5.46) we find

$$X^{-2}4\langle:\Delta\bar{a}_x^2:\rangle = 4I_{inc}/I_{coh} - N^{-1}4C\mu(\mu + 1)^{-1}(1 - X/Y) \ , \tag{5.65}$$

with I_{inc}/I_{coh} given in Eq.(5.47). This gives general expressions for antibunching and squeezing via Eqs.(5.54a) and (5.58a). As I_{inc}/I_{coh} is always positive, and vanishes for $X^2 = 0$, while $1 - X/Y = 2C/(1 + X^2 + 2C)$ is maximized for $X^2 = 0$, the maximum antibunching effect is given by

$$\lim_{X^2\to 0} g^{(2)}(0) - 1 = -N^{-1}4C\mu(\mu + 1)^{-1}2C(1 + 2C)^{-1} \ . \tag{5.66}$$

By comparison, squeezing vanishes for $X^2 = 0$, since for short counting times Q_x is proportional to the mean number of photons counted at the cavity output $\eta\kappa T'X^2n_{sat}$. Maximum squeezing must be found at some finite cavity intensity. The general expression derived from Eqs.(5.58a) and (5.65) is rather complicated; therefore, let us consider squeezing in the good cavity and bad cavity limits, as in the work by Lugiato and Strini (Lugiato and Strini, 1982). We obtain

$$Q_x(T') = -\eta\kappa T'2CX^2(1 - 2X^2)[(1 + X^2)^3 + 2C(1 - X^4)]^{-1} \ , \tag{5.67a}$$

$$Q_x(T') = -\eta\tfrac{1}{2}\gamma_{\|}T'2CX^2(1 - X^2)[(1 + X^2)^2 + 2C(1 - X^2)]^{-1} \ , \tag{5.67b}$$

respectively. Both of these expressions are maximized for large C. They give

$$Q_x(T')_{max} = -\eta\kappa T'[\tfrac{1}{2}(2 - \sqrt{3})] \ , \tag{5.68}$$

for $C \gg 1$ and $X^2 = 2 - \sqrt{3}$ in the good cavity limit, and

$$Q_x(T')_{max} = -\eta\tfrac{1}{2}\gamma_{\|}T' \ , \tag{5.69}$$

for $C >> 1$ and $X^2 \lesssim 1$ in the bad cavity limit. Correlation times for good and bad cavities will be characterized by κ^{-1} and $(\tfrac{1}{2}\gamma_{\|})^{-1}$, respectively. If we then compare Eqs.(5.68) and (5.69) for equal values of $\kappa T'$ and $\tfrac{1}{2}\gamma_{\|}T'$, the bad cavity is the better candidate for producing squeezing in absorptive bistability (Carmichael, 1986a). This is in contrast to the conclusion of Lugiato and Strini who make their comparison on the basis of $n_{sat}4\langle:\Delta\bar{a}_x^2:\rangle_s$, which is limited by photon numbers inside the cavity (Lugiato and Strini, 1982). Note that a judgement on the real maximum effect attainable is not possible on the basis of Eqs.(5.68) and (5.69) because, as short counting time results, they require $\kappa T'$ and $\tfrac{1}{2}\gamma_{\|}T'$ to be much smaller than unity. Note also that Eqs.(5.58)-(5.62) and Eqs. (5.67)-(5.69) are multiplied by a factor of 2 if a cavity with predominant loss at the output mirror is used rather than the symmetric cavity adopted here.

It will be noticed that Eqs.(5.67)-(5.69) do not exhibit the scaling with system size shown by Eqs.(5.66) and (5.47). This merits a brief comment. Squeezing does not scale with system size because it is measured in absolute photon number against an absolute quantum limit. Fluctuations in the unscaled field operators $\Delta a = a - \langle a\rangle_s$ and $\Delta a^\dagger = a^\dagger - \langle a^\dagger\rangle_s$ are constant for fixed C, μ and X and do not scale with system size (Eqs.(5.8) and (5.30)-(5.36)). Squeezing compares fluctuations in these operators against a fixed quantum limit $4\langle\Delta a_x^2\rangle = 1$ for a coherent state. On the other hand, both I_{inc}/I_{coh} and $g^{(2)}(0)-1$ are relative measures: they measure fluctuations of fixed size in Δa and $\Delta a^\dagger$ against a coherent background photon number $X^2 n_{sat}$ which increases with system size. The question with which a measurement of squeezing must be concerned is whether or not the quantum limit can be observed. As a crude illustration, suppose external fluctuations enter I_{inc} in Eq.(5.65) at a level of 0.1% of the coherent intensity. Then, in Eq. (5.58a) we will have

$$n_{sat}4\langle:\Delta\bar{a}_x^2:\rangle = 4(\langle\Delta a^\dagger\Delta a\rangle_s + 10^{-3}X^2 n_{sat}) - X^2(\mu+1)^{-1}(1-X/Y). \tag{5.70}$$

As n_{sat} increases the external fluctuations soon swamp the squeezing.

We return to photon antibunching for a final comment. For weak fields the drift matrix A (Eq.(5.28)) becomes rather simple and it is possible to find the elements of the correlation matrix required by Eq.(5.54b) analytically. This calculation gives (Carmichael, 1986a)

$$\lim_{X^2\to 0} g^{(2)}(t) = 1 - N^{-1}4C\mu(\mu + 1)^{-1}2C(1 + 2C)^{-1} \exp[-\tfrac{1}{4}(\mu + 1)\gamma_{||}t][\cosh(\Omega\gamma_{||}t) + (\mu + 1)(4\Omega)^{-1} \sinh(\Omega\gamma_{||}t)] , \quad (5.71)$$

with

$$\Omega = [(\mu - 1)^2/16 - \tfrac{1}{2}\mu C]^{\frac{1}{2}} . \quad (5.72)$$

In the good and bad cavity limits this reduces to the results obtained with adiabatic elimination of the atoms or the field (Casagrande and Lugiato, 1980):

$$\lim_{X^2\to 0} g^{(2)}(t) = 1 - n_{sat}^{-1}2C(1 + 2C)^{-1}\exp[-\kappa(1 + 2C)t] \quad (5.73)$$

for the good cavity limit, and

$$\lim_{X^2\to 0} g^{(2)}(t) = 1 - N^{-1}8C^2(1 + 2C)^{-1}\exp[-\tfrac{1}{2}\gamma_{||}(1 + 2C)t] \quad (5.74)$$

for the bad cavity limit. An interesting feature outside these limits is the possibility for an oscillatory response: at vanishing intensities and on resonance, so these are not the oscillations due to Rabi splitting or detuning, familiar from the behavior of a driven two-level atom. This oscillatory behavior is illustrated in Fig. 8. With $\mu = 1$ the frequency of the oscillation is obtained from Eqs.(5.33) and (3.31) as $\gamma_{||}(\tfrac{1}{2}\mu C)^{\frac{1}{2}} = \sqrt{N}g$. Ihis is just the level splitting for field and polarization oscillators coupled by the interaction Hamiltonian H_{AF} (Eq.(3.5)). A detailed discussion of this picture is given elsewhere (Carmichael, 1986a) where the oscillation shown by Eq.(5.71) is interpreted in terms of the

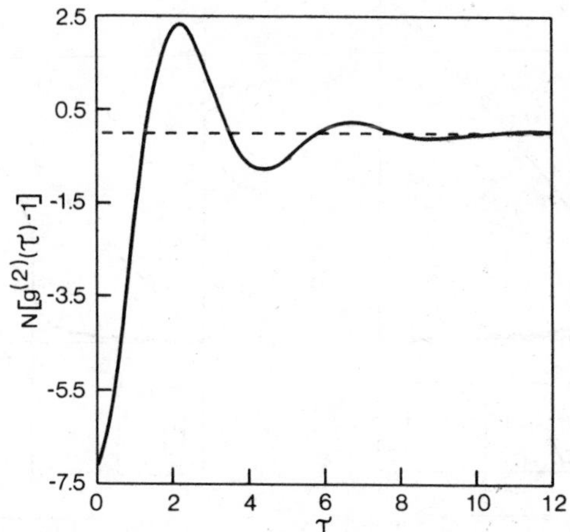

Fig. 8. Second-order correlation function of the transmitted light plotted from Eq.(5.71) for C = 4, $\delta = \phi = 0$, $\gamma_{\|} = 2\gamma_{\perp}$ and $\mu = 1$.

many-atom version of the so-called "vacuum Rabi splitting" reported for spontaneous emission from a single atom in a cavity (Sanchez-Mondragon et al., 1983; Agarwal, 1984).

6. NONLINEAR THEORY OF FLUCTUATIONS

We have seen that linearized theory gives diverging fluctuations at the turning points of bistability curves and at the critical point. In both situations nonlinearities are of primary importance. It is helpful to visualize what is happening in terms of a one-dimensional potential--although, of course, we cannot claim that fluctuations in optical bistability are always adequately described by such a simple model. In Fig. 9(a) we follow the development of a potential V(x) which might correspond to one of the bistable steady state curves plotted from Eq.(2.26). The deterministic steady states correspond to turning points of the potential ($V'(X) = 0$). Figure 9(b) illustrates the development of V(x) for a steady state curve passing through the critical point where the two local minima of the bistable potential coalesce. The turning points of bistability curves and the critical point all correspond to states with $V''(X) = 0$. For these states the quadratic term in

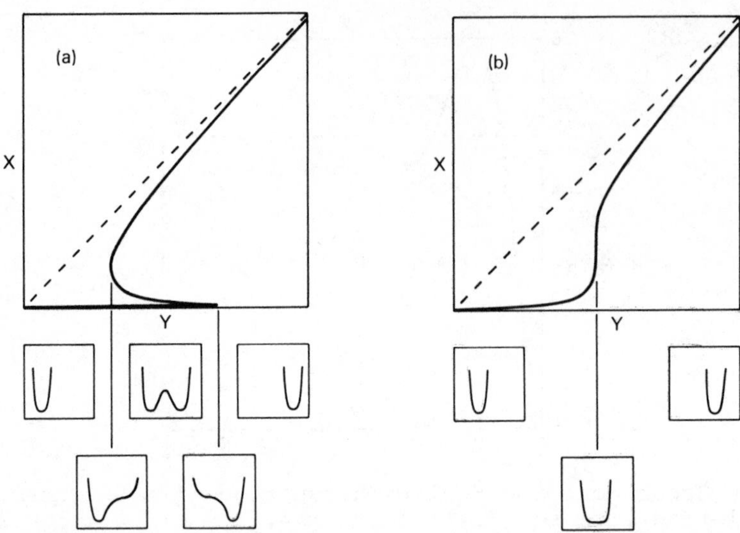

Fig. 9. Schematic representation of the one-dimensional potential model for absorptive bistability. The potential V(x) versus x is plotted for different values of Y: (a) for a bistable situation, C > 4; (b) for the critical case C = 4.

an expansion of V(x) around the steady state vanishes, and the parabolic approximation for the local potential shape becomes flat. A linearized theory, which makes such an approximation, then allows stochastic trajectories to wander without bound, while in reality they should be constrained by higher order terms in the potential expansion. In the linearized theory based on the positive P-representation the Fokker-Planck equation describes fluctuations in a ten-dimensional quadratic potential. Flattening of this potential then takes place along directions in ten dimensions defined by eigenvectors associated with the eigenvalues which vanish at instability points.

More generally, a nonlinear description is required, not just near instability points, but whenever fluctuations wander far from the deterministic steady state. In the bistable region, although they may be rare, transitions between states can occur even when operating far from the instability points. To calculate transition rates a nonlinear theory must be used. Also, linearized theory must eventually break down as the system size is reduced. According to Eqs.(5.30)-(5.36), fluctuations increase with decreasing system size. They can ultimately reach a level where the very existence of two distinct states in the bistable region must be questioned. A clear demonstration of the failure of linearized theory for small systems can be found in Eq.(5.66). Here, as N decreases $g^{(2)}(0)$ may become negative. But $g^{(2)}(0)$ is strictly a positive quantity. Thus, a nonlinear theory is required to treat quantum fluctuations in small systems. Exactly what constitutes "small" is, of course, one of the questions nonlinear theory must address: When do the results obtained with linearized theory begin to show significant inaccuracies?

Most work on the nonlinear treatment of quantum fluctuations in optical bistability is built around the one-dimensional picture outlined above. In the good cavity limit the Fokker-Planck equation in the Wigner representation is an equation in two dimensions, and, as mentioned earlier, it has positive definite diffusion even without linearization. If, for absorptive bistability, it is then assumed that the phase of the cavity field remains locked to that of the driving field, the two-dimensional Fokker-Planck equation is reduced to one dimension. Bonifacio et al. (1978) were the first to treat quantum fluctuations in absorptive bistability from this perspective. Although their original derivation of the one-dimensional Fokker-Planck equation was slightly in error (Drummond and Walls, 1981), this is corrected in later work with no qualitative change to the results (Lugiato, 1979; Lugiato, 1984). They construct the bimodal steady-state solution to the Fokker-Planck equation and evaluate the mean and variance for the field amplitude to demonstrate the first-order phase transition analogy for absorptive bistability. In extensions of this work, the one-dimensional Fokker-Planck equation in the Wigner representation has been used by a number of authors to calculate quantum induced transition rates between states in absorptive bistability (Schenzle and Brand, 1979; Farina et al., 1980; Hanggi et al., 1980; Bonifacio et al., 1981; Englund et al., 1981; Farina et al., 1981 England et al., 1982). Of course, these rates are extremely small in a system of macroscopic size, except very close to the deterministic instability points. These calculations can provide a basis for assessing the system size at which transitions might become important on an experimental time scale; although they are, of course, limited by the small noise assumptions implicit in the derivation of the Fokker-Planck equation. To this end, England et al. (1984) present a number of useful numerical results and make a nice comparison of various methods for calculating transition rates. The experimental observation of transitions between states would be a more attractive proposition if some signature of a purely quantum-dynamical process were present, as it is in photon antibunching and squeezing, for example. There is room for further theoretical work to resolve exactly what, if anything, is quantum mechanical about transitions in optical bistability, aside from the size of the diffusion coefficient. The terminology "quantum tunnelling" often used in reference to these transitions really calls for justification beyond anything provided in the literature to date. Where is the quantum coherence associated with a wavefunction residing simultaneously in two states?

A limited amount of work has gone beyond the one-dimensional approximation. Drummond and Walls (1980) have calculated all moments of the complex field amplitude in their nonlinear polarizability model for dispersive bistability using the complex P-representation (a close relative of the positive P-representation). Sakar and Satchell (1986) discuss transient behavior in multidimensional models for absorptive bistability, including some results obtained without adiabatic elimination of the atoms, using a Hamilton-Jacobi method. Graham and Schenzle (1981) also use a Hamilton-Jacobi equation to treat dispersive bistability in the good cavity limit; however, they drop the multiplicative part of the quantum noise.

If there is to be a vital experimental interest in our subject it is important to develop a theoretical approach which can deal with parameter values dictated by the convenience for experiments, rather than by the simplification brought by theoretical approximations. In this section we briefly explore the promise that our quantum-stochastic model based on the positive P-representation provides in this respect. This is our only option: recall that while the Wigner representation gives a Fokker-Planck equation with positive-definite diffusion after adiabatic elimination of the atoms, more generally it gives an equation which preserves positive definite diffusion only in its linearized form (section 4.2). We will review results obtained from numerical simulations of the full nonlinear stochastic differential equations in ten dimensions, Eqs.(4.42) (Carmichael et al., 1986). It appears from these results that there is a caveat which may limit the potential use of the positive P-representation. This arises when we address the question of structural stability in passing from the Maxwell-Bloch equations in five dimensions, Eqs.(2.30), to the ten-dimensional complexified version of these equations, Eqs. (4.29) (Carmichael et al., 1986; Sarkar et al., 1986).

6.1 Quantum-stochastic model in the positive P-representation

We will again focus on absorptive bistability ($\delta = \phi = 0$) with radiative damping ($\gamma_{\|} = 2\gamma_{\perp}$). Notice that the complex diffusion matrix is diagonal for these parameters (Eqs.(4.41)). Then, from Eqs.(4.39)-(4.43), an obvious choice for the matrix b gives the following set of stochastic differential equations:

$$d\bar{\alpha} = \mu(-\bar{\alpha} + 2C\bar{v} + Y)dt' , \tag{6.1a}$$

$$d\bar{\alpha}_* = \mu(-\bar{\alpha}_* + 2C\bar{v}_* + Y)dt' , \tag{6.1b}$$

$$d\bar{v} = \tfrac{1}{2}(-\bar{v} + \bar{\alpha}\bar{m})dt' + N^{-\frac{1}{2}}(\bar{\alpha}\bar{v})^{\frac{1}{2}}dW_3 , \tag{6.1c}$$

$$d\bar{v}_* = \tfrac{1}{2}(-\bar{v}_* + \bar{\alpha}_*\bar{m})dt' + N^{-\frac{1}{2}}(\bar{\alpha}_*\bar{v}_*)^{\frac{1}{2}}dW_4 , \tag{6.1d}$$

$$\begin{aligned} d\bar{m} = {} & (-\bar{m} - 1 - \tfrac{1}{2}\bar{\alpha}\bar{v}_* - \tfrac{1}{2}\bar{\alpha}_*\bar{v})dt' \\ & + (N/2)^{-\frac{1}{2}}(\bar{m} + 1 - \tfrac{1}{2}\bar{\alpha}\bar{v}_* - \tfrac{1}{2}\bar{\alpha}_*\bar{v})^{\frac{1}{2}}dW_5 , \end{aligned} \tag{6.1e}$$

where the dimensionless time t' is measured in atomic lifetimes:

$$t' = \gamma_{||} t \ .$$

These equations may be solved using a simple Euler algorithm to generate individual trajectories. For a time step $\Delta t'$, gaussian distributed random numbers with zero mean and variance $\Delta t'$ provide the noise sources

$$\Delta W_j(t') = \int_{t'}^{t'+\Delta t'} dW_j = W_j(t' + \Delta t') - W_j(t') \ , \qquad j = 3,4,5 \ . \tag{6.3}$$

From an ensemble of trajectories with the same initial condition stochastic averages $(\ )_{av}$ can be computed.

The results shown in Figs. 10 and 11 are for a system with $C = 4$, $\mu = \kappa/\gamma_\perp = 2\kappa/\gamma_{||} = 2$, and $N = 320$ ($n_s = N/4C\mu = 10$) (Carmichael et al., 1986). These parameters correspond to a small system which might reasonably be achieved in a scaled down version of the atomic beam experiments of Rosenberger et al. (1983, 1984). There are experimental difficulties with using fewer atoms. Note, also, that for small numbers of atoms the whole basis of the Fokker-Planck formalism must

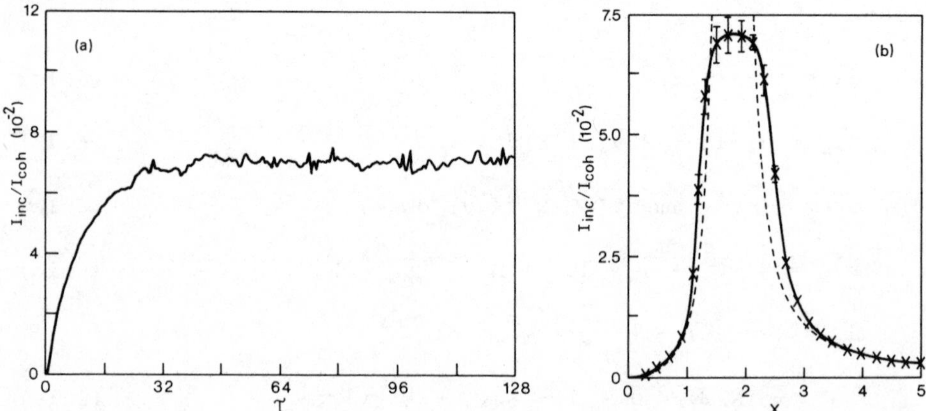

Fig. 10. Numerical simulations of Eqs. (6.1). I_{inc}/I_{coh} for $C = 4$, $\gamma_{||} = 2\gamma_\perp$, $\mu = 2$, and $N = 320$: (a) Evolution from the deterministic steady state to the stochastic stationary state for $X = 1.9$; (b) Behavior throughout the critical region in the stationary state.

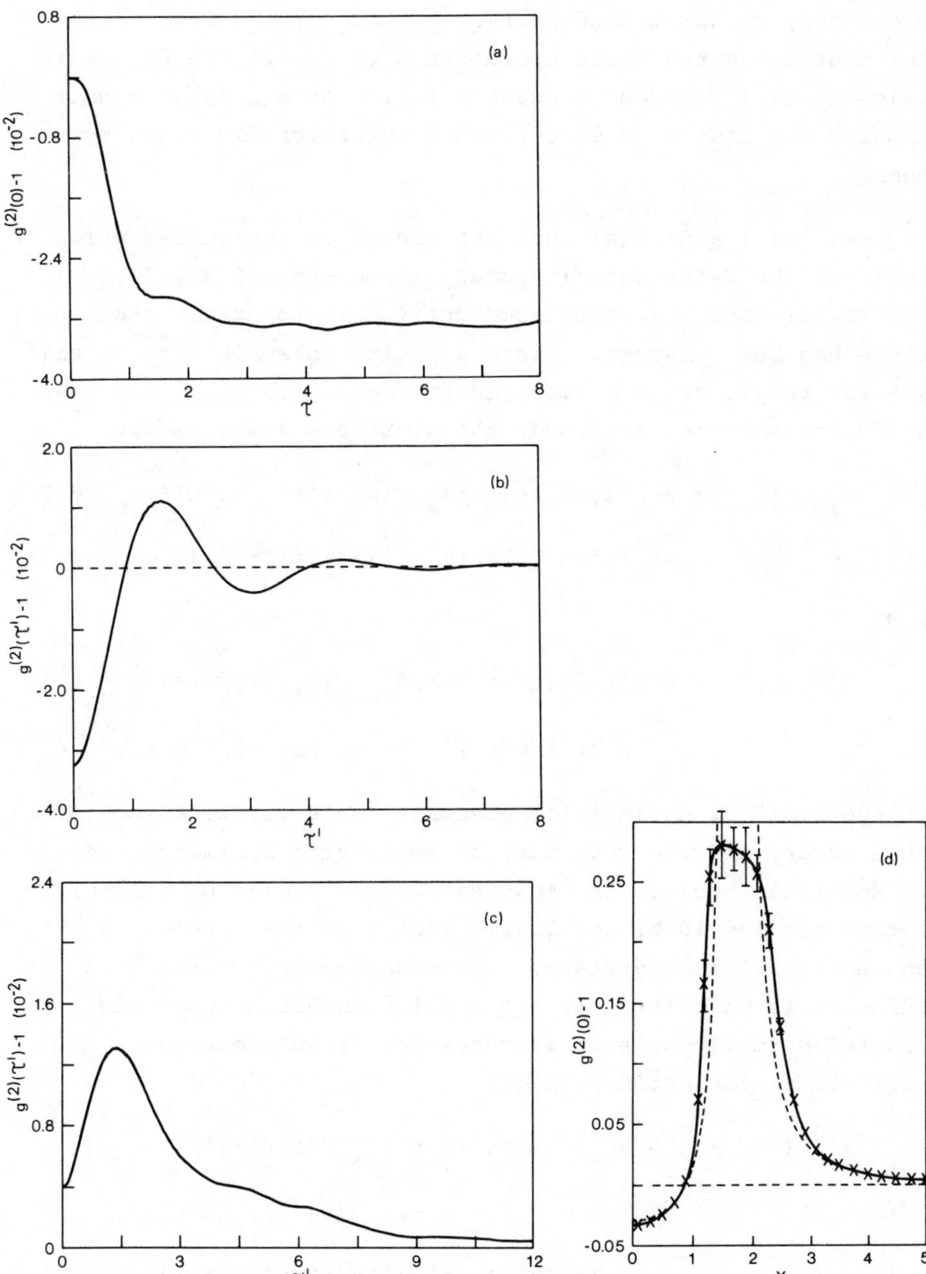

Fig. 11. Numerical simulations of Eqs.(6.1). Second-order correlation function for $C = 4$, $\gamma_{\parallel} = 2\gamma_{\perp}$, $\mu = 2$, and $N = 320$: (a) Evolution from the deterministic steady state to the stochastic stationary state for $X = 0.1$; (b) Behavior in the stationary state for $X = 0.1$; (c) Behavior in the stationary state for $X = 0.9$; (d) Behavior throughout the critical region in the stationary state.

eventually be questioned because of the large N requirement for truncating the shift operators $D^m_{\pm 1}$; recall the discussion below Eq.(4.7). Even an exact solution of Eqs.(6.1) cannot provide the last word on quantum fluctuations in small systems.

Figures 10(a) and 11(a) show transients for ensembles which begin in the deterministic steady state--Eqs.(5.3), (4.43), and (2.31)--and are integrated until the stochastic stationary state has been reached. Since a finite ensemble must be used (25,000 trajectories, except in the range $1.3 \leq X \leq 2.5$ where 4,000 trajectories are used) the plotted quantities are

$$I_{inc}/I_{coh} = \mathrm{Re}[(\bar{\alpha}_*(t')\bar{\alpha}(t'))_{av}/(\bar{\alpha}_*(t'))_{av}(\bar{\alpha}(t'))_{av}] - 1$$

$$\simeq \langle a^\dagger(t')a(t')\rangle/\langle a^\dagger(t')\rangle\langle a(t')\rangle - 1 , \qquad (6.4)$$

and

$$g^{(2)}(0) - 1 = \mathrm{Re}[(\bar{\alpha}_*(t')^2\bar{\alpha}(t')^2)_{av}/(\bar{\alpha}_*(t')\bar{\alpha}(t'))^2_{av}] - 1$$

$$\simeq \langle a^\dagger(t')^2a(t')^2\rangle/\langle a^\dagger(t')a(t')\rangle^2 - 1 . \qquad (6.5)$$

In the limit of an infinite ensemble the equalities should become exact, and the real part of stochastic averages need not be specified because the imaginary parts should be strictly zero. Figures 10(b) and 11(b)-11(d) show the results computed in the stationary state. The second-order correlation function is calculated from the relationship between time-ordered normally-ordered averages for field operators and multi-time stochastic averages:

$$g^{(2)}(\tau') - 1 = \mathrm{Re}[(\Lambda(t'+\tau')\Lambda(t'))_{av}/(\Lambda(t'+\tau'))_{av}$$

$$(\Lambda(t'))_{av}] - 1$$

$$\simeq \frac{\langle a^\dagger(t')a^\dagger(t'+\tau')a(t'+\tau')a(t')\rangle}{\langle a^\dagger(t'+\tau')a(t'+\tau')\rangle\langle a^\dagger(t')a(t')\rangle} - 1 , \qquad (6.6)$$

where $\Lambda = \bar{\alpha}_*\bar{\alpha}$, and t' is a time sufficiently long for the decay to the stationary state to be essentially complete. Note the correspondence between Fig. 11(b) and the oscillatory behavior given by linearized theory (Eq.(5.71) and Fig. 8). This weak-field result has $g^{(2)}(0) - 1 \simeq -0.0325$, in fairly close agreement with the value $-4/135 = -0.0296$ given by Eq. (5.66); the difference comes from the nonlinearity. For increasing intensities photon antibunching, as defined by Eq. (5.55), is reduced and eventually disappears. Antibunching in the sense of Eq.(5.56) can remain, however, after $g^{(2)}(0)$ has become positive. This is illustrated by Fig. 11(c). Figures 10(b) and 11(d) show the effects of nonlinearity in limiting fluctuations throughout the critical region in comparison with the divergence, at $X = \sqrt{3}$ $(Y = 3\sqrt{3})$, given by linearized theory. The agreement between the linear and nonlinear calculations is very good when fluctuations are small ($\sim$1%).

Further results from the numerical simulation of Eqs.(6.1) can be found elsewhere illustrating features such as critical slowing down, the scaling of critical fluctuations with system size and quantum induced transitions between states for $C > 4$ (Carmichael et al., 1986). These results are essentially proof in principle that our quantum stochastic model based on the positive P-representation can be used to extend the study of quantum fluctuations to moderately small systems, without restriction to the good or bad cavity limit, while retaining the full nonlinearity of the problem. There is a cloud hanging over this happy picture, however, which calls for a more thorough analysis of the formal basis for the positive P-representation and its properties under the approximations used to derive the Fokker-Planck equation. The positive P-representation shows rather questionable dynamical properties in the vicinity of bifurcation points and unstable states. We should not leave our discussion of nonlinear theory without a brief mention of this problem.

6.2 The positive P-representation and structural stability

The passage from a five-dimensional space to a ten-dimensional space in the positive P-representation does more than provide extra dimensinos to accommodate the quantum noise. It also profoundly alters the deterministic equations which govern the nonlinear dynamics of our model. This is illustrated by solutions in the steady state. If we drop the noise terms and set time derivatives to zero, Eqs.(4.29) (equivalently, Eq.(4.42)) have solutions:

$$\bar{\alpha}^s = Y[1 + i\phi + 2C(1 - i\delta)/(1 + \delta^2 + \Lambda_s)]^{-1} , \qquad (6.7a)$$

$$\bar{\alpha}^s_* = Y[1 - i\phi + 2C(1 + i\delta)/(1 + \delta^2 + \Lambda_s)]^{-1} , \qquad (6.7b)$$

$$\bar{v}^s = -\bar{\alpha}^s(1 - i\delta)/(1 + \delta^2 + \Lambda_s) , \qquad (6.7c)$$

$$\bar{v}^s_* = -\bar{\alpha}^s_*(1 + i\delta)/(1 + \delta^2 + \Lambda_s) , \qquad (6.7d)$$

$$\bar{m}^s = -(1 + \delta^2)/(1 + \delta^2 + \Lambda_s) , \qquad (6.7e)$$

where $\Lambda_s = \bar{\alpha}^s_*\bar{\alpha}^s$ satisfies the state equation

$$Y^2 = \Lambda_s\{[1+2C/(1+\delta^2+\Lambda_s)]^2+[\phi-2C\delta/(1+\delta^2+\Lambda_s)]^2\} . \qquad (6.8)$$

When Λ_s is real we recover Eqs.(2.26), (2.31) and (2.32), and the "physical" steady states which formed the basis for our linearized theory and gave the initial conditions for the numerical simulations of the last section. However, $\overline{\alpha_*\alpha}$ need not be real; extra dimensions have been introduced precisely so that $\bar{\alpha}_*$ need not remain complex conjugate to $\bar{\alpha}$. Then, we must always recognize the three solutions to Eq.(6.8), even when two of these are complex. Two branches of "nonphysical" steady states attach to the turning points of bistability curves as illustrated in Fig. (12); they correspond to solutions with $\bar{\alpha}^s_* \neq \bar{\alpha}^{s*}$, $\bar{v}^s_* \neq \bar{v}^{s*}$, and $\bar{m}^s \neq \bar{m}^{s*}$. Linearized theory in the positive P-representation gives the same results as theories based on the Wigner representation and quantum Langevin equations. There has been no call to worry about extra steady states when fluctuations are restricted to the vicinity of the physical steady states. However, what do we expect in nonlinear theory? Surely these extra states affect the dynamics close to bifurcation points where the physical and nonphysical steady states collide.

To assess this question we will consider the linear stability analysis. First, let us define a five-dimensional physical

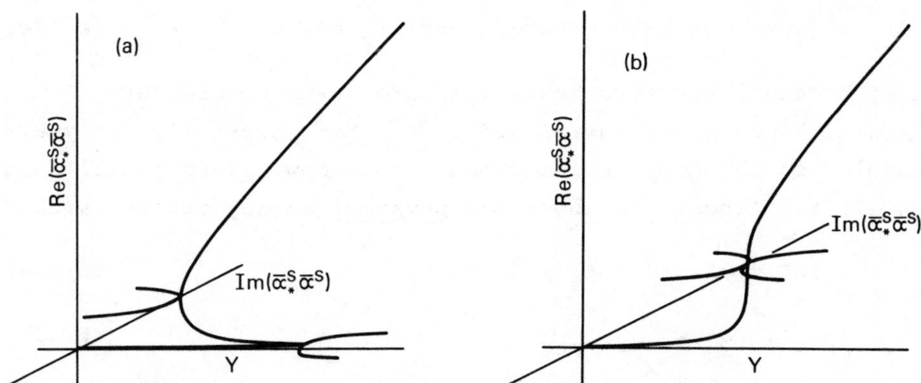

Fig. 12. Schematic representation of the new branches of nonphysical steady states introduced by the positive P-representation.

subspace (x, x^*, y, y^*, z) with

$$x = \tfrac{1}{2}(\bar{\alpha} + \bar{\alpha}^*_*) \;, \tag{6.9a}$$

$$y = \tfrac{1}{2}(\bar{v} + \bar{v}^*_*) \;, \tag{6.9b}$$

$$z = \tfrac{1}{2}(\bar{m} + \bar{m}^*) \;, \tag{6.9c}$$

and a nonphysical subspace (p, p^*, q, q^*, w) with

$$p = -i\tfrac{1}{2}(\bar{\alpha} - \bar{\alpha}^*_*) \;, \tag{6.10a}$$

$$q = -i\tfrac{1}{2}(\bar{v} - \bar{v}^*_*) \;, \tag{6.10b}$$

$$w = -i\tfrac{1}{2}(\bar{m} - \bar{m}^*) \;. \tag{6.10c}$$

Then our ten-dimensional complexified version of the Maxwell-Bloch equations has

$$\kappa^{-1}\dot{x} = -(1 + i\phi)x + 2Cy + Y \;, \tag{6.11a}$$

$$\gamma_\perp^{-1}\dot{y} = -(1 + i\delta)y + xz - pw \;, \tag{6.11b}$$

$$\gamma_\parallel^{-1}\dot{z} = -(z + 1) - \tfrac{1}{2}xy^* - \tfrac{1}{2}x^*y + \tfrac{1}{2}pq^* + \tfrac{1}{2}p^*q \;, \tag{6.11c}$$

and

$$\kappa^{-1}\dot{p} = -(1 + i\phi)p + 2Cq \ , \tag{6.12a}$$

$$\gamma_{\perp}^{-1}\dot{q} = -(1 + i\delta)q + xw + pz \ , \tag{6.12b}$$

$$\gamma_{\parallel}^{-1}\dot{w} = -w - \tfrac{1}{2}xq^* - \tfrac{1}{2}x^*q - \tfrac{1}{2}py^* - \tfrac{1}{2}p^*y \ . \tag{6.12c}$$

If p, q and w are zero Eqs.(6.11) are just the ordinary Maxwell-Bloch equations, Eqs.(2.30). More generally, they are coupled to the nonphysical dimensions. Now, if Eqs.(6.11) and (6.12) are linearized about the physical steady states, with

$$(x,x^*,y,y^*,z)^T = \vec{\xi}_s + \vec{\lambda}_p \ , \tag{6.13a}$$

$$(p,p^*,q,q^*,w)^T = \vec{\lambda}_{np} \ , \tag{6.13b}$$

we find

$$\dot{\vec{\lambda}}_p = A\vec{\lambda}_p \ , \tag{6.14a}$$

$$\dot{\vec{\lambda}}_{np} = A\vec{\lambda}_{np} \ , \tag{6.14b}$$

where A is the matrix defined by Eq.(2.35). The important point is that the same matrix A governs dynamics in the physical and nonphysical subspaces; eigenvalues of the linearized dynamics occur in pairs, one eigenvalue governing motion in the physical subspace and an identical eigenvalue governing motion in the nonphysical subspace. In particular, it follows that unstable physical steady states are always unstable in two, four, six, etc. directions, and one direction in each pair penetrates into the nonphysical subspace. The physical significance of this behavior is certainly questionable. Motion into nonphysical dimensions is not pathological in itself. Indeed, the positive P-representation provides the extra dimensions just for this purpose. However, we only expect to see noise driven motion into these dimensions. When this motion takes on a life of its own, as it does when the deterministic equations are unstable, it seems of suspect physical meaning. For example, might not trajectories in the unstable manifold lead to stable nonphysical attractors? The consequences of our observation are most disconcerting in chaotic systems, where chaotic trajectories constantly encounter instability (Sarkar et al., 1986).

The effects of this structural instability do show up in the numerical simulation of Eqs.(6.1) (Carmichael et al., 1986). However, they only seem to be a real cause for concern in systems small enough that the truncation of the displacement operators $D^m_{\pm 1}$ used to derive the Fokker-Planck equation is certainly suspect. Figure 13 is plotted from an average of 1000 trajectories and shows two anomalous features caused by large pulses generated in just two members of the ensemble. One of these pulses is shown in Fig. 13(b); it rises by more than an order of magnitude above the background fluctuations. This behavior is reminiscent of a superradiant pulse. We must not

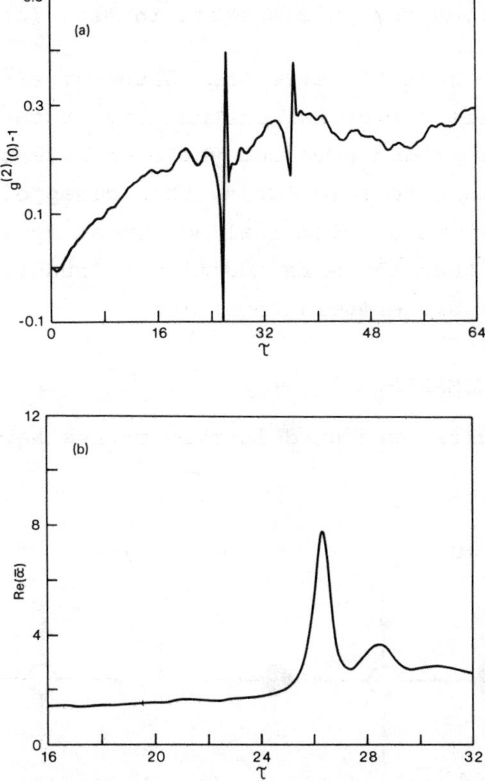

Fig. 13. The dynamical effects of structural instability: (a) Anomalous fluctuations in an average of 100 trajectories. (b) The single trajectory pulse responsible for the feature at $\tau \simeq 26$. For $C = 4$, $\gamma_{\parallel} = 2\gamma_{\perp}$, $\mu = 2$, $N = 320$, and $X = 1.3$.

be too ready to accept this appealing interpretation, however. The trajectory producing the pulse makes a wild excursion into the nonphysical space, with $\bar{\alpha}_* \simeq \bar{\alpha}$ and $\bar{v}_* \simeq \bar{v}$, rather than $\bar{\alpha}_* \simeq \bar{\alpha}^*$ and $\bar{v}_* \simeq \bar{v}^*$ (Carmichael et al., 1986). Its origin lies with the structure of stable and unstable manifolds in the vicinity of the critical point. Figure 14 illustrates the behavior of the manifolds associated with the two eigenvalues which vanish at the critical point. For $C = 4$, $X \lesssim \sqrt{3}$ ($Y \lesssim 3\sqrt{3}$) we have one stable physical steady state and two unstable nonphysical steady states, all very close together. At the critical point, $C = 4$, $X = \sqrt{3}$ ($Y = 3\sqrt{3}$), these steady states collide. An unstable manifold penetrating into the nonphysical subspace is left attached to the critical point, as illustrated by Fig. 14(b). Ejection along this unstable direction initiates the pulses shown in Fig. (13).

Unfortunately we have to leave this issue here. The full implications of this structural instability in the positive P-representation approach have not yet been assessed. Perhaps the difficulty is more stimulating than disappointing. Possibly the quantization of classical nonlinear dynamical systems is less trivial than the mere addition of quantum noise terms to the classical differential equations.

7. CONCLUDING REMARKS

Many of the results presented in this review have been in the

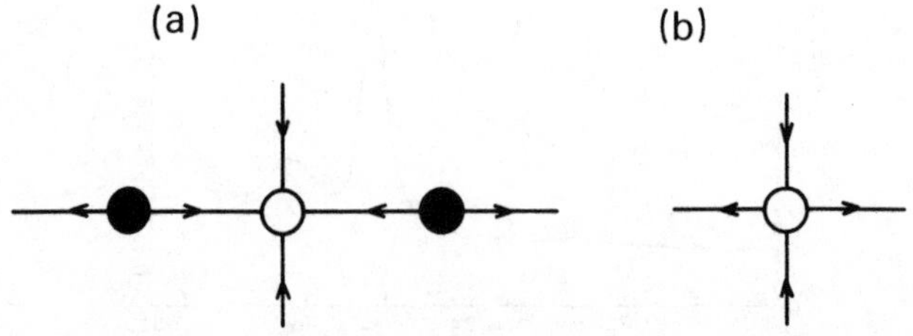

Fig. 14. Steady-state stability for the two eigenvalues which vanish at the critical point. Across the page motion is in a nonphysical dimension; up and down the page, in a physical dimension: (a) Two nonphysical saddles and one physical node for $C = 4$ and $X \lesssim \sqrt{3}$; (b) The critical point at $C = 4$ and $X = \sqrt{3}$.

literature now for a number of years. It is hoped that their presentation here, alongside discussion of some of the issues surrounding the model from which they are derived, has shown that the subject of quantum fluctuations in optical bistability holds the promise of interesting developments in the future to match those of the past.

Little has been said of experiments, but they will certainly play an important role in determining the future vitality of our subject. We can now realistically look forward to experimental measurements of quantum-statistical effects in optical bistability. The excellent quantitative agreement with theory that has now been achieved in experiments with sodium atomic beams--the published results (Rosenberger et al., 1983; Rosenberger et al., 1984) no longer do justice to the level of agreement most recently obtained in Prof. Kimble's laboratory--provides a strong stimulus for work in this direction. The key to such experiments will be the design of systems small enough and stable enough to make quantum fluctuations a dominant effect.

When we think of small systems interesting questions are also raised for theoretical studies. The treatment of quantum dissipative systems in optics using Fokker-Planck equation techniques has generally avoided the difficulties associated with non-positive-definite diffusion and use of the system-size expansion, having the a priori objective of reaching an essentially classical statistical description. It seems that the most interesting quantum mechanics is removed in the process. It becomes an interesting proposition to now face these issues: What are the precise limitations of the system-size expansion? How can the quantum dynamics represented by non-positive-definite diffusion be correctly taken into account? These questions are brought to focus in nonlinear systems, whose semiclassical descriptions exhibit bifurcations, self-oscillation, and chaos. We have seen that unexpected things can happen with the positive P-representation in a nonlinear theory, while this representation behaves quite satisfactorily in a linearized treatment. It is important that such issues be resolved if we are ultimately to understand the relationship between quantum mechanics and classical nonlinearity. One way to address this relationship is to study the passage from a quantized description for a few interacting atoms and photons to the classical limit for macroscopic systems. Quantum fluctuations can certainly destroy bistability for very small systems. How do the bistable states appear from a system dominated by quantum noise as system size is increased? The question of chaos in quantum systems is currently of wide interest. We have a classical nonlinear system which we claim we know how to quantize. We have not discussed its chaotic regime here, but our model can exhibit chaos. What does the quantum model do when its semiclassical version is chaotic? What is quantized chaos and how does it depend on system size? This question seems to be well posed. It avoids any debate over the existence and definition of quantum choas; we know something about classical chaos, and we know how to quantize optical systems. Something goes on in the quantized version of a chaotic system. Is it just a matter of adding noise terms to classical equations? At least for very small systems, surely it is not.

For a closing remark let us return to a point that was raised in the introduction and has probably been lost in the intervening discussion. Our model is very general. It is better viewed as an elementary model for the dynamical behavior of atoms inside cavities than a model peculiar to the phenomenon of optical bistability. Photon antibunching, squeezing, atom-atom correlation, vacuum Rabi splitting; these are all features we have met which have nothing in principle to do with bistability. To these we should add cavity enhanced spontaneous emission; there has not been space to include a section on the single-atom system which illustrates its role (Carmichael, 1985). Bistable behavior represents one result of instability and bifurcation in a model which also shows transitions to periodic and chaotic oscillatory states. Our ultimate objective should be to understand the full complement of quantum-dynamical and nonlinear-dynamical behavior exhibited by this model. If such a program can be completed, a large part of quantum optics will have been embraced.

ACKNOWLEDGEMENT

This paper is based upon work supported by the National Science Foundation under Grant No. PHY-8418070.

REFERENCES

Abraham, E. and Hassan, S.S. 1980, Opt. Commun. 35, 291-297
Abraham, E., Hassan, S.S., and Bullough, R.K., 1980, Opt. Commun. 33, 93-98
Abraham, E. and Smith, S.D. 1982, Rep. Prog. Phys. 45, 815-885
Agarwal, G.S., Narducci, L.M., Gilmore, R., and Feng, D.H., 1978, Phys. Rev. A 18, 620-634
Agarwal, G.S., Narducci, L.M., Feng, D.H., and Gilmore, R., 1980, Phys. Rev. A 21, 1029-1038
Agarwal, G.S. and Tewari, S.P. 1980, Phys. Rev. A 21, 1638-1647
Agarwal, G.S. 1982, Phys. Rev. A 26, 680-683
Agarwal, G.S. 1984, Phys. Rev. Lett. 53, 1732-1734
Agrawal, G.P. and Carmichael, H.J. 1979, Phys. Rev. A 19, 2074-2086
Agrawal, G.P. and Carmichael, H.J. 1980, Optica Acta 27, 651-660
Asquini, M.L., Lugiato, L.A., Carmichael, H.J., and Narducci, L.M., 1986, Phys. Rev. A 33, 360-374
Austin, J.W. and DeShazer, L.G. 1971, J. Opt. Soc. Am. 61, 650
Ballagh, R.J., Cooper, J., Hamilton, M.W., Sandle, W.J., and Warrington, D.M., 1981, Opt. Commun. 37, 143-148
Bonifacio, R., Schwendimann, P., and Haake, F., 1971, Phys. Rev. A 4, 302-313
Bonifacio, R. and Lugiato, L.A. 1976, Opt. Commun. 19, 172-176
Bonifacio, R. and Lugiato, L.A. 1978a, Phys. Rev. Lett. 40, 1023-1027
Bonifacio, R. and Lugiato, L.A. 1978b, Lett. Nuovo Cim. 21, 517-521
Bonifacio, R. and Lugiato, L.A. 1978c, Lett. Nuovo Cim. 21, 505-509
Bonifacio, R. and Lugiato, L.A. 1978d, Lett. Nuovo Cim. 21, 510-516
Bonifacio, R. and Lugiato, L.A. 1978e, Phys. Rev. A 18, 1129-1144
Bonifacio, R., Gronchi, M., and Lugiato, L.A., 1978, Phys. Rev. A 18, 2266-2279
Bonifacio, R., Gronchi, M., and Lugiato, L.A., 1979a, Nuovo Cim. B 53, 311-332
Bonifacio, R., Gronchi, M., and Lugiato, L.A., 1979b, Opt. Commun. 30, 129-133
Bonifacio, R., Lugiato, L.A., Farina, J.D., and Narducci, L.M. 1981, IEEE J. Quantum Electron. QE17, 357-365

Bowden, C.M. and Sung, C.C. 1979, Phys. Rev. A 19, 2392-2401
Bowden, C.M. 1981, Optical Bistability, Eds. C.M. Bowden, M. Ciftan, and H.R. Robl, Plenum, New York, pp 405-429
Bowden, C.M., Ciftan, M., and Robl, H.R., 1981, Eds., Optical Bistability, Plenum, New York
Bowden, C.M., Gibbs, H.M., and McCall, S.L., 1984, Eds., Optical Bistability 2, Plenum, New York
Carmichael, H.J. and Walls, D.F. 1973, J. Phys. A 6, 1552-1564
Carmichael, H.J. 1980a, J. Phys. B 13, 3551-3575
Carmichael, H.J. 1980b, Optica Acta 27, 147-158
Carmichael, H.J. and Agrawal, G.P. 1980, Opt. Commun. 34, 293-299
Carmichael, H.J. and Hermann, J.A. 1980, Z. Phys. B 38, 365-380
Carmichael, H.J. 1981, Z. Phys. B 42, 183-190
Carmichael, H.J. and Agrawal, G.P. 1981, Optical Bistability, Eds. C.M. Bowden, M. Ciftan, and H.R. Robl, Plenum, New York, pp 237-264
Carmichael, H.J., Snapp, R.R., and Schieve, W.C., 1982, Phys. Rev. A 26, 3408-3422
Carmichael, H.J. 1983, Laser Physics, Eds. J.D. Harvey and D.F. Walls, Springer, Berlin, pp 64-87
Carmichael, H.J., Walls, D.F., Drummond, P.D., and Hassan, S.S., 1983, Phys. Rev. A 27, 3112-3128
Carmichael, H.J. 1984, Phys. Rev. Lett. 52, 1292-1295
Carmichael, H.J. 1985, Phys. Rev. Lett. 55, 2790-2793
Carmichael, H.J. 1986a, Phys. Rev. A, to be published
Carmichael, H.J. 1986b, Quantum-Statistical Methods in Quantum Optics, Springer, Berlin, to be published
Carmichael, H.J., Satchell, J.S., and Sarkar, S., 1986, submitted to Phys. Rev. A
Casagrande, F. and Lugiato, L.A. 1980, Nuovo Cim. B 55, 173-190
Cresser, J.D., Häger, J., Leuchs, G., Rateike, M., and Walther, H., 1982, Dissipative Systems in Quantum Optics, Ed. R. Bonifacio, Springer, Berlin, pp 21-59
Drummond, P.D. and Carmichael, H.J. 1978, Opt. Commun. 27, 160-164
Drummond, P.D. 1979, Nonequilibrium Transitions in Quantum Optical Systems, D.Phil. thesis, University of Waikato, pp 59-72, 231-246
Drummond, P.D. and Gardiner, C.W. 1980, J. Phys. A 13, 2353-2368
Drummond, P.D. and Walls, D.F. 1980, J. Phys. A 13, 725-741
Drummond, P.D. 1981, IEEE J. Quantum Electron. QE17, 301-306
Drummond, P.D. and Walls, D.F. 1981, Phys. Rev. A 23, 2563-2579
Englund, J.C., Schieve, W.C., Zurek, W., and Gragg, R.F., 1981, Optical Bistability, Eds. C.M. Bowden, M. Ciftan, and H.R. Robl, Plenum, New York, pp 315-335
Englund, J.C., Schieve, W.C., Gragg, R.F., and Zurek, W., 1982, Instabilities, Bifurcations and Fluctuations in Chemical Systems, Eds. L.E. Reichl and W.C. Schieve, University of Texas Press, Austin
Englund, J.C., Snapp, R.R., and Schieve, W.C., 1984, Progress in Optics, Vol. XXI, Ed. E. Wolf, North Holland, Amsterdam,

pp 355-428
Farina, J.D., Narducci, L.M., Yuan, J.M., and Lugiato, L.A., 1980, Opt. Eng. 19, 469
Farina, J.D., Narducci, L.M., Yuan, J.M., and Lugiato, L.A., 1981, Optical Bistability, Eds. C.M. Bowden, M. Ciftan, and H.R. Robl, Plenum, New York, pp 337-352
Felber, F.S. and Marburger, J.H. 1976, Appl. Phys. Lett. 28, 731-733
Gardiner, C.W. 1983, Handbook of Stochastic Methods, Springer, Berlin, pp 96-98
Gardiner, C.W. and Savage, C.M. 1984, Opt. Commun. 50, 173-178
Gibbs, H.M., McCall, S.L., and Venkatesan, T.N.C., 1976, Phys. Rev. Lett. 36, 1135-1138
Gibbs, H.M. 1985, Optical Bistability: Controlling Light with Light, Academic Press, Orlando
Glauber, R.J. 1963a, Phys. Rev. Lett. 10, 84-86
Glauber, R.J. 1963b, Phys. Rev. 130, 2529-2539
Glauber, R.J. 1963c, Phys. Rev. 131, 2766-2788
Goy, P., Raimond, J.M., Gross, M., and Haroche, S., 1983, Phys. Rev. Lett. 50, 1903-1906
Graham, R. and Schenzle, A. 1981, Phys. Rev. A 23, 1302-1321
Grant, D.E. and Kimble, H.J. 1982, Opt. Lett. 7, 353-355
Grant, D.E. and Kimble, H.J. 1983, Opt. Commun. 44, 415-420
Gronchi, M. and Lugiato, L.A. 1978, Lett. Nuovo Cim. 23, 593-598
Gronchi, M. and Lugiato, L.A. 1980, Opt. Lett. 5, 108-110
Gronchi, M., Benza, V., Lugiato, L.A., Meystre, P., and Sargent III, M., 1981, Phys. Rev. A 24, 1419-1435.
Haken, H. 1970, Laser Theory, Encyclopedia of Physics, Vol. XXV/2C, Ed. L. Genzel, Springer, Berlin, pp 33-73
Hanggi, P., Bulsara, A.R., and Janda, R., 1980, Phys. Rev. A 22, 671-683
Hassan, S.S., Drummond, P.D., and Walls, D.F., 1978, Opt. Commun. 27, 480-484
Hassan, S.S., Bullough, R.K., Puri, R.R., and Lawande, S.V., 1980, Physica A 103, 213-225
Hermann, J.A. 1980, Optica Acta 27, 159-170
Hillery, M., O'Connell, R.F., Scully, M.O., and Wigner, E.P., 1984, Phys. Rep. 106, 121-167
Hopf, F.A. and Bowden, C.M. 1985, Phys. Rev. A 32, 268-275
Ikeda, K. 1979, Opt. Commun. 30, 257-261
Ikeda, K., Daido, H., and Akimoto, O., 1980, Phys. Rev. Lett. 45, 709-712
Ikeda, K. and Akimoto, O. 1982, Phys. Rev. Lett. 48, 617-620
Kaluzny, Y., Goy, P., Gross, M., Raimond, J.M., and Haroche, S., 1983, Phys. Rev. Lett. 51, 1175-1178
Kimble, H.J., Dagenais, M., and Mandel, L., 1977, Phys. Rev. Lett. 39, 691-695
Kimble, H.J., Grant, D.E., Rosenberger, A.T., and Drummond, P.D., 1983, Laser Physics, Eds. J.D. Harvey and D.F. Walls, Springer, Berlin, pp 14-63
Klauder, J.R. and Sudarshan, E.C.G. 1968, Fundamentals of Quantum Optics, Benjamin, New York, pp 178-201
Kleppner, D. 1981, Phys. Rev. Lett. 47, 233-236
Lax, M. 1963, Phys. Rev. 129, 2342-2348
Lax, M. 1967, Phys. Rev. 157, 213-231
Loudon, R. 1980, Rep. Prog. Phys. 43, 913-949

Lugiato, L.A. 1979, Nuovo Cim. B 50, 89-133
Lugiato, L.A. 1980a, Opt. Commun. 33, 108-112
Lugiato, L.A. 1980b, Lett. Nuovo Cim. 29, 375-380
Lugiato, L.A., Asquini, M.L., and Narducci, L.M., 1982a, Opt. Commun. 41, 450-454
Lugiato, L.A., Narducci, L.M., Bandy, D.K., and Pennise, C.A., 1982b, Opt. Commun. 43, 281-286
Lugiato, L.A., Casagrande, F., and Pizzuto, L., 1982c, Phys. Rev. A 26, 3438-3458
Lugiato, L.A. and Strini, G. 1982, Opt. Commun. 41, 67-70
Lugiato, L.A. 1984, Progress in Optics, Vol. XXI, Ed. E. Wolf, North Holland, Amsterdam, pp 69-216
Lugiato, L.A., Horowicz, R.J., Strini, G., and Narducci, L.M., 1984, Phys. Rev. A 30, 1366-1376
Lugiato, L.A. and Narducci, L.M. 1984, Coherence and Quantum Optics, Eds. L. Mandel and E. Wolf, Plenum, New York, pp 941-955
Lugiato, L.A. 1986, Phys. Rev. A, to be published
Mandel, L. 1982, Phys. Rev. Lett. 49, 136-138
Marburger, J.H. and Felber, F.S. 1978, Phys. Rev. A 17, 335-342
McCall, S.L. 1974, Phys. Rev. A 9, 1515-1523
McCall, S.L. and Gibbs, H.M. 1980, Opt. Commun. 33, 335-339
Meschede, D., Walther, H., and Müller, G., 1985, Phys. Rev. Lett. 54, 551-554
Meystre, P. 1978, Opt. Commun. 26, 277-280
Mollow, B.R. 1969, Phys. Rev. 188, 1969-1975
Moloney, J.V. and Gibbs, H.M. 1982, Phys. Rev. Lett. 48, 1607-1610
Moloney, J.V., Sargent III, M., and Gibbs, H.M., 1982, Opt. Commun. 44, 289-292
Narducci, L.M., Gilmore, R., Feng, D.H., and Agarwal, G.S., 1978a, Opt. Lett. 2, 88-90
Narducci, L.M., Feng, D.H., Gilmore, R., and Agarwal, G.S., 1978b, Phys. Rev. A 18, 1571-1576
Orozco, L.A., Rosenberger, A.T., and Kimble, H.J., 1984, Phys. Rev. Lett. 53, 2547-2550
Paul, H. 1982, Rev. Mod. Phys. 54, 1061-1102
Puri, R.R. and Lawande, S.V. 1979, Phys. Lett. A 72, 200-202
Puri, R.R. and Lawande, S.V. 1980, Physica A 101, 599-612
Puri, R.R., Lawande, S.V., and Hassan, S.S., 1980, Opt. Commun. 35, 179-184
Puri, R.R., Hildred, G.P., Hassan, S.S., and Bullough, R.K., 1984, Optical Bistability 2, Eds. C.M. Bowden, H.M. Gibbs, and S.L. McCall, Plenum, New York, pp 471-478
Raimond, J.M., Goy, P., Gross, M., Fabre, C., and Haroche, S., 1982a, Phys. Rev. Lett. 49, 117-120
Raimond, J.M., Goy, P., Gross, M., Fabre, C., and Haroche, S., 1982b, Phys. Rev. Lett. 49, 1924-1927
Reid, M.D. and Walls, D.F. 1985, Phys. Rev. A 32, 396-401
Rogonov, N.N. and Semenov, V.E.S. 1980, Opt. Spectrosc. 48, 59-63
Rosenberger, A.T., Orozco, L.A., and Kimble, H.J., 1983, Phys. Rev. A 28, 2569-2572
Rosenberger, A.T., Orozco, L.A., and Kimble, H.J., 1984, Fluctuations and Sensitivity in Nonequilibrium Systems, Eds. W. Horsthemke and D.K. Kondepudi, Springer, Berlin,

pp 62-69
Roy, R. and Zubairy, M.S. 1980a, Opt. Commun. 32, 163-168
Roy, R. and Zubairy, M.S. 1980b, Phys. Rev. A 21, 274-280
Sanchez-Mondragon, J.J., Narozhny, N.B., and Eberly, J.H., 1983, Phys. Rev. Lett. 51, 550-553
Sandle, W.J. and Gallagher, A. 1980, J. Opt. Soc. Am. 70, 656
Sandle, W.J. and Gallagher, A. 1981, Phys. Rev. A 24, 2017-2028
Sargent III, M., Zubairy, M.S., and DeMartini, F., 1983, Opt. Lett. 8, 76-78
Sargent III, M., Holm, D.A., and Zubairy, M.S., 1985, Phys. Rev. A 31, 3112-3131
Sarkar, S. and Satchell, J.S. 1986, Phys. Rev. A, to be published
Sarkar, S., Satchell, J.S., and Carmichael, H.J., 1986, submitted to J. Phys. A
Schenzle, A. and Brand, H. 1979, Opt. Commun. 31, 401-407
Schumaker, B.L. 1984, Opt. Lett. 9, 189-191
Seidel, H., U.S. Patent 3,610,731, filed May 19, 1969, granted October 5, 1971
Senitzky, I.R. 1960, Phys. Rev. 119, 670-679
Senitzky, I.R. 1961, Phys. Rev. 124, 642-648
Senitzky, I.R. 1972, Phys. Rev. A 6, 1171-1174
Spiller, E. 1971, J. Opt. Soc. Am. 61, 669
Spiller, E. 1972, J. Appl. Phys. 43, 1673-1681
Stenholm, S., Holm, D.A., and Sargent III, M., 1985, Phys. Rev. A 31, 3124-3131
Sudarshan, E.C.G. 1963, Phys. Rev. Lett. 10, 277-279
Szöke, A., Daneu, V., Goldhar, J., and Kurnit, N.A., 1969, Appl. Phys. Lett. 15, 376-379
Tavis, M. and Cummings, F.W. 1968, Phys. Rev. 170, 379-384
Tavis, M. and Cummings, F.W. 1969, Phys. Rev. 188, 692-695
Vaidyanathan, A.G., Spencer, W.P., and Kleppner, D., 1981, Phys. Rev. Lett. 47, 1592-1595
Walls, D.F., Drummond, P.D., Hassan, S.S., and Carmichael, H.J., 1978, Prog. Theor. Phys. 64, 307-320
Walls, D.F. 1979, Nature 280, 451-454
Walls, D.F. 1983, Nature 306, 141-146
Walls, D.F. and Milburn, G.J. 1983, Quantum Optics, Experimental Gravity and Measurement Theory, Eds. P. Meystre and M.O. Scully, Plenum, New York, pp 209-247
Weidlich, W. and Haake, F. 1965a, Z. Phys. 185, 30-47
Weidlich, W. and Haake, F. 1965b, Z. Phys. 186, 203-221
Weyer, K.G., Wiedenmann, H., Rateike, M., MacGillivray, W.R., Meystre, P., and Walther, H., 1981, Opt. Commun. 37, 426-430
Wherrett, B.S. and Smith, S.D. 1984, Eds., Bistability, Dynamical Nonlinearity and Photonic Logic, The Royal Society, London
Wigner, E.P. 1932, Phys. Rev. 40, 749-759
Willis, C.R. 1977, Opt. Commun. 23, 151-154
Willis, C.R. 1978, Opt. Commun. 26, 62-65
Willis, C.R. and Day, J. 1979, Opt. Commun. 28, 137-141
Xiao, M., Kimble, H.J., and Carmichael, H.J., 1986, J. Opt. Soc. Am. A 2, p 87
Yuen, H.P. and Chan, V.W.S. 1983, Opt. Lett. 8, 177-179

QUANTUM THEORY OF OPTICAL BISTABILITY FOR SMALL SYSTEMS

J S SATCHELL AND SARBEN SARKAR

1. INTRODUCTION

Optical bistability (OB) is a subject which may well be coming of age. There have been and continue to be ingenious new schemes for OB but there is now a growing interest in having the closest possible correspondence between experiments and the simplest rigorous theories. This theoretical work, just as for the laser, has based itself on two level atoms. Several groups (Sandle and Gallagher, 1981; Weyer et al, 1981; Grant and Kimble, 1982; Rosenberger et al, 1983) have been active in producing OB experiments which approximate two level atom schemes. Semi-classical features such as the steady-state hysteresis cycle in absorptive bistability have been observed rather convincingly (Grant and Kimble, 1982; Rosenberger et al, 1983). It is a natural progression to wish to examine the detailed quantum statistical mechanical predictions of the theory. In experiments there are usually many sources of fluctuations other than spontaneous emission. However such fluctuations can be described by a classical stochastic process which implies certain inequalities for correlation functions. Under some circumstances fluctuations due to the spontaneous emission violate these inequalities. The phenomenon of antibunching (Kimble et al, 1977; Jakeman et al, 1977; Paul, 1982) is an example. In the theory of communications (Middleton, 1960) the intensity autocorrelation function $G^{(2)}(\tau)$ always satisfies

$$G^{(2)}(o) \geqslant G^{(2)}(\tau) \quad \text{for all } \tau \tag{1}$$

In anti-bunching this inequality does not hold since it is a quantum mechanical (ie non-classical) effect. Antibunching is possible in OB. The predicted effects are usually very small since the magnitude scales as the inverse of the saturation photon number n_o or, equivalently, as the inverse of the number of atoms, N. However, in principle, the effects can be big for sufficiently small numbers of atoms and high co-operativity. We will consider a ring cavity shown in Fig.1, and first a good cavity, which is perhaps experimentally simpler.

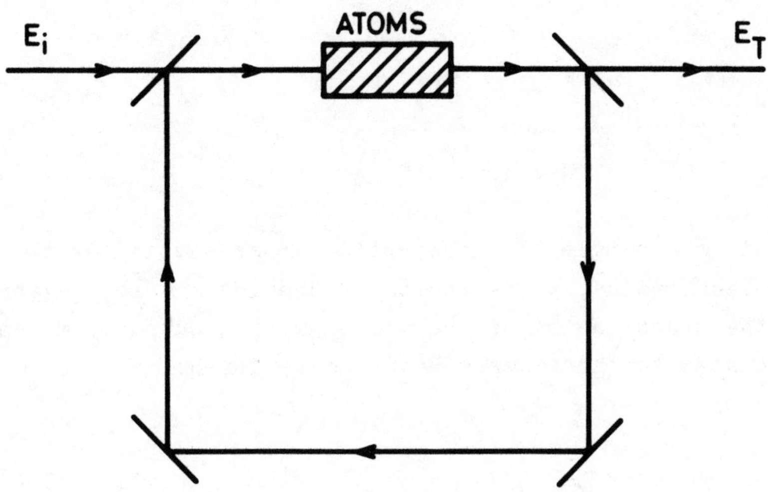

FIGURE 1. A ring cavity with incident electric field E_i and transmitted field E_T.

In the good cavity limit (Drummond and Walls, 1981) the linearised theory gives

$$G^{(2)}(\tau) - 1 = -\frac{1}{n_o}\frac{2C}{1+2C}\,e^{-\kappa(1+2C)\tau} \tag{2}$$

for very small values of the input field and pure radiative damping.

Now

$$C = \frac{\alpha L}{T} \tag{3}$$

and

$$\alpha = N\,\sigma/\pi r^2 \ell \tag{4}$$

Here σ is the single atom absorption cross-section, r the atomic beam radius, ℓ the length of the interaction region and T the transmission of the mirror at the output. Moreover κ is related to the finesse F of the cavity by

$$\kappa = \frac{C\pi}{\mathcal{L}F} \tag{5}$$

where $\mathcal{L}$ is the cavity round trip length. From Eqn. (2) it is clear that C has to be small in order that the effective time scale is not too short. However the prefactor (2C/1+2C) dictates that C should not be too small. For calcium and $\kappa = 5 \times 10^7$ sec^{-1} it is not unreasonable (theoretically) to have effects of the order of 10%. In general to design systems with small N we can see from Eqns. (3) and (4) that we require

i. large σ and hence to use resonance lines of the atoms and avoid broadening by using an atomic beam.

ii. small T and to utilise very high reflectivity mirrors. It is possible to obtain reflectivities of 99.99%.

iii. small r which is obtained through tightly focussed or very small cavities.

Currently experiments to measure antibunching are being designed both at Austin and Malvern.

The rigorous theories using two level atoms are based on master equations similar to those obtained for the laser (Haken, 1976). For a collection of homogeneously broadened atoms interacting on resonance with a single quantised cavity mode (in Fig.1) the electric field operator, at time t and a point x, is

$$\underline{E}(x,t) = i\,\underline{e}\left(\frac{\hbar\omega_o}{2\epsilon V}\right)^{\frac{1}{2}} [a(t)\,e^{ikx} - a^+(t)\,e^{-ikx}] \qquad (6)$$

where a^+ and a are photon creation and annihilation operators, $\underline{e}$ is a polarisation vector, V is the quantisation volume, ω the resonant frequency, $k = \omega/c$ (c being the speed of light), and ϵ is the vacuum permittivity. The atoms are described by Pauli spin matrices

$$\sigma_{\pm}^{(j)} \quad , \quad \sigma_3^{(j)} \qquad\qquad j = 1, \ldots, N$$

where

$$\begin{aligned} [\sigma_{\pm}^{(j)}, \sigma_3^{(k)}] &= \mp\, \sigma_{\pm}^{(j)}\, \delta_{jk} \\ [\sigma_+^{(j)}, \sigma_-^{(k)}] &= 2\, \sigma_3^{(j)}\, \delta_{jk} \end{aligned} \qquad (7)$$

Collective atomic motion is described by the operators

$$J_{\pm} = \sum_{j=1}^{N} e^{\pm ikx} \sigma_{\pm}(j)$$
$$J_3 = \sum_{j=1}^{N} \sigma_3(j) \qquad (8)$$

In most situations the driving field $\underline{E}_d$ is well described classically by

$$\underline{E}_d(\underline{x},t) = \underline{e}\,\beta\, e^{i(\omega t-k.x)} + c.c. \qquad (9)$$

where β is a field amplitude. The master equation for the density matrix ρ of atoms plus field, in a frame rotating at frequency ω, is given by

$$\begin{aligned}\dot{\rho} &= \alpha[a^+-a,\rho] + g[a^+J_+-aJ_-,\rho] \\ &+ \tfrac{1}{2}\gamma_\| \sum_{j=1}^{N} (2\sigma_-(j)\rho\sigma_+(j) - \sigma_+(j)\sigma_-(j)\rho-\rho\sigma_+(j)\sigma_-(j)) \\ &+ 2\gamma_o \sum_{j=1}^{N} (\sigma_3(j)\rho\sigma_3(j)-\rho) + \kappa(2a\rho a^+-a^+a\rho-\rho a^+a)\end{aligned} \qquad (10)$$

where the radiative lifetime is $\gamma_\|^{-1}$, γ_o is the atomic decay rate due to collisions and g is the atom field coupling constant ($g = (\omega\mu^2/2\hbar\varepsilon V)^{\frac{1}{2}}$) with μ the atomic dipole moment strength

$$\alpha = -i\,(2eV/\hbar\omega)^{\frac{1}{2}}\, e^{i\phi_T} \frac{\sqrt{TF}}{\pi}\,\beta \qquad (11)$$

ϕ_T is the phase change at the input mirror on transmission. We let $\rho_A(t)$ be the density matrix traced over field variables. If the single atom (or incoherent) atomic decay is small compared to the collective effects

$$\gamma_\| \,,\ (\tfrac{1}{2}\gamma_\|+\gamma_o) \ll gN^{\frac{1}{2}} \,,\ gN^{\frac{1}{2}}/\log N \qquad (12)$$

and of a photon once produced escapes from the cavity so rapidly that is does not interact with an atom so that

$$gN^{\frac{1}{2}}\ ,\ gN^{\frac{1}{2}}/\log N << \kappa \tag{13}$$

then $\rho_A(t)$ satisfies (Bonifacio et al, 1971; Carmichael, 1981)

$$\begin{aligned}\dot{\rho}_A(t) = -i\hat{\Omega}\ [J_+ + J_-, \rho_A] + \frac{g^2}{\kappa}\ (2J_-\rho_A J_+ - J_+J_-\rho_A - \rho_A J_+ J_-) \\ + \frac{\gamma_\parallel}{2} \sum_{j=1}^{N} (2\sigma_-^{(j)}\rho_A\sigma_+^{(j)} - \sigma_+^{(j)}\sigma_-^{(j)}\rho_A - \rho_A\sigma_+^{(j)}\sigma_-^{(j)}) \\ + \tfrac{1}{2}\ \gamma_0 \sum_{j=1}^{N} (2\sigma_3^{(j)}\rho_A\sigma_3^{(j)} - \tfrac{1}{2}\ \rho_A)\end{aligned} \tag{14}$$

and $\hat{\Omega} = g\alpha/\kappa$. The master equation (ME) can be written in terms of a c-number representation (Casagrande and Lugiato, 1980; Drummond and Walls, 1981; Carmichael et al, 1983) using a normally ordered characteristic function

$$\chi_N(j_1, j_2, j_3, \xi_1, \xi_2) = \mathrm{tr}(\rho\ e^{ij_2J_+}\ e^{ij_3J_3}\ e^{ij_1J_-}\ e^{i\xi_2 a^+}\ e^{i\xi_1 a}) \tag{15}$$

where j_1, j_2, j_3, ξ_1 and ξ_2 are independent complex variables. The quasi-distribution function $P(J_-, J_+, J_3, \alpha_1, \alpha_2)$ in the positive P representation (Drummond and Gardiner, 1980) is introduced via

$$\begin{aligned}\chi_N(j_1, j_2, j_3, \xi_1, \xi_2) = \int d^2j_1 \int d^2j_2 \int d^2j_3 \int d^2\xi_1 \int d^2\xi^2 \\ P(\tilde{J}_-, \tilde{J}_+, \tilde{J}_3, \alpha_1, \alpha_2) \\ e^{i\xi_1\alpha_1}\ e^{i\xi_2\alpha_2}\ e^{i\tilde{J}_- j_1}\ e^{i\tilde{J}_+ j_2}\ e^{i\tilde{J}_3 j_3}\end{aligned} \tag{16}$$

where $\hat{J}_-$, $\hat{J}_+$, $\hat{J}_3$, α_1 and α_2 are independent complex variables. P satisfies a generalised Fokker-Planck equation

$$\dot{P} = (L_A + L_F + L_{AF} + L_{F}\,\alpha)\,P \tag{17}$$

where

$$L_A = \gamma_\perp \left(\frac{\partial}{\partial \hat{J}_-}\hat{J}_- + \frac{\partial}{\partial \hat{J}_+}\hat{J}_+\right)$$

$$+ \gamma_\parallel \left(e^{\frac{\partial}{\partial \hat{J}_3}} - 1\right)\left(\frac{N}{2} + \hat{J}_3\right)$$

$$L_F = \kappa\left(\frac{\partial}{\partial \alpha_1}\alpha_1 + \frac{\partial}{\partial \alpha_2}\alpha_2\right)$$

$$L_{AF} = -g\left(\frac{\partial}{\partial \alpha_1}\hat{J}_- + \frac{\partial}{\partial \alpha_2}\hat{J}_+\right)$$

$$-g\left(\left(e^{\frac{\partial}{\partial \hat{J}_3}} - 1\right)\hat{J}_+ + 2\frac{\partial}{\partial \hat{J}_-}\hat{J}_3 + \frac{\partial^2}{\partial \hat{J}_-^2}\hat{J}_-\right)\alpha_1$$

$$-g\left(\left(e^{\frac{\partial}{\partial \hat{J}_3}} - 1\right)\hat{J}_- + 2\frac{\partial}{\partial \hat{J}_+}\hat{J}_3 + \frac{\partial^2}{\partial \hat{J}_+^2}\hat{J}_+\right)\alpha_2$$

$$L_{F\alpha} = -\alpha\left(\frac{\partial}{\partial \alpha_1} + \frac{\partial}{\partial \alpha_2}\right)$$

and we have taken $\gamma_o = 0$ for simplicity. It is not usually possible to solve Eqn. (17). In order to understand its nature we will consider a simpler example. If we examine an equation similar to (14) with detuning but without single atom decays viz

$$\dot{\rho}_A = -i\Omega\,[J_+ + J_-, \rho_A] + i\delta_o[J_3, \rho_A]$$

$$+ \gamma\,(2J_-\rho_A J_+ - J_+J_-\rho - \rho J_+J_-) \qquad (18)$$

then it is known that the steady state solution for N = 1 is given by

$$\rho_{SS} = (2+(1-\Delta^2)|\hat{h}|^{-2})$$

$$\cdot\,(1+(1+\Delta)\hat{h}^{-1}\,J_+ + (1-\Delta)\hat{h}^{*-1}\,J_- + (\hat{h}^*\hat{h})^{-1}(1-\Delta^2)J_-J_+) \qquad (19)$$

where

$$\hat{h} = \frac{i\hat{\Omega}}{\gamma}$$

and

$$\Delta = \frac{i\delta_o}{\gamma}$$

The related characteristic function $\chi(j_1, j_2, j_3)$ for N = 1 satisfies

$$\chi(j_1, j_2, j_3) = (2+(1-\Delta^2)|\hat{h}|^{-2})^{-1}$$

$$\left\{e^{ij_3/2} + e^{-ij_3/2}\,(i(1-\Delta)\hat{h}^{*-1}\,j_2 + 1 \right. \qquad (20)$$

$$\left. + (\hat{h}\hat{h}^*)^{-1}(1-\Delta^2) - j_1 j_2 + i(1+\Delta)\hat{h}^{-1} j_1\right\}$$

It is easy to deduce that the positive P function is

$$
\begin{aligned}
P(\tilde{J}_-,\tilde{J}_+,\tilde{J}_3) = {} & (2 + (1-\Delta^2)\ |\hat{h}|^{-2})^{-1} \\
& \Big\{ \delta^{(2)}(\tilde{J}_3-\tfrac{1}{2})\ \delta^{(2)}(\tilde{J}_-)\ \delta^{(2)}(\tilde{J}_+) \\
& - \delta^{(2)}(\tilde{J}_3+\tfrac{1}{2})((1-\Delta)\ \hat{h}^{*-1} \left(\frac{\partial}{\partial \tilde{J}_+}\ \delta(\tilde{J}_+)\right)\ \delta(\tilde{J}_-) \\
& + \frac{1+\Delta}{g}\ \delta(\tilde{J}_+)\ \frac{\partial}{\partial \tilde{J}_-}\ \delta(\tilde{J}_-) \\
& - (1 + (\hat{h}\hat{h}^*)^{-1}\ (1-\Delta^2))\ \delta^{(2)}(\tilde{J}^-)\ \delta^{(2)}(\tilde{J}^+) \\
& - \frac{\partial}{\partial \tilde{J}^-}\ \delta(\tilde{J}^-)\ \frac{\partial}{\partial \tilde{J}^+}\ \delta(\tilde{J}^+) \Big\}
\end{aligned}
\tag{21}
$$

It is the non-polynomial form of the dependence of χ on j_3 which gives rise to the δ-functions in the inversion. This feature disappears when the standard expansion to a second order polynomial in the j_3 is made; the usual Fokker-Planck expression is obtained thus. When more than one atom is considered, the rigorous distribution over the inversion variable has support on only the discrete set of states $J_3 = -\tfrac{1}{2}N$, $-\tfrac{1}{2}N+1$, ..., $\tfrac{1}{2}N-1$, $\tfrac{1}{2}N$. In effect the Fokker-Planck equation (FPE) approximates this distribution by its smooth envelope. Such a description would be expected to break down for small N. Further insight may be obtained from a linearised analysis of fluctuations around the steady state solutions of the drift in the bad cavity Fokker-Planck equation. Such a calculation has been done (Carmichael et al, 1983). It can be shown that

$$
\langle (J_+ - \langle J_+ \rangle)(J_- - \langle J_- \rangle) \rangle = \frac{1}{2N}\ \frac{X^2}{1+X^2}\ \frac{X^2}{\frac{Y}{X}\left(\frac{Y}{X} + X^2\left(2 - \frac{Y}{X}\right)\right)}
\tag{22}
$$

where

$$
Y = \frac{2\hat{\Omega}}{(\gamma_\parallel\ \gamma_\perp)^{\frac{1}{2}}}
$$

and

$$X = \frac{2}{(\gamma_\| \gamma_\perp)^{\frac{1}{2}}} \left(\hat{\Omega} - \frac{ig^2}{\kappa} <S_->\right)$$

Moreover

$$Y = X\left(1 + \frac{2C}{1+X^2}\right) \tag{23}$$

which is the standard relation between the input and intra-cavity fields. The ratio r of the expectation of the incoherent intensity over the coherent intensity is

$$r = \frac{4C^2}{N} \frac{X^2}{1+X^2} \frac{1}{\frac{Y}{X}\left(\frac{Y}{X} + X^2\left(2 - \frac{Y}{X}\right)\right)} \tag{24}$$

This simplifies for small X

$$r \sim \frac{1}{N} \left(\frac{2C}{1+2C}\right)^2 X^2 \tag{25}$$

and so r is small when N = 1 (if we assume the validity of the theory for N = 1). For $X^2 \gg 2C > 1$ we have

$$r \sim \frac{4C^2}{N} \frac{1}{X^2} \tag{26}$$

Moreover $r \to \infty$ for C = 4 as $X \to 3^{\frac{1}{2}}$ and $Y \to 3^{3/2}$. These calculations indicate that fluctuation effects are enhanced when X is large ($\gg (2C)^{\frac{1}{2}}$ and $C > \frac{1}{2}$) and we are near the critical point at C = 4. The linearised calculation is breaking down in a neighbourhood of the critical point.

It is of course possible to study the FPE using a different quasi-probability distribution such as the Wigner representation which gives symmetrically ordered expectation values (Lugiato, 1979). The diffusion matrix involved is however not always positive definite. In the region where a stochastic differential equation formulation of the FPE is allowed, exact agreement, even in principle, should not be expected between the positive P and Wigner approaches since the detailed nature of the truncation to obtain the FPEs are different in the two cases. Preliminary calculations indicate that at the critical point ($C = 4$ and $Y = 3\sqrt{3}$) there may be differences in results using the two FPEs.

Apart from OB, special cases of Eqn. (14) are of interest in the study of co-operative effects due to the coupling of two-level atoms to a common radiation field. In particular we can have co-operative resonance fluorescence where the N two-level atoms are confined to a volume less than a resonant wavelength cubed and coherently driven on resonance. However we shall here be explicitly only considering applications to OB. We will examine the small N limit of the ME (which is valid for arbitrary N) and also attempt to determine the smallest value of N for which the Fokker-Planck equation is a good description. The value of N depends on the process under consideration, and, where linearised analysis breaks down in principle, qualitative differences between the ME and (nonlinear) FPE analysis is found.

2. THE MASTER EQUATION

The master equation in (14) implies that

$$\frac{d}{dt}\langle J_-\rangle = 2i\hat{\Omega}\langle J_3\rangle + \frac{2g^2}{\kappa}\langle J_3 J_-\rangle - \gamma_\perp\langle J_-\rangle \tag{27}$$

$$\frac{d}{dt}\langle J_3\rangle = i\hat{\Omega}\left(\langle J_-\rangle - \langle J_+\rangle\right) - \frac{2g^2}{\kappa}\langle J_+ J_-\rangle - \gamma_\parallel\left(\frac{N}{2} + \langle J_3\rangle\right) \tag{28}$$

Clearly the equations are not closed. Similarly we have

$$\frac{d}{dt}\langle\sigma_-^{(j)}\rangle = 2i\hat{\Omega}\langle\sigma_3^{(j)}\rangle + \frac{2g^2}{\kappa}\langle\sigma_3^{(j)}S_-\rangle - \gamma_\perp\langle\sigma_-^{(j)}\rangle \tag{29}$$

$$\frac{d}{dt}\langle\sigma_3^{(j)}\rangle = i\hat{\Omega}(\langle\sigma_-^{(j)}\rangle - \langle\sigma_+^{(j)}\rangle) - \frac{g^2}{\kappa}(\langle\sigma_+^{(j)}S_-\rangle + \langle S_+\sigma_-^{(j)}\rangle) - \gamma_\parallel(\tfrac{1}{2} + \langle\sigma_3^{(j)}\rangle) \tag{30}$$

Existing treatments, if they wish to proceed with these equations, assume the scaling of collective fluctuations with respect to N, and make suitable decorrelations[13]. In Reference 12 only unlike atom correlation functions were decorrelated. Such an approach is in the spirit of the Fokker-Planck equation and is unlikely to be satisfactory for N small. Hence we will proceed differently, and will check the validity of the decorrelation approximation a posteriori. Our approach will be exact but, naturally, will only be practicable for a restricted range of N values. Once a basis for states has been chosen the density matrix equation yields a coupled set of ordinary linear differential equations. For two-level atoms an obvious basis is a direct product of elementary spinors $|i_j\rangle$

$$|\underline{i}\rangle = \bigotimes_{j=1}^{N} |i_j\rangle \tag{31}$$

where $i_j = 1$, or -1, ie the jth atom is in the excited or ground state. Unfortunately the number of such states scales as 2^N. This in turn implies that the density matrix has 2^{2N} elements. Even for the largest computers available it would be currently difficult to treat $N > 6$. In order to consider questions such as the validity of the Fokker-Planck equation or decorrelations, it is clearly completely inadequate to calculate only for such N. For collective resonance fluorescence, however, Dicke state bases are possible but these are not useful when there are incoherent decay processes. Fortunately Eqn. (14) has permutation symmetry between the atoms, ie all atoms are treated in the same way. This allows us to devise a canonical form for a density matrix element.

A general density matrix element $<\underline{i}|\rho|\underline{j}>$ ($\equiv \rho_{\underline{i}\underline{j}}$) can be written as

$$\rho_{\underline{i}\underline{j}} = \rho(n;\underline{\tilde{j}}) \tag{32}$$

where there is a permutation σ such that

$$\sigma(\underline{i}) = \underline{\tilde{i}}$$

and $\tilde{i}_j = 1$ if $j \leqslant n$

$= -1$ otherwise

and $\underline{\tilde{j}} = \sigma(\underline{j})$

We can standardise the form of $\rho_{\underline{i}\underline{j}}$ further by applying a permutation π such that

$$\pi(\underline{\tilde{i}}) = \underline{\tilde{i}}$$

and $\pi(\underline{\tilde{j}}) = \underline{j}'$

where $j'_\ell = 1$ if $\ell \leqslant m$

$= -1$ if $m < \ell \leqslant n$

$= 1$ if $n < \ell \leqslant n+m'$

$= -1$ if $n+m' < \ell \leqslant N$

Finally we will write

$$\rho(n;m,m') = \rho_{\sigma\pi(\underline{i})\ \sigma\pi(\underline{j})} \tag{33}$$

This is our canonical density matrix element. In terms of these elements

$$
\begin{aligned}
\dot{\rho}(n;mm') = & -i\hat{\Omega}((n-m)\ \rho(n-1;m,m') + m\ \rho(n-1;m-1,m'+1) \\
& + (N-n-m')\ \rho(n+1;m,m') + m'\ \rho(n+1;m+1,m'-1) \\
& - m\ \rho(n;m-1,m') - m'\ \rho(n;m,m'-1) \\
& - (n-m)\ \rho(n;m+1,m') - (N-n-m')\ \rho(n;m,m'+1)) \\
& + \gamma((N-n-m')\ \rho(n+1;m+1,m') \\
& - \tfrac{1}{2}(n+m+m')\ \rho(n;m,m')) \\
& + \frac{g^2}{\kappa}\ (2((N-n-m')\ \rho(n+1;m+1,m') \\
& + m'(n-m)\ \rho(n+1;m+2,m'-1) \\
& + (N-n-m')(n-m+m')\ \rho(n+1;m+1,m') \\
& + (N-n-m')(N-n-m'-1)\ \rho(n+1;m,m'+1)) \\
& - (m'm + (N-n-m')(n-m) + m(n-m) + m'(N-n-m') \\
& + n+m+m')\ \rho(n;m,m') \\
& - 2m(N-n-m')\ \rho(n;m-1,m'+1) \\
& - 2m'(n-m)\ \rho(n;m+1,m'-1)) \qquad (34)
\end{aligned}
$$

Given an n there are (n+1)(N-n+1) different pairs (m,m') that are possible. Consequently the total number of states in the (n;m,m') basis for the density matrix elements is

$$\sum_{n=0}^{N} (n+1)(N-n+1) = \frac{1}{6}(N+1)(N+2)(N+3) \qquad (35)$$

This should be contrasted with 2^{2N} for the number of density matrix elements required if the basis is used naively.

If we wish to treat the master equation when the field has not been adiabatically eliminated, then the above basis has to be supplemented with extra labels corresponding to the photon number states. The matrix element

$$\rho^{(n;mm')}_{\overline{n}'\overline{n}} = \langle\overline{n}'|\langle\sigma\pi(\underline{i})|\ \rho|\sigma\pi(\underline{j})\rangle|\overline{n}\rangle \tag{36}$$

and $|\overline{n}\rangle$, $|\overline{n}'\rangle$ are photon number states. However, here, $\overline{n}$ and $\overline{n}'$ no longer have an upper-bound, and so the basis is not finite. In many situations a truncation of the photon number states $\overline{n}$ and $\overline{n}'$ above a certain integer is justified; if the saturation photon number is about .5, then it is adequate to truncate the basis by restricting $\overline{n}$ and $\overline{n}'$ to be less than 5. This reduces the number of atoms that we can usefully treat by approximately three. Hence if we leave the bad cavity situation, we are limited to results for atom numbers of up to three (with our computers).

In order to compare our ME solutions with those derived from FPEs, it is useful to express the latter in terms of an equivalent set of Ito stochastic differential equations. Given a multivariable FPE

$$\frac{\partial p(\underline{x},t)}{\partial t} = -\sum_i \frac{\partial}{\partial x_i}[A_i(\underline{x},t)\ p(\underline{x},t)]$$

$$+\tfrac{1}{2}\sum_{i,j,\ell}\frac{\partial^2}{\partial x_i \partial x_j}(B_{i\ell}(\underline{x},t)\ B_{j\ell}(\underline{x},t)\ p(\underline{x},t)) \tag{37}$$

we can associate[14] the stochastic differential equations

$$dx_i = A_i(\underline{x},t)dt + B_{ij}(\underline{x},t)\ dW_j(t) \tag{38}$$

where $dW_j(t)$ are independent Wiener processes.

For the bad cavity FPE the appropriate set of Ito stochastic differential equations is

$$dz_- = (-\tfrac{1}{2}z_- + 2Cz_-z_3 + \tfrac{1}{2}Yz_3)\,dt + \left(z_-\left(\frac{2C}{N}z_- + \frac{Y}{2N}\right)\right)^{\frac{1}{2}} dW_1 \tag{39}$$

$$dz_+ = (-\tfrac{1}{2}z_+ + 2Cz_+z_3 + \tfrac{1}{2}Yz_3)\,dt + \left(z_+\left(\frac{2C}{N}z_+ + \frac{Y}{2N}\right)\right)^{\frac{1}{2}} dW_2 \tag{40}$$

$$dz_3 = -(\tfrac{1}{2} + z_3 + 4Cz_-z_+ + \frac{Y}{2}(z_-+z_+))\,dt + \frac{1}{N^{\frac{1}{2}}}(\tfrac{1}{2} + z_3 - 4Cz_+z_- - \frac{Y}{2}(z_+ + z_-))^{\frac{1}{2}} dW_3 \tag{41}$$

Here

$$z_\pm = \frac{\tilde{J}_\pm}{2^{\frac{1}{2}}N}, \quad z_3 = \frac{\tilde{J}_3}{N} \quad \text{and} \quad Y = \frac{2^{3/2}\hat{\Omega}}{\gamma}.$$

The solution of these equations can be obtained as a Taylor series (Satchell and Sarkar, 1986). The simplest algorithm may be derived by retaining terms which contribute to the means and variances to order t. The analogue of this for ordinary differential equations is the forward Euler method. Such a numerical simulation of an ensemble of trajectories allows us to estimate the moments of observable quantities in a fully nonlinear manner. The approximations inherent in the FPE reveal themselves as trajectories that leave the physical region. In particular if the inversion variable for a particular member of the ensemble leaves the physical region $(-\frac{1}{2},\frac{1}{2})$, the trajectory has a tendency to run off to infinity.

This is perhaps not surprising since the FPE is not physically sensible outside this region. Different strategies can be adopted to deal with such trajectories. It is possible to adopt boundary conditions so that trajectories on crossing the boundary of $[-\frac{1}{2},\frac{1}{2}]$ are bounced back into the physical region in some fashion. Alternatively, as we have done, we can reject such trajectories entirely. For very small systems such a procedure leads to an unacceptably high rejection rate of trajectories, and a correct treatment requires the use of the master equation.

The differential equations (34) can be solved for as many as 45 atoms (on the computer available to us). Once we have the steady state values for $\rho(n;m,m')$ we can evaluate the steady state moments of observables. For sufficiently small N ($N \leqslant 12$) we can use a different technique. The master equation can be written as

$$\dot{\rho}(n;m,m') = \sum_{u,v,v'} L(n;m,m'|u;v,v')\,\rho(u;v,v') \tag{42}$$

where the 'super operator' $L(n;m,m'|u;v,v')$ is the Liouvillian. L is a matrix and can be diagonalised directly for $N \leqslant 12$.

There are theorems (Spohn, 1980) which state that equations such as (14) have a unique steady state solution, but we also know that the semi-classical state equation (23) shows hysteresis for $C > 4$. In fact the 'bistable' states are actually metastable and fluctuations induce transitions between states. This is a very slow process for large systems (Gardiner, 1983) which is typical of tunnelling. The theory of first passage times used in these discussions is not appropriate to master equations in the basis that we are using. We wish to suggest that a necesary condition for the existence of bistability in a quantum mechanical system is the existence of a relaxation time that is much longer than any of the natural timescales in the problem (Sarkar and Satchell, 1986). This is not a sufficient condition. The semiclassical analysis indicates that we find long timescales for $C = 4-\varepsilon$ (where ε is small) due to critical slowing down. When a potential picture is valid, critical slowing down is implied by the extreme flatness of the potential for C values near 4 and y near $3^{3/2}$. For fixed C, critical slowing down is independent of N, whereas the quantum mechanical transition rate has a strong dependence of N. $G^{(2)}(o)$ is also useful for distinguishing between the possibilities. Fluctuations and, hence, $G^{(2)}(o)$ are much greater when there is genuine bistability.

The smallest eigenvalue of the Liouvillian gives us a simple measure for the existence of bistability. From the linearised analysis, for C values for which bistability exists, we expect to get largest effects of fluctuations for large X and C. We find that there is a large C limit where the eigenvalue is independent of C for a given N. The slowest timescale is found for the driving field which is linearly related to C. For two atoms the eigenvalue is greater than $\gamma_{\|}$, and so bistability is not possible. With three atoms we find that the smallest eigenvalue has a minimum of $.83\ \gamma_{\|}$. Again it is not possible to argue for bistability. Figure 2 shows the results for N up to 12. We can see that the rate of quantum mechanical transitions is dropping with increasing N. The value of $\Omega\kappa/g^2$ for the optimum effect is also increasing with N. Moreover $G^{(2)}(o)$ for these cases is very large. For eight atoms with C = 20 and Y = 14.14 $G^{(2)}(o)$ is 428. This behaviour is consistent with bistability. For 12 atoms we see that timescales are sufficiently well separated so that distinct metastable states may be considered to exist. For systems with more than twelve atoms we can, in principle, use a different technique. It is possible, using the quantum regression theorem (Lax, 1967), to examine the two time intensity correlation function $G^{(2)}(t)$ for large t, and thus determine the largest timescale.

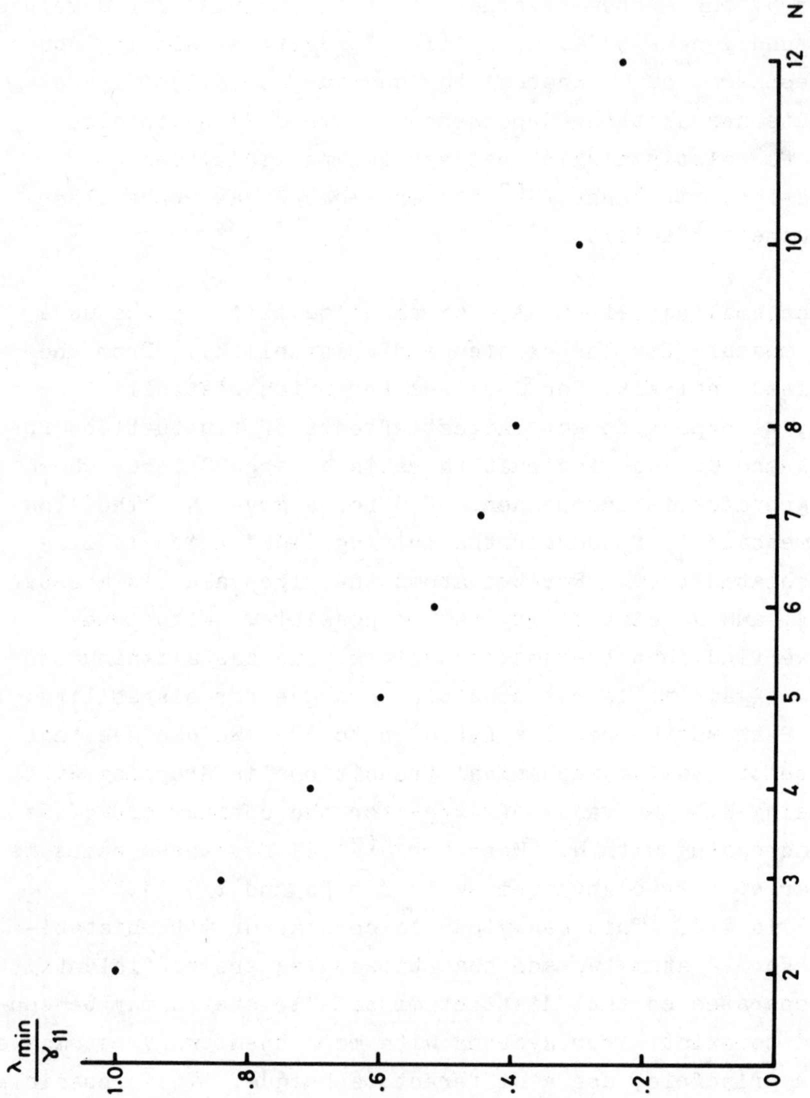

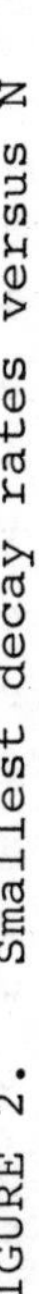

FIGURE 2. Smallest decay rates versus N

From an experimental point of view the good cavity is marginally of more interest than the bad cavity. The general ME in (10) requires the introduction of photon states as in Eqn. (36). If the saturation photon number is small, a truncation of the number of photon states up to an occupation number $\hat{n}$ may be valid. The cases N = 2 and $\hat{n}$ = 6, and N = 3 and $\hat{n}$ = 5, with $\kappa/\gamma_\parallel$ = .1 have been studied. There is no evidence for bistability. The smallest eigenvalue is rather larger than κ for values of the driving field for which bistability may be expected. Even from this limited calculation we see that bistability is harder to observe in the good cavity case.

The potential difficulties for the FPE have been discussed in the introduction. We now need to examine how real these difficulties are and in what circumstances they are present. At the critical point (C=4 and Y = $3\sqrt{3}$) fluctuations are so large that the linearised theory breaks down completely, and so this is a particularly apt place to compare the ME with the FPE. We will be interested in field statistics, but, in the bad cavity limit, the field has been adiabatically eliminated and so we need to express the moments of the field in terms of the atomic operators. From Bonifacio et al (1971) we know that

$$\langle (a^+)^n \, a^m \rangle \;=\; \langle \left(\frac{i\hat{\Omega}}{g} - \frac{g}{\kappa} J^+\right)^n \left(-\frac{i\hat{\Omega}}{g} - \frac{g}{\kappa} J^-\right)^m \rangle \qquad (43)$$

The field statistics of particular interest for us will be $g^{(2)}(o)$ and $\langle a^+a \rangle$. For $g^{(2)}(o)$ the master and Fokker-Planck equations agree well as can be seen in Fig.3a. However from this it is not possible to conclude that all quantities calculated for this same choice of parameters will agree to the same extent. A calculation of $\langle a^+a \rangle$ is shown in Fig.3b. The horizontal line is the deterministic value. The ME and FPE have qualitatively different trends. The fact that $\langle a^+a \rangle$ from the FPE is dropping below 3 does not indicate non-classical statistics since $\langle a \rangle \langle a^+ \rangle$ also falls as N decreases and

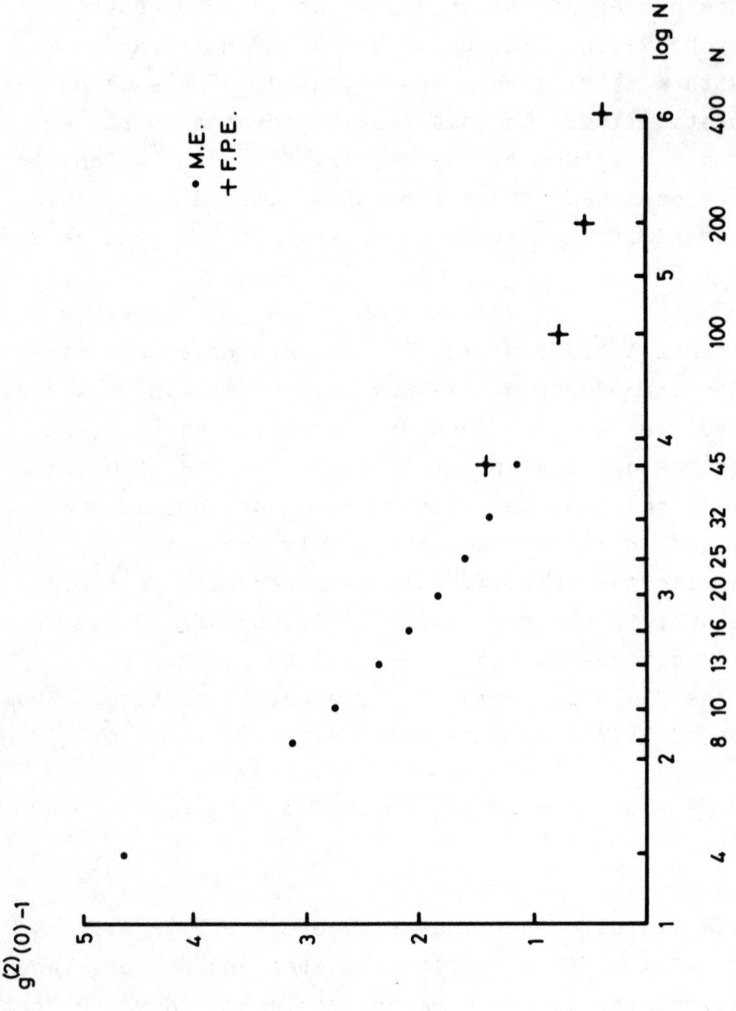

FIGURE 3a. Behaviour of $g^{(2)}(o)$ at the critical point

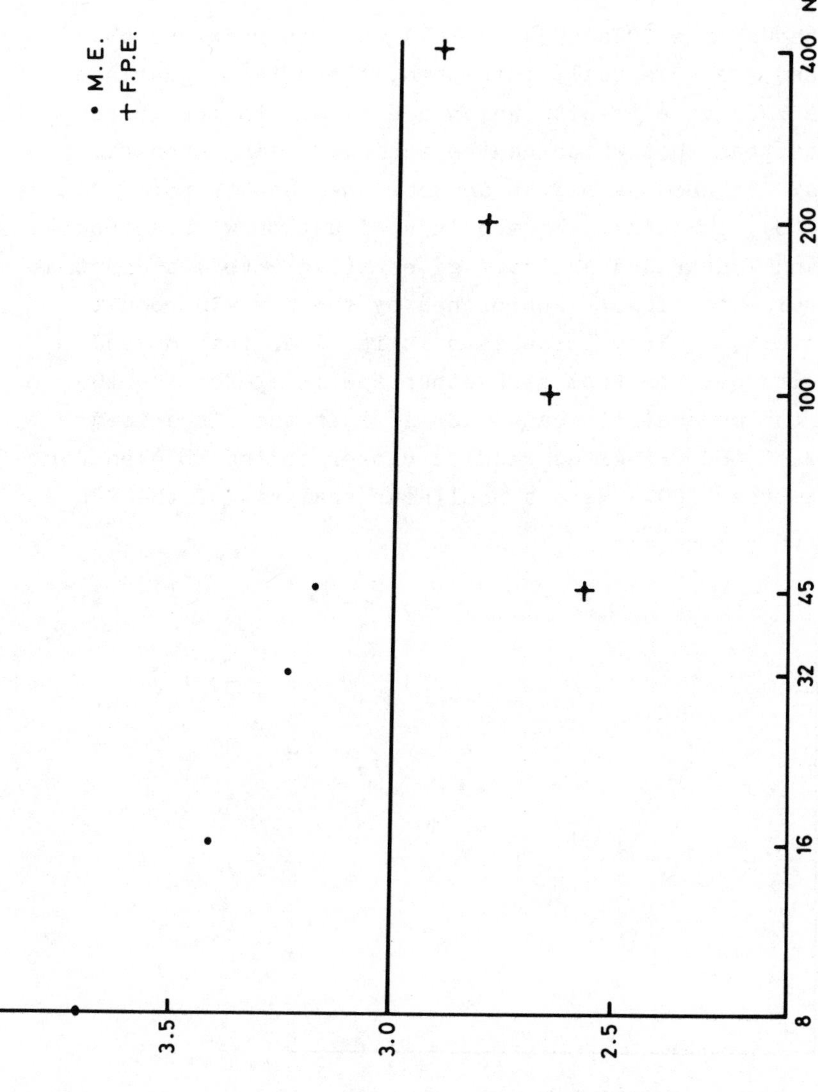

Figure 3b. Expectation of a scaled intensity at the critical point

$(\langle a^{+}\rangle - \langle a\rangle\langle a^{+}\rangle)$ is positive. Both the ME and FPE show this drop in $|\langle a\rangle|$ and this is shown in Fig. 3c. It is difficult to make general statements concerning the validity of the FPE without reference to the actual expectation values being calculated.

Although the SDEs (39) to (41) have corrections to linearised analysis built into them, there is no guarantee that the SDEs give results which are closer to the correct (ME) ones than that which can be achieved using linearised analysis. Figures 4a and 4b do show the results for $N(1-g^{(2)}(o))$ and $\langle a^{+}a\rangle$. We are in a situation with antibunching, and linearised analysis gives $N(1-g^{(2)}(o))$ a constant (whose value is closely approached by the N = 400 result from the FPEs). From inspection it is clear that ME and FPE results deviate from each other specially for N = 100. However the more significant fact is that the linearised analysis of the FPE gives results closer to the ME even for N smaller than 100. Here a nonlinear analysis of the FPE has not been fruitful.

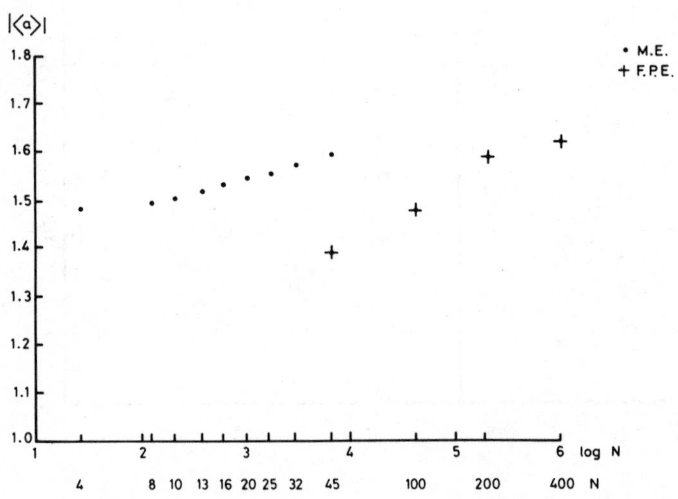

Figure 3c. Deviation of $|\langle a\rangle|$ from the deterministic value

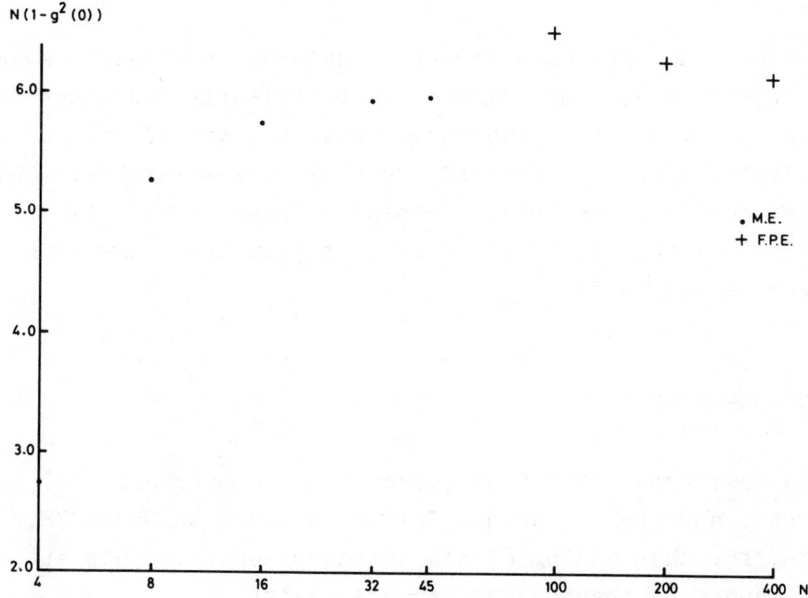

FIGURE 4a. $N(1-g^{(2)}(o))$ versus N for C = 2 and y = $\sqrt{2}$

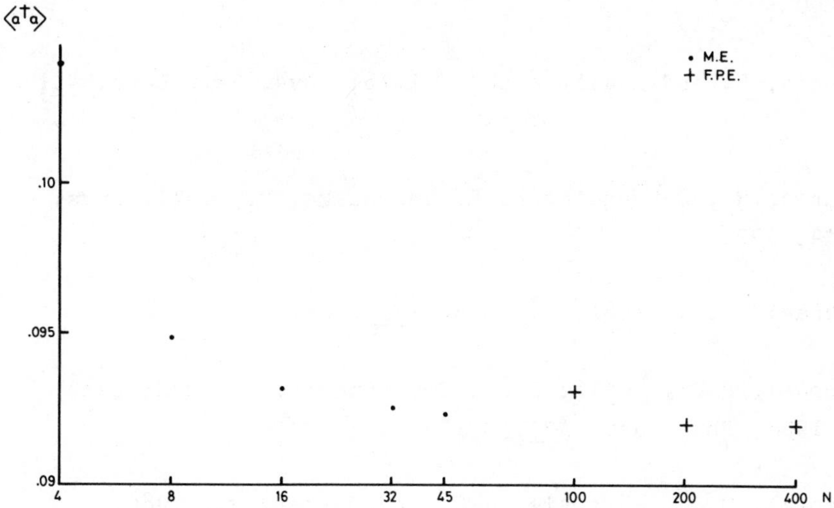

FIGURE 4b. $\langle a^{+}a\rangle$ versus N for C = 2 and y = $\sqrt{2}$

CONCLUSIONS

The straightforward direct solution of the master equation, when permutation symmetry is efficiently implemented, allows the study of bistability, metastability of states and the validity of the associated Fokker-Planck solutions. Our methods provide a non-trivial class of "exactly soluble" models in quantum optics which are, in principle, amenable to experimental test.

ACKNOWLEDGEMENTS

We have benefitted from discussions with Dr H J Carmichael and Prof H Haken. One of us (JSS) thanks RSRE, Malvern, for the funding of his Research Assistantship at the Clarendon Laboratory, Oxford University.

REFERENCES

Bonifacio, R. and Lugiato, L.A., 1978, Phys. Rev. Lett. 40, 1023.

Bonifacio, R., Schwendimann, P. and Haake, F., 1971, Phys. Rev. A4, 302.

Carmichael, H.J., 1981, Z. Phys. B42, 183.

Carmichael, H.J., Walls, D.F., Drummond, P.D. and Hassan, S.S., 1983, Phys. Rev. A27, 3112.

Carmichael, H.J., Satchell, J.S. and Sarkar, S., 1986, Nonlinear Analysis of Quantum Fluctuations in Absorptive Bistability.

Drummond, P.D. and Gardiner, C.W., 1980, J. Phys. A13, 2353.

Drummond, P.D. and Walls, D.F., 1981, Phys. Rev. A23, 2563.

Gardiner, C.W., 1983, Handbook of Stochastic Methods, Springer-Verlag.

Grant, D.E. and Kimble, H.J., 1982, Optics Lett. 7, 353.

Haken, H., 1970, Laser Theory, Encycl. of Phys. XXV/2C, Springer-Verlag, Berlin.

Kimble, H.J., Dagenais, M. and Mandel, L., 1977, Phys. Rev. Lett. 39, 691.

Lax, M., 1967, Phys. Rev. 157, 213.

Lugiato, L.A., 1979, Nuovo Cim. B50, 89.

Meystre, P. and Walther, H., 1981, Opt. Comm. 37, 426.

Middleton, D., 1960, An Introduction to Statistical Communication Theory, McGraw-Hill, New York.

Narducci, L.M., Gilmore, R., Feng, D.H. and Agarwal, G.S., 1978, Opt. Lett. 2, 88.

Paul, H., 1982, Rev. Mod. Phys. 54, 1061.

Puri, R.R. and Lawande, S.V., 1979, Phys. Lett. A72, 200.

Rosenberger, A.T., Orozco, L.A. and Kimble, H.J., 1983, Phys. Rev. A28, 2569.

Sandle, W.J. and Gallagher, A., 1981, Phys. Rev. A24, 2017.

Sarkar, S. and Satchell, J.S., 1986, Optical Bistability with Small Numbers of Atoms, RSRE, Malvern, Preprint.

Satchell, J.S. and Sarkar, S., 1986, Quantisation of Limit Cycles in the P-representation of a Dissipative Driven Anharmonic Oscillator, J. Phys. A (to be published).

Spohn, H., 1980, Rev. Mod. Phys. 52, 569.

Weyer, K.G., Widenmann, H., Rateike, M., MacGillivray, W.R., Meystre, P. and Walther, H., 1981, Opt. Comm. 37, 426.

NOISE EFFECTS IN OPTICAL BISTABILITY

L A LUGIATO, G BROGGI AND A COLOMBO

1. INTRODUCTION

The macroscopic description of the dynamical behaviour in nonlinear, dissipative systems admits two distinct levels, deterministic and stochastic. The latter description takes into account the sources of noise that can affect the evolution of the system.

Let us consider first the systems which, at deterministic level, admit a single stable attractor. In the stochastic description the standard picture is that the probability distribution of the intensive macroscopic observables of the system is a single peak with a mean value which follows the deterministic motion, and a width inversely proportional to the square root of the system size (Van Kampen 1961).

The picture is more complicated when the system, at deterministic level, has more than one stable attractor in its phase space; for the sake of definiteness let us assume that there are two attractors, which corresponds to a bistable situation. In this case the steady-state probability distribution has usually two peaks which, for a low noise level, are centered near the stable stationary solutions of the deterministic theory. This configuration of the probability distribution is called bimodal. Clearly, the probability distribution becomes bimodal also in the transient approach towards the steady-state configuration.

In the stochastic description of dissipative systems, the most interesting situations are those in which anomalous fluctuations lead to deviate from this general picture. A whole class of situations of this type is associated with the rise of instabilities, which lead to qualitative transitions in the behaviour of the system (Haken 1977). When the control parameter is near the critical value which corresponds to the onset of the instabilities, the width of the stationary probability distribution does no longer scale as the inverse square root of the system size. Another example of deviation from the general picture given above are the so-called noise-induced transitions (Horsthemke and Lefever 1983) in which, as a consequence of multiplicative noise, the number of peaks in the steady-state distribution is different from the number of attractors in the deterministic picture.

The study of the role of noise in quantum optical systems has attracted a continuous interest since the very early days of Laser Physics (Glauber 1963 *a* and *b*; Haken 1970; Sargent, Scully, Lamb 1974; Arecchi, Degiorgio and Querzola 1967, Davidson and Mandel 1967; Pike 1969). The interest focussed mainly on quantum mechanical fluctuations, however it must be stressed that in laser-type systems classical noise is also interesting, because most often it overshadows quantum noise, and can give rise to effects quite similar to those which are caused by quantum noise. The only exception are, of course, nonclassical effects as photon antibunching, sub-Poissonian photon statistics and squeezing.

In this paper we illustrate two phenomena in which the probability distribution, which describes the statistical behaviour of a bistable optical system, exhibits a strongly anomalous behaviour.

1) If the input intensity in a bistable system is switched on stepwise to a value slightly beyond the bistable region, the deterministic time evolution exhibits critical slowing down behaviour (Bonifacio and Lugiato 1978*a*, Bonifacio and Meystre 1978; Benza and Lugiato 1979). In the stochastic description, the probability distribution becomes two-peaked during the transient approach towards the single attractor. This phenomenon, called transient bimodality, arises also in a variety of physical, chemical, biological systems, whenever the deterministic time evolution presents a slow lethargic stage followed by a fast explosive stage (Baras *et al.* 1983, Frankowicz and Nicolis 1983, Frankowicz *et al.* 1984). In the case of optical bistability, it can originate both from quantum and from classical noise. We compare our theoretical predictions (Broggi and Lugiato 1984 *a* and *b*, Broggi *et al.* 1985) with the results of the experiments performed at the Univerity of Hannover (Mitscke *et al.* 1985, Lange *et al.* 1985).

2) We show that a special type of multiplicative noise can destroy completely the expected two-peaked character of the steady-state distribution in a bistable system (Lugiato *et al.* 1985). This situation arises for istance in dispersive optical bistability, when the fluctuations in the frequency of the incident field become dominant over the other sources of noise. An important feature of this phenomenon is that here the influence of multiplicative noise leads to dramatic qualitative effects even for low noise

levels. This is in contrast with the case of noise induced transitions, which arise only in presence of a high noise level, *i.e.* when the stochastic term in the dynamical equation becomes comparable with the deterministic term.

2. THE MEAN FIELD MODEL OF OPTICAL BISTABILITY

Let us first remind what is optical bistability (Lugiato 1984). The coherent beam of a laser, operating in a stationary regime, is injected into an optical cavity resonant or nearly resonant with the frequency of the incident light. The incident beam is in part trasmitted and in part reflected by the cavity. When this is empty, the trasmitted intensity I_T is simply proportional to the incident intensity I_I. On the other hand, if the cavity is filled with material resonant or nearly resonant with the frequency of the input field (Fig. 1*a*), the trasmitted intensity becomes in general a nonlinear function of the incident intensity. By suitably varying the external parameters of the system, one obtains that the steady-state curve of I_T as a function of I_I becomes S-shaped (Fig. 1*b*). The segment with negative slope is unstable, hence one obtains a hysteresis cycle with two distinct states of trasmission. This phenomenon, called optical bistability (OB), is attracting noteworthy attenction in view of the perspective of realizing practical optical memories, transistors, and eventually optical computers (Abraham and Smith 1982, Gibbs 1985).

Clearly, the role of noise is especially important in the case of OB, because it can induce spontaneous random jumps between the two branches of the hysteresis cycle. In other words, the two branches are not absolutely stable but metastable. Of course, these jumps are an undesired effect if we want to use these systems as optical memories. There are essentially three sources of noise: *a*) external noise, *e.g.* fluctuations in the amplitude and in the frequency of the incident field, *b*) thermal noise and *c*) quantum noise, which roughly speaking arises via spontaneous emission.

We will consider one of the simplest models of optical bistability, the so called "*mean field model*" (Bonifacio and Lugiato 1976 and 1978*b*, Hassan *et al.* 1978). Here, the words "*mean field*" indicate that the spatial variation of the electric field along the sample is neglected. This model gives an adeguate description of the dynamics of the system when this operates in a single mode regime. The material is schematized as a collection of identical two-level atoms with transition frquency ω_a, interacting with a single cavity mode with frequency ω_c.

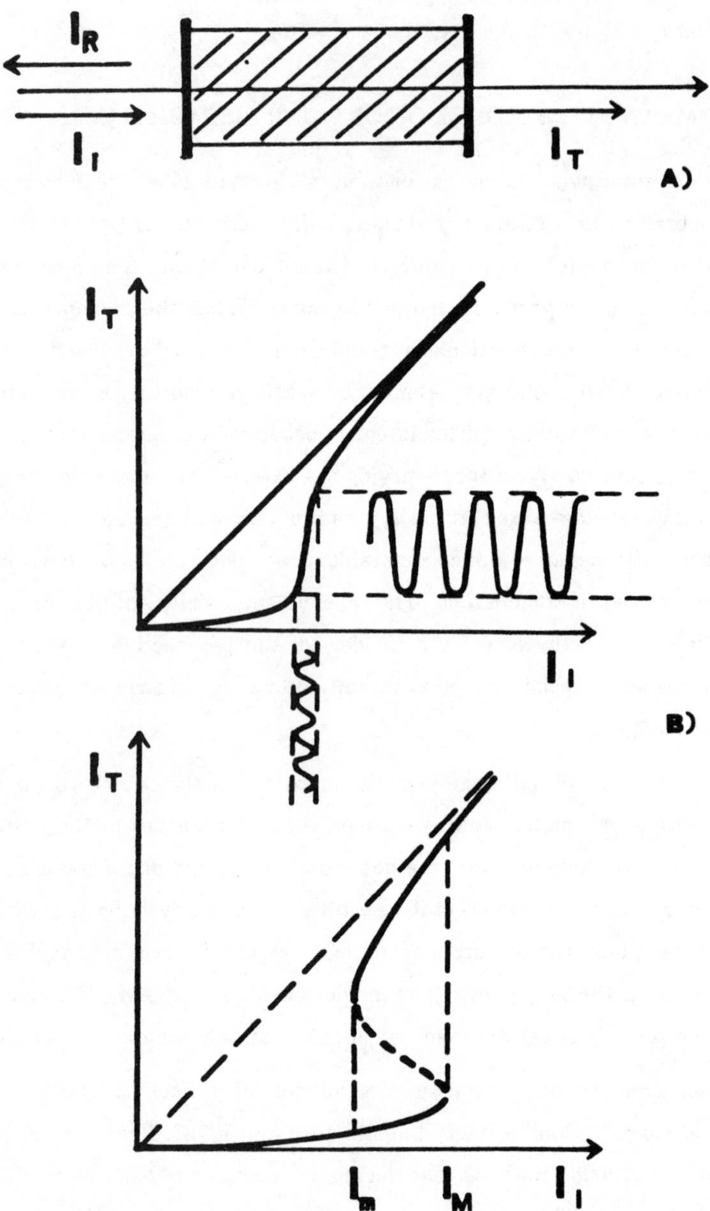

Fig. 1: *a*) I_I, I_T and I_R are respectively the incident, trasmitted and reflected intensity. *b*) Steady-state hysteresis cycle of trasmitted *vs.* incident intensity. The part of the curve with negative slope is unstable

The dynamical equations read

$$\frac{dx}{dt} = -\kappa\left[x(1+i\theta) - y + 2CP\right], \tag{1.1}$$

$$\frac{dP}{dt} = \gamma_\perp\left[xD - (1+i\Delta)P\right], \tag{1.2}$$

$$\frac{dD}{dt} = -\gamma_\parallel\left[\frac{1}{2}(xP^* + x^*P) + D - 1\right], \tag{1.3}$$

where x is the normalized slowly-varying envelope of the electric field inside the cavity, y is the normalized amplitude of the incident field, which is taken real and positive for definiteness. The atomic variables P and D are respectively the normalized dielectric polarization of the medium, and the normalized difference between the population of the lower and the upper state of the two-level atoms. The field damping rate κ arises from the fact that the mirrors are semitrasparent and therefore the photons escape from the cavity, and is defined by

$$\kappa = cT/2L, \tag{2}$$

where T is the trasmissivity coefficient of the mirrors, and L is the cavity length. The atomic damping rates $\gamma_\perp$ and $\gamma_\parallel$ arise from spontaneous emission and from collisions between atoms, and the atomic system is assumed to be homogeneously broadened. The cavity detuning parameter θ and the atomic detuning parameter Δ are defined respectively as follows:

$$\theta = \frac{\omega_c - \omega_o}{\kappa} \quad , \quad \Delta = \frac{\omega_a - \omega_o}{\gamma_\perp}, \tag{3}$$

where ω_o is the frequency of the incident field. Last but not least, the bistability parameter C is given by

$$C = \frac{\alpha L}{T}, \tag{4}$$

where α is the absorption coefficient of the material per unit length.

In the following we shall analyze the set of equations (1) in various limits, with the system subjected to different sources of noise.

3. PURELY ABSORPTIVE CASE. GOOD CAVITY LIMIT. AMPLITUDE NOISE.

We consider first the case of exact resonance between the incident field, the cavity and the atoms, i.e. $\theta = \Delta = 0$. In this situation, in which the behaviour of the system is determined solely by absorption, one has <u>purely absorptive bistability</u>.

Furthermore, we assume that the field damping constant κ is the slowest rate, *i.e.*

$$\kappa \ll \gamma_\perp, \gamma_\| \tag{5}$$

This is usually called "*good cavity limit*" because the cavity linewidth κ is much smaller than the atomic linewidth $\gamma_\perp$. The assumption (5) allows us to eliminate adiabatically the atomic variables (Haken 1977, Lugiato *et al.* 1984). By setting $dP/dt = dD/dt = 0$ in Eqs. (1.2) and (1.3), one obtains the expression of P and D as a function of the field variable x, which in the good cavity limit governs the dynamic of the system. On substituting the expression of P in Eq. (1.1), we obtain the following self-contained equation of motion for x:

$$\frac{dx}{dt} = -\kappa\left(x - y - \frac{2Cx}{1+x^2}\right). \tag{6}$$

The steady-state behaviour is described by the equation

$$y = x + \frac{2Cx}{1+x^2} \tag{7}$$

which, for a given y, admits either one or three solutions. The steady-state curve of x *vs.* y, obtained from Eq. (7), exhibits a hysteresis cycle for $C > 4$ (Fig. 2). For $y < y_m$ and $y > y_M$ Eq. (7) admits one stationary solution $\bar{x}$, whereas for $y_m < y < y_M$ it admits three stationary solutions $x_1 < x_2 < x_3$, of which x_1 and x_3 are stable and belong to the hysteresis cycle, whereas x_2 is unstable and therefore physically not realizable.

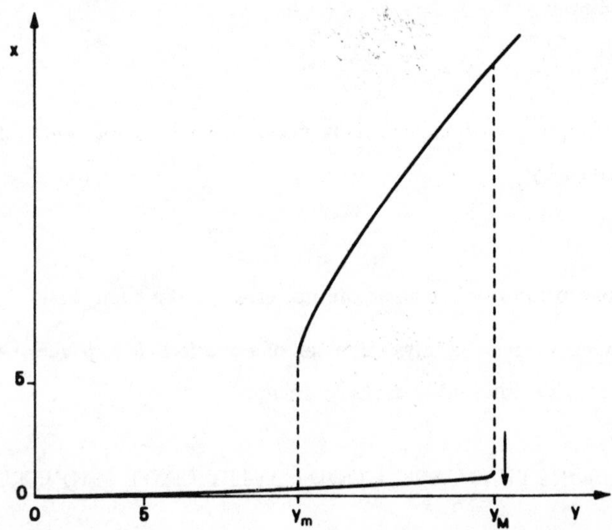

Fig. 2: Purely absorptive bistability. Hysteresis cycle of normalized trasmitted field x as a function of the normalized incident field y for $C = 20$. The arrow indicates a value of the incident field slightly larger than the up-switching threshold $y_M = 21.0264$

It is convenient to introduce the normalized time

$$\tau = \kappa t \tag{8}$$

and the potential $V(x,y)$ defined as follows:

$$V(x,y) = \int dx \left(x - y + \frac{2Cx}{1+x^2} \right) = \frac{x^2}{2} - yx - C\,ln(1+x^2), \tag{9}$$

so that Eq. (6) takes the compact form

$$\frac{dx}{d\tau} = -\frac{\partial V(x,y)}{\partial x} \tag{10}$$

Next, we assume that the amplitude y of the input field fluctuates in time. Precisely, we replace in Eq. (6) $y \to y + \delta y(t)$, where the fluctuating part $\delta y(t)$ undergoes a stationary Gaussian process with

$$< \delta y(t) >= 0 \qquad , \qquad < \delta y(t)\, \delta y(t') >= 2q\delta(t-t'). \tag{11}$$

Hence the noise is white, and the parameter q measures the level of noise. In such a way, Eq. (6) becomes a Langevin equation, which is equivalent to the following Fokker-Planck equation (FPE) (Gardiner 1983) for the probability distribution $P(x,\tau)$ of the trasmitted field x:

$$\frac{\partial P(x,\tau)}{\partial \tau} = \frac{\partial}{\partial x} \left(x - y + \frac{2Cx}{1+x^2} \right) P(x,\tau) + q \frac{\partial^2 P(x,\tau)}{\partial x^2}. \tag{12}$$

The drift term arises from the deterministic part of the Langevin equation, whereas the diffusion term originates from the stochastic term.

It is immediate to calculate the stationary solution ($\partial P(x,\tau)/\partial\tau = 0$) of Eq. (12), which has the form

$$P_{st}(x,y) = \mathcal{N}\, exp \left[-\frac{1}{q} V(x,y) \right], \tag{13}$$

where $\mathcal{N}$ is the normalization constant such that $\int P_{st}(x,y)\, dx \equiv 1$. On comparing with the Landau theory of equilibrium phase transitions, we see from Eq. (13) that the potential V plays the role of a generalized free energy in this system, which is driven far from thermal equilibrium.

From Eqs. (9) and (7) we see that the extrema of the function $V(x)$ coincide with the stationary solutions of the deterministic theory; from Eq. (13) we have that the maxima (minima) of $P_{st}(x)$ coincide with the minima (maxima) of $V(x)$.

The shape of the potential V, and of the associated stationary probability distribution P_{st}, depends on the value y of the input field. For $y < y_m$ and $y > y_M$

(see Fig. 2) V has a single minimum exactly at the deterministic stationary solution $\bar{x}$, and P_{st} has a single peak centered at $\bar{x}$. For $y_m < y < y_M$ V has two minima in correspondence with the deterministic stable stationary solutions x_1 and x_3, and a local maximum in correspondence with the unstable solution x_2 (Fig. 3*a*). In this case, P_{st} has two distinct peaks centered at x_1 and x_3. This bimodal structure of the stationary probability distribution is the very signature of optical bistability from a statistical viewpoint. However, it is tremendously difficult to observe experimentally this stationary configuration, because it is attained only for extremely long times. In fact, in order to approach this distribution the system must have the time to perform transitions from one metastable state to the other, and the lifetime of the two metastable states is exceedingly long. The average transition time can be calculated using a classic method introduced by Kramers (Kramers 1940). Assume that initially the system sits at the bottom of the lefthand well in Fig. 3*a*; in order to perform a transition to the more stable righthand well it must overcome the potential barrier centered at the local maximum of the potential. Now, Kramers's method says that the average transition time is proportional to $exp\,[\Delta V/q]$, where ΔV is the height of the barrier, defined as the difference between the values of V at the local maximum and at the lefthand minimum. Because q is small, the transition time turns out to be enormous. The only exception is when the value of the input field is very near the boundary of the bistable region, so that also the potential barrier ΔV is small.

Hence, it is not surprising that the only observation of a two-peaked stationary distribution in the framework of OB was not obtained using an all-optical system as that described in Sec. 2, but a hybrid electro-optical system in which fluctuations are artificially and skillfully introduced and controlled (McCall *et al.* 1982).

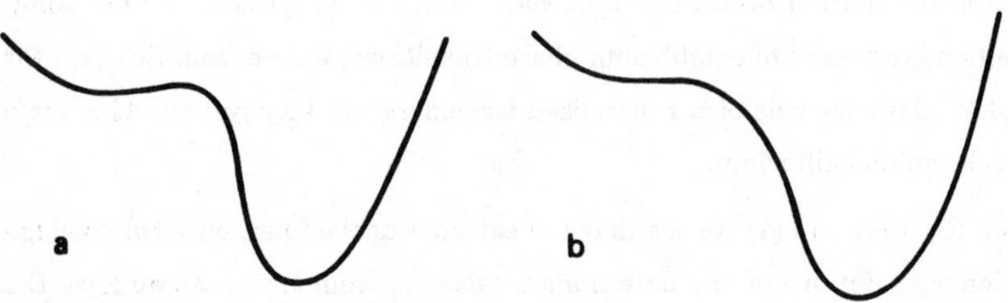

Fig. 3: *a*) Qualitative shape of the potential V in the bistable region. *b*) Qualitative shape of the potential V for a value of y corresponding to the arrow in Fig. 2

4. CRITICAL SLOWING DOWN AND TRANSIENT OPTICAL BIMODALITY

Let us now show that there is also another possibility of obtaining a bimodal probability distribution in OB, and this time the bimodality does not appear at steady-state, but in the transient, in such a way that it is amenable to experimental observation.

Precisely, let us proceed as follows. We switch on the incident field stepwise to a value that lies slightly beyond the bistable region (Fig. 2). The deterministic theory predicts the time evolution shown in Fig. 4: a long lethargic stage of slow evolution is followed by a rapid switching to the stationary state. This critical slowing down behaviour was first predicted by Bonifacio and Lugiato (1976 and 1978*a*) and has been the object of intense theoretical study (Bonifacio and Meystre 1978, Benza and Lugiato 1979; Mandel and Erneux 1982, Grynberg and Cribier 1983). Its experimental observation has been attained in a variety of systems: electro-optical (Garmire *et al.* 1979), microwave (Barbarino *et al.* 1982), sodium atomic beams (Grant and Kimble 1983), sodium vapour (Mitscke *et al.* 1983) and rubidium (Cribier *et al.* 1983).

It is easy to understand the origin of this phenomenon. We have seen before that in the bistable region the generalized free energy V has two minima. For a value of y as indicated by the arrow in Fig. 2, we are out of the bistable region and therefore V presents only one minimum. However, we are near the boundary of the bistable region, and therefore there is a remainder of the other minimum which just disappeared, and which leaved a flat part in the potential, as we see in Fig. 3*b*. When, during the deterministic time evolution, the system arrives at the flat part of V, the evolution slows down and becomes quick again only when the system reaches the edge of the potential well.

Let us now see the time evolution as predicted by the stochastic theory, and obtained by numerically solving the FPE (12) (Broggi and Lugiato 1984 *a* and *b*, Broggi *et al.* 1985). The probability distribution is initially a delta function $P(x,0) = \delta(x)$, because for $\tau < 0$ the incident field is switched off. The distribution drifts to the right and, because there is diffusion, it immediately acquires a nonzero width (curve *a* in Fig. 5). Later, the distribution forms a long tail towards the right (curve *b*), and subsequently it develops a second peak (curve *c*). Finally, the first peak disappears (curves *d*,*e*) and $P(x,\tau)$ approaches the stationary one-peaked configuration. In fact, in this case the steady-state distribution has one peak because we are out of the bistable region.

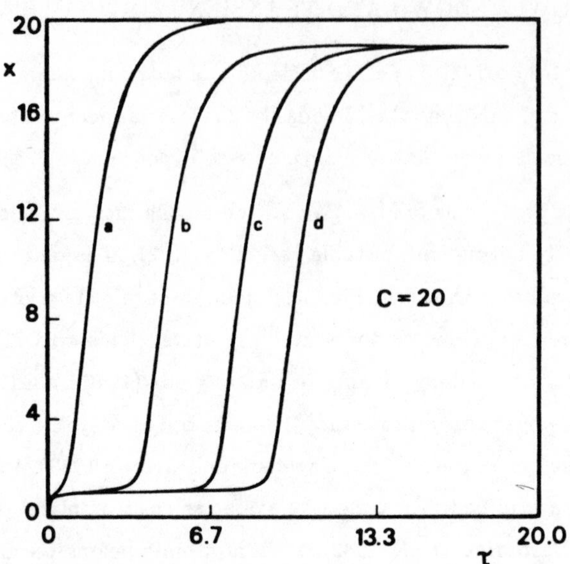

Fig. 4: Time evolution of the trasmitted field for $C = 20$ and $a)$ $y = 22$, $b)$ $y = 21.1$, $c)$ $y = 21.05$, $d)$ $y = 21.04$

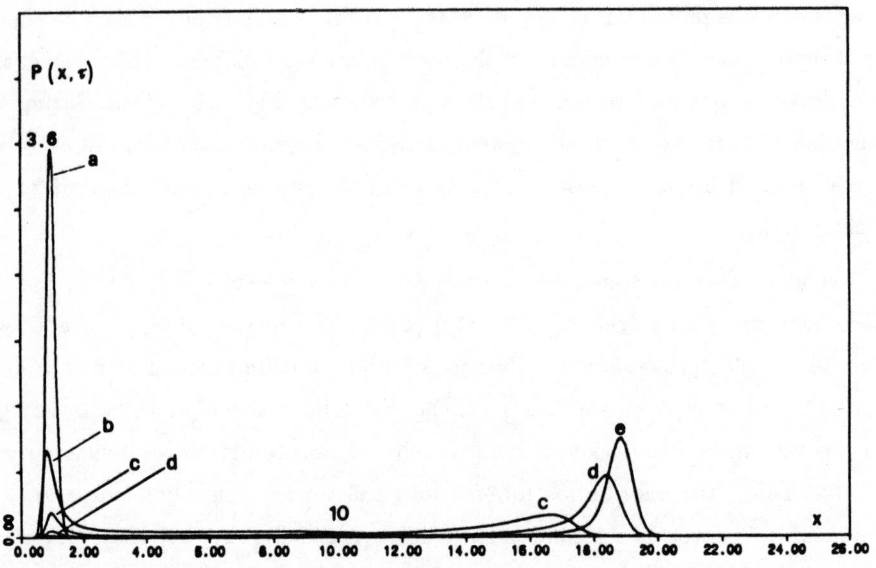

Fig. 5: The probability distribution $P(x, \tau)$ is shown for $C = 20, y = 21.04, q = 0.1$ and $a)$ $\tau = 0.2$, $b)$ $\tau = 1.8$, $c)$ $\tau = 3.6$, $d)$ $\tau = 5.4$, and $e)$ $\tau = 7.2$

This type of time evolution of $P(x,\tau)$ is pictorially represented by the three-dimensional diagram shown in Fig. 6, from which we see that the double-peaked character of the probability distribution persists for a sizable amount of time.

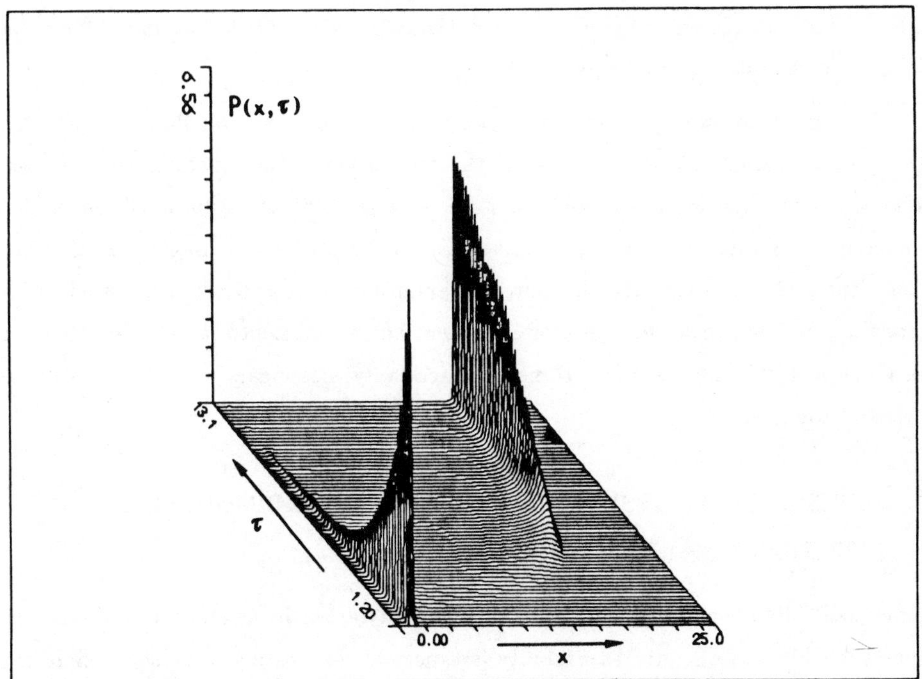

Fig. 6: Time evolution of the probability distribution $P(x,\tau)$ for $C = 20, y = 21.1, q = 0.005$

This phenomenon, in which the distribution becomes bimodal during the transient approach to a one-peaked distribution, has been called transient bimodality. Also the reason of this phenomenon becomes clear if we consider the role of the potential V shown in Fig. 3*b*. When the distribution arrives at the flat part of the potential, due to the slowing down it sits there for a long time almost without any movement in its center of mass. However, because of diffusion the distribution broadens with time and, due to the asymmetry of the potential, it develops a tail in the direction of the well. As soon as the leading edge of the tail reaches the boundary of the potential well, it is quickly trasferred to the bottom, thereby giving rise to the second peak. Only later the remaining mass of the probability distribution falls into the well, thus restoring a one-peaked distribution.

It must be stressed that the phenomenon of transient bimodality is not restricted to the case of optical bistability, but is of general nature. In fact, it was first predicted by Nicolis and collaborators for explosive chemical reactions and combustion (Baras *et al.* 1983, Frankowicz and Nicolis 1983,Frankowicz *et al.* 1984), and arises universally in all systems with a deterministic time evolution which presents a long induction stage followed by a fast switching stage.

The explanation of transient bimodality, that we gave before, shows clearly that this phenomenon is not at all linked to the special kind of fluctuations that we considered, which corresponds to additive noise. Also multiplicative noise will do the job perfectly. A special case of multiplicative noise is quantum noise (Lugiato *et al.* 1984, Carmichael *et al.* 1983), and therefore also quantum noise can produce transient bimodality, at least provided the atomic system has a sufficiently small size. In fact, as we show in the next section, transient bimodality disappears when the noise level becomes too low.

5. EXPERIMENTAL OBSERVATION OF TRANSIENT BIMODALITY. FLUCTUATIONS IN THE SWITCHING TIME.

As we said, the phenomenon of transient bimodality arises in several types of systems. However, the case of optical bistability is especially interesting because it offers the possibility of a precise experimental observation. This goal has been achieved by Lange and his collaborators using optically pumped sodium vapour (Mitscke *et al.* 1985, Lange *et al.* 1985). Figure 7 shows a typical result of this refined experiment. The inset illustrates how the experiment itself was performed: at time zero, the input field is switched on, and one finds the critical slowing down and the switching. At a fixed time delay T one records the value of the trasmitted intensity. One repeats this procedure several thousands of times, thus obtaining the probability distribution of the output intensity at time T. This distribution is shown in the figure for various values of the delay. The behaviour is qualitatively very similar to that predicted by our theoretical model (compare with Fig. 6), even if the physical conditions of the experiment do not match this model, because they do not correspond to the good cavity limit, but to the opposite bad cavity limit. This agreement under different physical conditions confirms the universality of the phenomenon of transient bimodality.

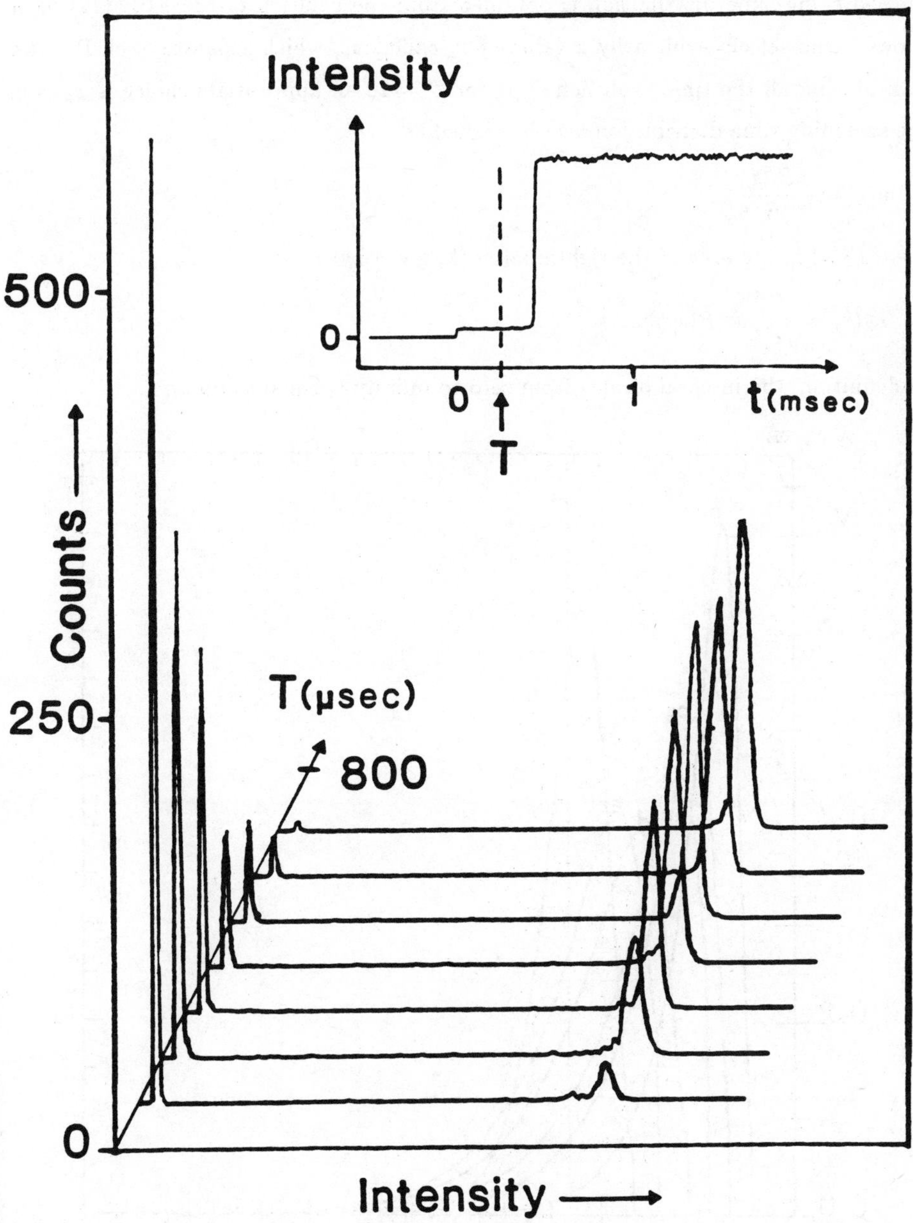

Fig. 7: Measured histograms of output intensities for different times T after switch-on of the incident light (from Mitschke *et al.* 1985)

If one looks at the switching time one finds, both theoretically and experimentally, that it undergoes large fluctuations. Let us consider first our theoretical predictions. The switching time distribution is obtained from the solution of the FPE (12) as it follows. One selects arbitrarily a value of x, called x_s, which separates well the two peaks during all the time evolution; *e.g.*, for $C = 20$ an appropriate choice is $x_s = 8$. The switching time distribution $v(\tau)$ is defined as

$$v(\tau) = \frac{dP_2(\tau)}{d\tau},$$

where $P_2(\tau)$ is the area of the righthand peak, given by

$$P_2(\tau) = \int_{x_s}^{\infty} dx\, P(x, \tau).$$

By definition, the integral of $v(\tau)$ from zero to infinity is equal to unity.

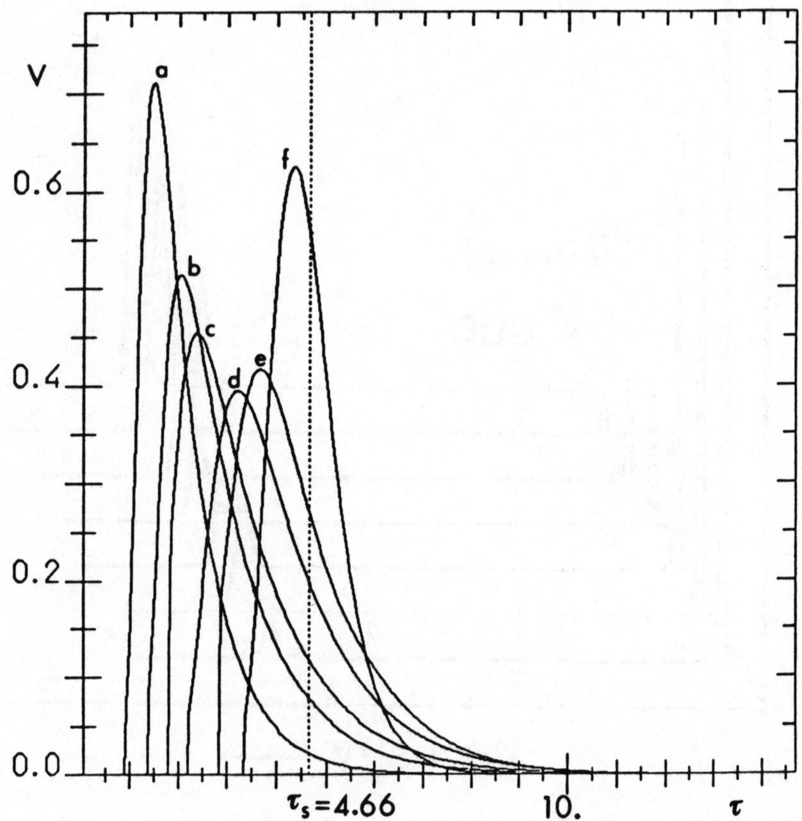

Fig. 8: Switching time distributions for $C = 20, y = 21.1$ and *a*) $q = 0.5$, *b*) $q = 0.1$, *c*) $q = 0.05$, *d*) $q = 0.01$, *e*) $q = 0.005$, *f*) $q = 0.001$

Figure 8 shows the switching time distribution for several values of the noise level q. The vertical line indicates the deterministic value of the switching time, *i.e.* the time τ_s at which the solution $x(\tau)$ of Eq. (6) for $x(0) = 0$ is equal to x_s. For relatively large noise levels the most probable switching time is much shorter than the deterministic value τ_s. As the noise level is reduced, the switching time distribution broadens, while the most probable switching time increases. This broadening is accompained by an enhancement of the phenomenon of transient bimodality, in the sense that the time interval, in which the distribution $P(x, \tau)$ is bimodal, becomes larger. For $q = 10^{-2}$ we find an inversion of tendency: on decreasing further the noise level, the most probable switching time continues to grow, but the switching time distribution now becomes narrower and more symmetical, until it approaches a delta function centered at the deterministic value of the switching time. This narrowing is accompained by a gradual disappearence of the phenomenon of transient bimodality, until, when the noise level is small enough, the distribution $P(x, \tau)$ remains always a single peak which trivially follows the deterministic time evolution.

An interesting feature of this analysis is that, even if transient bimodality disappears when noise becomes too small, it persists for quite low values of the noise level. This fact was relevant with respect to ensuring the feasibility of an experimental observation.

Figure 9 shows the switching time distribution as observed in the Hannover experiment. The bottom curve is obtained for a noise level which corresponds to the natural noise of the system, while the other curves are obtained by adding an increasing controlled amount of noise in the intensity of the input field. In agreement with Fig. 8, the switching time distribution broadens as the noise level is reduced.

In this system, the natural noise imposes a lower bound to the noise level. For this reason, Lange and collaborators performed also another experiment (Mitscke *et al.* 1985) using, instead of an optical system, an electronic bistable system in which the noise level can be controlled to a much higher degree. The behaviour of the distribution $P(x, \tau)$ turns out to be qualitatively similar to that in the optical system, which gives a further confirmation of the universality of the phenomenon of transient bimodality. Figure 10 exhibits the behaviour of the switching time distribution, as the noise level in the electronic system is reduced. The distribution broadens until there is a reversal of tendency, for which the distribution narrows and becomes more symmetrical, again in full qualitative agreement with theory.

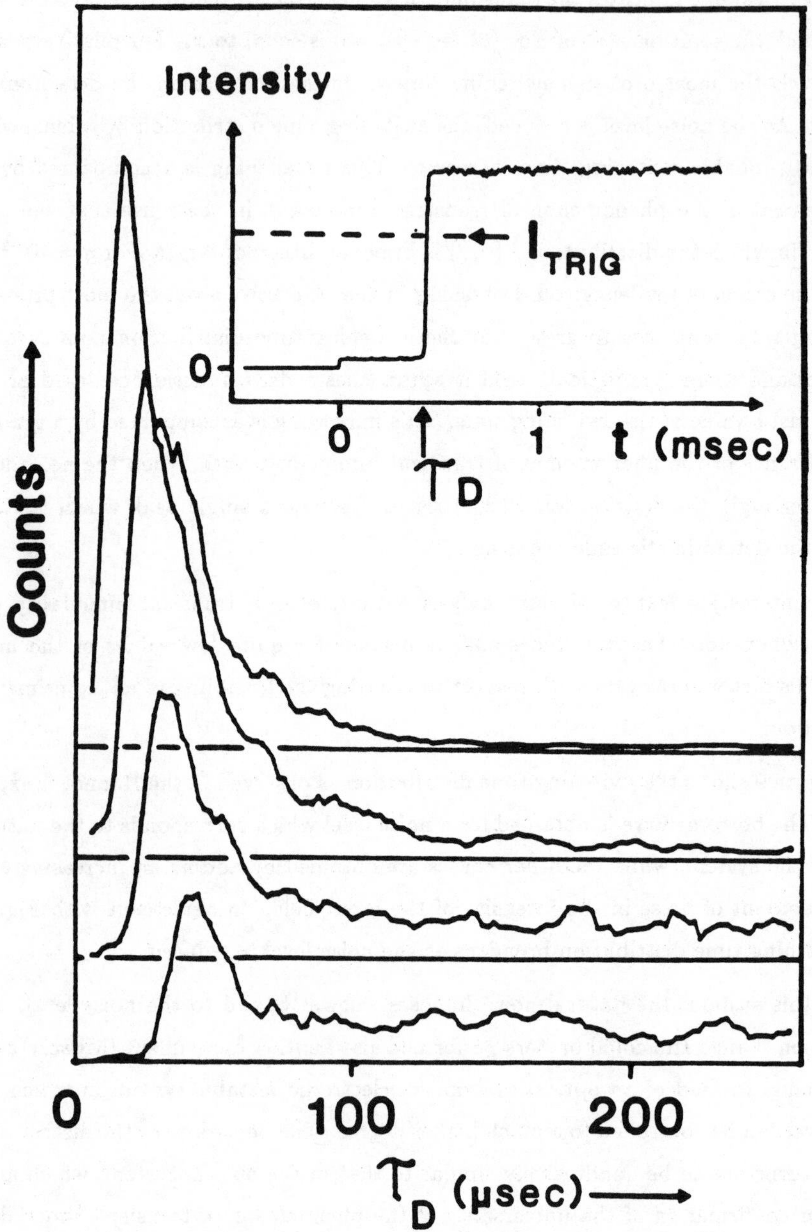

Fig. 9: Measured histograms of switching time for different noise levels, optical system (from Mitschke *et al.* 1985)

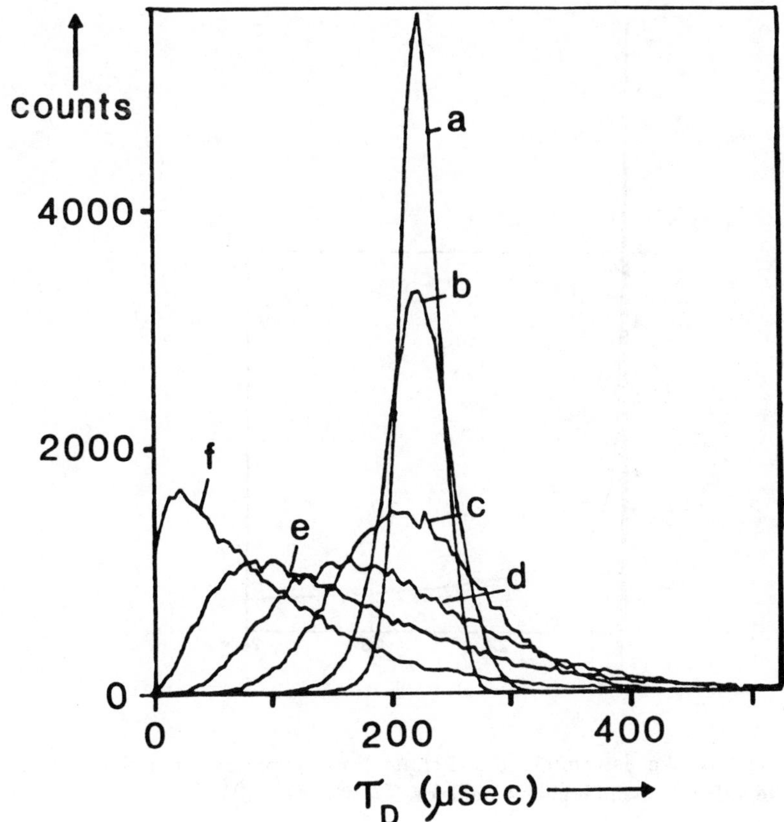

Fig. 10: Measured histograms of switching time for different noise levels, electronic system (from Mitschke *et al.* 1985)

6. CRITICAL SLOWING DOWN AND NOISE

After discussing the dependence on the noise level, another aspect that is of interest is the dependence on the difference $(y - y_M)$ between the value y of the input field and the bistability threshold, or critical value, y_M (see Fig. 2). This is connected with scaling laws in the critical slowing down.

It is well known that according to the deterministic theory the switching time diverges when the critical value is approached (see Figs. 4 and 11). Grynberg and Cribier (1983) analyzed this divergence and found that it scales as $(y - y_M)^{-1/2}$. If one includes noise, the picture changes qualitatively because the divergence disappears.

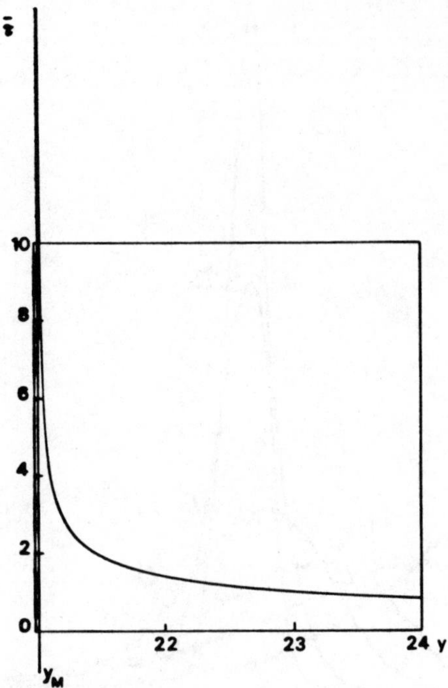

Fig. 11: Plot of the deterministic switching time (expressed in units κ^{-1}) as a function of the value y of the incident field for $C = 20$ (Eq. (5))

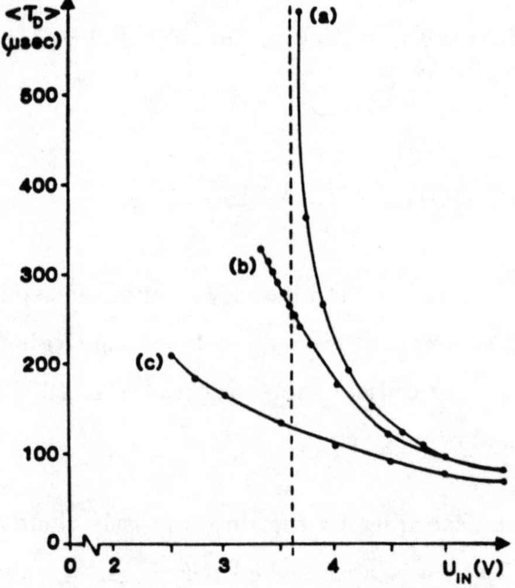

Fig. 12: Measured average switching time versus input voltage for different noise levels, electronic system (from Mitschke *et al.* 1985)

This is clearly shown by the experimental plot of the average switching time shown in Fig. 12, obtained in Hannover using the electronic system. In curve *a* the noise level is so low that the situation cannot be distinguished from the deterministic case, and in fact the curve indicates something similar to a divergence as the critical value is approached. Curve *b* and *c* correspond to an increasing amount of noise in the input voltage, and we see that not only there is no divergence, but the curve of the average switching time continues smoothly into the bistable region. This is due to the fact that in presence of noise the two branches of the hysteresis cycle are metastable. When we are near the bistability threshold y_M and the noise level is large enough, the system tends to switch to the upper branch because the lifetime of the lower trasmission state is not too long. Hence the transition can be observed and the curve of the average switching time can be continued into the bistable region. Again the experimental findings are in qualitative agreement with theory, as we see from Fig. 13.

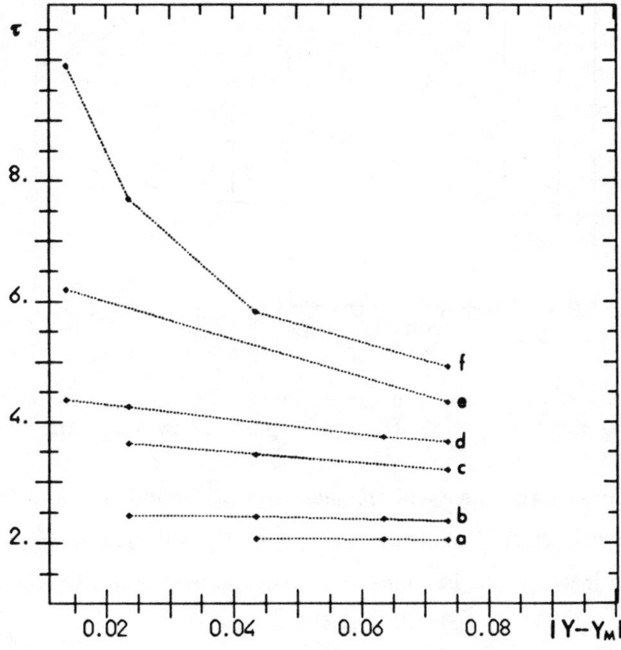

Fig. 13: Most probable delay time as a function of $(y - y_M)$ for *a*) $q = 0.1$, *b*) $q = 0.05$, *c*) $q = 0.01$, *d*) $q = 0.005$, *e*) $q = 0.001$, *f*) $q = 0$ (deterministic case)

However, we must not forget that the average switching time is an insufficient information because of the large fluctuations in the switching time itself. For this reason in Fig. 14 we indicate also the variance of the switching time distribution, and it is evident that this variance increases when the critical value is approached. This feature is relevant in connection with the experiments on critical slowing down performed up to now, that were carried out without being aware of the fact that the large fluctuations in the switching time are intrinsic to the critical slowing down process, and are not due to imperfections in the experimental apparatus.

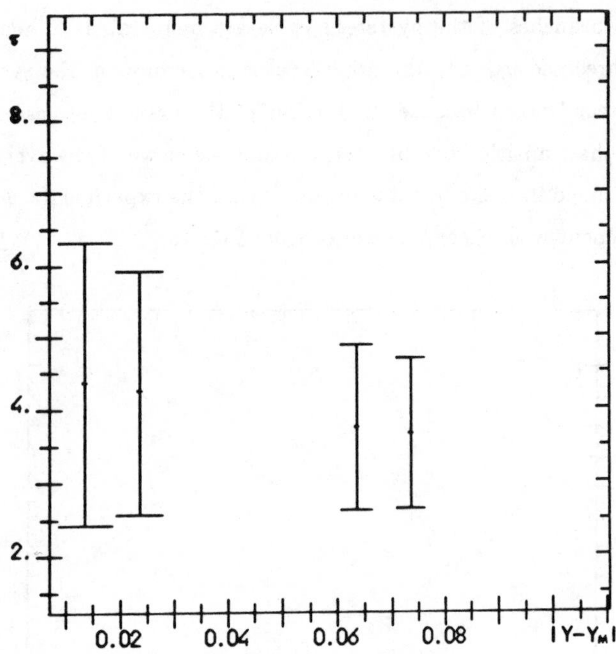

Fig. 14: Most probable delay time as a function of $(y - y_M)$ for $q = 0.005$. The vertical bar represents the FWHM of the corresponding switching time cuves

Further details on the theory of transient optical bimodality, and the discussion of the relations with the phenomena, can be found in (Broggi *et al.* 1985). These results may be of interest also in connection with the practical problem of optimizing the switching time process in optically bistable devices.

7. ADIABATIC ELIMINATION OF THE ATOMIC POLARIZATION. PURELY DISPERSIVE CASE, FREQUENCY NOISE.

We now turn to analyze Eqs. (1) under conditions quite different from those considered in Sec. 3. First of all, we assume the limit

$$\gamma_\perp \gg \kappa, \gamma_\parallel \tag{14}$$

in which we can adiabatically eliminate the polarization by setting $dP/dt = 0$ in Eq. (1.2). Thus, we obtain the following two equations for x and D:

$$\frac{dx}{dt} = -\kappa\left[x(1+i\theta) - y + \frac{2C}{1+i\Delta}xD\right], \tag{15.1}$$

$$\frac{dD}{dt} = -\gamma_\parallel\left[\frac{|x|^2 D}{1+\Delta^2} + D - 1\right]. \tag{15.2}$$

In the case of purely dispersive bistability, almost all the atoms are in the lower level (Lugiato 1984). Hence we set $D = 1 - n$, where $n \ll 1$ is proportional to the nonlinear part of the refractive index of the material. By neglecting the nonlinear absorption term in Eq. (15.1) and setting $D = 1$ in the first term in the righthand side of Eq. (15.2), our set of dynamical equations takes the form

$$\frac{dx}{dt} = -\kappa\left[x(1+2Cb) - y + ix\bar{\theta} - iaxn\right], \tag{16.1}$$

$$\frac{dn}{dt} = -\gamma_\parallel\left(n - b|x|^2\right), \tag{16.2}$$

where we set

$$a = -\frac{2C\Delta}{1+\Delta^2} \quad , \quad b = \frac{1}{1+\Delta^2} \quad , \quad \bar{\theta} = \theta + a \tag{17}$$

Next, assuming $a > 0$ we introduce the trasformations

$$x = \left(\frac{1+2Cb}{ab}\right)^{1/2}\bar{x} \quad , \quad y = \frac{(1+2Cb)^{3/2}}{(ab)^{1/2}}\bar{y},$$

$$n = \frac{1+2Cb}{a}\bar{n} \quad , \quad \bar{\theta} = (1+2Cb)\,\bar{\bar{\theta}}, \tag{18}$$

by which we obtain a set of equations with only four parameters $\bar{\bar{\theta}}, \kappa, \gamma_\parallel, \bar{y}$:

$$\frac{d\bar{x}}{dt} = -\kappa\left\{\bar{x} - \bar{y} + i(\bar{\bar{\theta}} - n)\bar{x}\right\}, \tag{19.1}$$

$$\frac{d\bar{n}}{dt} = -\gamma_\parallel\left(\bar{n} - |x|^2\right). \tag{19.2}$$

As shown in (Lugiato and Horowicz 1985) the model (19) can also describe OB in a Kerr medium or in a semiconductor device. In the following we omit the bars and write x, y, n, θ instead of $\bar{x}, \bar{y}, \bar{n}, \bar{\bar{\theta}}$.

Now we assume the condition

$$\kappa \gg \gamma_\parallel, \tag{20}$$

which is typical of the miniaturized optically bistable devices which utilize semiconductors (Abraham and Smith 1982, Gibbs 1985). Hence by setting $dx/dt = 0$ in Eq. (19.1) we obtain

$$|x|^2 = \frac{y^2}{1+(n-\theta)^2}, \tag{21}$$

while Eq. (19.2) becomes

$$\frac{dn}{dt} = -\gamma_{\|}\left[n - \frac{y^2}{1+(n-\theta)^2}\right]. \tag{22}$$

In these conditions, the behaviour of the system is governed by the variable n. At steady-state, as we see from Eq. (19.2) n coincides with the normalized trasmitted intensity $|x|^2$, and its stationary values are determined by the equation

$$y^2 = n\left[1+(n-\theta)^2\right], \tag{23}$$

as we obtain from Eq. (22). For $\theta > \sqrt{3}$ Eqs. (23) and (22) predicts bistability. (Fig. 15)

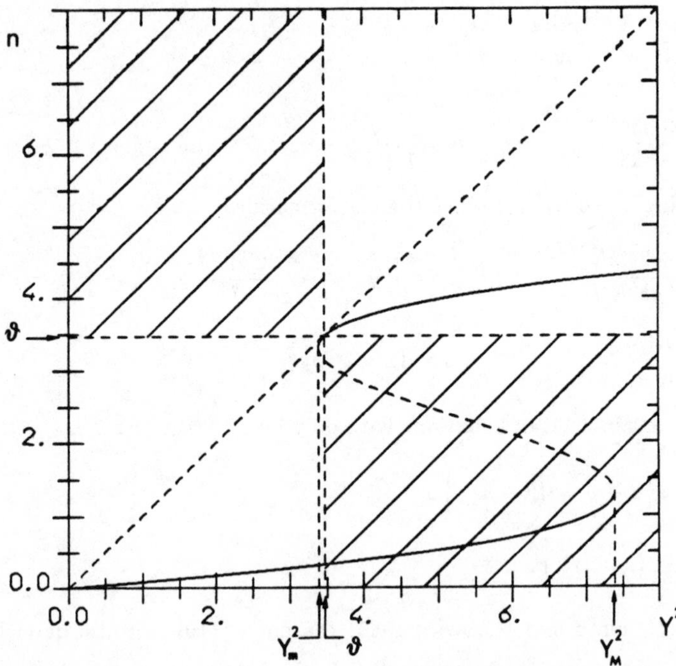

Fig. 15: Purely dispersive bistability stationary curve for $\theta = 2\sqrt{3}$. At steady state, the normalized transmitted intensity coincides with n. The negative slope part of the curve is unstable. The hatched regions indicate the range of n in correspondence of which the stationary probability distribution $P^f_{st}(n)$ vanishes, for each value of the input intensity y^2.

At this point, we assume that the frequency of the incident field fluctuates in time. The correct procedure to incorporate in this model the frequency fluctuations, as well as other types of noise, is shown in (Lugiato and Horowicz 1985). Here we give an intuitive nonrigorous treatment, which however leads to the correct stochastic equation. This is obtained by treating the detuning parameter θ as a stochastic variable, *i.e.* setting $\theta \longrightarrow \theta + \delta\theta(t)$, where the fluctuating part $\delta\theta(t)$ undergoes a stationary Gaussian stochastic process with

$$< \delta\theta(t) >= 0,$$

$$< \delta\theta(t)\, \delta\theta(t') >= 2\, \frac{q_f}{\kappa}\, \delta(t - t'), \tag{24}$$

where q_f is the adimensional frequency noise parameter. Accordingly, we replace θ by $\theta + \delta\theta(t)$ in Eq. (22), which therefore becomes a stochastic equation. We expand the function $[\,1 + (\,n - \theta - \delta\theta(t)\,]^{-1}$ in power series of $\delta\theta$ and neglect all the terms nonlinear in $\delta\theta$, because they give very small contributions. *I.e.*, we substitute in Eq. (22)

$$\frac{1}{1 + (n - \theta)^2} \longrightarrow \frac{1}{1 + (n - \theta)^2} - \frac{2\,(n - \theta)}{[\,1 + (n - \theta)^2\,]^2}\, \delta\theta(t), \tag{25}$$

obtaining the stochastic equation with multiplicative noise

$$\dot{n} = -\gamma_{\|} \left(n - \frac{y^2}{1 + (n - \theta)^2} \right) + 2\gamma_{\|} y^2\, \frac{n - \theta}{[1 + (n - \theta)^2\,]^2}\, \delta\theta(t). \tag{26}$$

Finally, by using the Stratonovich rule (Gardiner 1983)Eq. (6) is translated into the following Fokker-Planck equation (FPE) for the probability distribution $P(n, t)$ of the variable n:

$$\frac{\partial P(n, t')}{\partial t'} = \left\{ \frac{\partial}{\partial n} \left(n - \frac{y^2}{1 + \sigma^2} \right) + d\frac{\partial^2}{\partial n^2}\, \frac{\sigma^2}{(1 + \sigma^2)^4} \right\} P(n, t'), \tag{27}$$

where we defined

$$t' = \gamma_{\|}\, t, \tag{28}$$

$$\sigma = n - \theta, \tag{29}$$

$$d = 4q_f \frac{\gamma_{\|}}{\kappa} y^4, \tag{30}$$

and we neglect the correction to the drift term which arises in presence of multiplicative noise, because it is proportional to $q_f \ll 1$. The inclusion of this correction does not change at all the following results.

If, instead of the frequency noise, we consider thermal noise in the material, the dynamics of the system is governed by a stochastic equation obtained by including an additive noise term in Eq. (22) (Lugiato and Horowicz 1985). This leads to the FPE with costant diffusion coefficient q_{th}:

$$\frac{\partial P(n,\tau)}{\partial \tau} = \left\{ \frac{\partial}{\partial n}\left(n - \frac{y^2}{1+\sigma^2}\right) + q_{th}\frac{\partial^2}{\partial n^2} \right\} P(n,\tau). \tag{31}$$

8. DESCRIPTION OF THE RESULTS

Let us consider first the stationary solution of the thermal FPE (31), which is given by

$$P_{st}^{th}(n,y) = \mathcal{N}\, exp\left(-V_{th}(n,y)\right),$$
$$V_{th}(n,y) = \frac{1}{q_{th}}\left\{\sigma\left(\theta + \frac{\sigma}{2}\right) - y^2\, tg^{-1}\sigma\right\}, \tag{32}$$

where $\mathcal{N}$ is the normalization costant. For $y_m^2 < y^2 < y_M^2$ $P_{st}^{th}(n,y)$ has two distinct peaks centered at the two stable stationary solutions which corresponds with the given value of y^2 (see Fig. 15). On the contrary, for $y^2 < y_m^2$ and $y^2 > y_M^2$ the distribution $P_{st}^{th}(n,y)$ is one-peaked.

On the other hand, the stationary solution of the FPE (27) for frequency noise is as follows (Lugiato *et al.* 1985). Defining the function

$$f(n,y) = \frac{(1+\sigma^2)^4}{\sigma^2}\, exp\left\{-V_f(n,y)\right\}, \tag{33}$$

$$V_f(n,y) = \frac{1}{d}\left\{\frac{\sigma^8}{8} + \theta\frac{\sigma^7}{7} + \frac{2}{3}\sigma^6 + \frac{4\theta - y^2}{5}\sigma^5 + \frac{3}{2}\sigma^4 + (2\theta - y^2)\sigma^3\right.$$
$$\left. +2\sigma^2 + (4\theta - 3y^2)\sigma + ln|\sigma| - (\theta - y^2)\frac{1}{\sigma}\right\}, \tag{34}$$

we have

$$\begin{aligned}
&\text{a)}\quad \text{for } y^2 < \theta \qquad P_{st}^f(n,y) = \begin{cases} \mathcal{N}' f(n,y) & \text{for } n < \theta, \\ 0 & \text{for } n \geq \theta, \end{cases} \\
&\text{b)}\quad \text{for } y^2 = \theta \qquad P_{st}^f(n,y) = \delta(n-\theta), \\
&\text{c)}\quad \text{for } y^2 > \theta \qquad P_{st}^f(n,y) = \begin{cases} 0 & \text{for } n \leq \theta, \\ \mathcal{N}' f(n,y) & \text{for } n > \theta. \end{cases}
\end{aligned} \tag{35}$$

The region for which $P_{st}^f(n,y)$ vanishes is shown in Fig. 15, for all values of y^2. The peculiar structure of the stationary distribution (35) arises from the fact that the diffusion coefficient in the FPE (27) vanishes as σ^2 for $\sigma = 0$ ($i.e.$ $n = \theta$), which originates a divergence in $V_f(n,y)$. This feature produces a steady-state solution that is not analytical in n, but is continuos with all its derivatives.

The special property of distribution (35), which distinguishes it from the thermal case of Eq. (32), as well as from all other distributions which describe a bistable system, is now the following. For $d \lesssim 5$, the distribution (35) has one peak even in the range $\theta \leq y^2 < y_M^2$, in which the deterministic theory predicts bistability.

Let us explain the physical reasons which give rise to the configuration (35). The key role is played by the point $\sigma = 0$, which in the deteministic steady-state theory corresponds to the situation $n = \theta = y^2 = |x|^2$, as we see from Eqs. (29), (23), and (21). Hence this is the "*bleaching point*" in which all the incident light is trasmitted. The frequency fluctuations are governed by the trasmission curve $\left[1+(n-\theta)^2\right]^{-1}$, as we see from Eq. (25). At the point $n - \theta = 0$ this curve attains its maximum (Fig. 16), and therefore the effect of frequency fluctuations vanishes to first order (*i.e.*, in the linear approximation with respect to the fluctuating quantity $\delta\theta(t)$).

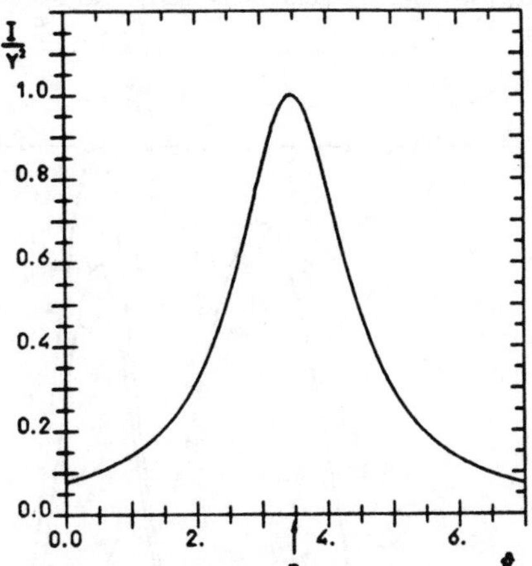

Fig. 16: The transmission of the cavity, defined by the ratio of the intensity $|x|^2$ of the transmitted beam to that of the incident beam y^2, is given by the function $\left[1+(n-\theta)^2\right]^{-1}$ (see Eq. (21)). In correspondence with the maximum $\theta = n$ the system is insensitive to the fluctuations of θ (*i.e.*, to the frequency noise).

From Eq. (27), we see that the point $\sigma = 0$ plays the role of a "*Maxwell devil*" in this problem. In fact, let us consider the probability current in the FPE (27):

$$j(n,t') = \left(\frac{y^2}{1+\sigma^2} - n\right) P(n,t') - d\frac{\partial}{\partial n}\frac{\sigma^2}{(1+\sigma^2)^4}P(n,t'). \qquad (36)$$

For $\sigma = 0$ $(n = \theta)$ it becomes

$$j(\theta,t') = \left(y^2 - \theta\right) P(\theta,t'),$$

hence for $y^2 > \theta$ the flux is always towards the right, whereas for $y^2 < \theta$ it is towards the left. This one-sided flux across the point $n = \theta$ biases the evolution of the probability distribution and, for long times, gives rise to the configuration described by Eq. (35). Completely different is the situation of the thermal FPE (31), for which the sign of the flux is determined by the shape of the probability distribution, for any value of n. *I.e.* the flux is to the right or to the left, according to the configuration of $P(n, \tau)$.

In the case of pure frequency noise for $\theta \leq y^2 < y_M^2$ we are sure that, if we wait long enough, the system performs a transition to the upper branch and will never come back, whereas in the case of thermal noise there is always some probability of performing the opposite transition. However, it must be noted that for small noise level the transition from the lower to the upper branch takes an extremely long time also in the case of pure frequency noise. For this reason, in order to show the approach to steady-state of the probability distribution, we choose relatively large values of the noise parameter.

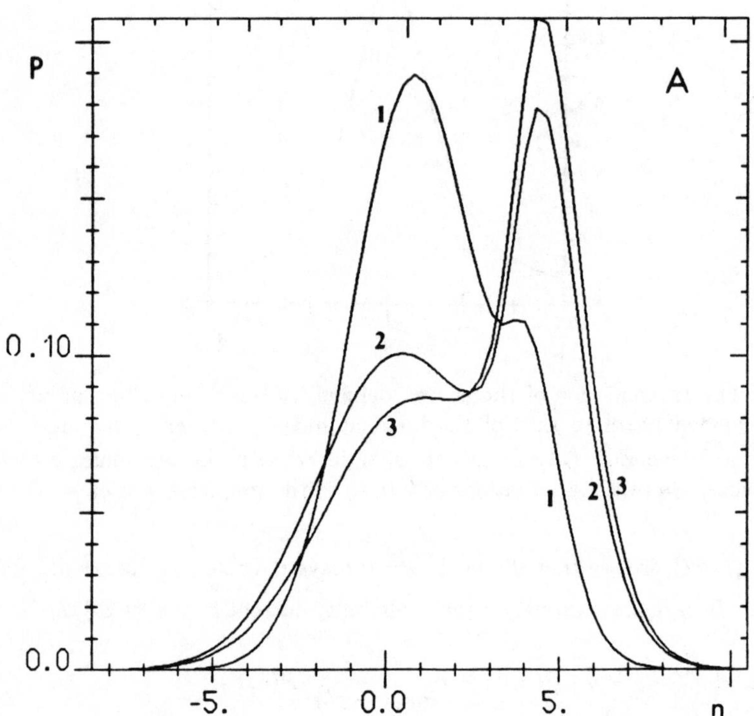

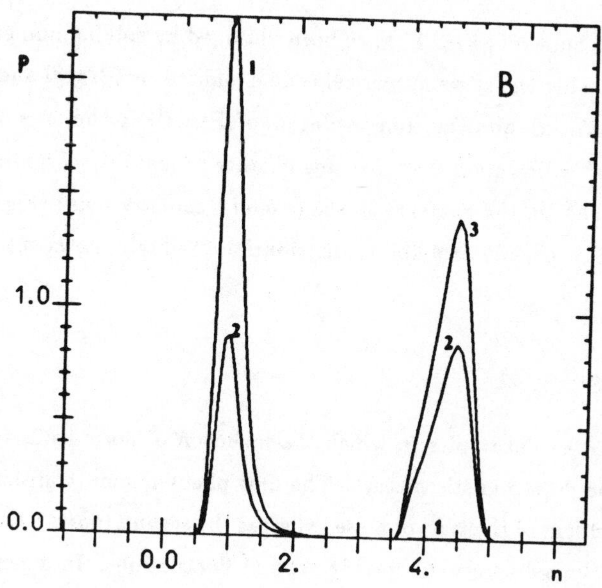

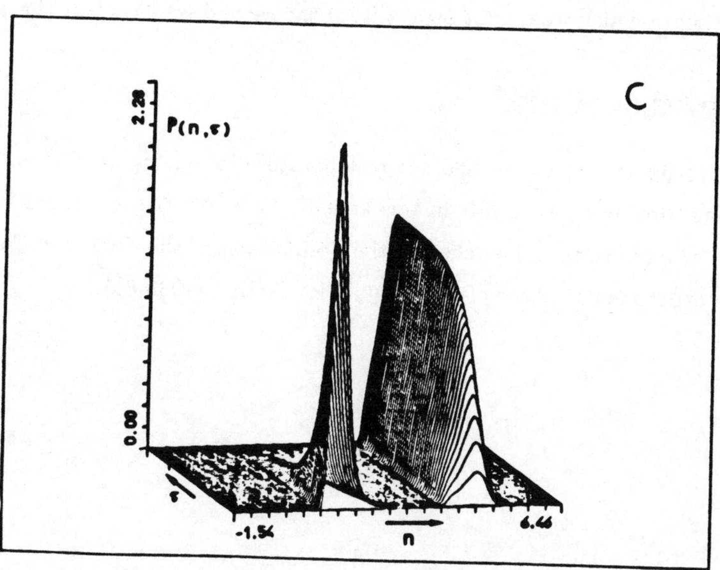

Fig. 17: Time evolution of the probability distribution for $\theta = 2\sqrt{3}$, $y^2 = 7$ and with the initial condition $P(n,0) = \delta(n - \bar{n})$, where $\bar{n} = 0.969$ is the lower-branch stationary value given by the deterministic theory (see Fig. 15). *a*) : solution of Eq. (31) (thermal noise) for $q = 5$ and 1) $\tau = 0.464$, 2) $\tau = 2.177$, 3) $\tau = 15.173$. *b*) : solution of Eq. (27) (frequency noise) for $a = 5$ and 1) $\tau = 5.00$, 2) $\tau = 154.705$, 3) $\tau = 1132.562$. In both *a*) and *b*), curve 3 practically coincides with the stationary probability distribution. *c*) 3-dimensional diagram of the evolution of the probability distribution $P(n,\tau)$ for the case of frequency noise, $a = 5$

The results shown in Fig. 17 have been obtained by solving numerically the FPEs (27) and (31), using the same numerical code exploited in (Broggi and Lugiato 1984 *a* and *b*). Fig. 17*a* exhibits the time evolution of $P(n,t')$ in the case of thermal noise, for a value of y^2 which gives a sizable steady-state probability of finding the system in the lower branch. On the contrary in the case of frequency noise (Fig. 17 *b* and *c*) for the same value of y^2, the probability distribution gradually flows entirely to the right of the point $n = \theta$.

9. CONCLUSION

We have shown two examples in which the inclusion of noise leads to dramatic deviations from the deterministic picture. The first phenomenon (transient bimodality) is largely independent of the type of noise, whereas the second (destruction of steady-state bimodality) is linked to a very specific type of fluctuations. In a sense, the two phenomena are opposite to each other, because in the first we find a bimodal distribution in a situation in which we do not expect it, in the second we have just the contrary.

AKNOWLEDGEMENTS

We are grateful to W. Lange and his collaborators for giving us the permission of reproducing some of their results in this article. This research was carried out in the framework of an Operation launched by the Commission of the European Communities under the experimental phase of the Stimulation Action (1983-85).

REFERENCES

Abraham, F. and Smith, S.D., 1982, *Rep. Progr. Phys.* **45**, 815-885

Arecchi, F.T., Degiorgio, V. and Querzola, B., 1967, *Phys. Rev. Lett.* **19**, 20-24

Baras, F., Nicolis, G., Malek Mansour, M. and Turner, J.W., 1983 *J. Stat. Phys.* **32**, 1-14

Barbarino, S., Gozzini, A., Maccarrone, F.,Longo, I., and Stampacchia, R., 1982, *Nuovo Cimento* **71**B, 183-191

Benza, V. and Lugiato, L.A. 1979, *Lett. Nuovo Cimento* **26**, 405-410

Bonifacio, R. and Lugiato, L.A., 1976 *Opt. Commun.* **19**, 172-175

Bonifacio, R., and Lugiato, L.A., 1978*a* *Phys. Rev. A* **18**, 1129-1140

Bonifacio, R., and Lugiato, L.A., 1978*b* *Lett. Nuovo Cimento* **21**, 517-521

Bonifacio, R. and Meystre, P., 1978 *Opt. Commun.* **29**, 131-135

Broggi, G., and Lugiato, L.A., 1984*a* *Phys. Rev. A* **29**, 2949-2953

Broggi, G. and Lugiato, L.A., 1984*b* *Phyl. Trans. R. Soc. Lond. A* **313**, 425-429

Broggi, G., Lugiato, L.A. and Colombo, A. 1985 *Phys. Rev. A* in press

Carmichael, H.J., Walls, D.F., Drummond, P.D. and Hassan, S.S., 1983, *Phys. Rev. A* **27**, 3112-3128

Cribier, S., Giacobino, E., and Grynberg, G., *Opt. Commun.* **47**, 170-174

Davidson, F., and Mandel, L., 1967 *Phys. Lett.* **25** A, 700-703

Frankowicz, M., and Nicolis, G., 1983 *J. Stat. Phys.* **33**, 595-603

Frankowicz, M., Malek-Mansour, M., and Nicolis, G., 1984 *Physica* **12**A, 237-250

Gardiner, C.W., 1983, "*Handbook of Stochastic Methods*", in Springer Series in Synergetics, ed. by H. Haken, Vol. 13, Springer-Verlag, Berlin

Garmire, E., Marburger, J.H., Allen, S.D. and Winful, H.G., 1979, *Appl. Phys. Lett.* **34**, 374-377

Gibbs, H.M., 1985 "*Optical Bistability: Controlling Light by Light*", Academic Press, New York

Glauber, R.J., 1963*a*, *Phys. Rev.* **130**, 2529-2540

Glauber, R.J., 1963*b*, *Phys. Rev.* **131**, 2766-2771

Grant, D.E. and Kimble, H.J., 1983, *Opt. Commun.* **44**, 415-420

Grynberg, G., and Cribier, S., 1983 *J. de Physique Lett.* **44**, 4449-4454

Haken, H., 1977, "*Synergetics - An Introduction*", Springer-Verlag, Berlin

Haken, H., 1970, "*Laser Theory*", in "Light and Matter Ic", ed. by L. Genzel, Handbuch der Physik, Vol. XXV/2c, Springer-Verlag, Berlin

Hassan, S.S., Drummond, P.D. and Walls, D.F., 1978, *Opt. Commun.* **27**, 480-484

Horsthemke, W. and Lefever, R., 1983, "*Noise-Induced Transitions*", Springer Series in Synergetics, ed. by H. Haken, Vol. 15, Springer-Verlag, Berlin

Kramers, H.A., 1940, *Physica* **7**, 284-298

Lange, W., Mitschke, F., Deserno, R. and Mlynek, J., 1985 *Phys. Rev. A* in press

Lugiato, L.A., Mandel, P. and Narducci, L.M., 1984, *Phys. Rev. A*, 1438-1462

Lugiato, L.A., 1984, "*Theory of Optical Bistability*", in "Progress in Optics", Vol. XXI, ed. by E. Wolf, North-Holland, Amsterdam, p. 69-216

Lugiato, L.A. and Horowicz, R.J., 1985, *J. Opt. Soc. Am. B* **2**, 971-981

Lugiato, L.A., Colombo, A., Broggi, G. and Horowicz, R.J., 1985, submitted for publication

Mandel, P. and Erneux, T., 1982, *Opt. Commun.* **42**, 362-366

McCall, S.L., Gibbs, H.M., Hopf, F.A., Kaplan, D.L. and Ovadia, S., 1982 *Appl. Phys. B* **28**, 99-106

Mitschke, F., Deserno, R., Mlynek, J., and Lange, W., 1983 *Opt. Commun.* **46**, 135-140

Mitschke, F., Deserno, R., Mlynek, J., and Lange, W., 1985 *IEEE. J. Quant. Electron.*, September 1985

Pike, R.E., 1969, in "*Quantum Optics*", Proceedings of the International School of Physics "Enrico Fermi", Course XLII, ed. by R.J. Glauber, Academic Press, New York, p. 160-175

Sargent,III M., Scully, M.O. and Lamb, W.E. Jr., 1974, "*Laser Physics*", Addison-Wesley, Reading, MA

Van Kampen, N.G., 1961, *Can. J. Phys.* **30**, 551-564

A REVIEW OF INSTABILITIES IN OPTICAL BISTABILITY

P MANDEL

1. INTRODUCTION

Optical bistability (OB) is the property displayed by a class of optical systems whose output intensity is not uniquely determined by the control parameters. Depending on its preparation the system can transmit a large or a small fraction of the input intensity. Only nonlinear systems may have such an hysteresis. Associated with the nonlinearity and the multiplicity of solutions is the question of stability. It arises whenever two solutions (steady or time-dependent) coincide. In this contribution we review the various instabilities known in a particular class of optically bistable systems.

In the same way as laser physics has its roots in maser physics, OB originates from experiments performed with masers. It was demonstrated by Gerritson (1963) that a maser with an intracavity saturable absorber has a domain of bistability. Lasher (1964) suggested that the same idea be applied to lasers,i.e., to lasers with a saturable absorber. His theoretical calculations were soon tested (Nathan et al.,1965) using a pulsed GaAs diode laser with active and passive segments, at cryogenic temperature (see also Basov, 1968, for a similar work). In view of the difficulties related to the cw diode laser technology, Lasher'scheme came to a dead-end. It is only recently (Harder et al.,1982) that bistability was observed with GaAs semiconductor in burried heterostructures and with cw outputs. OB in an active cavity, as proposed by Lasher, rests on a competition between the usual nonlinear gain due to the amplifying medium and the nonlinear losses induced by the saturable absorber whic make the cavity decay rate intensity-dependent. It turns out that this question is best analyzed with long (1 to 3 meters) gas lasers which display a wealth of instabilities. The literature on lasers with saturable absorbers is too large to be quoted here. We just mention two recent works (Arimondo et al., 1983 and Mandel et al.,1984) and a review (Erneux, 1985) and refer to their list of references for an introduction to this field.

An alternative scheme was proposed by Szöke et al.(1969) in which the amplifying and absorbing media of the Lasher scheme are placed in distinct resonnant cavities. Thus the active medium, in its cavity, is now a simple laser whose cw beam is sent into a cavity containing only the saturable absorber. The output of this last cavity displays OB as a function of the input laser intensity. Experiments were first reported on Na vapour (Gibbs et al., 1976), then on liquids and liquid crystals (Bischofberger et al., 1978) and finally on semiconductors (Miller et al., 1979 and Gibbs et al., 1979). Since then, a fairly large and sustained effort from experimentalists and theoreticians has been devoted to an analysis of OB. To a large extent, it is justified by the potential applications of OB to all-optical signal processing. An international research progra on OB has been set up in the EC countries (Swinnerton-Dyer, 1984) and large programs exist in the US. As a result OB is by now the chapter of quantum optics which is best understood apart from the laser itself. However this review will not deal with the applied side of OB. It will concentrate on the problem of instabilities which is of interest to applied and fundamental physicists, though for opposite reasons. The point of view adopted in this review is one of fundamental physics. In order to have a comprehensive review we must reduce the scope of this contribution. Hence we shall limit ourselves to OB arising in a passive unidirectional ring cavity in which the saturable medium is modelled by homogeneously broadened two-level atoms interacting only with the field and via one-photon transitions.

No comprehensive review of OB exists as of now. However, a high concentration of informations can be found in proceedings of conferences (Bowden et al., 1981 and 1984; Smith et al., 1985), in two special issues of IEEE Journal of Quantum Electronics (Smith, 1981; Garmire, 1985) and in review articles (Abraham et al., 1982; Bonifacio et al., 1982; McCall et al., 1982; Lugiato, 1983 and 1984; Carmichael, 1983; Englund et al.,1984).

In the review by Englund et al.(1984) an important result has been proved: it is not possible to derive analytically in closed form the characteristic equation of the linear stability analysis of OB in the general case. By general case we imply dispersive OB outside the mean field limit. Hence asymptotic analyses are required to get analytical results. As shown by Carmichael (1983) the adiabatic elimination of the atomic polarization is sufficient to lead to a transcendental characteristic equation from which all results obtained at that time could be recovered. We shall therefore refer the reader to these two rewiew articles and deal here only with results which have been obtained since then.

2. MATHEMATICAL FORMULATION

This section is intended only to set up the framework for the following sections. We consider a ring cavity defined by n mirrors M_n ($n \geq 3$). Two consecutive mirrors, M_1 and M_2, are the input and output coupling mirrors with identical reflection coefficients R and identical transmission coefficients $T=1-R$. The remaining mirrors are perfectly reflecting ($T=0$). The distance between M_1 and M_2 is L. The nonlinear medium, of volume V and length L, is placed between M_1 and M_2. It consists of N honogeneously broadened two-level atoms with a transition frequency ω_a and a dipole moment of modulus μ. The total length of the cavity is L' and ω_c is the cavity eigenfrequency nearest to the incident field frequency ω_o. The input field amplitude is E_i and its arbitrary phase is set equal to zero so that E_i is real and positive. The envelope of the field inside the cavity, $E(z,t)$, is a solution of the coupled Maxwell-Bloch equations (in units $h=2\pi$):

$$\frac{\partial E}{\partial t} + c\frac{\partial E}{\partial z} = -gE \tag{2.1}$$

$$\frac{\partial P}{\partial t} = -\{\gamma_p + i(\omega_a - \omega_o)\}P + \mu ED \tag{2.2}$$

$$\frac{\partial D}{\partial t} = -\gamma_d(D - N/2) - \frac{\mu}{2}(EP^* + E^*P) \tag{2.3}$$

where the atom-field coupling constant g is defined by $g=4\pi\omega_o\mu/V$, γ_p is the decay rate of the atomic polarization μP and γ_d is the decay rate of D which is half the difference between the population of the lower and the upper levels. Alternative notations for γ_p are $\gamma_\perp$ or $1/T_2$ and alternative notations for γ_d are $\gamma_\parallel$ or $1/T_1$. Eqs.(2.1)-(2.3) are to be solved with the boundary condition

$$E(0,t) = T^{1/2}E_i + Re^{-i\delta}E(L, t - \frac{L'-L}{c}) \tag{2.4}$$

where δ is the cavity detuning

$$\delta = \tau(\omega_c - \omega_o)\ , \quad -\pi \leqq \delta \leqq \pi \tag{2.5}$$

and τ the cavity round trip time

$$\tau = L'/c \tag{2.6}$$

Finally we introduce the unsaturated absorption coefficient

$$\alpha = \mu gN/2c\gamma_p(1+\Delta^2) = \alpha_o/(1+\Delta^2) \tag{2.7}$$

the atomic detuning

$$\Delta = (\omega_a - \omega_o)/\gamma_p \tag{2.8}$$

the reduced cavity detuning

$$\theta = \delta/T \tag{2.9}$$

the field decay rate

$$\kappa = T/\tau \tag{2.10}$$

and the cooperation or bistability parameter

$$C = \alpha L/2T \tag{2.11}$$

As mentioned in the introduction, it is not possible to carry out analytically the linear stability analysis of Eqs.(2.1)-

(2.4). This justifies the introduction of asymptotic expansions. They are based on one or a combination of the following limits.

1. Adiabatic elimination of one or two of the three variables E, P and D (Lugiato et al.,1984a).
2. Single mode of oscillation in the cavity.
3. Absorptive OB ($\Delta=0$) in a tuned ($\delta=0$) or a detuned ($\delta\neq0$) cavity.
4. Neglect of transverse effects.
5. When the spatial variation of the field inside the cavity is very weak, the mean field limit (Lugiato, 1984) may be introduced as an asymptotic expansion based on

$$\alpha L\to0,\ T\to0,\ \delta\to0,\quad C \text{ and } \theta \text{ finite} \tag{2.12}$$

In the mean field limit, Eqs.(2.1)-(2.4) reduce to (Lugiato, 1984):

$$\frac{\partial E}{\partial t'}+c\frac{L}{L'}\frac{\partial E}{\partial z}=\kappa\{-(1+i\theta)E+Y-2CP\} \tag{2.13}$$

$$\frac{\partial P}{\partial t'}=\gamma_p\{-(1+i\Delta)P+FD\} \tag{2.14}$$

$$\frac{\partial D}{\partial t'}=\gamma_d\{-D+1-\frac{1}{2}(EP^*+E^*P)\} \tag{2.15}$$

$$t'=t+\frac{L'-L}{c}\,\frac{z}{L} \tag{2.16}$$

$$E(0,t')=E(L,t') \tag{2.17}$$

with a trivial rescaling of E,P,D and E_i to absorb the constants N and μ.

3. CORRESPONDENCE PRINCIPLES

In this section we review two papers which deal with classes of equivalence between instabilities. The first paper on this subject is due to Carmichael (1984) who compares multimode instabilities in a tuned cavity ($\delta=0$) and in a cavity tuned midway between cavity resonances ($\delta=-\pi$).In the tuned cavity there can be bistability. In the case of midtuning, there is never bistability, the relation between input and cavity intensities being linear. However, in the limit $\gamma_p\tau\to\infty$ and

$\gamma_d \tau \to \infty$ there is a one to one correspondence between the eigenvalues of the linear stability analysis of the tuned and the midtuned cavities. That is, the stability of the tuned cavity also rules the stability of the midtuned cavity. When there is optical bistability, the middle branch is always unstable. Consequently the condition for the existence of optical bistability is also the condition for the existence of a finite domain of instability in the midtuned cavity. Furthermore when an instability appears on the branches with positive slope in the tuned cavity, a similar instability will appear in the midtuned case.

A second paper, by Lugiato et al.(1985), establishes two correspondence principles. In this case the restriction is the mean field limit but the results cover the possibility of inhomogeneous broadening and transverse beam profile. Furthermore, they are not limited to optical bistability. The correspondence principles again follow from a careful discussion of the stability conditions derived from the linear stability analysis. The first principle is general and states that for a given system a single mode instability necessarily implies the existence of a multimode instability in a sufficiently good cavity ($\kappa \ll \gamma_p + \gamma_d$). The second principle is somewhat more restricted. It states that a multimode instability in the good cavity limit can, but need not, imply a single mode instability in the bad cavity limit. The origin of this asymmetry is easily understood. In the multimode case, an instability sets up as soon as one of the modes (of which there is an infinite number) becomes unstable. Which mode becomes unstable is not important since a single unstable mode will drive all other modes via nonlinear mode coupling. On the contrary, single mode instability needs more stringent conditions to express that precisely the mode under study becomes unstable. The symmetry would be restored if we impose a particular mode of the multimode cavity to be unstable.

4. LINEAR STABILITY ANALYSIS

In this section we begin to review the instabilities which have not been covered by Carmichael (1983) and by Englund et al. (1984). The first paper is by Asquini et al.(1985) and deals with the instabilities of the nonresonant modes in the mean field limit. Two adiabatic eliminations of atomic variables are also performed. The macroscopic polarization is eliminated by the requirement that the decay rates verify the inequalities

$$\gamma_p >> \gamma_d \ , \quad \gamma_p \tau >> 1 \tag{4.1}$$

This leads to a set of multimode rate equations which are projected on the infinite set of cavity modes. For each cavity mode the contribution to the atomic inversion is further eliminated by the requirement

$$\gamma_d >> \kappa, \ \kappa\tau << 1 \tag{4.2}$$

A linear stability analysis of the remaining equations gives the conditions for the occurence of Hopf bifurcations. When $\Delta=-\theta$, the following results have been proved analytically for $C>4(1+\Delta^2)$ which is the condition for optical bistability:
1.When $\Delta^2<1$ there is no Hopf bifurcation.
2.When $\Delta^2>1$ there are instabilities on both the upper and lower branches. The domain on instability is bounded from above by X_+^2 and from below by X_-^2 with

$$X_\pm^2/(1+\Delta^2)=\frac{C}{2|\Delta|}-1\pm\frac{C}{2|\Delta|}\{1-\frac{4|\Delta|(1+|\Delta|)^2}{C(1+\Delta^2)}\}^{1/2} \tag{4.3}$$

where X^2 is the cavity intensity in steady state scaled to the saturation intensity. This domain includes the middle branch. All X^2 between X_+^2 and X_-^2 are unstable. Numerically it has been proved that this instability persists for $\Delta\neq-\theta$ and remains even for $\Delta\theta>0$. On the contrary only when $\Delta\theta>0$ another domain of instability can appear, which may coexist with the previous one. It affects only the upper branch and does not, in general, begin at the limit point. A difference between these two domains of instability is that the first domain is linked to optical bistability and vanishes when $C<4(1+\Delta^2)$. On the contrary the second domain may remain even when the condition for optical bistability is not met.

Another stability analysis has been performed by Erneux et al. (1985) in the single-mode and mean field limits for a poor cavity:

$$\gamma_p \cong \gamma_d \cong \kappa \tag{4.4}$$

but in the limit of large C. Although this analysis was carried out without knowledge of the results due to Asquini et al.(1985) and to Lugiato et al.(1985), it happens to be a counterpart of the previous analysis in the sense of the correspondence principles. The method used by Erneux et al.(1985) is a multiple time-scale analysis in which time-dependent solutions are sought which have the same scaling in C as the steady solutions on the upper and lower branches. On the lower branch a finite domain of instability, extending till the limit point, exists if

$$\Delta\theta<0, \quad |\theta|>\frac{1+\gamma_p(1+\Delta^2)\kappa^{-1}}{|\Delta|}, \quad Y^2>Y_c^2=\frac{4\Gamma\gamma_d\kappa^{-1}C^2}{(\Gamma+\gamma_d\kappa^{-1})^2} \tag{4.5}$$

$$\text{with} \quad \Gamma=\frac{(1+\gamma_p/\kappa)^2+(\theta+\Delta\gamma_p/\kappa)^2}{-\Delta\theta-1-\gamma_p(1+\Delta^2)/\kappa}>0$$

The bifurcation equations from which (4.5) is obtained have the rare property of being linear. As a result they only locate the instability point but give no indication on the nature of the bifurcation. A numerical integration of Eqs.(2.13)-(2.17) for a limited set of parameters reveals that at Y_c a jump to the upper branch occurs. This result suggests that a subcritical Hopf bifurcation is the instability at Y_c. This should not be confused with anomalous switching in dispersive optical bistability (Hopf et al.,1979; Lugiato et al.,1982a). Although the result (jump to the upper branch) is the same, anomalous switching results from an instability under large fluctuations whereas (4.5) results from a standard linear stability analysis which assumes infinitesimal perturbations. For the upper branch a finite domain of instability limited by Hopf bifurcations exists when the steady intensity $I=C^2X^2$ verifies the two conditions

$$X^2>2/(2+\gamma_d/\kappa),$$

$$-X^4\{\theta^2+(1+\gamma_d/\kappa)^2\}+X^2\{3(1+\gamma_d/\kappa)+\theta^2+\Delta\theta(2+\gamma_d/\kappa)\}$$
$$-2\{1+\theta\Delta+\gamma_d(1+\Delta^2)/\kappa\}<0 \qquad (4.6)$$

The subcritical instability at Y_c on the lower branch is a new result. For the upper branch, the existence of a domain of instability in the single mode theory was already known (Lugiato et al.,1982b; Narducci et al., 1983) though it was not characterized analytically. It should be emphasized that in Lugiato et al.(1982b) the limit $\gamma_p>>\kappa,\gamma_d$ is first taken to eliminate the polarization. The resulting equations are then studied numerically in the limit C>>4. On the contrary, in Erneux et al.(1985) the limit C>>4 is studied but with equal orders of magnitude for the three decay rates. If the limit $\gamma_p>>\kappa,\gamma_d$ is then introduced it leads to divergencies, indicating the necessity of a new asymptotic expansion. In other words, the two limits $C\to\infty$ and $\gamma_p/\kappa\to\infty$ do not commute in this problem and the behavior of $C\kappa/\gamma_p$ (which may either remain finite or diverge) will lead to different results. This non-commutation of two limits is more a general rule than an exception.

5. NONLINEAR MAPS

An important difference between the general equations of optical bistability, Eqs.(2.1)-(2.4), and their mean field limit, Eqs.(2.13)-(2.17), resides in the boundary conditions. In the mean field limit, a suitable change of coordinates leads to the isochronous condition (2.17). Outside the mean field limit, the field E(0,t) is related to the field E(L,t-τ+L/c). This introduces a delay, τ, which is the cavity round-trip time. A very original approach to optical bistability has been proposed by Ikeda et al.(1979, 1980) in which the delayed boundary condition (2.4) is incorporated in the Eqs.(2.1)-(2.3) in some suitable limit. This leads to a set of coupled nonlinear delay-differential equations, the delay being τ. The success of this approach is due to the fact that in the limit of a large delay (compared to the material relaxation times γ_p and γ_d) the equations may be reduced to purely discrete equations, i.e., to nonlinear maps. The theory of nonlinear maps has recently been the subject of intense research so that a large body of methods and, above all, universal scaling laws are available (see, e.g., Collet et al.,1983).

To determine the linear stability of a steady state solution, it is entirely equivalent to use either directly the map or the differential equations and the limits which lead to the map. Although there were some doubts about this equivalence, it has been explicitly demonstrated by Lugiato et al.(1982c). The power of the description in terms of maps, and to some extend of delay-differential equations as well, is that it provides quite easily relevant informations about the nature and the stability of the time-dependent solutions beyond the threshold of instability for the steady solution.

A review of the stability properties of Eqs.(2.1)-(2.4) has been published by Englund et al.(1984). Here we shall review only later papers. Under the assumption $\gamma_p >> \gamma_d, \kappa$ the macroscopic polarization can be adiabatically eliminated from Eqs.(2.1) -(2.4). As shown by Ikeda (1979) these equations can then be transformed into

$$\frac{d\phi(t)}{dt} = -\gamma_d(\phi(t) - \tfrac{1}{2}) - \frac{1}{2\alpha L}|E'(t-\tau)|^2(e^{2\alpha L\phi(t)} - 1) \tag{5.1}$$

$$E'(t) = \sqrt{T}E_i' + RE'(t-\tau)e^{\alpha L\phi(t)}e^{i\{\alpha L\Delta\phi(t) + kL'\}} \tag{5.2}$$

where $\phi(t) = \frac{1}{L}\int_0^L D(t-\tau+z'/c, z')dz'$

$$E'(t) = |E(t,0)|/E_s, \quad E_i' = E_i/E_s, \quad E_s^2 = \frac{\gamma_p\gamma_d}{\mu^2}(1+\Delta^2)$$

In the absence of nonlinearities, the steady solution of (5.1) is $\phi = 1/2$. The so-called dispersive limit rests on the neglect of nonlinear absorption. It is assumed to be justified when $\Delta^2 >>1$. In this limit a trivial rescaling of the variables leads to (Ikeda et al., 1980):

$$\gamma_d^{-1}\frac{dX(t)}{dt}=-X(t)+|E(t-\tau)|^2 \tag{5.3}$$

$$E(t)=A+BE(t-\tau)\exp\{i(X(t)-X_0)\} \tag{5.4}$$

where A^2 is proportional to the input intensity and $B=Re^{-\alpha L/2}$. Finally, in the limit $B\to 0$ but $A^2B=O(1)$ we obtain

$$\gamma_d^{-1}\frac{dX(t)}{dt}=-X(t)+A^2+A^2B\cos\{X(t-\tau)-X_0\}$$

A significant amount of research has concentrated on a particular member of this class of equations

$$\gamma_d^{-1}\frac{dX(t)}{dt}=-X(t)+a-b\sin X(t-\tau) \tag{5.5}$$

The reason for analyzing this particular equation is that beside its relevance in all-optical bistable systems (Nakatsuka et al.,1983; Harrison et al.,1983) it is easy to set up a bistable electro-optical analog circuit which models (5.5) (Gibbs et al., 1981; Okada et al.,1981; Neyer et al., 1982, Hopf et al., 1982; Petersen et al., 1984). Furthermore in the limit $G=\gamma_d\tau\to\infty$ the delay-differential equation (5.5) reduces to a nonlinear map:

$$X_{n+1}=a-b\sin X_n\equiv f(X_n,a,b) \tag{5.6}$$

An erroneous interpretation of this map has led to many misconceptions about the nature of its solutions. Although it is true that near one maximum of $\sin X$ the map can be approximated by the logistic map $Y_{n+1}=\mu Y_n(1-Y_n)$ if $a=b$, the Ikeda map (5.6) is much richer because it is a two-dimensional multiple extrema map. Hence there is no reason to expect that (5.6) will closely follow only the Feigenbaum scenario.

The key to analyze (5.6) is to realize that a period-n solution of the map is also a steady solution of the nth composition of the map $f^{(n)}(X,a,b)=f(f^{(n-1)}(x,a,b),a,b)$ with $f^{(1)}=f$. Accordingly the steady solution of (5.6) is called the period-1 solution and verifies

$$X=a-b\sin X \tag{5.7}$$

which has an infinite number of branches. A linear stability analysis amounts to seek solutions of (5.6) in the form $X_n=X+\varepsilon X'\exp(\lambda n)+O(\varepsilon^2)$. This leads to the characteristic equation

$$f'(X,a,b)=e^{\lambda}=-b\cos X$$

The steady solution is stable as long as $|e^{\lambda}|<1$. Consequently there are two types of instability boundaries:
(i)When $\lambda=0$ modulo 2π (tangent boundary), $f'=+1$ and a limit point occurs where a pair of stable and unstable solutions emerge.
(ii)When $\lambda=i\pi$ modulo 2π (harmonic boundary), $f'=-1$ and a Hopf bifurcation occurs with exact period doubling.
This extends directly to period-n solutions which verify $X=f^{(n)}(X,a,b)$ and have harmonic boundaries of stability

$$X=f^{(n)}(X,a,b), \quad f'^{(n)}(X,a,b)=+1 \tag{5.8}$$

corresponding to a period doubling and tangent boundaries

$$X=f^{(n)}(X,a,b), \quad f'^{(n)}(X,a,b)=-1 \tag{5.9}$$

corresponding to domains of bistability for the period-n solutions. It is now an easy matter to plot the stability boundaries in the plane (a,b) and to predict, for any variation of a and b, the sequence of bifurcations and bistabilities (Mandel et al., 1983). Such a diagram is useful to predict, for instance, bistability between period-n solutions, between period-n and chaotic solutions or crises of the strange attractor.

In the same spirit, Moloney (1984) and Hammet et al.(1985) have studied the nonlinear map which derives from (5.2), further assuming that (i) γ_d is also much larger than κ so that $D(z,t)=(1+\Delta^2)/(1+\Delta^2+|E(z,t)|^2)$, (ii) the dispersive limit $(\Delta^2>>1)$

holds so that nonlinear absorption is negligible and (iii) the z dependence of D(z,t) is sufficiently weak to be neglected. The resulting map is

$$g_{n+1}=T^{1/2}E_i+Rg_n\exp\{i(\delta-\frac{p}{1+|g_n|^2})\} \tag{5.10}$$

where $g_n=E'(n\tau)/\Delta$ and $p=\alpha L\Delta=\alpha_o L/\Delta$. For this two-dimensional map (g is complex) the authors also stress the need to describe the stability boundaries in the parameter plane (E_i,p). Furthermore they discuss a class of boundaries which cannot be obtained from a linear stability analysis of (5.10). This new class of period-n solutions (n=6 is treated in details whereas n=8 and 10 are mentioned) and their period doubling cascades arise via the Newhouse sink phenemenon (Newhouse, 1980).

When γ_p and γ_d are both much greater than κ, the atomic variables in (2.1)-(2.4) can be adiabatically eliminated to yield

$$\frac{\partial E}{\partial t}+c\frac{\partial E}{\partial z}=\chi(|E|^2)E \tag{5.11}$$

$$E(0,t)=T^{1/2}E_i+Re^{-i\delta}E(L,t-\frac{L'-L}{c}) \tag{5.12}$$

A change of variables reduces (5.11) to an ordinary differential equation which is easily integrated. In this way the problem is again reduced to a nonlinear map for the complex field. It was analyzed by Haus et al.(1984) with an emphasis on the universal scaling laws which rule the subharmonic cascades to chaos. A similar analysis to determine Feigenbaum's universal constants is found in Carmichael et al.(1982).

It should be borne in mind that the maps described in this section are two-parameter multiple extrema maps (in one or two dimensions). Besides the scalar Feigenbaum scaling, these maps also exhibit vector scaling involving two new universal constants (Chang et al., 1981; Schell et al., 1983) or alternative "universal" routes to chaos (Mackay et al., 1984). In addition, period-n windows of chaotic dynamical systems also have their own scaling laws (Yorke et al., 1985).

Two papers deal with the influence of noise on the maps considered in this section. A paper by Zardecki (1982) studies numerically how additive noise increases the fractal dimension of the stange attractor of the map (5.4) when $1/\gamma_d=0$. Another paper, by Derstine et al.(1982), studies experimentally how multiplicative noise truncates the subharmonic cascade to chaos

6. DELAY-DIFFERENTIAL EQUATIONS

In this section we investigate the influence of the differential terms which are neglected to obtain discrete maps. In more physical terms, it amounts to analyze the influence of the cavity round trip time which is assumed to diverge in order to obtain the discrete map. We can state easily the difficulty in relation with Eq.(5.5):

$$G^{-1}\frac{dX(t)}{dt}=-X(t)+a-b\sin X(t-1) \tag{6.1}$$

where $G=\gamma_d\tau$. The limit $G\to\infty$ is a singular limit since it affects the derivative. That is, the solution $X(t;G)$ of (6.1) is not an analytic function of $1/G$ in the limit $1/G\to 0$. This is clearly displayed in the work of Chow et al.(1983) who constructed explicitly the first periodic solution of (6.1) in the limit $1/G\to 0$. Experimentally it is clear that when the delay is reduced, a wealth of bifurcation sequences still exists as shown in hybrid electro-optical bistable circuits modelling Eq.(6.1) by Petersen et al.(1984) for $G\cong 1$ and by Okada et al.(1981) with $G<1$. In the absence of delay Eq.(6.1) can no longer lead to chaos. However, it remains possible to set up hybrid electro optical bistable devices with no delay but with another type of nonlinearity which still display a subharmonic cascade to chaos (Mitschke et al.,1984).

For Eq.(6.1) the first theoretical paper to indicate that a finite G is likely to modify the bifurcation diagram is by Ikeda et al.(1982). They study numerically Eq.(6.1) with $G=40$ and $a=b$. The bifurcation parameter is a. They indicate that just beyond the first chaos threshold, the solution is a square wave of period $T \cong 2\tau$ with small-scale chaos added to it. This square wave undergoes a bifurcation sequence with periods T/3, T/5, T/7, ... ending abruptly with fully developed chaos. Each transition is first order with an hysteresis domain. The sequence is finite for finite G. As G increases, the number of periods T/(2n+1) in the sequence increases and the distance between two consecutive bifurcations decreases. Although such a sequence has been observed in a hybrid electro-optical device (Derstine et al., 1983) the results of Ikeda et al.(1982) and the experimental confirmation suffer from a bias because the studies were performed by linearly increasing in time the bifurcation parameter. As recently shown by Kapral et al. (1985) this bias delays the bifurcations and truncates bifurcation sequences.

In three papers, Gao et al. (1983a, 1983b, 1984) have reported results from a numerical analysis of (6.1) for $G \lesssim O(1)$. From their results we can conclude that an infinite sequence of bifurcations to chaos exists down to at least G=1/2. Very near the onset of oscillation, the power spectrum has a dominant peak corresponding to a single unstable eigenvalue of the linearized problem. However they point out that in this case the power spectrum of the periodic solutions also displays peaks of small amplitudes which correspond to the imaginary parts of the stable eigenvalues. These frequencies may be of importance since they correspond to a set of incommensurate frequencies.

These results are clarified by an analytic study of (6.1) due to Nardone et al.(1985). They discuss a striation process of the parameter plane (a,b). Each harmonic boundary (5.8) is split into an infinite set of secondary boundaries. Crossing the first member of an infinite sequence results in a quasi-period doubling (it is an exact doubling only if 1/G=0). Crossing the secondary boundaries of the sequence results in the progressive squaring of the periodic solution. Likewise, each tangent boundary (5.9) is also split into an infinite set of secondary boundaries. Crossing the first tangent boundary of an infinite sequence results in bistability (for steady or periodic solutions). Crossing the following boundaries of th sequence modifies the transient oscillatory evolution. Futhermore it is easily proved that the first periodic solution (corresponding to a Hopf bifurcation of the steady state) has a period $2\pi/\omega$ which varies continuously between 2 when $1/G=0$ and 4 when $G=0$. It is given by the first non-zero root of the dispersion relation

$$\mathrm{tg}\,\omega=-\omega/G$$

The following odd roots of the dispersion relation correspond to the first sequence of secondary harmonic boundaries whereas the even roots correspond to the first sequence of secondary tangent boundaries.

Finally we mention that Le Berre et al.(1985) have studied numerically Eqs.(5.1) and (5.2) without further approximation. In particular they stress the influence of the atomic detuning. For $G=10$, $R=0.95$ and $\chi=\alpha L(\Delta-i)=6\pi-2i$ there are chaotic attractors but no subharmonic cascades. Changing $\mathrm{Re}\chi$ into 7π , however, restores the period doubling cascade whereas $\mathrm{Re}\chi=\pi$ gives bistability but no instability.

Bar-Joseph et al.(1983) and Silberberg et al.(1984) have analyzed in detail the characteristic equation which is obtained from a linear stability analysis of Eqs.(5.3) and (5.4). They point out that an identical equation is obtained by ascribing the origin of instabilities to a four-wave mixing process (see also Moloney et al.,1984; Firth et al.,1984). The roots of the characteristic equation are studied numerically in the limit $B\to1$ and the period of oscillation for the first periodic solution is analyzed versus the cavity detuning. When $\gamma_d\tau\to\infty$ the period of oscillation is around 2τ only if $\delta\cong\pi$. Otherwise the period is around γ_d^{-1} with a continuous transition between the two values. When $\gamma_d\tau\to0$ the domain of instability is sharply reduced, but around $\delta=0$ period γ_d^{-1} solutions are again found.

7. TRANSVERSE EFFECTS

A set of experiments, performed on Na atoms, has been aimed at carefully testing the predictions of the two-level atom theory (Kimble et al.,1984; Rosenberger et al., 1985; Sandle et al., 1981; Weyer et al.,1981). Here the emphasis is on determining the steady solution and its instabilities without fiddling around with adjustable constants. One of the difficulties is that the input beam has, like in any laser, a transverse intensity profile. The influence of this transverse variation on the steady state and on the onset of bistability has been studied analytically (Ballagh et al.,1981; Drummond, 1981; Lugiato et al.,1983a and 1983b) for a cavity limited by spherical mirrors. This last point means that the TEM_{oo} mode structure of the input field may be assumed to be compatible with the filled ring cavity mode structure, which is surely not the case with plane mirrors. Under this assumption (called the one transverse mode assumption) and in the mean field limit the steady state equation becomes

$$Y^2=X^2\{(1+\frac{2C}{X^2}g(X^2))^2+(\theta-\frac{2C\Delta}{X^2}g(X^2))^2\} \tag{7.1}$$

$$g(X^2)=\ln\{\frac{1+\Delta^2+X^2}{1+\Delta^2+X^2\exp(-2R^2/w_o^2)}\} \tag{7.2}$$

where Y^2 and X^2 are the input and cavity intensities, respectively, scaled to the saturation intensity, w_o is the beam waist and R the radius of the cylindrical material medium. In the limit

$$R/w_o\to 0,\quad C\to\infty,\quad CR^2/w_o^2 \text{ finite}$$

the plane wave steady state is recovered. Lugiato et al.(1983a) have studied the stability of the steady solutions (7.1)-(7.2) in the absorptive mean field and good cavity limits. They were able to prove analytically that in the limit $R/w_o\to\infty$ all instabilities on the positive slope branches disappear. These results were extended by a numerical analysis for dispersive optical bistability without adiabatic elimination of the atomic variables (Lugiato et al.,1984b). Here,on the contrary, the authors conclude that the instability domain is enhanced when $R/w_o \to \infty$. Hence the disappearance of the plane wave instabilities is limited to the resonant case. This is confirmed by further results obtained by Asquini et al.(1985).

Another study of the influence of transverse effects is due to Moloney et al.(1982). The study is entirely numerical and is aimed at determining the corrections to the Ikeda map (5.10) when the transverse beam profile is retained. Their results indicate that when $R/w_o \neq 0$ new domains of instability appear and that the bifurcation sequence obtained from the map (5.10) can be drastically modified by quasiperiodic and frequency-locked solutions when the Fresnel number $F=\pi w_o^2/\lambda L$ is small.

A remarkable result has been proved by McLaughlin et al.(1983). The generalization of the map (5.10) which includes transverse effects is, in scaled variables,:

$$2i\frac{\partial G_n}{\partial \zeta}+\frac{\partial^2 G_n}{\partial y^2}=\frac{G_n}{1+2G_nG_n^*} \tag{7.3}$$

$$G_n(y,o)=T^{1/2}E_i(y)+Re^{ikl}G_{n-1}(y,p) \tag{7.4}$$

where ζ is the longitudinal coordinate, y the transverse coordinate and $p=\alpha_o L/\Delta$. The map (5.10) is simply the y-independent solution of (7.3) and (7.4) when $E_i(y)=E_i$. However, two results can be proved when the input field has a radial intensity distribution:

1.In the limit of weak nonlinearities (i.e. for a Kerr medium) the right-hand side of (7.3) reduces to $G_n-2G_nG_n^*$. Then (7.3) has a transverse soliton solution

$$G(y,\zeta)=P(\lambda y,\lambda)\exp\{i(\lambda^2-1)\zeta/2+i\gamma\} \tag{7.5}$$

$$P(\lambda y,\lambda)=\lambda \operatorname{sech}(\lambda y) \tag{7.6}$$

2.For the more general Eq.(7.3) an exact soliton solution is not known. Nevertheless an analytic expression (which has to be solved numerically) gives the condition for the emergence of a soliton.

Finally the authors show by direct numerical integration of (7.3) and (7.4) that higher transverse solitary waves can occur.

ACKNOWLEDGMENTS

This work was supported by the FNRS(Belgium), the Stimulation Action program of the European Community and the Association Euratom-Etat Belge.

REFERENCES

Abraham, E. and Smith, S. D. 1982, Rep.Prog.Phys. 45, 815

Arimondo, E., Casagrande, F., Lugiato, L. A. and Glorieux, P. 1983, Appl. Phys. 30B, 57

Asquini, M. L., Lugiato, L. A., Carmichael, H. J. and Narducci, L. M. 1985,Off-resonant mode instabilities in mixed absorptive and dispersive optical bistability, preprint

Ballagh, R. J., Cooper, J., Hamilton, M. W., Sandle, W. J. and Warrington, D. M. 1981, Optcs Comm. 37,143

Bar-Joseph, I. and Silberberg, Y. 1983, Optics Comm. 48,53

Basov, N. G. 1968, IEEE J. Quant.Electron. QE-4,855

Bischofberger, T. and Shen, Y. R. 1978, Appl.Phys.Lett.32,156

Bonifacio, R. and Lugiato, L. A. 1982, in Topics in current physics, vol.27, 61, Springer Verlag, Heidelberg

Bowden, C. M., Ciftan, M. and Robl, H. R.(ed.)1981, Optical Bistability, Plenum Press, New York

Bowden, C. M., Gibbs, H. M. and McCall, S. L. (ed.) 1984, Optical Bistability 2, Plenum Press, New York

Carmichael, H. J., Snapp, R. R. and Schieve, W. C. 1982,Phys. Rev. 26A, 3408

Carmichael, H. J. 1983, in Lecture notes in physics, vol.182, 64, Springer Verlag, Heidelberg

Carmichael, H. J. 1984, Phys.Rev.Lett. 52, 1292

Chang, S.-J., Wortis, M. and Wright, J. A. 1981, Phys.Rev. 24A, 2669

Chow, S. N. and Mallet-Paret, J. 1983, in Coupled nonlinear oscillators, 7, North-Holland, Amsterdam

Collet, P. and Eckmann, J. P. 1983, Itarated map on the interval as dynamical systems, Birkhäuser Verlag, Basel

Derstine, M. W., Gibbs, H. M., Hopf, F. A. and Kaplan, D.L. 1982, Phys.Rev. 26A, 3720

Derstine, M. W., Gibbs, H. M., Hopf, F. A. and Kaplan, D. L. 1983, Phys.Rev. 27A, 3200

Drummond, P. D. 1981, IEEE J. Quant. Electron. QE-17, 301

Englund, J. C., Snapp, R. R. and Schieve, W. C. 1984, in Progress in optics, vol. XXI, 335, North-Holland, Amsterdam

Erneux, T. 1985, Analytic studies of a laser with a saturable absorber, preprint
Erneux, T. and Mandel, P. 1985, Dispersive optical bistability: stability of the steady states, preprint
Firth, W. J., Wright, E. M. and Cummins, J. D. 1984, in Bowden et al.,111
Gao, J. Y., Yuan, J. M. and Narducci, L. M. 1983a,Optics Comm. 44, 201
Gao, J. Y., Narducci, L. M., Schulman, L. S., Squicciarini, M. and Yuan, J. M. 1983b, Phys. Rev. 28A, 2910
Gao, J. Y., Narducci, L. M., Sadiky, H., Squicciarini, M. and Yuan, J. M. 1984, Phys. Rev. 30A, 901
Garmire, E. (ed.) 1985, Special issue of IEEE J. Quant. Electro QE-21
Gibbs, H. M., McCall, S. L. and Venkatesan, T. N. C. 1976, Phys Rev.Lett. 36, 1135
Gibbs, H. M., McCall, S. L., Venkatesan, T. N. C., Gossard, A. C., Passner, A. and Wiegman, W. 1979, Appl. Phys.Lett. 35,451
Gibbs, H. M., Hopf, F. A., Kaplan, D. L. and Shoemaker 1981, Phys. Rev. Lett. 46, 474
Gerritsen, H. J. 1963, Proc. IEEE 51, 934
Hammel, S. M., Jones, C. K. R. T. and Moloney, J. V. 1985, J. Opt. Soc. Am. 2B,552
Harder, Ch., Lau, K. Y. and Yariv, A. 1982, IEEE J. Quant. Electron. QE-18, 1351
Harrison, R. G., Firth, W. J., Emshary, C. A. and Al-Saidi, I. A. 1983, Phys. Rev. Lett. 51, 562
Haus, J. W., Bowden, C. M. and Sung, C. C. 1984, J. Opt. Soc. Am. 1B, 742
Hopf, F. A., Meystre, P., Drummond, P. D. and Walls, D. F. 1979 Optics Comm. 31, 245
Hopf, F. A., Kaplan, D. L., Gibbs, H. M. and Shoemaker, R. L. 1982, Phys. Rev. 25, 2172
Ikeda, K. 1979, Optics Comm. 30, 257
Ikeda, K., Daido, H. and Akimoto, O. 1980, Phys. Rev. Lett. 45, 709
Ikeda, K., Kondo, K. and Akimoto, O. 1982, Phys. Rev. Lett. 49, 1467
Kapral, R. and Mandel, P. 1985, Phys. Rev. 32A, 1076
Kimble, H. R., Rosenberger, A. T. and Drummond, P. D. 1984, in Bowden et al.,1
Lasher, G. J. 1964, Solid-Statte Electronics, 7, 707
Le Berre, M., Ressayre, E., Tallet, A. and Gibbs, H. M. 1985, Non-Feigenbaum route to chaos of a nonlinear ring cavity with infinite temporal dimension, preprint
Lugiato, L. A. 1983, Contemp. Physics 24, 333
Lugiato, L. A. 1984, in Progress in optics, vol. XXI, 69, North-Holland, Amsterdam
Lugiato, L. A., Milani, M. and Meystre, P. 1982a, Optics Comm. 40, 307
Lugiato, L. A., Narducci, L. M., Bandy, D. K. and Pennise, C.A. 1982b, Optics Comm. 43, 281
Lugiato, L. A., Asquini, M. L. and Narducci, L. M. 1982c, Optic Comm. 41, 450
Lugiato, L. M. and Milani, M. 1983a, Z. Physik 50B, 171
Lugiato, L. M. and Milani, M. 1983b, Optics Comm. 46, 57
Lugiato, L. M., Mandel, P. and Narducci, L. M. 1984a, Phys. Rev 29A, 1438

Lugiato, L. A., Horowicz, R. J., Strini, G. and Narducci, L. M. 1984b, Phys. Rev. 30A,1366
Lugiato, L. A. and Narducci, L. M. 1985, Phys. Rev. 32A, 1576
Mackay, R. S. and Tresser, C. 1984, J. Physique Lett. 45, L-741
Mandel, P. and Erneux, T. 1984, Phys. Rev. 30A,1893
Mandel, P. and Kapral, R. 1983, Optics Comm. 47, 151
McCall, S. L. and Gibbs, H. M. 1982, in Topics in current physics, vol. 27, 93, Springer Verlag, Heidelberg
McLaughlin, D. W., Moloney, J. V. and Newell, A. C. 1983, Phys. Rev. Lett. 51, 75
Miller, D. A. B., Smith, S. D. and Johnston, A. 1979, Appl. Phys. Lett. 35, 658
Mitschke, F. and Flüggen, N. 1984, Z. Physik 35B, 59
Moloney, J. V. 1984, Optics Comm. 48, 435
Moloney, J. V., Hopf, F. A. and Gibbs, H. M. 1982, Phys. Rev. 25A, 3442
Moloney, J. V., Hammel, S. and Jones, C. R. T. 1984, in Bowden et al., 87
Nakatsuka, I., Asaka, S., Itoh, H., Ikeda, K. and Matsuoka, M. 1983, Phys. Rev. Lett. 50, 109
Nardone, P., Mandel, P. and Kapral, R. 1985, Analysis of a delay-differential equation in optical bistability, preprint
Narducci, L. M., Bandy, D. K., Pennise, C. A. and Lugiato, L. A. 1983, Optics Comm. 44, 207
Nathan, M. I., Marinace, J. C., Rutz, R. F., Michel, A. E. and Lasher, G. L. 1965, J. Appl. Phys. 36, 473
Newhouse, S. E. 1980, in Progress in mathematics, vol. 8,1, Birkhäuser, Basel
Neyer, A. and Voges, E. 1982, IEEE J. Quant. Electron. QE-18, 2009
Okada, M. and Takizawa, K. 1981, IEEE J. Quant. Electron. QE-17, 2135
Petersen, P. M., Ravn, J. N. and Skettup, T. 1984, IEEE J. Quant. Electron. QE-20, 690
Rosenberger, A. T., Orozco, L. A. and Kimble, H. J. 1985, Intrinsic dynamical instability in optical bistability with two-level atoms, preprint
Sandle, W. J. and Callagher, A. 1981, Phys. Rev. 24A, 2017
Schell, M., Fraser, S. and Kapral, R. 1983, Phys. Rev. 28A,373
Silberberg, Y. and Bar-Joseph, I. 1984, J. Opt. Soc. Am. 1B,662
Smith, P. W. (ed.)1981, Special issue of IEEE J. Quant. Electron. QE-17
Smith, S. D., Miller, A. and Wherrett, B. S. (ed.) 1985, Optical bistability, dynamical nonlinearity and photonic logic, The Royal Society, London; also published in Phil. Trans. R. Soc. Lond. 313A, 187
Swinnerton-Dyer, P. 1984, Phil.Trans.R.Soc.Lond.313A,191
Szöke, A., Daneu, V., Goldhar, J. and Kunit, N. A. 1969, Appl. Phys. Lett. 15,376
Weyer, K. G., Wiedemann, M. W., Rateike, M., McGillavray, W.R., Meystre, P. and Walther, H. 1981, Optics Comm. 37, 426
Yorke, J. A., Grebogi, C., Ott, E. and Tedeschini-Lalli, L. 1985, Phys. Rev. Lett. 54, 1095
Zardecki, A. 1982, Phys. Lett. 90A, 274

PHOTONS AND INTERFERENCE

E R PIKE AND SARBEN SARKAR

1. INTRODUCTION

Although a satisfactory general theory of photodetection (Glauber, 1963) was formulated over twenty years ago, there has been, from time to time, revival of interest in formulating the concept of a localised photon independent of the detection process. This in itself is not absurd since in non-relativistic quantum mechanics we regularly introduce the wavefunction of a particle (moving in a potential well, say), although it, by itself, is not the quantity that is directly measured. For a photon the arguments of Landau and Peierls (1930), Newton and Wigner (1949) and Wightman (1962) on the non-existence of a local wavefunction discouraged further discussion for a time. However, Jauch and Piron (1967) and Amrein (1969) returned to these issues and introduced concepts more general than that of a wavefunction. Others have also considered the problem (Mandel (1966), Tatarski (1972) and Cook (1982a, 1982b)). On the experimental side Burnham and Weinberg (1970) and Friberg et al (1985a, 1985b) showed that using parametric down conversion it is possible to produce closely correlated photon pairs which may be in most respects replicas of each other, and with suitable gates and shutters this allows, in principle, convenient exploration of the properties of localised single photon states. Some experiments along these lines have been published recently (Jakeman and Walker (1985), Pike (1986)). In such experiments we enter realms where implicit or inadvertent use of classical ideas of local properties may not be permissible.

It is for this reason that it is opportune to examine the issues of localisation which are involved critically. Glauber's annihilation photodetection operator may then be one of the possible operators which may be applied to a correctly formulated single photon state.

In order to set the scene we will consider what is involved in describing a highly idealised experiment to measure the interference of light from two separate sources S_1 and S_2 (see Figure 1) which each give off one photon during the same short time period. Classically, to describe interference we would add the electric fields $\underline{E}_1$ and $\underline{E}_2$ at the screen emanating from S_1 and S_2, and then square the modulus of this sum. The cross-term obtained as a result of this squaring gives rise to the effect of interference.

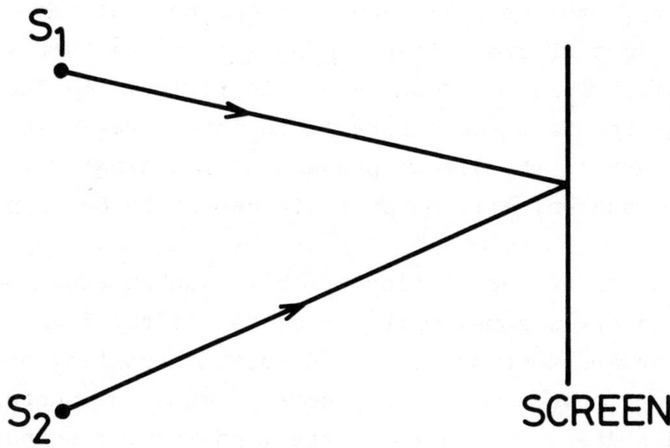

Figure 1. Idealised Interference Experiment

In quantum mechanics on the other hand the radiation field is described by states in an abstract Hilbert space. When this space is described in terms of the occupation numbers of states of given momentum and polarisation, it is called a Fock space (Dirac (1978)). In terms of this description, the fact that there are two sources of light requires that we have a direct product of states of the Fock space. Moreover the quantum mechanical description requires the existence of only one electric field operator E, albeit normally expressed as a Fourier sum, since the information about the existence of two sources is contained exclusively in the nature of the Fock space states to which they give rise. Any interference effects are then determined by the matrix elements

$$\langle \text{product state} \,|E^{(-)}(x)\, E^{(+)}(x)|\, \text{product state}\rangle$$

where x is a spatial co-ordinate in the field along the screen and $E^{(+)}$ and $E^{(-)}$ are the positive and negative frequency parts of the electric field operator (Glauber 1963). It is only in the special case when the product state is a direct product of two suitable coherent states that we obtain the classical rule of adding electric fields. In fact this, as well as the fact that there is no interference when the product state is the direct product of two single photon wavepacket states, will be shown in detail in Section 6.

In the above description of the gedanken experiment we have implicitly assumed that the states in the Fock space are straightforward to construct. In fact, in most theoretical treatments little attempt is generally made to construct localised states. Plane waves are used whether or not they are appropriate. In experiments of the Burnham and Weinberg kind, a wavepacket description is necessary. However, for an electric field, we cannot construct an arbitrarily localised wavepacket since this is not possible for a transverse vector field. This extra condition of transversality is at the root of the photon localisation problems and led to the introduction of the concept of a "generalised imprimitivity" by Jauch and Piron (1967) and Amrein (1968), and we will re-interpret many of their ideas from the 'practical' point of view of constructing theoretically sound wavepackets for the photon. Hence it is clear that even to get a correct description of the above simple gedanken experiment we need to change radically the way of thinking which is natural in classical optics; in particular, the addition of electric fields to represent a superposition will not be allowed in an arbitrary case.

The plan of this paper follows. In section 2 we outline the quantisation of the electromagnetic field in the radiation gauge as well as in a class of closely related physical gauges. In section 3 we summarise the main results of the representation theory of the Poincaré group for both massive and massless particles. In section 4 an approach due to Newton and Wigner (1949) and Wightman (1962) for the study of localisability of elementary systems is given. The approach is illustrated by the example of the Klein-Gordon particle (Schweber, 1966). The Tatarski-Cook continuity equation is examined with special stress on gauge invariance. Arguments are given for the impossibility of a conventional localisation of the photon. The section concludes with a simple but relevant example of a generalised imprimitivity due to Jauch and Piron (1967). In section 5 the photon is treated more rigorously and it is shown to be 'weakly localisable' (Jauch and Piron, 1967; Amrein, 1967). A rigorous new non-local continuity equation is also presented. In section 6 the relevance of the new quantities introduced for the theory of photodetection (Glauber, 1963) is given. In particular, using creation operators for true single photon wavepackets (which may be been formed with the help of generalised imprimitivities), criteria for interference between two light beams containing small numbers of photons is formulated.

2. QUANTISATION OF FREE ELECTROMAGNETIC FIELDS

We will briefly outline the quantisation of the electromagnetic field partly to establish notation. We will follow very closely the treatment of Kibble (1968).

In quantum optics it is usual to quantise in the Coulomb gauge because the physical degrees of freedom of the photon are manifest (Schwinger 1948, 1962). We will present the quantisation in this gauge in a way which superficially 'appears' more covariant and then we will introduce other gauges which are simply related to the Coulomb gauge potentials but satisfy the Lorentz gauge condition. Such a formulation has the advantage of not needing to work with Gupta-Bleuler indefinite metrics (Gupta 1950, Bleuler 1950) since again only the two helicity states of the photon are introduced. This general philosophy will be explained fully below.

The electromagnetic field is described by the vector Heisenberg operator $A_\mu(x)$. x denotes (t, $\underline{x}$), ie time and space. (We will throughout try not to distinguish space and time indices explicitly, and $\mu = 0$ will be the time index whereas $\mu = 1,2,3$ will denote the spatial indices.) The Minkowski metric will be taken to be $g_{oo} = 1$, $g_{ii} = -1$ for $i = 1,2,3$. It is useful to introduce the electromagnetic field tensor $F_{\mu\nu}$

$$F_{\mu\nu} = \partial_\mu A_\nu - \partial_\nu A_\mu \tag{1}$$

The electric field E_k and magnetic field B_k are

$$\begin{aligned} E_k &= F_{ko} \\ B_k &= \tfrac{1}{2}\,\varepsilon_{ijk}\,F_{jk} \end{aligned} \tag{2}$$

In keeping with standard theoretical practice we are choosing units such that $c = \hbar = 1$, c being the speed of light and

($2\pi\hbar$) Planck's constant. Keeping to a 4-vector and tensor formulation we believe helps in the study of the questions raised in this paper. The Lagrangian density L(x) for the electromagnetic field is given by

$$L(x) = -\tfrac{1}{4} F^{\mu\nu} F_{\mu\nu} \tag{3}$$

The Euler-Lagrange equations give

$$\partial_\nu F^{\mu\nu} = 0 \tag{4}$$

In this context gauge invariance of the second kind means that Eqn. (4) is invariant under

$$A_\mu(x) \to A_\mu(x) + \partial_\mu \lambda(x) \tag{5}$$

for any scalar function $\lambda(x)$. The longitudinal component of $A_\mu(x)$ is not a dynamical variable. It is fixed in the Coulomb gauge by

$$\partial_k A^k = 0 \tag{6}$$

Equation (4) then reduces to

$$\partial^2 A_\mu(x) = 0 \tag{7}$$

$A_\mu(x)$ is Hermitian and so $A_\mu(p)$, its Fourier transform (with 4 momentum p) has the form

$$\tilde{A}_\mu(p) = 2\pi\ \delta(p^2)\ [\theta(p^\circ)\ C_\mu(\underline{p}) + \theta(-p^\circ)\ C_\mu(-\underline{p})^+]$$

where θ is the Heaviside function and δ is the Dirac δ-function and where

$$C_\mu(p) = \sum_{\lambda=1}^{2} e_\mu^{(\lambda)}(p)\ a_\lambda(\underline{p}) \tag{9}$$

with

$$e_o^{(\lambda)} = 0 \quad , \quad e_\mu^{(\lambda)} e^{(\lambda')\mu} = \delta_{\lambda\lambda'}$$

$$p^\mu e_\mu^{(\lambda)}(p) = 0 \; , \; \sum_{\lambda=1}^{2} e_i^{(\lambda)}(p) e_j^{(\lambda)}(p) = \delta_{ij} - \frac{p_i p_j}{p_o^2} \; (\equiv \gamma_{ij}(\underline{p})) \qquad (10)$$

(The summation convention is adopted for repeated 4 indices which are written with Greek letters. Latin letters will denote the 'spatial' indices.)

The $e^{(\lambda)}(\underline{p})$ are the polarisation vectors and $a_\lambda(p)$ is the creation operator for a single photon state of momentum p and helicity λ satisfying

$$[a_\lambda(p), a^+_{\lambda'}(\underline{p}')] = (2\pi)^3 \, 2p^\circ \, \delta_{\lambda\lambda'} \, \delta^{(3)}(\underline{p}-\underline{p}') \qquad (11)$$

with $p^\circ = |\underline{p}|$. It is important to note that the appearance of p° or otherwise in Eqn. (11) is a matter of convention (Bjorken and Drell, 1965). With the convention of Eqn. (11)

$$[c_\mu(\underline{p}), c_\nu(\underline{p}')^+] = \gamma_{\mu\nu}(\underline{p}) \, (2\pi)^3 \, 2p^\circ \, \delta^{(3)}(\underline{p}-\underline{p}') \qquad (12)$$

The projector $\gamma_{\mu\nu}(\underline{p})$ is non-zero only if $\mu, \nu \neq 0$. Moreover

$$A_\mu(x) = \int \frac{d^3p}{(2\pi)^3 2p_o} \left(c_\mu(\underline{p}) e^{ip.x} + c^+_\mu(p) e^{-ip.x}\right) \qquad (13)$$

If we had chosen the convention in Eqns. (11) and (12) without the p° then

$$\hat{A}_\mu(x) = \int \frac{d^3p}{(2\pi)^3 2p_o^{\frac{1}{2}}} \left(\hat{c}_\mu(\underline{p}) e^{ip.x} + \hat{c}^+_\mu(p) e^{-ip.x}\right) \qquad (14)$$

where the '^' denotes the operators in the new convention. The attraction of the convention in Eqns. (11) and (13) is that the integration measure

$$\int \frac{d^3\underline{p}}{p_o} \tag{15}$$

is Lorentz invariant. Only physical states are involved since λ = 1,2 (ie there are only two helicities). It is possible to go to a 'Lorentz type' gauge with

$$\partial_\mu A^\mu(x) = 0 \tag{16}$$

via a gauge transformation. In particular we define

$$\tilde{c}_\mu(p) = c_\mu(\underline{p}) + p_\mu b^\nu c_\nu(\underline{p}) \tag{17}$$

where b^ν is some 4 component c-number. It is then easy to show that

$$[\tilde{c}_\mu(p), \tilde{c}_\nu^+(p')] = (2\pi)^3 2p_o \delta^{(3)}(\underline{p}-\underline{p}') \tilde{\gamma}_{\mu\nu}(\underline{p}) \tag{18}$$

and

$$\begin{aligned}\tilde{\gamma}_{\mu\nu}(p) = \gamma_{\mu\nu}(\underline{p}) + p_\nu(b^\rho)^* \gamma_{\mu\rho}(\underline{p}) + p_\mu b^\lambda \gamma_{\lambda\nu}(\underline{p}) \\ + p_\mu p_\nu b^\lambda(b^\rho)^* \gamma_{\lambda\rho}(\underline{p})\end{aligned} \tag{19}$$

Again $\tilde{\gamma}_{\mu\nu}$ is a 2 dimensional projector and

$$\begin{aligned} p^\mu \tilde{\gamma}_{\mu\nu}(p) &= 0 \\ s^\mu \tilde{\gamma}_{\mu\nu}(p) &= 0 \end{aligned} \tag{20}$$

for some vector s^μ.

Clearly by construction

$$p^\mu \ c_\mu(p) \ = \ 0 \tag{21}$$

The physical Fock space can still be made out of the Coulomb vacuum $|o\rangle$ and the operators $a_\lambda(p)$. $|o\rangle$ satisfies

$$c_\mu(p) \ |o\rangle \ = \ 0 \tag{22}$$

The most general one photon state is characterised by a photon wavepacket $f_\mu(\underline{p})$ for which

$$\begin{aligned} f^\circ &= 0 \\ p_j f^j &= 0 \end{aligned} \tag{23}$$

This just reflects the transverse nature of the electromagnetic field. We will discuss later the limit to which a photon can be localised for a given f_μ. The Hilbert space state corresponding to this wavepacket is

$$|1,f\rangle \ = \ f^+ \ |o\rangle \tag{24}$$

where

$$f^+ \ = \int \frac{d^3p}{(2\pi)^3 2p^o} \ c_\mu^+(\underline{p}) f^\mu(\underline{p}) \tag{25}$$

The f^+ satisfy the commutation relations

$$[f^-,g^+] \ = \int \frac{d^3p}{(2\pi)^3 2p^o} \ f_\mu^*(p) \ g^\mu(p) \tag{26}$$

where $f^+|o\rangle$ and $g^+|o\rangle$ are any photon wavepackets, and $f^- = (f^+)^*$.

(In the class of gauges of Eqn. (17) a similar construction goes through except that

$$\tilde{\gamma}_{\mu\nu}(\underline{p})\ f^{\nu}(\underline{p}) \quad = \quad f_{\mu}(\underline{p}) \tag{27}$$

instead of Eqn. (23).) The one photon state is of course gauge invariant since any gauge transformation has the form

$$c_{\mu}(\underline{p}) \rightarrow c_{\mu}(\underline{p}) + p_{\mu}\ \lambda(\underline{p})$$

and

$$p^{\mu}\ f_{\mu}(\underline{p}) \quad = \quad 0 \tag{28}$$

An n photon Fock state is given by

$$|n;f_1,\ldots,f_n\rangle \quad = \quad f_1^{+}\cdots\cdots f_n^{+}\ \ |o\rangle \tag{29}$$

3. REPRESENTATIONS OF THE POINCARÉ GROUP

In order to understand the concept of generalised imprimitivity as applied to a photon (Jauch and Piron 1967, Amrein 1968) it is necessary to recall some of the main results for the representations of the Poincaré group (Wigner 1939, Bargmann and Wigner 1948).

The Poincaré group is the semidirect product of the translation group T(4) and the orthochronous homogeneous Lorentz group SO(3,1). A useful representation of the group is in terms of 2x2 complex matrices, since x_{μ} can be associated with a 2x2 matrix as follows

$$x_{\mu} \rightarrow \underline{\underline{x}} \quad = \quad x^{\mu}\sigma_{\mu} \quad = \quad \begin{pmatrix} x_o - x_3 & -x_1 + ix_2 \\ -x_1 - ix_2 & x_o + x_3 \end{pmatrix} \tag{30}$$

and

$$\sigma_0 = \begin{pmatrix} 1 & 0 \\ 0 & 1 \end{pmatrix}, \sigma_1 = \begin{pmatrix} 0 & 1 \\ 1 & 0 \end{pmatrix}, \sigma_2 = \begin{pmatrix} 0 & -i \\ i & 0 \end{pmatrix}, \sigma_3 = \begin{pmatrix} 1 & 0 \\ 0 & -1 \end{pmatrix} \quad (31)$$

which are the Pauli matrices. A translation by a vector a_μ can be represented as

$$\underline{\underline{x}} \rightarrow \underline{\underline{x}} + \underline{\underline{a}} \quad (32)$$

A general homogeneous Lorentz transformation can be written as

$$\underline{\underline{x}} \rightarrow U(\Lambda)\, \underline{\underline{x}}\, U^+(\Lambda) \quad (33)$$

where

$$U(\Lambda) = \exp[-\tfrac{1}{2}i(\underline{\theta}+i\underline{\zeta}).\underline{\sigma}] \quad (34)$$

with $\underline{\theta}$ and $\underline{\zeta}$ two real 3-vectors. ($\underline{\theta} = 0$, $\underline{\zeta} \neq 0$) corresponds to boosts and ($\underline{\zeta} = 0$, $\underline{\theta} \neq 0$) corresponds to rotations. The importance of the Poincaré group for particles is that any non-composite system transforms as a unitary representation of the Poincaré group. The Lie algebra (Bjorken and Drell, 1965) for the Poincare group is

$$\begin{aligned} [P_\mu, P_\nu] &= 0 \\ [J_{\mu\nu}, P_\lambda] &= i(\eta_{\nu\lambda}P_\mu - \eta_{\mu\lambda}P_\nu) \\ [J_{\kappa\lambda}, J_{\mu\nu}] &= i(\eta_{\lambda\mu}J_{\kappa\nu} + \eta_{\kappa\nu}J_{\lambda\mu} - \eta_{\kappa\mu}J_{\lambda\nu} - \eta_{\lambda\nu}J_{\kappa\mu}) \end{aligned} \quad (35)$$

P_μ is the generator of translations while the antisymmetric operator $J_{\mu\nu}$ is the generator of rotations and Lorentz boosts. The Casimir operators (Wigner 1939, Bargmann and Wigner 1948) (which form the set of operators which commute with all operators) are

$$P^2 = P^\mu P_\mu$$

and

$$S^2 = S^\mu S_\mu \tag{36}$$

$$S_\kappa = -\tfrac{1}{2} \varepsilon_{\kappa\lambda\mu\nu} J^{\lambda\mu} P^\nu/(P^2)^{\frac{1}{2}} \tag{37}$$

S_κ behaves as a 4-vector under homogeneous Lorentz transformations but is invariant under translations. From the standard theory of Lie groups (Wybourne 1974) the irreducible representation are labelled by the eigenvalues of S^2 and P^2. A representation is given by the states $|p,j,\lambda\rangle$ in a Hilbert space such that

$$\underline{P}^2 \; |p,j,\lambda\rangle = p^2 \; |p,j,\lambda\rangle$$

$$S^2 \; |p,j,\lambda\rangle = s^2(j)|p,j,\lambda\rangle \tag{38}$$

$$S_o \; |p,j,\lambda\rangle = w_o(\lambda)|p,j,\lambda\rangle$$

where $s(j)$ and $w_o(\lambda)$ are functions whose precise form depends on p^2. The theory of induced representations (Mackey 1978) has been used to obtain such representations. p^2 is of course the square of the mass of the one particle state. The mass m can be positive or zero. It might seem that for the photon we can restrict ourselves to m=0 representations. Although this is true, the generalised imprimitivities that we will develop later depend on a certain isomorphism with the representation of massive states having spin one.

If $U(b,\Lambda)$ is the unitary operator corresponding to a translation b and homogeneous Lorentz transformation Λ, then for a massless particle of helicity λ (Amrein 1969)

$$U(b,\Lambda)|\Lambda^{-1}p,|\lambda|,\lambda\rangle = e^{ip.b} \delta_{\lambda\lambda'}(\chi(p,\Lambda))^{2\lambda} \; |p,|\lambda|,\lambda'\rangle \tag{39}$$

where

$$\chi(p,\Lambda) = \frac{(|\underline{p}| + p^3)u^* - (p^1 - ip^2)w^*}{|(|\underline{p}|+p^3)u^* - (p^1-ip^2)w^*|}$$

and

$$\Lambda = \begin{pmatrix} v & w \\ t & u \end{pmatrix} \in SL(2,C)$$

with $vu - tw = 1$.

(We recall that SL(2,C) is the covering group of the Lorentz group.) For the massless case (and the massless case only) λ plays the role of a Casimir label, and $S_\mu = \lambda P_\mu/(P^2)^{\frac{1}{2}}$.

We will now consider the representations of a massive particle. There are two unitarily equivalent representations which are useful. We proceed as follows.

Let $\hat{p} = (m,o,o,o)$ and L_p be a Lorentz boost such that

$$L_p \hat{p} = p \tag{40}$$

There is an arbitrariness in the choice of L_p which will lead to the 'canonical' formalism or the 'helicity' formalism depending on the precise form of L_p. It is necessary to work with both formalisms since Mackey's theorem on ordinary imprimitivities (Mackey 1978, Wightman 1962) is most naturally stated in the canonical formalism, but the isomorphism theorem (Amrein 1969) between massive and massless representations for particles of given j can only be stated in the helicity formalism. Given a choice of L_p

$$U(b,\Lambda)|p,j,\lambda\rangle = |\Lambda p,j,\lambda'\rangle e^{i\Lambda p.a} D(j)^{\lambda}_{\lambda'}(\hat{\Lambda}(\Lambda,p)) \tag{41}$$

where

$$\hat{\Lambda}(\Lambda,p) = L^{-1}_{\Lambda p} \Lambda L_p$$

and $D^{(j)}$ are the finite dimensional representations of SU(2,C) well-known from the theory of angular momentum (Rose 1957).

If

$$L_p^{(c)} = (2m(p^\circ+m))^{-\frac{1}{2}} \begin{pmatrix} m+p^\circ+p^3 & p^1-ip^2 \\ p^1+ip^2 & m+p^\circ-p^3 \end{pmatrix} \tag{42}$$

then we have the canonical formalism.

If

$$L_p^{(h)} = \tfrac{1}{2}(m(p^\circ+m)(|\underline{p}|+p^3)|\underline{p}|)^{-\frac{1}{2}} \begin{pmatrix} |\underline{p}|+p^3 & -p^1+ip^2 \\ p^1+ip^2 & |\underline{p}|+p^3 \end{pmatrix} \begin{pmatrix} m+p^\circ+|\underline{p}| & 0 \\ 0 & m+p^\circ-|\underline{p}| \end{pmatrix} \tag{43}$$

then we have the helicity formalism. Equation (43) just represents a boost (of the form in Eqn. (42)) from $\hat{p}$ to $(p^\circ,o,o,|\underline{p}|)$ followed by a rotation of $(p^\circ,o,o,|\underline{p}|)$ into p.

The D^J matrices satisfy the recursion relations

$$D^J_{\alpha\beta}(B) = \sum_{\gamma,\delta,\varepsilon,\chi} C(J-\tfrac{1}{2},\tfrac{1}{2},J|\gamma,\delta,\alpha)C(J-\tfrac{1}{2},\tfrac{1}{2},J|\varepsilon,\chi,\beta)D^{J-\frac{1}{2}}_{\gamma\varepsilon}(B)B_{\delta\chi} \tag{44}$$

with

$$D^o(B) = 1 \quad \text{and} \quad D^{\frac{1}{2}}_{\alpha\beta}(B) = B_{\alpha\beta}$$

The Cs are the Clebsch-Gordan coefficients (Rose 1957). From the addition of the spin of two spin ½ particles we know that

$$C(\tfrac{1}{2},\tfrac{1}{2},1|\tfrac{1}{2},\tfrac{1}{2},1) = C(\tfrac{1}{2},\tfrac{1}{2},1|-\tfrac{1}{2},-\tfrac{1}{2},-1) = 1$$
$$C(\tfrac{1}{2},\tfrac{1}{2},1|\tfrac{1}{2},-\tfrac{1}{2},o) = C(\tfrac{1}{2},\tfrac{1}{2},1|-\tfrac{1}{2},\tfrac{1}{2},o) = 2^{-\frac{1}{2}} \qquad (45)$$

and, for the case of interest to us, these are the only Clebsch-Gordan coefficients that would be required for explicit calculations of wavepackets.

4. IMPRIMITIVITIES AND LOCALISABILITY

If for a particle there is a position operator x_{op} then all eigenstates $|x\rangle$ of x_{op} are localised states of the particle and if the state of the particle is $|\psi\rangle$ then

$$\int_{\Omega} |\langle\psi|x\rangle|^2 \, d\mu(x) \qquad (46)$$

is the probability of finding the particle in a region Ω where $\mu(x)$ is the Lebesgue measure (Reed and Simon 1972). It is interesting to note that there has been no mention of the measuring process and (46) is just a statement of the Born interpretation of the wavefunction (Landau and Lifshitz, 1959; Wheeler and Zurek, 1983). For relativistic particles the arguments are more subtle.

In order to make progress on the question of localisability it is necessary to be precise about our requirements. Our aim will, of course, always be to find the analogue of $|x\rangle$ in the relativistic situation. Newton and Wigner (1949) required:

i. that the set of all states localised at time 0 at $\underline{x} = 0$ forms a linear manifold invariant under spatial rotations about the origin, spatial inversions and time inversions;

ii. that if a state $|\psi_{\underline{x}}\rangle$ is localised at some point $\underline{x}$, then a spatial displacement shall make it orthogonal to the set of states localised at the point $\underline{x}$;

iii. that the infinitesimal operators of the Lorentz group are meaningful when applied on the 'localised' states. This is a regularity condition.

As an example of states satisfying the Newton-Wigner criteria we will study the massive Klein-Gordon equation (Schweber, 1966)

$$(\Box + m^2)\, \phi(x) = 0 \tag{47}$$

We will consider the positive energy solutions which form a linear vector space. A Hilbert space structure can then be introduced on defining a scalar product (at a time t)

$$\langle\phi|\psi\rangle_t = i\int d^3x(\tilde{\phi}^*(x)\, \partial_0\, \tilde{\psi}(x) - (\partial_0\tilde{\phi}^*(x))\, \tilde{\psi}(x)) \tag{48}$$

In momentum space

$$\langle\phi|\psi\rangle_t = \int \frac{d^3k}{k_0}\, \phi^*(\underline{k})\, \psi(\underline{k}) \tag{49}$$

where $k_0 = (\underline{k}^2+m^2)^{\frac{1}{2}}$ and

$$\tilde{\phi}(x) = \frac{1}{(2\pi)^{3/2}}\int\frac{d^3k}{k_0}\, e^{-ik.x}\, \phi(\underline{k}) \tag{50}$$

The naive position operator is, surprisingly, unsatisfactory

$$\begin{aligned}\langle\psi|\underline{x}\phi\rangle_t &= i\int\frac{d^3p}{p_o}\,\psi^*(\underline{p})\,\underline{\nabla}_p\,\phi(\underline{p}) \\ &= \int\frac{d^3p}{p_o}\left[\left(-i\nabla_p + \frac{i\underline{p}}{\underline{p}^2+m^2}\right)\psi^*(\underline{p})\right]\phi(\underline{p}) \qquad (51) \\ &\neq \langle\underline{x}\psi,\phi\rangle_t\end{aligned}$$

A measurable property of the system should be expressible as a self-adjoint operator (Dirac 1978). $\underline{x}$ is not, and so it is not acceptable as a position operator. It is easy to show that a suitable operator is

$$\underline{x}_{op} = i\nabla_{\underline{p}} - \tfrac{1}{2}\,\frac{i\underline{p}}{\underline{p}^2+m^2} \qquad (52)$$

and a typical wavefunction is

$$\psi_{\underline{x}_o}(\underline{k}) = (2\pi)^{-3/2}\,e^{-i\underline{k}.\underline{x}_o}\,k_o^{\frac{1}{2}} \qquad (53)$$

with eigenvalue $\underline{x}_o$. In configuration space

$$\tilde{\psi}_{\underline{x}_o}(\underline{x}) \propto \left(\frac{m}{r}\right)^{5/4} H^{(1)}_{5/4}(imr) \qquad (54)$$

where $H^{(1)}$ is a Hankel function and

$$r = |\underline{x} - \underline{x}_o|.$$

These states are exactly the localised states since

$$\langle\psi_{\underline{x}_o}|\psi_{\underline{x}'_o}\rangle = \delta(\underline{x}_o-\underline{x}'_o) \qquad (55)$$

However in configuration space they look like extended objects, but this is just a reflection of the fact that $\underline{x}$ is not a position operator. As $m\to\infty$ we recover the naive picture. The operators $\underline{x}_{op}$ have the relevant good properties such as

$$[(x_{op})_i \, , \, (x_{op})_j] = 0$$
$$[(x_{op})_\ell \, , \, (P)_j] = i\delta_{\ell j} \tag{56}$$

From our earlier heuristic considerations we expect to be unable to treat the photon in the same way as the Klein-Gordon particle. The non-existence of (46) can be seen in a related difficulty of constructing probability densities and currents for the photon. Tatarski (1972) and Cook (1982a, 1982b) have constructed operators

$$b_\mu(\underline{x},t) = \frac{1}{(2\pi)^{3/2}} \int \frac{d^3k}{k_o} \, e^{ik.x} \, k_o^{\frac{1}{2}} \, C_\mu(k) \tag{57}$$

Now we introduce

$$n(\underline{x},t) = b_\mu^+(\underline{x},t) \, b^\mu(\underline{x},t)$$
$$= \frac{1}{(2\pi)^3} \iint \frac{d^3k}{k_o} \frac{d^3k'}{k'_o} \, e^{-i(k-k').x} (k_o k'_o)^{\frac{1}{2}} C_\mu^+(k) C^\mu(k') \tag{58}$$

and

$$\underline{j}(\underline{x},t) = \frac{1}{(2\pi)^3} \iint \frac{d^3k}{k_o} \frac{d^3k'}{k'_o} \frac{(k_o k'_o)^{\frac{1}{2}}}{k_o + k'_o} (\underline{k}+\underline{k}') C_\mu^+(k) C^\mu(k') \tag{59}$$

It is easy to show that

$$\frac{\partial}{\partial t} n(\underline{x},t) + \nabla \cdot \underline{j}(\underline{x},t) = 0 \tag{60}$$

Under a gauge transformation (28)

$$n(\underline{x},t) \rightarrow \frac{1}{(2\pi)^3}\iint \frac{d^3k}{k_o}\frac{d^3k'}{k'_o} e^{-i(k-k').x}(k_o k'_o)^{\frac{1}{2}} (C^+_\mu(k) + k_\mu\lambda^*(k))(C^\mu(k') + k'^\mu\lambda(k'))$$

$$= n(\underline{x},t) + \frac{1}{(2\pi)^3}\iint \frac{d^3k}{k_o}\frac{d^3k'}{k'_o} e^{-i(k-k').x}(k_o k'_o)^{\frac{1}{2}} \left\{\lambda^*(k)k_\mu C^\mu(k') + \lambda(k')k'^\mu C^+_\mu(k) + k_\mu k'^\mu\lambda^*(k)\lambda(k')\right\} \quad (61)$$

It is clear that $n(\underline{x},t)$ is not gauge invariant. A similar calculation shows that $\underline{j}(\underline{x},t)$ is not gauge invariant. As Tatarski has pointed out, $(n,\underline{j})$ is not a 4-vector and so the continuity equation is not Lorentz invariant. The lack of gauge invariance makes Eqn. (63) not useful. Since localisability is a concept valid at a particular time, the lack of Lorentz covariance is not so crucial.

It is clear that a somewhat more powerful approach needs to be brought to the problem. First we will give Wightman's reformulation of the ideas of Newton and Wigner. Instead of localisability at a point Wightman (1962) considered the probability of localisability in a region based on the concept of imprimitivity for the Euclidean group. The operators $\{E(S)\}$ are a system of imprimitivity for the unitary representation $U(\underline{a},A)$ of $\mathcal{E}_3$ (the three dimensional Euclidean group) with base R^3 if

a. for every set S obtained from the complements and finite union of cubes, E(S) is a projection operator. Moreover the expectation value

$$\langle E(S)\rangle \;=\; \text{probability of the particle being in S}$$

b. $E(S_1 \cap S_2) = E(S_1)E(S_2)$

c. $E(S_1 \cup S_2) = E(S_1) + E(S_2) - E(S_1 \cap S_2)$

d. $E(\bigcup_{i=1}^{n} S_i) = \sum_{i=1}^{n} E(S_i)$

e. $E(R^3) = 1$, $E(\phi) = 0$ (ϕ is the empty set)

f. $E(AS+\underline{a}) = U(\underline{a},A)\, E(S)\, U(\underline{a},A)^{-1}$

Here $AS+\underline{a}$ is set obtained from S by first rotating all points (around the origin) determined by A and then translating by the vector $\underline{a}$. $S_1 \cup S_2$ denotes all points either in S_1 or S_2. $S_1 \cap S_2$ denotes points which are in both S_1 and S_2. From (b) it is clear that

$$E(S_1)E(S_2) = E(S_2)E(S_1) \tag{62}$$

It is possible to introduce a co-ordinate operator (Wightman 1962) and that all components of the operator commute follows from Eqn. (62). Wightman has shown that it is not possible to find non-trivial E(S) for the Hilbert space of photon states. The difficulty with the photon (as far as its localisability is concerned) is that although it has spin 1 it has only two (rather than 3) degrees of freedom. This is echoed by the electric field which needs to be transverse to the direction of propagation of the field. It is useful to examine a two dimensional example. We will consider a 'transverse' wavepacket

$$h_j(\underline{x}) = \int d^2k\, f_j(\underline{k})\, e^{i\underline{k}\cdot\underline{x}} \qquad j = 1,2 \tag{63}$$

with $\underline{k} = (k_1,k_2)$, $\underline{x} = (x_1,x_2)$

Since $k_j f_j(k) = 0$ (64)

we have

$$\underline{f}(\underline{k}) = (-k_2,k_1)g(\underline{k}) \tag{65}$$

for some function $g(\underline{k})$.

Hence

$$(h_1(\underline{x}), h_2(\underline{x})) = \frac{1}{i}\left(-\frac{\partial}{\partial x_2}, \frac{\partial}{\partial x_1}\right)\int d^2k\, g(\underline{k})\, e^{i\underline{k}\cdot\underline{x}} \qquad (66)$$

Consequently it is impossible to find a solution

$$h_1(\underline{x}),\ h_2(\underline{x}) \propto \delta(x_1)\,\delta(x_2) \qquad (67)$$

This is seen from inspection. The transversality condition (64) is the cause of the difficulty.

Jauch and Piron (1967) have relaxed condition (b). As a result the expectation of E(S) is no longer the probability of finding the state localised in S. The expectation of E(S) is still non-negative and has many of the properties of a probability. Following Jauch and Piron (1967) we will give a simple concrete example of a generalised imprimitivity. Even if the attractions of a generalised probability are debatable, the theory does allow the construction of in some sense the best possible localised wavefunction in a given region compatible with transversality conditions. This would then allow an explicit construction of the states given in (24) and (29). Moreover an eigenstate of E(S) represents a photon whose probability of being outside S is zero.

We will consider all 4 component functions $f_\mu(p)$ which satisfy $p^\mu f_\mu(p) = 0$ and transform under the Poincaré group as

$$[U(a,\Lambda)f]^\mu\,(p) = e^{-ia.p}\,\Lambda^\mu{}_\nu f^\nu(\Lambda^{-1}p) \qquad (68)$$

Two functions which differ by a gauge transformation will be considered to be equivalent. The equivalence class of f_μ will be denoted by $\hat{f}_\mu$, and in particular we will choose the representative member such that $f_o(p) = 0$. An inner product on the equivalence class (or gauge invariant) space will be given by

$$\langle \underline{f} | \underline{g} \rangle = \int d^3p \; \underline{f}^*(\underline{p}) \cdot \underline{g}(\underline{p}) \tag{69}$$

where purely for convenience we have absorbed a factor of $p_o^{-\frac{1}{2}}$ in $\underline{f}$ and $\underline{g}$. The representatives of the equivalence class have only nonzero spatial components and so for convenience are denoted by three vectors. The transversality condition becomes

$$\underline{p} \cdot \underline{f} = 0 \tag{70}$$

The functions which satisfy Eqn. (70) and have a finite inner product form a Hilbert space h. This can be extended to h^+ by dropping the condition (70). We define the projection P from h^+ to h

$$[P\underline{f}](k) = \underline{f}(k) - \frac{\underline{k}}{k_o^2} \underline{k} \cdot \underline{f}(k) \tag{71}$$

From Eqn. (68) it is easy to see that

$$[P, U(\underline{a}, \Lambda)] = 0 \tag{72}$$

In order to define imprimitivities we need to be in configuration space which is accomplished by Fourier transforming

$$F : \underline{f}(\underline{p}) \rightarrow \tilde{\underline{f}}(\underline{x}) = \frac{1}{(2\pi)^{3/2}} \int d^3p \; e^{i\underline{p}\cdot\underline{x}} \; \underline{f}(p) \tag{73}$$

It is straightforward to define an ordinary imprimitivity on the functions in h^+

$$E(S) = F^{-1} \chi(S) F \tag{74}$$

where

$$\chi(S)\,\underline{\tilde{f}}(\underline{x}) = \underline{\tilde{f}}(\underline{x}) \quad \text{if} \quad \underline{x} \in S$$
$$= 0 \quad \text{otherwise}$$

A generalised imprimitivity is given by

$$E(S) \cap P$$
$$\neq E(S)\,P \qquad \text{in general since } [E(S),P] \neq 0 \tag{75}$$

($E(S) \cap P$ is the projection on to the largest subspace which is contained in the ranges of $E(S)$ and P.) Basically localising in a region S using an ordinary imprimitivity destroys the transversality and so the generalised imprimitivity demands both localisation and transversality. This is clearly seen from

$$E(S) \cap P = \lim_{n\to\infty} (E(S)\,P)^n \tag{76}$$

5. WEAK LOCALISATION

A generalised imprimitivity will be constructed for the photon (Amrein, 1969). h^+ will denote the Hilbert space of states in the irreducible representation for the mass m and spin 1 particle given by (41) and (43) with j = 1.

$h_{\pm 1}$ will denote the subspaces of h^+ with states having $\lambda = \pm 1$. The key fact used to construct a generalised imprimitivity for the photon is the isomorphism between $h_{\pm 1}$ with the Hilbert space of the representation vectors of the massless particle with helicities ± 1 in Eqn. (39). This follows from

$$(\chi(p,\Lambda))^{2\lambda} = D^{|\lambda|}_{\lambda\lambda}\left(L^{(h)^{-1}}_{\Lambda p}\,\Lambda\, L^{(h)}_{p}\right) \tag{77}$$

$P^{\pm}$ will denote the projection of h^{+} on to h_{+} or h_{-} and is the analogue of the P of Eqn. (71).

The construction of the generalised imprimitivity proceeds in much the same way as given in the last section. If the one particle state given in Eqn. (24) is decomposed into its positive and negative helicity components it is weakly localised in S if

$$(\breve{E}(S)\ \hat{f}(\underline{p}))_{+} = \sum_{\beta=-1}^{1} D^{1}_{1\beta}\ (O_p^{-1})(\underline{p}^2)^{\frac{1}{4}}\ F^{-1}\ E(S)\ F\ \Sigma_{\beta}(\underline{p})$$

$$(\breve{E}(S)\ \hat{f}(\underline{p}))_{-} = \sum_{\beta=-1}^{1} D^{1}_{-1\beta}\ (O_p^{-1})(\underline{p}^2)^{\frac{1}{4}}\ F^{-1}\ E(S)\ F\ \Sigma_{\beta}(\underline{p}) \qquad (78)$$

where $\hat{f}(\underline{p}) = (f_{+}(\underline{p}),f_{-}(\underline{p}))$ and f_{+} and f_{-} are linear combinations of f_{μ} in Eqn. (25) necessary to make positive and negative helicities. Rougly speaking $F^{-1}E(S)F$ gives the projection on to S and the D factors fix the helicities (or photon nature) of the state, ie play the role of P.

Also

$$\Sigma_{\beta}(\underline{p}) = \frac{1}{|\underline{p}|^{\frac{1}{2}}}\ (D^{1}_{\beta 1}(O_p)\ f_{+}(\underline{p}) + D^{1}_{\beta -1}(O_p)\ f_{-}(\underline{p})) \qquad (79)$$

with

$$O_p = (2|\underline{p}|(|\underline{p}|+p^3))^{-\frac{1}{2}} \begin{pmatrix} |\underline{p}|+p^3 & -p^1+ip^2 \\ p^1+ip^2 & |p|+p^3 \end{pmatrix} \qquad (80)$$

the rotation taking $(p°,o,o,|\underline{p}|)$ to p. It is easy to show that

$$(\breve{E}(S))^2\ \hat{f} = \breve{E}(S)\ \hat{f} \qquad (81)$$

and if $S_1 \cap S_2 = \emptyset$ then

$$\tilde{E}(S_1)\,\tilde{E}(S_2)\,\hat{f} = 0 \tag{82}$$

From Eqn. (84) we can see that given an arbitrary state $\hat{f}$, $\tilde{E}(S)\hat{f}$, provided it is non-zero, gives a weakly localised state. In this way we have a means of producing localised photon wavepackets. It is obvious from the formulae (78) and (79) that if the region S is very much larger in dimensions than the wavelengths involved in the one photon state the generalised imprimitivity becomes essentially an ordinary imprimitivity. For any dimensions of S of the order of a wavelength (or less) there is a leakage of the configuration space wavepacket outside S in those directions.

The $\tilde{E}(S)$ are self-adjoint operators and according to Dirac (1978) they are in principle measurable. We do not know of a practicable scheme, although it is possible that some sort of state preparation (such as going through a slit) may correspond to $\tilde{E}(S)$. For a generalised imprimitivity for a sequence of disjoint sets S_i

$$\sum_i \tilde{E}(S_i) \leqslant \tilde{E}(\bigcup_i S_i) \tag{83}$$

(A projection P is greater than an operator O if the range of P is larger than that of O, and P and O agree on the intersection of their ranges.) The fact that the equality does not always hold means that the expectation of $\tilde{E}(S)$ in a state is not strictly a probability; we may refer to it as a generalised probability since $\tilde{E}(R^3)$ is 1 and for two sets S_1 and S_2 whose closures are disjoint

$$\tilde{E}(S_1 \cup S_2) = \tilde{E}(S_1) + \tilde{E}(s_2) \tag{84}$$

This is illustrated in Figure 2 with the function space example of generalised imprimitivity given in the last section.

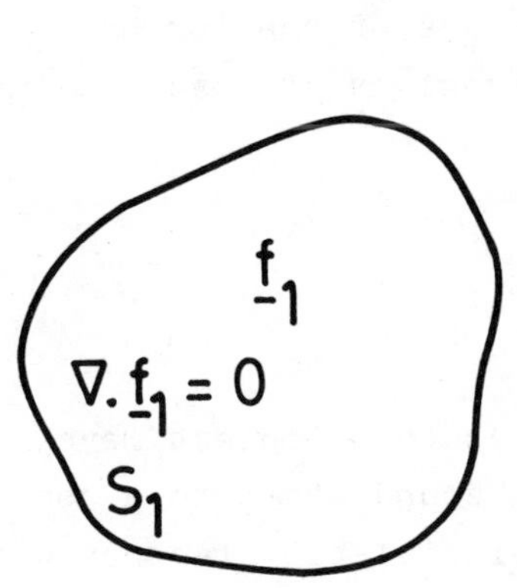

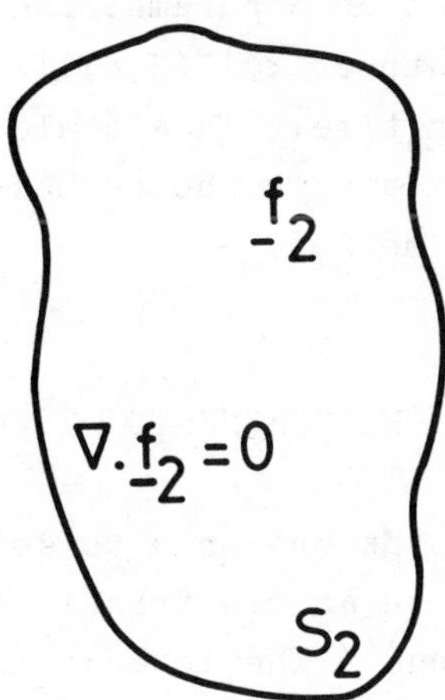

$$\underline{f} = \underline{f}_1 + \underline{f}_2$$

$$\nabla . \underline{f} = \nabla . \underline{f}_1 + \underline{\nabla} . \underline{f}_2 = 0$$

Figure 2. Projections in a Function Space

Clearly if the S_1 and S_2 can touch then $\underline{f}$ has more possibilities than just being $(\underline{f}_1, 0) \oplus (0, \underline{f}_2)$.

Given a fixed region S, for any one photon state $|1,f\rangle$

$$\frac{d}{dt}\langle 1,f|\breve{E}(S)|1,f\rangle = \langle 1,f|\breve{J}(S)|1,f\rangle \qquad (85)$$

where

$$\breve{J}(S)\hat{f}(p) = i(p_o\breve{E}(S) + \breve{E}(S)p_o)\hat{f}(p) \qquad (86)$$

Equation (85) is just an expression of the evolution of a wavepacket solution of the free electromagnetic field. It does not have an analogue in terms of local densities. We suggest that it may be the best that can be done as regards a continuity equation for the photon. Moreover by considering a set of S parametrised by time, $\{S(t)\}$ it is possible to construct $\langle\psi|E(S(t))|\psi\rangle$ where $|\psi\rangle$ the wavepacket also evolves with time. This could provide a measure of the 'probability' of a single photon being in certain regions of space during its motion.

6. MEASUREMENTS

Experiments to detect a photon at the simplest level excite an electron in an atom from a bound state to a continuum. The interaction of a non-relativistic electron with the electromagnetic field is described by the Hamiltonian

$$H_1 = H^{atom} + H^{field} - \underline{d} \cdot \underline{E}(\underline{o})^{\perp} \tag{87}$$

where $\underline{d} = e\underline{r}$, $\underline{o}$ is the 'position' of the nucleus of the atom, $\underline{E}^{\perp}$ is the transverse part of the electromagnetic field, and $\underline{r}$ the position operator for the electron. It is also possible to use the Hamiltonian

$$H_2 = H^{atom} + H^{field} - \tfrac{1}{2}\frac{e}{m}(\underline{p}.\underline{A}(\underline{r})+\underline{A}(\underline{r}).\underline{p}) + \tfrac{1}{2}\frac{e^2}{m}\underline{A}(\underline{r}).\underline{A}(\underline{r}) \tag{88}$$

In the electric dipole approximation H_1 and H_2 are connected by a canonical transformation (Göppert-Mayer 1931, Richards 1948). It is well-known that (in the rotating wave approximation) the probability $P(\underline{r},t,\Delta t)$ that a photoelectron is detected over the time interval from t to $t+\Delta t$ is

$$P(\underline{r},t,\Delta t) = \int_0^{\Delta t} dt' \int_0^{\Delta t} dt'' \langle E_i^{(-)}(\underline{r},t+t')E_j^{(+)}(\underline{r},t+t'')\rangle k_{ji}(t'-t'') \tag{89}$$

and $k_{ji}(t)$ is the response function of the detector (Glauber, 1963). For a pure state $|1,f\rangle$ by inserting a complete set of states in (89) we find the quantities

$$<o|E_j^{(+)}(\underline{r},t)|1,f> \tag{90}$$

These may be regarded as one photon 'detection' wavefunctions (Titulaer amd Glauber 1966) but, however, to not satisfy our objective, stated in the Introduction, of finding an analogue of a wavefunction. From Eqns. (2) and (13) we know that

$$E_k(x) = i\int \frac{d^3p}{(2\pi)^3 2p_o} \left\{(p_k C_o(\underline{p}) + p_o C_k(\underline{p}))\, e^{ip.x} - (p_k C_o^+(\underline{p}) + C_k^+(\underline{p})p_o)\, e^{-ip.x}\right\} \tag{91}$$

We shall from now on concentrate on the Coulomb gauge (for which $C_o = 0$). The question of interest to us is the nature of the interference pattern between two beams of light each with a small number of photons. As a simple example we will consider our introductory problem of the interference of two beams each with one photon. The wavepacket in one beam is determined by $f^+_{(1)}|o>$ and the wavepacket in another by $f^+_{(2)}|o>$. The corresponding Fock state $|\psi>$ is

$$|\psi> = f^+_{(1)}\, f^+_{(2)}\, |o> \tag{92}$$

We are interested in interference phenomena which are determined by

$$<\psi|\, E_i^{(-)}(x')\, E_j^{(+)}(x'')\, |\psi>$$

where $x' = (\underline{r},t+t')$ and $x'' = (\underline{r},t+t'')$. This becomes

$$\int d\Omega(p) \int d\Omega(p')\, <o|\, f^-_{(1)}\, f^-_{(2)}\, C_i^+(p)\, p_o\, C_j(p')\, p_o'\, f^+_{(1)} f^+_{(2)}|o>\, e^{-ip.x'}\, e^{ip'.x''} \tag{93}$$

where

$$d\Omega(p) = \frac{d^3p}{(2\pi)^3 2p_o}$$

For situations where $f_{(1)}$ and $f_{(2)}$ do not have wavevectors in common (Burnham and Weinberg 1970) (93) reduces to

$$\int d\Omega(k_{(2)}) \int d\Omega(k'_{(2)}) f^*_{(2)i}(k_{(2)}) f_{(2)j}(k'_{(2)})\, e^{ik'_{(2)}.x''}\, e^{-ik_{(2)}.x''}$$

$$+ \qquad (94)$$

$$\int d\Omega(k_{(1)}) \int d\Omega(k'_{(1)}) f^*_{(1)i}(k_{(1)}) f_{(1)j}(k'_{(1)})\, e^{ik'_{(1)}.x''}\, e^{-ik_{(1)}.x'}$$

(94) does not show any interference. The only $\underline{r}$ dependence is determined by the shape of the individual beams.

We can however have non-trivial interference when $|\psi\rangle$ is given by

$$|\psi\rangle = (\alpha_{(1)}(f^+_{(1)})^n + \beta_{(1)}(f^+_{(1)})^{n+1})$$

$$\cdot\ (\alpha_{(2)}(f^+_{(2)})^n + \beta_{(2)}(f_2^{(+)})^{n+1})\ |o\rangle \qquad (95)$$

(where $\alpha_{(j)}$ and $\beta_{(j)}$ ($j = 1,2$) are normalisation constants). For simplicity of presentation we shall consider n=1 but our analysis goes through in a straightforward way for general n. The analogue of Eqn. (93) is then

$$\int d\Omega(p) \int d\Omega(p')\, e^{-ip.x'}\, e^{ip'.x''}$$

$$\langle o| \{ (\alpha^*_{(1)} f^-_{(1)} + \beta^*_{(1)}(f^-_{(1)})^2)(\alpha^*_{(2)} f^-_{(2)}$$

$$+ \beta^*_{(2)})(f^-_{(2)})^2) C^+_i(p) p_o C_j(p') p'_o \ \cdot\ (\alpha_{(1)} f^+_{(1)}$$

$$+ \beta_{(1)}(f^+_{(1)})^2)(\alpha_{(2)} f^+_{(2)} + \beta_{(2)}(f^+_{(2)})^2)\}\ |o\rangle \qquad (96)$$

Since

$$[C_j(p'),\ C_\mu^+ (p_1)\ C_\mu^+ (p_2)]$$

$$= C_\mu^+ (p_1)\ \gamma_{j\mu} (p')\ (2\pi)^3\ 2|\underline{p}'|\ \delta^{(3)}(\underline{p}'-\underline{p}_2) \tag{97}$$

$$+ C_\mu^+ (p_2)\ \gamma_{j\mu} (p')\ (2\pi)^3\ 2|\underline{p}'|\ \delta^{(3)}(\underline{p}'-\underline{p}_1)$$

It is easy to see that

$$C_j(p')f^+_{(1)}\ |o\rangle = 2f_{(1)j}(p')\ f^+_{(1)}\ |o\rangle \tag{98}$$

We can see that Eqn. (96) contains interference between the two beams since the action of $C_j(p')$ on the ket can give rise to the state $f^+_{(1)}f^+_{(2)}\ |o\rangle$ from Eqn. (98), whereas similarly $C_j^+(p)$ can give $\langle o|f^-_{(1)}f^-_{(2)}$, by destroying a photon in the operator $(f^-_{(2)})^2$; so one photon has been destroyed in each packet. This contribution to Eqn. (66) gives

$$4|\beta_{(1)}|^2|\alpha_{(2)}|^2 \int d\Omega(p) \int d\Omega(p')p_o p_o'\ e^{-ip.x'}\ e^{ip'.x''}$$

$$f_{(1)j}(p')f^*_{(2)i}(p) \tag{99}$$

and we have taken normalisations so that $f_{(2)}{}^*f_{(2)} = f_{(1)}{}^*f_{(1)} = 1$. It should be clear that the above argument goes through not only for the state in Eqn. (95) but for other states which contain it as a part. These calculations generalise those of earlier plane wave formulations (Titulaer and Glauber, 1966; Walls, 1977).

The results of classical optics, however, can be derived if we take for $|\psi\rangle$

$$|\psi\rangle = f^+_{(1)c}\ f^+_{(2)c}\ |o\rangle \tag{100}$$

where we define

$$f^+_{(1)c} = e^{f^+_{(1)}-f^-_{(1)}} \tag{101}$$

and

$$f^*_{(2)}f_{(1)} = 0$$
$$f^*_{(1)}f_{(1)} = f^*_{(2)}f_{(2)} = 1 \qquad (102)$$

It is easy to see that

$$[f^+_{(1)c}, f^+_{(2)c}] = 0 \qquad (103)$$

The operators f^+_c generate wavepacket coherent states (Kibble, 1968), which are no longer one photon states. Moreover

$$[f^+_c, C_\mu(k)] = -f_\mu(k)f^+_c \qquad (104)$$

and so

$$f^- f^+_{(1)c}|o\rangle = f^+_{(1)c}|o\rangle\,(f^* f_{(1)}) \qquad (105)$$

The matrix element determining interference phenomena now has the form

$$\int d\Omega(p) \int d\Omega(p') \, \langle o|f^-_{(1)c}f^-_{(2)c}C^+_i(p)p_o C_j(p')p'_o \, e^{-ip.x'} \, e^{ip'.x''} \, f^+_{(1)c} \, f^+_{(2)c} \, |o\rangle \qquad (106)$$

From Eqn. (104) we can show that

$$C_j(p')f^+_{(1)c}f^+_{(2)c}|o\rangle = (f^+_{(1)c}f^+_{(2)c}|o\rangle)(f_{(1)j}(p') + f_{(2)j}(p')) \qquad (107)$$

and we have emerging the addition of wavepacket amplitudes which are just proportional to those for the electric field. On substituting this result in (106) we have

$$\int d\Omega(p) \int d\Omega(p') p_o (f^*_{(1)i}(p) + f^*_{(2)i}(p)) p'_o (f_{(1)j}(p') + f_{(2)j}(p'))\, e^{-ip.x'}\, e^{ip'.x''} \tag{108}$$

which is the classical prescription of first adding electric fields and then taking the product with the complex conjugate sum. The usual form of interference follows. A less general form of this argument has appeared before (Titulaer and Glauber 1966, Walls 1977).

We will now come to the intriguing question of detecting a photon outside its region of localisation Δ and we will re-derive the convolution integral obtained by Amrein (1969) by a more direct argument. The theoretical analysis that we have presented shows that a photon state which is an eigenstate of $\tilde{E}(\Delta)$ can be sensibly regarded as localised in Δ. The representation of the state in configuration space does not have compact support; but the photodetection process probes the configuration space properties of the state and so there is a probability of photodetection outside the region Δ. Explicitly if the localised state is given by

$$\int \frac{d^3p}{(2\pi)^3 2p_o} (f_+(p)\, a^+(p,2) + f_-(p)\, a^+(p,1)) \tag{109}$$

then the detection probability is related to the matrix element

$$-i \int \frac{d^3p}{(2\pi)^3 2}\, e^{ip.x}\, p_o^{\frac{1}{2}} \left(e_j^{(2)}(p) \frac{f_+(p)}{p_o^{\frac{1}{2}}} + e_j^{(1)}(p) \frac{f_-(p)}{p_o^{\frac{1}{2}}} \right)$$

$$= -i \int \frac{d^3p}{(2\pi)^3 2}\, e^{ip.x}\, p_o^{\frac{1}{2}}\, \mathcal{E}(p) \tag{110}$$

where x is the location of the detector. The matrix element has the form of a convolution in real space and the jth component of the field associated with the state is given in momentum space by the combination

$$e_j^{(2)}(p)\ \frac{f_+(p)}{p_o^{\frac{1}{2}}} + e_j^{(1)}(p)\ \frac{f_-(p)}{p_o^{\frac{1}{2}}} \qquad (111)$$

(The functions $f_\pm(p)$ should not be confused with the operators $f^\pm$.)

The three dimensional Fourier transform of $(p_o)^{\frac{1}{2}}$ has an asymptotic $r^{-7/2}$ behaviour for large distances r and provided the field $\underline{\tilde{\mathcal{E}}}(x)$ falls off much more sharply than this, the matrix elements for points far from Δ will be dominated by the $r^{-7/2}$ behaviour. An experimental measurement of such a power law would be of great interest and should not be impractical now that we may isolate single photon wavepackets by nonlinear optical techniques. Similarly, experiments on single photon wavepacket interference could be possible and valuable.

7. CONCLUSIONS

We have examined various approaches for understanding 'localised' photon states, giving special emphasis to helicity structure and gauge invariance. The concept of generalised imprimitivities is shown to be useful for constructing single photon localised wavepackets, and the associated creation operators are necessary to understand interference of general light fields.

REFERENCES

Amrein, W.O., 1969, Helv. Phys. Acta 42, 149.

Bargmann, V. and Wigner, E.P., 1948, Proc. Nat'l Acad. Sci. 34, 211.

Bjorken, J.D. and Drell, S.D., 1965, Relativistic Quantum Fields, McGraw-Hill.

Bleuler, K., 1950, Helv. Phys. Acta 23, 567.

Burnham, D.C. and Weinberg, D.L., 1970, Phys. Rev. Lett. 25, 84.

Cook, R.J., 1982a, Phys. Rev. A25, 2164.

Cook, R.J., 1982b, Phys. Rev. A26, 2754.

Dirac, P.A.M., 1978, The Principles of Quantum Mechanics, Oxford University Press.

Feynman, R.P., 1962, Quantum Electrodynamics, Benjamin, New York.

Friberg, S., Hong, C.K. and Mandel, L., 1985a, Phys. Rev. Lett. 54, 2011.

Friberg, S., Hong, C.K. and Mandel, L., 1985b, Opt. Commun. 54, 311.

Glauber, R.J., 1963, Phys. Rev. 131, 2766.

Goppert-Mayer, M., 1931, Am. Phys. (Leipzig) 9, 273.

Gupta, S.N., 1950, Proc. Phys. Soc. (London) A63, 681.

Jakeman, E. and Walker, J.G., 1985, Opt. Comm. 55, 219.

Jauch, J.M. and Piron, C., 1967, Helv. Phys. Acta 40, 559.

Jauch, J.M. and Rohrlich, F., 1976, The Theory of Photons and Electrons, Springer-Verlag.

Kibble, T.W.B., 1968, Coherent States and Infrared Divergencies, Lectures presented at the Theoretical Physics Institute, Univ. of Colorado.

Landau, L.D. and Lifshitz, E.M., 1959, Quantum Mechanics, Pergamon Press, London.

Mackey, G.W., 1978, Unitary Group Representations in Physics, Probability and Number Theory, Benjamin.

Mandel, L., 1966, Phys. Rev. 144, 1071.

Newton, T.D. and Wigner, E.P., 1949, Rev. Mod. Phys. 21, 400.

Pike, E.R., 1986, Proc. Conf. in honour of R J Glauber's 60th birthday, Academic Press, New York.

Reed, M. and Simon, B., 1972, Methods of Modern Mathematical Physics, Vol. 1: Functional Analysis, Academic Press.

Richards, P.I., 1948, Phys. Rev. 73, 254.

Rose, M.E., 1957, Elementary Theory of Angular Momentum, Wiley, New York.

Schweber, S.S., 1966, An Introduction to Relativistic Quantum Field Theory, Harper.

Schwinger, J., 1948, Phys. Rev. 74, 1439.

Schwinger, J., 1962, Phys. Rev. 127, 324.

Tatarski, V.I., 1972, Sov. Phys. JETP 36, 1097.

Titulaer, U.M. and Glauber, R.J., 1966, Phys. Rev. 145, 1041.

Walls, D.F., 1977, Am. J. Phys. 45, 952.

Wheeler, J.A. and Zurek, W.H., 1983, Quantum Theory and Measurement, Princeton Univ. Press.

Wightman, A.S., 1962, Rev. Mod. Phys. 34, 845.

Wigner, E.P., 1939, Ann. Math. 40, 149.

Wybourne, B.G., 1974, Classical Groups for Physicists, Wiley.

QUANTUM EFFECTS IN SPONTANEOUS PARAMETRIC DOWN-CONVERSION OF LIGHT

L MANDEL

1. INTRODUCTION

The process of spontaneous parametric down-conversion of light is one of several interesting phenomena in quantum optics in which a nonclassical field is created as a result of some nonlinear interaction. In this process an incident pump photon of energy $\hbar\omega_0$ interacts with a nonlinear dielectric so as to fission into two lower frequency signal and idler photons of energies $\hbar\omega_1$ and $\hbar\omega_2$. The resulting electromagnetic field exhibits several explicitly quantum mechanical features, such as a highly correlated two-photon state, sub-Poissonian statistics and squeezing. Several of these can be put to the experimental test relatively easily. As we show below, in the process of spontaneous parametric down-conversion one also comes close to realizing a localized one-photon state, which may make it possible to test certain paradoxical non-local aspects of quantum electrodynamics. Finally, some quantum features lend themselves to an interesting application to optical communication, which makes the phenomenon attractive for further exploration.

The subject of parametric amplification and oscillation has been treated numerous times in the past by Louisell (1960), Louisell et al (1961), Gordon et al (1963), Mollow and Glauber (1967a, 1967b), Mollow (1967, 1973), Giallorenzi and Tang (1968), Kleinman (1968), Freund and Levine (1969), Zel'dovich and Klyshko (1969), Tucker and Walls (1969), Stoler (1970, 1971, 1974), Raiford (1970, 1974), Smithers and Lu (1974), Yariv (1975), Mista et al (1977a, 1977b), Drummond et al (1979), Chmela (1979, 1981a, 1981b), Neumann and Haug (1979), Milburn and Walls (1981, 1983), Wódkiewicz and Zubairy (1983), Friberg and Mandel (1984), Graham

(1984), Hillery and Zubairy (1984), Meystre et al (1984), Hong and Mandel (1985), with emphasis on different aspects of the problem. In most of these treatments the field was resolved into a few discrete modes. However, time-resolved information about correlation functions requires a multimode treatment [see Giallorenzi and Tang (1968), Kleinman (1968), Mollow (1973), Hong and Mandel (1985)], which also allows better contact with experiment to be made. The process of spontaneous down-conversion or frequency splitting of light was first demonstrated experimentally by Burnham and Weinberg (1970) with uv light. Later experiments made use of visible light as pump [see Montgomery and Giallorenzi (1973)] and also showed that X-rays could similarly be down-converted [see Eisenberg and McCall (1971), Danino and Freund (1981)]. More recently the distribution of time intervals between signal and idler photons was measured with sub-nanosecond time resolution [see Friberg et al (1985a)], and the intensity dependence of the cross-correlation function was investigated [see Friberg et al (1985b)]. It was also shown experimentally that the photon statistics are sub-Poissonian, and that one comes close to realizing a localized one-photon state [see Hong and Mandel (1986)].

2. THEORY WITH TWO OR THREE DISCRETE MODES

In order to describe the phenomenon we start in the simplest possible way, by treating the process as a parametric interaction between incident pump photons $(\underline{k}_0,\omega_0)$ and signal $(\underline{k}_1,\omega_1)$ and idler photons $(\underline{k}_2,\omega_2)$ in some non-linear crystal. When the phase matching condition is satisfied, both energy and momentum conservation hold to a good approximation,

$$\left.\begin{aligned} \underline{k}_0 &= \underline{k}_1 + \underline{k}_2 \\ \omega_0 &= \omega_1 + \omega_2 \quad . \end{aligned}\right\} \qquad (1)$$

The total energy of the coupled system of three modes may be expressed in the form [see Graham (1984)]

$$\hat{H} = H(\hat{a}_0,\hat{a}_0^\dagger,t) + \hbar\omega_1\hat{n}_1 + \hbar\omega_2\hat{n}_2 + i\hbar g(\hat{a}_1^\dagger\hat{a}_2^\dagger\hat{a}_0 - \text{h.c.}) \quad . \qquad (2)$$

Here $\hat{a}_0,\hat{a}_1,\hat{a}_2$ are time-dependent photon annihilation operators for the three modes, H is the (possibly time-dependent) energy of the pump mode alone, and g is the parametric mode coupling constant. We label all Hilbert space operators by the caret ^.

It follows immediately from the Heisenberg equations of motion that

$$\frac{d}{dt}(\hat{n}_1-\hat{n}_2) = \frac{1}{i\hbar}[\hat{n}_1-\hat{n}_2,\hat{H}] = 0 \quad , \tag{3}$$

where $\hat{n}_j = \hat{a}_j^\dagger\hat{a}_j$ (j=1,2) is the photon number operator, so that $\hat{n}_1-\hat{n}_2$ is a constant of the motion, and

$$\hat{n}_1(t)-\hat{n}_2(t) = \hat{n}_1(0)-\hat{n}_2(0) \quad . \tag{4}$$

This shows that signal and idler and photons are always created in pairs, as implied by the interaction in Eq. (2). The probability of detecting a photon of mode j at time t is proportional to $\langle\hat{n}_j(t)\rangle$, and the joint probability of detecting both a signal and idler photon is proportional to the normally ordered correlation $\langle:\hat{n}_1(t)\hat{n}_2(t):\rangle$ [see Glauber (1963a, 1963b)]. As operators belonging to different modes commute at the same time, we have

$$\begin{aligned}\langle:\hat{n}_1(t)\hat{n}_2(t):\rangle &= \langle\hat{n}_1(t)\hat{n}_2(t)\rangle \\ &= \langle\hat{n}_1^2(t)\rangle-\langle\hat{n}_1(t)\hat{n}_1(0)\rangle+\langle\hat{n}_1(t)\hat{n}_2(0)\rangle \quad ,\end{aligned} \tag{5}$$

where the last line follows with the help of Eq. (4).

In order to evaluate the expectations we have to specify the quantum state. We shall suppose that the signal and idler modes are initially in the vacuum state $|vac_1\rangle,|vac_2\rangle$. Then since $\hat{a}_1(0)|vac_1\rangle = 0 = \hat{a}_2(0)|vac_2\rangle$, we have from Eq. (5)

$$\langle:\hat{n}_1(t)\hat{n}_2(t):\rangle = \langle\hat{n}_1^2(t)\rangle \quad ,$$

and with the help of the commutation relations it then follows that [see Graham (1984)]

$$\langle:\hat{n}_1(t)\hat{n}_2(t):\rangle = \langle:\hat{n}_1^2(t):\rangle + \langle\hat{n}_1(t)\rangle \quad . \tag{6}$$

It is not difficult to show that this relation implies that the down-converted light has no classical description. For this purpose we first make a diagonal coherent state representation of the density operator $\hat{\rho}$ [see Glauber (1963a, 1963b), Sudarshan (1963), Klauder (1966)],

$$\hat{\rho} = \int \phi(\{v\}) |\{v\}\rangle\langle\{v\}| d\{v\} \quad , \tag{7}$$

in which $\phi(\{v\})$ is a phase space quasiprobability density. Next we make use of the optical equivalence theorem [see Sudarshan (1963), Klauder (1966)] to express the expectations of the three normally ordered operators in Eq. (6) as c-number averages with the phase space density ϕ as weight function. We then obtain the relation

$$\langle|v_1 v_2|^2\rangle_\phi = \langle|v_1|^4\rangle_\phi + \langle|v_1|^2\rangle_\phi \quad , \tag{8}$$

where v_1, v_2 are the right eigenvalues of $\hat{a}_1(t), \hat{a}_2(t)$ in the coherent state $|\{v\}\rangle$. By contrast we note that for classical random variables v_1, v_2 the Schwarz inequality requires

$$\langle|v_1|^2|v_2|^2\rangle_\phi \leqq \sqrt{\langle|v_1|^4\rangle\langle|v_2|^4\rangle} = \langle|v_1|^4\rangle \tag{9}$$

when the averages for modes 1 and 2 are equal. Comparison of Eqs. (8) and (9) shows that the signal and idler modes violate the Schwarz inequality, so that $\phi(\{v\})$ cannot be a probability density, and the down-converted field is not describable by a classical ensemble [see Graham (1984)]. Indeed, $\langle|v_1|^2\rangle_\phi$ is generally much larger than $\langle|v_1|^4\rangle_\phi$ because the down-converted field is weak, so that the violation can be very large. The joint probability of detecting both a signal and an idler photon is then much larger than the joint probability of detecting two signal photons or two idler photons. There are no classical light beams that exhibit this kind of behavior.

Although these conclusions follow directly from the Hamiltonian, we cannot evaluate the various expectations without integrating the equations of motion for the system. Here we shall further simplify the problem, by taking the pump mode to be classical and monochromatic, with complex amplitude

$$\hat{a}_0(t) \to v_0 \, e^{-i\omega_0 t} = |v_0| e^{-i(\omega_0 t - \theta)} \quad , \tag{10}$$

so that H becomes a c-number in Eq. (2). The Heisenberg equations for $\hat{a}_1(t)$ and $\hat{a}_2(t)$ then take the form

$$\dot{\hat{a}}_j(t) = \frac{1}{i\hbar}[\hat{a}_j(t),\hat{H}] \qquad (j,k = 1,2)$$

$$= -i\omega_j\hat{a}_j(t) + gv_0\, e^{i\omega_0 t}\, \hat{a}_k^\dagger(t) \ , \quad (j \neq k) \quad ,$$

and they are readily integrated. We obtain the well-known solutions [see Mollow (1967), Graham (1984)]

$$\hat{a}_j(t) = [\hat{a}_j(0)\cosh(g|v_0|t) + \hat{a}_k^\dagger(0)e^{i\theta}\sinh(g|v_0|t)]e^{-i\omega_j t} \quad . \tag{11}$$

$$(j,k = 1,2) \quad (j \neq k) \quad .$$

From these we find for the expectations in the vacuum state

$$\langle\hat{a}_j^N(t)\rangle = 0 \quad , \quad N=1,2 \ ,\ldots. \quad (j=1,2) \tag{12}$$

$$\langle\hat{n}_j(t)\rangle = \langle\hat{a}_j^\dagger(t)\hat{a}_j(t)\rangle = \sinh^2(g|v_0|t) \quad , \quad (j=1,2) \tag{13}$$

$$\langle:\hat{n}_1(t)\hat{n}_2(t):\rangle = [\cosh^2(g|v_0|t) + \sinh^2(g|v_0|t)]\sinh^2(g|v_0|t) \tag{14}$$

$$\langle:\hat{n}_j^2(t):\rangle = 2\sinh^4(g|v_0|t) \quad , \quad (j=1,2). \tag{15}$$

So long as $g|v_0|t \ll 1$, $\langle\hat{n}_j(t)\rangle \ll 1$, and $\langle\hat{n}_j(t)\rangle$ grows quadratically in time. The down-conversion is then predominantly spontaneous. However, once $\langle\hat{n}_j(t)\rangle$ exceeds unity, stimulated down-conversion predominates, and $\hat{n}_j(t)$ grows exponentially in time. In practice the effective interaction time t is the transit time of the light through the non-linear crystal, which is usually extremely short, so that $\langle\hat{n}_j(t)\rangle$ is generally much less than 1. Under those conditions we see from Eqs. (13) to (15) that

$$\langle:\hat{n}(t_1)\hat{n}_2(t):\rangle \gg \langle:\hat{n}_1^2(t):\rangle \tag{16}$$

and

$$\langle:\hat{n}_1(t)\hat{n}_2(t):\rangle \approx \langle\hat{n}_1(t)\rangle = \hat{n}_2(t)\rangle = \sinh^2(g|v_0|t) \quad . \tag{17}$$

Hence the joint probability of obtaining both a signal and idler photon equals the probability of obtaining a signal or idler photon alone. This shows once again that the signal and idler photor are produced together. Also from Eqs. (13) and (15) we note that

$$\langle(\Delta\hat{n}_j(t))^2\rangle - \langle\hat{n}_j(t)\rangle = \sinh^4(g|v_0|t) \quad , \quad j=1,2 \tag{18}$$

so that the down-converted photons of each mode obey thermal statistics and are approximately Poissonian when $\langle\hat{n}_j(t)\rangle \ll 1$. Finally, we observe that if we introduce a normalized correlation function λ_{12} by writing

$$\lambda_{12} = \frac{\langle:\hat{n}_1(t)\hat{n}_2(t):\rangle}{\langle\hat{n}_1(t)\rangle\langle\hat{n}_2(t)\rangle} - 1 \quad ,$$

then from Eqs. (13) and (14)

$$\lambda_{12} = \coth^2(g|v_0|t) \approx \frac{1}{\langle\hat{n}_j(t)\rangle} \gg 1, \quad j=1,2 \quad , \tag{19}$$

so that λ_{12} can be very large, as well as being inversely proportional to the photon number.

3. A CLASSICAL FIELD EXHIBITING CORRELATED PULSES

An interesting question that arises here is to what extent this last behavior could be modelled classically. We have already seen that the classical inequality for correlated light intensities is violated, so that Eq. (8), for example, cannot be derived by a classical argument. Nevertheless, it is not difficult to construct a classical field consisting of two light beams with the properties that when the beams fall on two photodetectors, whenever one detector registers a response then with high probability the other one does also. For this purpose we consider a classical light source that emits a succession of identical short light pulses of duration δt at random, at an average rate R. If a light beam derived from this source is split into two by a half-silvered mirror, we obtain two light beams with intensities $I_1(t)$ and $I_2(t)$ whose time developments are identical, as shown in Fig. 1.

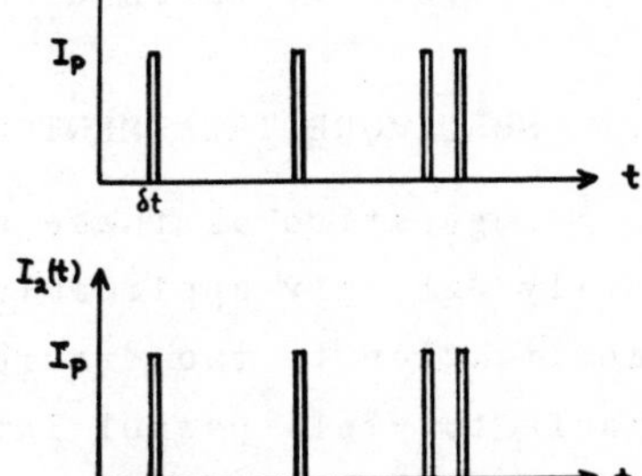

Fig. 1. A classical pair of light intensities that would give rise to strongly correlated photodetections at two detectors.

When these beams fall on two efficient photodetectors, then with high probability photodetections will be registered by both detectors simultaneously within δt.

To make these considerations a little more quantitative, let us take the light pulses to be rectangular, of peak intensity I_p and duration δt. Then the average light intensities are

$$\langle I_j \rangle = I_p R \delta t \quad , \qquad (j=1,2) \tag{20}$$

and the two-time intensity correlation function is given by

$$\left.\begin{aligned} \langle I_j(t) I_k(t+\tau) \rangle &= I_p^2 [R(\delta t-\tau) + R^2(\delta t)^2] \text{ for } 0 \leq \tau \leq \delta t \\ &= I_p^2 [R^2(\delta t)^2] \text{ for } \tau > \delta t \quad . \\ &\qquad (j,k=1,2) \end{aligned}\right\} \tag{21}$$

We note that $\langle I_1(t) I_1(t+\tau) \rangle = \langle I_2(t) I_2(t+\tau) \rangle = \langle I_1(t) I_2(t+\tau) \rangle$, so that the Schwarz inequality is satisfied. The normalized intensity cross-correlation function $\lambda_{12}(0)$ is given by

$$\begin{aligned} \lambda_{12}(0) &= \frac{\langle I_1(t) I_2(t) \rangle}{\langle I_1 \rangle \langle I_2 \rangle} - 1 \\ &= \frac{1}{R\delta t} \quad , \end{aligned} \tag{22}$$

and, like λ_{12} given by Eq. (19), it is inversely proportional to the light intensity and can be very large. Indeed, if measurements made in a time interval δt are compared, then λ_{12} given by Eq. (22) is larger than λ_{12} given by Eq. (19) by the number of photons in each light pulse. The absolute value of λ_{12} therefore provides a test of the quantum character of the field. Only in the limit in which each 'classical' light pulse contains one photon on the average, would the two values of λ_{12} coincide, which evidently takes us out of the strictly classical domain.

4. MULTIMODE TREATMENT WITH CONTINUOUS EXPANSION

Suggestive as these conclusions are, they are not necessarily directly applicable to an experiment, because of the idealization to two discrete modes. Moreover, the treatment is unable to yield useful information about the time-dependence of correlation functions. We therefore re-examine the problem with the help of a multimode expansion of the quantum electromagnetic field [see Hong and Mandel (1985)], with the assumption that the incoming field is again a classical monochromatic plane wave

$$\underline{V}(\underline{r},t) = \underline{V}\, e^{i(\underline{k}_0 \cdot \underline{r} - \omega_0 t)} \quad . \tag{23}$$

We shall use a discrete mode decomposition in a space L^3 with periodic boundary conditions, although sums will be replaced by integrals in later stages of the calculation, so that we are effectively dealing with a continuum of modes. The Hamiltonian will be taken to be of the form [see Kleinman (1968), Hong and Mandel (1985)]

$$\begin{aligned}\hat{H} = \sum_{\underline{k},s} \hbar\omega(\underline{k})\hat{n}_{\underline{k}s} + \frac{1}{L^3}\int_V \sum_{\underline{k}',s'} \sum_{\underline{k}'',s''} \\ \times \chi_{ij\ell}[\omega(\underline{k}_0),\omega(\underline{k}'),\omega(\underline{k}'')]\hat{a}^\dagger_{\underline{k}'s'}\hat{a}^\dagger_{\underline{k}''s''} \\ \times (\underline{\varepsilon}^*_{\underline{k}'s'})_i(\underline{\varepsilon}^*_{\underline{k}''s''})_j V_\ell\, e^{i[(\underline{k}_0-\underline{k}'-\underline{k}'')\cdot\underline{r}-\omega(\underline{k}_0)t]}\, d^3x \\ + \text{h.c.}\end{aligned} \tag{24}$$

This contains a natural generalization of the interaction used previously in Eq. (2). Here $\underline{\varepsilon}_{\underline{k}s}$ is a unit polarization vector, s is a polarization index (s=1,2), and the integral extends over the volume V of the nonlinear crystal. $\chi_{ij\ell}$ is a 2nd order susceptibility tensor for the crystal, which in general depends on the three wave vectors $\underline{k}_0$, $\underline{k}'$, $\underline{k}''$, but has the symmetry

$$\chi_{ij\ell}[\omega(\underline{k}_0),\omega(\underline{k}'),\omega(\underline{k}'')] = \chi_{ij\ell}[\omega(\underline{k}_0),\omega(\underline{k}''),\omega(\underline{k}')] \quad . \tag{25}$$

For a uniform cubic crystal in the form of a rectangular parallelepiped centered at $\underline{r}_0$ with sides ℓ_1,ℓ_2,ℓ_3 the volume integral yields

$$\prod_{m=1}^{3}\left[\frac{\sin\frac{1}{2}(\underline{k}_0-\underline{k}'-\underline{k}'')_m\ell_m}{\frac{1}{2}(\underline{k}_0-\underline{k}'-\underline{k}'')_m}\right] \quad .$$

The Heisenberg equation of motion for the annihilation operator $\hat{a}_{\underline{k}s}(t)$ can be written down immediately, as before. It is a little more convenient to work with the slowly varying variables

$$\hat{A}_{\underline{k}s}(t) \equiv \hat{a}_{\underline{k}s}(t)e^{i\omega(\underline{k})t} \quad . \tag{26}$$

If we assume that the interaction time Δt is the brief propagation time of the light through the crystal in the ℓ_3 direction, then the Heisenberg equation of motion is readily integrated, and we can write, to a good approximation [see Hong and Mandel (1985)],

$$\begin{aligned}\hat{A}_{\underline{k}s}(\Delta t) = \hat{A}_{\underline{k}s}(0) + \frac{1}{i\hbar L^3}\sum_{\underline{k}',s'}\chi_{ij\ell}[\omega(\underline{k}_0),\omega(\underline{k}),\omega(\underline{k}')] \\ \times \hat{A}^{\dagger}_{\underline{k}'s'}(0)[(\underline{\varepsilon}^*_{\underline{k}s})_i(\underline{\varepsilon}^*_{\underline{k}'s'})_j + (\underline{\varepsilon}^*_{\underline{k}'s'})_i(\underline{\varepsilon}^*_{\underline{k}s})_j] \\ \times V_\ell \frac{\sin\frac{1}{2}[\omega(\underline{k}_0)-\omega(\underline{k})-\omega(\underline{k}')]\Delta t}{\frac{1}{2}[\omega(\underline{k}_0)-\omega(\underline{k})-\omega(\underline{k}')]} \\ \times \prod_{m=1}^{3}\frac{\sin\frac{1}{2}(\underline{k}_0-\underline{k}-\underline{k}')_m\ell_m}{\frac{1}{2}(\underline{k}_0-\underline{k}-\underline{k}')_m} \\ \times e^{i(\underline{k}_0-\underline{k}-\underline{k}')\cdot\underline{r}_0}\, e^{-\frac{1}{2}i[\omega(\underline{k}_0)-\omega(\underline{k})-\omega(\underline{k}')]\Delta t} \quad ,\end{aligned} \tag{27}$$

and

$$\hat{A}_{\underline{k}s}(t) = \hat{A}_{\underline{k}s}(\Delta t) \text{ for } t \geq \Delta t \quad . \tag{28}$$

The down-converted light is allowed to fall on two detectors centered at positions $\underline{r}_1$ and $\underline{r}_2$, that respond to a limited set of modes $[\underline{k}]_1$ and $[\underline{k}]_2$, which might be determined by filter It is therefore convenient to introduce a response or detector operator $\hat{W}$ defined by

$$\hat{\underline{W}}(\underline{r}_j,t) = \frac{1}{L^{3/2}}\sum_{[\underline{k}]_j,s}\hat{A}_{\underline{k}s}(t)\underline{\varepsilon}_{\underline{k}s}\, e^{i[\underline{k}\cdot\underline{r}_j-\omega(\underline{k})t]} \quad , \quad j=1,2. \tag{29}$$

If the detector j has an entrance aperture of area δA_j and is open for a short time δt that is much greater than the reciprocal bandwidth, then the probability of detecting a photon with a perfect detector is given by

$$P_j = \int_{\delta V_j}\langle\hat{\underline{W}}^{\dagger}(\underline{r},t)\cdot\hat{\underline{W}}(\underline{r},t)\rangle d^3x \quad , \qquad j=1,2 \quad . \tag{30}$$

Here δV_j is a volume in the form of a cylinder of base δA_j and height $c\delta t$, which is the volume of space effectively explored by the detector in time δt. Similarly, the joint probability $P_{12}(\tau)$ of detecting a photon at detector 1 at time t within δt, and at detector 2 at a later time $t+\tau$ within δt, is given by

$$P_{12}(\tau) = \int_{\delta V_1}\int_{\delta V_2} \langle \hat{W}_p^\dagger(\underline{r},t)\hat{W}_q^\dagger(\underline{r}',t+\tau)\hat{W}_q(\underline{r}',t+\tau)\hat{W}_p(\underline{r},t)\rangle \times d^3x\, d^3x' \quad . \tag{31}$$

By substituting for $\hat{A}_{\underline{k}s}(t)$ from Eqs. (27) and (28) in Eq. (29), and then using the result in Eqs. (30) and (31), we can calculate these probabilities on the assumption that the initial quantum state is the vacuum. After some straightforward but somewhat lengthy manipulation we arrive at the result [see Hong and Mandel (1985)]

$$\begin{aligned} P_1 = \frac{1}{\hbar^2}\int_{\delta V_1} d^3x_1 \frac{1}{(2\pi)^6}\int d^3k \int_{[\underline{k}']_1} d^3k \sum_{s,s'} \\ \times \chi^*_{ij\ell}[\omega(\underline{k}_0),\omega(\underline{k}),\omega(\underline{k}')]\chi_{uvw}[\omega(\underline{k}_0),\omega(\underline{k}),\omega(\underline{k}')] \\ \times [(\underline{\varepsilon}_{\underline{k}s})_i(\underline{\varepsilon}_{\underline{k}'s'})_j + (\underline{\varepsilon}_{\underline{k}'s'})_i(\underline{\varepsilon}_{\underline{k}s})_j] \\ \times [(\underline{\varepsilon}^*_{\underline{k}s})_u(\underline{\varepsilon}^*_{\underline{k}'s'})_v + (\underline{\varepsilon}^*_{\underline{k}'s'})_u(\underline{\varepsilon}^*_{\underline{k}s})_v]V^*_\ell V_w \\ \times \int_0^{\Delta t} dt''\, e^{i(\underline{k}_0-\underline{k}-\underline{k}')\cdot[\underline{r}_1-\underline{r}_0-\underline{\kappa}'u(t-t'')]} \\ \times e^{i[\omega(\underline{k}_0)-\omega(\underline{k})-\omega(\underline{k}')][(1/2)\Delta t-t'']} \\ \times U(\underline{r}_1-\underline{r}_0-\kappa'u(t-t'')|\ell_1\ell_2\ell_3) \\ \times \frac{\sin\frac{1}{2}[\omega(\underline{k}_0)-\omega(\underline{k})-\omega(\underline{k}')]\Delta t}{\frac{1}{2}[\omega(\underline{k}_0)-\omega(\underline{k})-\omega(\underline{k}')]} \prod_{m=1}^{3}\left[\frac{\sin\frac{1}{2}(\underline{k}_0-\underline{k}-\underline{k}')_m\ell_m}{\frac{1}{2}(\underline{k}_0-\underline{k}-\underline{k}')_m}\right]. \end{aligned} \tag{32}$$

Here $\underline{\kappa}\equiv\underline{k}/|\underline{k}|$, $\underline{u}$ is the group velocity $\underline{\nabla}_{\underline{k}}\omega(\underline{k})$, and the function $U(\underline{r}|\ell_1\ell_2\ell_3)$ is unity if $\underline{r}$ falls within the crystal volume $\ell_1\ell_2\ell_3$, and zero otherwise. The expression for the joint detection probability $P_{12}(0)$ for conjugate photons turns out to be identical, provided the two detectors are appropriately located, except that the integral over d^3k is limited to the set $[\underline{k}]_2$ that is conjugate to $[\underline{k}]_1$. However, because of the sinc factors, the significant contributions to the $\underline{k}$-integral in Eq. (32) come from

$$\left.\begin{aligned} \underline{k} &= \underline{k}_0-\underline{k}_1'+0(1/\ell) \\ \omega(\underline{k}) &= \omega(\underline{k}_0)-\omega(\underline{k}')+0(1/\Delta t) \quad , \end{aligned}\right\} \tag{33}$$

so that the integration is effectively limited to the set $[\underline{k}]_2$ in any case. It then follows that to a very good approximation

$$P_1 = P_{12}(0) = P_2 \quad . \tag{34}$$

The time range of $P_{12}(\tau)$ is found to be approximately δt, provided δt exceeds the reciprocal bandwidth, and it is not determined by the propagation time of the light through the crystal or by the coherence time of the pump beam, which was effectively infinite in our calculation [see Hong and Mandel (1985)].

The requirement that δt exceed the reciprocal acceptance bandwidth and that the detection volumes $\delta V_1, \delta V_2$ be chosen appropriately, ensures that the photons of a conjugate pair are not missed in time or space by one or the other detector. This leads to an interesting paradox when the detector areas δA and the detection time intervals δt are both very small. In that case one would expect the integral in Eq. (30) to reduce to

$$P_j = \langle \underline{\hat{W}}^\dagger(\underline{r}_j,t) \cdot \underline{\hat{W}}(\underline{r}_j,t)\rangle c\delta A\ \delta t \qquad (j=1,2) \quad . \tag{35}$$

Superficially it might appear that the double integral in Eq. (31) would then also reduce to

$$P_{12}(0) = \langle \hat{W}_p^\dagger(\underline{r}_1,t)\hat{W}_q(\underline{r}_2,t)\hat{W}_q(\underline{r}_2,t)\hat{W}_p(\underline{r}_1,t)\rangle c^2(\delta A)^2(\delta t)^2 \quad .$$

However, in that case Eq. (34) could not possibly hold, because P_1, P_2 would be of order $\delta A\delta t$, whereas P_{12} would be of order $(\delta A\delta t)^2$. The resolution of the paradox depends on the fact that the two-point correlation function under the integral in Eq. (31) is highly peaked in space and time. If we write it in the form

$$\langle \hat{W}_p^\dagger(\underline{r}_1,t_1)\hat{W}_q^\dagger(\underline{r}_1+\underline{r},t_1+t)\hat{W}_q(\underline{r}_1+\underline{r},t_1+t)\hat{W}_p(\underline{r}_1,t_1)\rangle \quad ,$$

with $\underline{r} \equiv \underline{r}_2-\underline{r}_1$, $t \equiv t_2-t_1$, then the $\underline{r}_1, t_1$ integrations can be approximated by $\delta A\ \delta t$, but the variation with $\underline{r}, t$ is so rapid that such an approximation is not valid in the remaining integrals.

In an actual experiment the quantum efficiencies of the two detectors will of course have some values α_1, α_2 less than 100%. Moreover, the collection efficiencies along the paths leading to detectors 1 and 2 may not be equal either. We can allow for this possibility by introducing a quantity β_1 which is the probability that when a signal or idler photon is detected at detector 1, then the conjugate photon reaches detector 2, and similarly for β_2. Then Eq. (34) has to be replaced by

$$\alpha_1\beta_1 P_2 = P_{12} = \alpha_2\beta_2 P_1 \quad . \tag{36}$$

Because of the equalities (34) and (36), the process of spontaneous parametric down-conversion comes very close to giving us a one-photon state in which the photon is localized in both space and time (albeit still within a region whose linear dimensions are large compared with the wavelength and in a time interval long compared with the optical period). The detection of a signal photon at some place at some time implies the presence of an idler photon at a certain place at a certain time. Although many common processes, like the decay of an excited atom, give rise to one-photon states after a time, the resulting photon is generally distributed over space and time rather than being localized.

Ideally, the probability p(n) that n idler photons are located in a certain region of space and in a short time interval, conditioned on the detection of a signal photon, would be given by

$$p(n) = \delta_{n1} \quad . \tag{37}$$

However, because of the limited collection and quantum efficiencies of the photodetector the measured photon counting distribution P(n) in the idler channel will differ from p(n). If η is the probability that an idler photon conjugate to the detected signal photon is registered, then P(n) is related to p(n) by the Bernoulli convolution

$$P(m) = \sum_{n=m} p(n) \binom{n}{m} \eta^{m} (1-\eta)^{n-m} \quad . \tag{38}$$

In principle, this relation can be inverted to yield p(n) from measurements of P(m). In practice, corrections for background have to be made.

5. CORRELATION EXPERIMENTS

Some of the foregoing conclusions have recently been tested by two-time photoelectric correlation measurements [see Friberg et al (1985a, 1985b)]. Figure 2 gives an outline of one experiment. Down-conversion into the visible was achieved by allowing the 351.1 nm line of an argon laser to fall on a 6 cm long crystal of KDP, whose optical axis makes an angle of 50.35° with the normal to the crystal face. The down-converted light emerges in cones with their axes aligned with the laser beam. Light at the conjugate wavelengths of 660 nm and 750 nm was selected to fall on two microchannel plate photodetectors, whose output pulses were standardized by two constant fraction discriminators and then directed to the start and stop inputs of a time-to-digital converter (TDC). This measures and digitizes the time interval between the two pulses and sorts the events into groups of range 50 psec. Because of the 100 psec transit time spreads of the electrons in the microchannel plate detectors, even strictly coincident photon pairs would result in a distribution of about 150 psec width centered on zero. In the experiment the distribution was displaced from zero to 1000 psec by delaying the stop pulse an extra nanosecond.

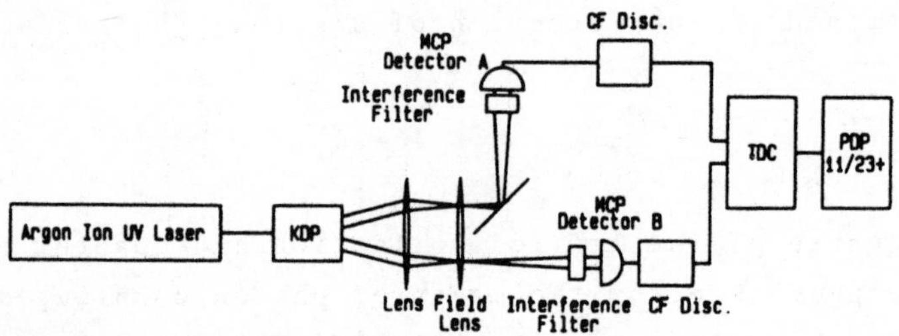

Fig. 2. Outline of the apparatus for measuring time intervals between signal and idler photons.
[Reproduced from S. Friberg, C. K. Hong and L. Mandel, Phys. Rev. Lett. 54, 2011 (1985).]

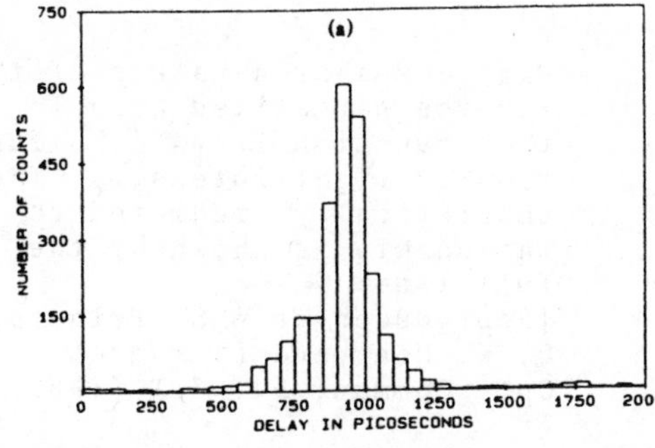

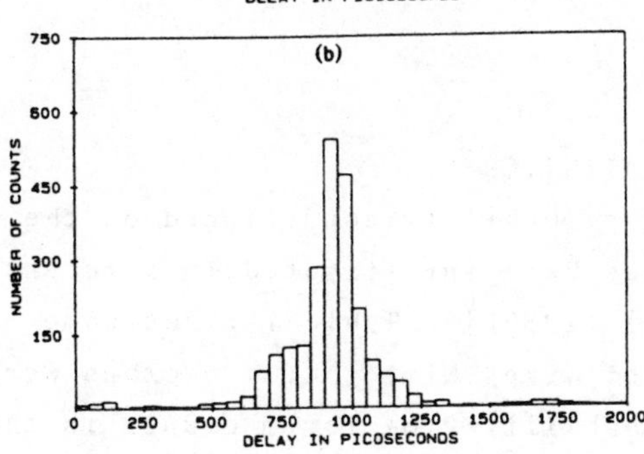

Fig. 3 Measured distributions of time intervals between signal and idler photons (a) for a single mode pump laser; (b) for a multimode pump laser. [Reproduced from S. Friberg, C. K. Hong and L. Mandel, Phys. Rev. Lett. 54, 2011 (1985).]

A typical set of experimental results is shown in Fig. 3. The standard deviations of the approximately Gaussian distributions are about 175 psec, which is close to the instrumental spread. This is significantly less than the 400 psec transit time of the light through the crystal, even before instrumental effects connected with detector electronics are subtracted out. Moreover, the distributions obtained with a pump beam in the form of a single-mode laser of coherence time greater than 40 nsec (Fig. 3a), and a multimode laser with coherence time of about 100 psec (Fig. 3b), were virtually identical. This shows that the down-converted photons are produced 'simultaneously', and both the propagation time and the coherence time of the pump light play a negligible role.

Figure 4 shows a plot of the reciprocal normalized correlation function λ_{12} derived from the measured coincidence detection probabilities as a function of the down-converted light intensity [see Friberg et al (1985b)]. The predicted inverse relationship is confirmed, and the constant of proportionality is found to be consistent with Eq. 19).

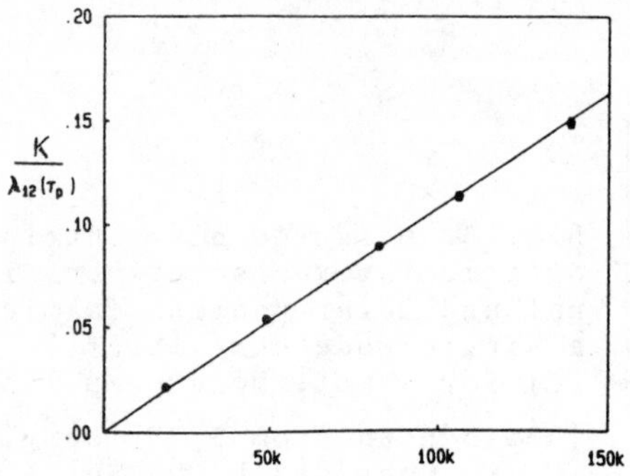

Fig. 4 Measured values of the inverse normalized correlation function $1/\lambda_{12}$ as a function of light intensity. The theoretically predicted relationship is shown by the full line.
[Reproduced from S. Friberg, C. K. Hong and L. Mandel, Opt. Commun. 54, 311 (1985).]

6. MEASUREMENTS OF PHOTON STATISTICS

The idler photon counting probability conditioned on the detection of a signal photon has been investigated in a recent experiment [see Hong and Mandel (1986)]. Table 1 gives some measured values of the number of times N(m) that m photons were counted, and the resulting probability P(m) for m counts in the idler channel, in a 20 nsec interval, conditioned on the detection of a signal count. It is apparent that P(m) is negligibly small except for m=0,1. Even for the raw data represented by the measured probability P(m), without any corrections applied, we find that the counting statistics are sub-Poissonian with parameter

$$Q \equiv \frac{\langle(\Delta m)^2\rangle - \langle m\rangle}{\langle m\rangle} = -.0024 \pm .0001 \quad . \tag{39}$$

But Eq. (38) can readily be inverted to yield p(n) from P(m),

$$\left.\begin{aligned} p(0) &= 1 - P(1)/\eta \\ p(1) &= P(1)/\eta \quad . \end{aligned}\right\} \tag{40}$$

Observed Data	Counting Probabilities	Background Probabilities	Derived Photon Probabilities
N = 6,000,000			
N(0) = 5,985,901	P(0) = N(0)/N = .99765	$p_B(0) \gtrsim 1$	$p(0) \gtrsim -.17$ to $.04$
N(1) = 14,098	P(1) = N(1)/N = .00235	$p_B(1) \gtrsim 8.8\times10^{-5}$	$p(1) \gtrsim 1.06\pm10\%$
N(2) = 1	$P(2) = N(2)/N < 10^{-6}$	$p_B(2) \gtrsim 0$	$p(2) \gtrsim 0$

Table 1. Measured Photon Counting Distributions.

In practice the situation is made a little more complicated by the presence of background counts, and Eq. (40) has to be replaced by [see Hong and Mandel (1986)]

$$\left.\begin{aligned} p(0) &= 1-p(1) \\ p(1) &= [P(1)p_B(0)-P(0)p_B(1)]/\eta\theta\; p_B^2(0) \quad . \end{aligned}\right\} \quad (41)$$

Here $p_B(n)$ is the probability that there are n background counts during the measurement interval, and θ is the fraction of counts in the signal channel initiated by a signal photon. The measured background probabilities $p_B(n)$ are also shown in Table 1, together with the derived probabilities p(n). The latter are plotted in Fig. 5. The indicated uncertainties are not statistical, but stem largely from uncertainties in the value of η. It will be seen that we have a very good approximation to a localized one-photon state. Clearly such a field is highly nonclassical.

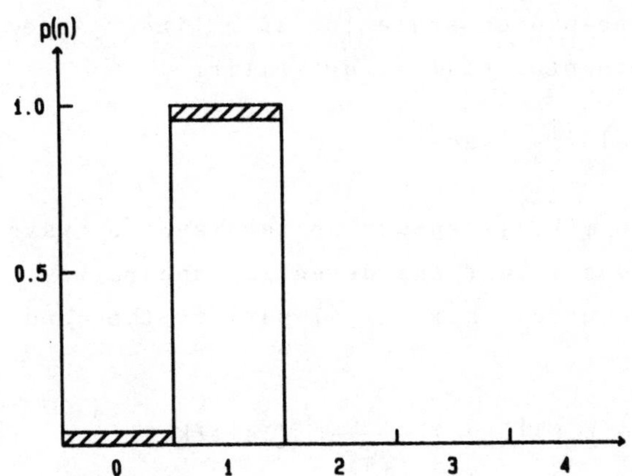

Fig. 5 Derived probability distribution of idler photons conditioned on the detection of a signal photon.

[Reproduced from C. K. Hong and L. Mandel, Phys. Rev. Lett. 56, 58 (1986).]

7. PHOTON LOCALIZATION AND PHOTOELECTRIC MEASUREMENT

Although a photon has no position in a precise sense [see Newton and Wigner (1949)], a photoelectric measurement nevertheless appears to come close to localizing the photon in space-time to a certain extent. However, in attempting to interpret the measurement in this way we encounter a paradox, because of the non-local relationship between electromagnetic field vectors.

Despite the absence of a position operator for photons, one

can associate an approximate position with a photon through the use of a localized photon number operator [see Mandel (1966), Jauch and Piron (1967), Amrein (1969)]. For this purpose we let $\hat{\underline{V}}(\underline{r},t)$ be a configuration space photon annihilation operator

$$\hat{\underline{V}}(\underline{r},t) \equiv \frac{1}{L^{3/2}} \sum_{[\underline{k},s]} \hat{a}_{\underline{k}s}\, \underline{\varepsilon}_{\underline{k}s}\, e^{i(\underline{k}\cdot\underline{r}-\omega t)} . \tag{42}$$

Then the operator associated with the number of photons in the volume V at time t is defined by [see Mandel (1966)]

$$\hat{n}(V,t) \equiv \int_V \hat{\underline{V}}^{\dagger}(\underline{r},t)\cdot\hat{\underline{V}}(\underline{r},t)\,d^3x \quad , \tag{43}$$

with the understanding that the dimensions of V are large compared with optical wavelengths. Let us use this to calculate the average number of idler photons in some small volume V, conditioned on the detection of a signal photon.

We first need to assign a quantum state to the idler photon. We shall represent the one-photon state $|\phi\rangle$ as a linear superposition of one-photon momentum states, by writing

$$|\phi\rangle = \frac{1}{L^{3/2}} \sum_{\underline{k},s} \phi(\underline{k},s)\, \hat{a}^{\dagger}_{\underline{k}s}\, |vac\rangle \quad , \tag{44}$$

where the weight function $\phi(\underline{k},s)$ depends on the range of wave vectors $\underline{k}$ and polarizations s that the detection apparatus accepts. Normalization requires that $\phi(\underline{k},s)$ satisfy the condition

$$\langle\phi|\phi\rangle = 1 = \frac{1}{L^3} \sum_{\underline{k},s} |\phi(\underline{k},s)|^2 \to \sum_s \frac{1}{(2\pi)^3} \int d^3k\, |\phi(\underline{k},s)|^2 , \tag{45}$$

but $\phi(\underline{k},s)$ could otherwise be of any form. From Eqs. (43) and (44) we then obtain for the average number of photons in V at time t,

$$\begin{aligned}\langle\phi|\hat{n}(V,t)|\phi\rangle &= \frac{1}{L^6} \int_V \sum_{\underline{k},s} \sum_{\underline{k}',s'} \phi^*(\underline{k},s)\phi(\underline{k}',s') \\ &\qquad e^{i[(\underline{k}'-\underline{k})\cdot\underline{r}-(\omega'-\omega)t]}\, d^3x \\ &= \int_V d^3x\, |\phi(\underline{r},t)|^2 \quad , \end{aligned} \tag{46}$$

$$\underline{\Phi}(\underline{r},t) \equiv \frac{1}{L^3}\sum_{\underline{k},s} \phi(\underline{k},s)\underline{\varepsilon}_{\underline{k}s}\, e^{i(\underline{k}\cdot\underline{r}-\omega t)}$$
$$\rightarrow \sum_s \frac{1}{(2\pi)^3}\int d^3k\; \phi(\underline{k},s)\underline{\varepsilon}_{\underline{k}s}\, e^{i(\underline{k}\cdot\underline{r}-\omega t)} \quad . \tag{47}$$

Evidently $\underline{\Phi}(\underline{r},t)$ is the wave function of the photon in state $|\phi\rangle$ [see Jauch and Piron (1967)], obtainable by projecting $|\phi\rangle$ on to the position $\underline{r}$ one-photon state $\hat{\underline{V}}^{\dagger}(\underline{r},t)|vac\rangle$,

$$\underline{\Phi}(\underline{r},t) = \langle vac|\hat{\underline{V}}(\underline{r},t)|\phi\rangle \quad , \tag{48}$$

but with the interpretation of $\underline{\Phi}(\underline{r},t)$ restricted to volume integrals as in Eq. (46). On the basis of the foregoing discussion in Section 6, we expect $\underline{\Phi}(\underline{r},t)$, conditioned on the detection of a signal photon, to be strongly peaked both in position and in time, to reflect the localized form of the idler photon.

Let us now calculate the probability that this localized photon is detected at position $\underline{r}$ at time t with the help of a photoelectric detector. We shall suppose that the electron-field interaction occurs through the electric field. Then the probability P of photoelectric detection at time t is proportional to the expectation of $\hat{\underline{E}}^{(-)}(\underline{r},t)\cdot\hat{\underline{E}}^{(+)}(\underline{r},t)$ [see Glauber (1963a, 1963b)] integrated over the detector area and the short measurement time interval. As before we represent this by an equivalent integral over a volume V, and write

$$P = K\int_V \langle\phi|\hat{\underline{E}}^{(-)}(\underline{r},t)\cdot\hat{\underline{E}}^{(+)}(\underline{r},t)|\phi\rangle d^3x \quad , \tag{49}$$

where $\hat{\underline{E}}^{(+)}(\underline{r},t)$ is the positive frequency part of the electric field, and is given by

$$\hat{\underline{E}}^{(+)}(\underline{r},t) = \frac{1}{L^{3/2}}\sum_{\underline{k},s} i\left(\frac{\hbar\omega}{2\varepsilon_0}\right)^{\frac{1}{2}} \hat{a}_{\underline{k}s}\,\underline{\varepsilon}_{\underline{k}s}\, e^{i(\underline{k}\cdot\underline{r}-\omega t)} \quad . \tag{50}$$

From Eqs. (44) and (50) we find

$$\hat{\underline{E}}^{(+)}(\underline{r},t)|\phi\rangle = \frac{1}{L^3}\sum_{\underline{k},s} i\left(\frac{\hbar\omega}{2\varepsilon_0}\right)^{\frac{1}{2}} \phi(\underline{k},s)\underline{\varepsilon}_{\underline{k}s}\, e^{i(\underline{k}\cdot\underline{r}-\omega t)}|vac\rangle$$
$$\rightarrow \sum_s \frac{1}{(2\pi)^3}\int d^3k\; i\left(\frac{\hbar\omega}{2\varepsilon_0}\right)^{\frac{1}{2}} \phi(\underline{k},s)\underline{\varepsilon}_{\underline{k}s}\, e^{i(\underline{k}\cdot\underline{r}-\omega t)}|vac\rangle \quad , \tag{51}$$

and from Eq. (49), with the help of the notation

$$\underline{\Psi}(\underline{r},t) \equiv \sum \frac{1}{(2\pi)^3} \int d^3k \; i\left(\frac{\hbar\omega}{2\varepsilon_0}\right)^{\frac{1}{2}} \phi(\underline{k},s) \; e^{i(\underline{k}\cdot\underline{r}-\omega t)} \underline{\varepsilon}_{\underline{k}s} \;, \tag{52}$$

$$P = K \int_V \langle vac|\underline{\Psi}^*(\underline{r},t)\cdot\underline{\Psi}(\underline{r},t)|vac\rangle d^3x = K \int_V |\underline{\Psi}(\underline{r},t)|^2 \, d^3x \,. \tag{53}$$

By definition, the Fourier transform of $\underline{\Psi}(\underline{r},t)$ differs by the factor $i(\hbar\omega/2\varepsilon_0)^{\frac{1}{2}}$ (and possibly by other factors characteristic of the detector) from the Fourier transform of $\underline{\Phi}(\underline{r},t)$. It follows that $\underline{\Psi}(\underline{r},t)$ and $\underline{\Phi}(\underline{r},t)$ are related through the convolution

$$\underline{\Psi}(\underline{r},t) = \int d^3x' \; G(\underline{r}-\underline{r}')\underline{\Phi}(\underline{r}',t) \;, \tag{54}$$

in which the Green function $G(\underline{r})$ is the three-dimensional Fourier transform of $i(\hbar\omega/2\varepsilon_0)^{\frac{1}{2}}$,

$$G(\underline{r}) = \frac{i}{(2\pi)^3} \int d^3k \; \left(\frac{\hbar\omega}{2\varepsilon_0}\right) \; e^{i\underline{k}\cdot\underline{r}} \;. \tag{55}$$

The form of $G(\underline{r})$ was investigated by Amrein (1969), who showed that for large r

$$|G(\underline{r})| \approx \frac{3\sqrt{\hbar c}}{16\pi^{3/2}\sqrt{\varepsilon_0}\; r^{7/2}} \;. \tag{56}$$

Because of the $r^{-7/2}$ dependence on distance it is possible for $|\underline{\Psi}(\underline{r},t)|^2$ to be non-negligible at positions where $|\underline{\Phi}(\underline{r},t)|^2$ is negligibly small. In other words, although the photon may be localized at one place, it could be detected photoelectrically elsewhere. This very unparticle-like aspect of photons might possibly be explorable experimentally through the process of parametric down-conversion, if $\underline{\Phi}(\underline{r},t)$ is physically meaningful.

8. AN APPLICATION TO OPTICAL COMMUNICATION

The simultaneous creation of a signal and idler photon, which are distinguishable by their wavelengths, suggests an interesting application to an optical communication channel [see Mandel (1984)]. By modulating the pump beam and detecting both down-converted photons in coincidence, it is possible to achieve considerable discrimination against unwanted photons. Indeed, it is possible to create a situation in which a single detector receives no intelligible information because the signal is lost in the noise, and yet two detectors working in coincidence can achieve almost error-free communication.

Let suffices 1 and 2 label parameters in the signal and idler channel, respectively. When the pump beam is on, the rate R_c at which photons are detected in coincidence at the two detectors within some resolving time $\delta\tau$ is given by

$$R_c = R_1\alpha_2 + [(1-\alpha_2)R_1+r_1][(1-\alpha_1)R_2+r_2]\delta\tau \quad . \tag{57}$$

Here R_1,R_2 are the counting rates of the two detectors due to conjugate signal and idler photons, r_1,r_2 are the corresponding counting rates attributable to all other causes, and α_1,α_2 are the two detection efficiencies. The first term on the right of Eq. (57) is due to the arrival of conjugate photon pairs, whereas the other term represents accidental coincidences within the resolving time $\delta\tau$. Because $\delta\tau$ can be made very short, it is possible for the first term to exceed the second greatly, so that the coincidence counting rate is dominated by correlated signal and idler photons, and falls drastically when the pump beam is turned off.

Figure 6 shows the outline of a recent experiment in which the feasibility of using the down-converted photons in this way in a communication channel was tested [see Hong et al (1985)].

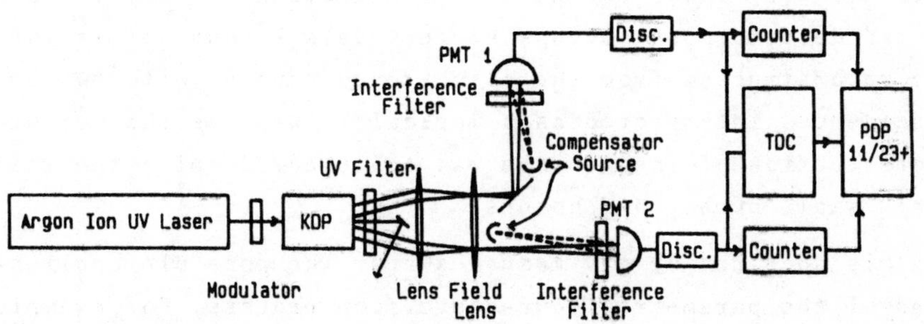

Fig. 6. Outline of a pilot experiment for using coincident signal and idler photons in an optical communication channel. [Reproduced from C. K. Hong, S. R. Friberg and L. Mandel, Appl. Opt. 24, 3877 (1985).]

A simple binary communication channel was formed by on-off modulation of the pump beam by a shutter, so that 'on' corresponds to message symbol 1 and 'off' corresponds to 0, and each symbol has the same duration T. Down-converted beams consisting of signal and idler photons were transmitted separately and detected by two photomultipliers, whose output pulses in each interval T were counted. At the same time the pulses from the two detectors were fed to a time-to-digital converter that effectively counted the number of coincidences in time T with some instrumental resolving time of 3 or 4 nsec. Two or more coincidences were interpreted as a logical 1, while zero or one coincidence was interpreted as logical 0. In order deliberately to make the counts registered by the single detectors as unintelligible as possible, two compensator light sources with Poisson statistics were turned on in each beam whenever the pump beam was turned off. By careful adjustment of the compensator sources it was possible to make the average photon counting rate at each detector independent of whether a 0 or 1 was being transmitted. The individual photon counts registered by each detector then contained no useful information. Only simultaneously detected photon pairs conveyed information.

The results of the experiment are shown in Fig. 7. (a) shows the transmitted message consisting of an alternation of 1 and 0, and (b) and (c) give the number of photon counts registered by the signal and idler detectors in successive counting intervals T. These counts simply exhibit random Poissonian fluctuations. (d) shows the number of coincidences registered by the two detectors in successive intervals T, and (e) is the message reconstructed from the coincidence counts, with two or more coincidences interpreted as a logical 1. We note that error-free reconstruction of the message was achieved, despite the relatively small number of photons.

This does not by any means exhaust the possible applications of the parametric down-conversion process. For example, in the degenerate limit the down-converted field exhibits squeezing [see Stoler (1970, 1971), Milburn and Walls (1981, 1983), Wódkiewicz and Zubairy (1983), Friberg and Mandel (1984), Meystre et al (1984)], which also has potential for optical

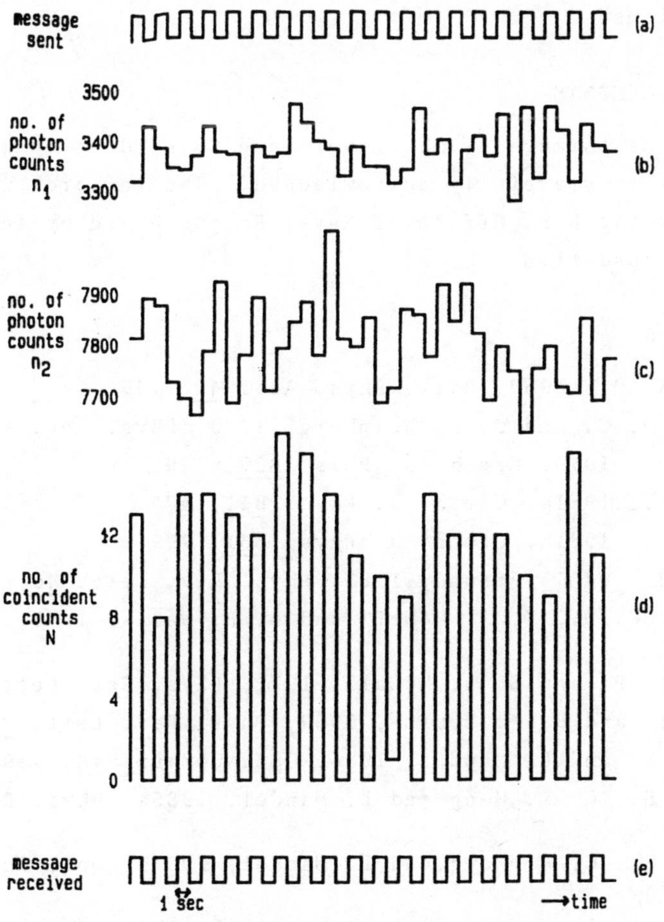

Fig. 7. Results of the communication experiment. (a) transmitted message; (b) number of photon counts registered by detector 1 in successive time intervals T; (c) number of photon counts registered by detector 2 in successive time intervals T; (d) number of coincidences registered by the two detectors in successive time intervals T; (e) reconstructed message.
[Reproduced from C. K. Hong, S. R. Friberg and L. Mandel, Appl. Opt. 24, 3877 (1985).]

communication, and a combination of feedback with down-conversion has recently been proposed for the generation of light with sub-Poisson statistics [see Jakeman and Walker (1985)]. It is clear from the foregoing that the process is not only interesting for fundamental reasons, because it generates highly nonclassical two-photon states, but that it may also have more practical applications.

ACKNOWLEDGEMENTS

The assistance of S. Friberg and C. K. Hong in most of this work is gratefully acknowledged. The research was supported by the U.S. Office of Naval Research and by the National Science Foundation.

REFERENCES

Amrein, W. O., 1969, Helv. Phys. Act. 42, 149.

Burnham, D. C. and D. L. Weinberg, 1970, Phys. Rev. Lett. 25, 84.

Chmela, P., 1979, Czech. J. Phys. B29, 129.

Chmela, P., 1981a, Czech. J. Phys. B31, 977.

Chmela, P., 1981b, Czech. J. Phys. B31, 999.

Danino, H. and I. Freund, 1981, Phys. Rev. Lett. 46, 1127.

Drummond, P. D., K. J. McNeil and D. F. Walls, 1979, Opt. Commun. 28, 255.

Eisenberg, P. and S. L. McCall, 1971, Phys. Rev. Lett. 26, 684.

Freund, I. and B. F. Levine, 1969, Phys. Rev. Lett. 23, 854.

Friberg, S. and L. Mandel, 1984, Opt. Commun. 48, 439.

Friberg, S., C. K. Hong and L. Mandel, 1985a, Phys. Rev. Lett. 54, 2011.

Friberg, S., C. K. Hong and L. Mandel, 1985b, Opt. Commun. 54, 311.

Giallorenzi, G. T. and C. L. Tang, 1968, Phys. Rev. 166, 225.

Glauber, R. J., 1963a, Phys. Rev. 130, 2529.

Glauber, R. J., 1963b, Phys. Rev. 131, 2766.

Gordon, J. P., W. H. Louisell and L. R. Walker, 1963, Phys. Rev. 129, 481.

Graham, R., 1984, Phys. Rev. Lett. 52, 117.

Hillery, M. and M. S. Zubairy, 1984, Phys. Rev. A 29, 1275.

Hong, C. K. and L. Mandel, 1985, Phys. Rev. A 31, 2409.

Hong, C. K. and L. Mandel, 1986, Phys. Rev. Lett. 56, 58.

Hong, C. K., S. Friberg and L. Mandel, 1985, Appl. Opt. 24, 3877.

Jakeman, E. and J. G. Walker, 1985, Opt. Commun. 55, 219.

Jauch, J. M. and C. Piron, 1967, Helv. Phys. Act. 40, 559.

Klauder, J. R., 1966, Phys. Rev. Lett. 16, 534.

Kleinman, D. A., 1968, Phys. Rev. 174, 1027.

Louisell, W. H., 1960, *Coupled Mode and Parametric Electronics* (New York: Wiley).

Louisell, W. H., A. Yariv and A. E. Siegman, 1961, Phys. Rev. 124, 1646.

Mandel, L., 1966, Phys. Rev. 144, 1071.

Mandel, L., 1984, J. Opt. Soc. Am. B 1, 108.

Meystre, P., K. Wódkiewicz and M. S. Zubairy, 1984, in *Coherence and Quantum Optics V*, eds. L. Mandel and E. Wolf, 761 (New York: Plenum).

Milburn, G. and D. F. Walls, 1981, Opt. Commun. 39, 401.

Milburn, G. and D. F. Walls, 1983, Phys. Rev. A 27, 392.

Mista, L., V. Perinova and J. Perina, 1977a, Act. Phys. Pol. A51, 739.

Mista, L., V. Perinova and J. Perina, 1977b, Act. Phys. Pol. A52, 425.

Mollow, B. R., 1967, Phys. Rev. 162, 1256.

Mollow, B. R., 1973, Phys. Rev. A 8, 2684.

Mollow, B. R. and R. J. Glauber, 1967a, Phys. Rev. 160, 1076.

Mollow, B. R. and R. J. Glauber, 1967b, Phys. Rev. 160, 1097.

Montgomery, G. P. and T. G. Giallorenzi, 1973, Phys. Rev. B 8, 808.

Neumann, R. and H. Haug, 1979, Opt. Commun. 31, 267.

Newton, T. D. and E. P. Wigner, 1949, Rev. Mod. Phys. 21, 400.

Raiford, M. T., 1970, Phys. Rev. A 2, 1541.

Raiford, M. T., 1974, Phys. Rev. A 9, 2060.

Smithers, M. and E.Y.C. Lu, 1974, Phys. Rev. A 10, 187.

Stoler, D., 1970, Phys. Rev. D 1, 3217.

Stoler, D., 1971, Phys. Rev. D 4, 1927.

Stoler, D., 1974, Phys. Rev. Lett. 33, 1397.

Sudarshan, E.C.G., 1963, Phys. Rev. Lett. 10, 277.

Tucker, J. and D. F. Walls, 1969, Phys. Rev. 178, 2036.

Wódkiewicz, K. and M. S. Zubairy, 1983, Phys. Rev. A 27, 203.

Yariv, A., 1975, *Quantum Electronics*, 2nd Ed., Ch. 17 (New York: Wiley).

Zel'dovich, B. Ya and D. W. Klyshko, 1969, Zh. Eksp. Teor. Fiz. Pis. Red. 9, 69 [Sov. Phys. JETP Pis. Red. 9, 40 (1969)].

THE USE OF PHOTO EVENT TRIGGERED OPTICAL SHUTTERS TO GENERATE SUB-POISSONIAN PHOTOELECTRON STATISTICS

E JAKEMAN

1. INTRODUCTION

Many of the contributions in this volume testify to the growing interest non-classical states of light. Since the early theoretical proposals Stoler (1974), Simaan and Loudon (1975), Yuen (1976) and others there h been a steady growth of physical insight and understanding of the subjec and of how non-classical states can be generated and observ experimentally. This has culminated in the first reported observations squeezing (Slusher 1986), perhaps one of the more difficult phenomena interpret , but one which holds some promise for the improvement heterodyne measurements (Caves, 1981). Any doubts as to the existence the more readily appreciated photon antibunching were dispelled some yea ago by the pioneering experiments of Kimble, Dagenais and Mandel (1977 More recently, observation of the related phenomena of sub-Poissoni Photon Statistics has been reported (Short and Mandel 1983, Teich and Sal 1985) and a number of different proposals for experiments generating th kind of light have now appeared in the literature (Saleh and Teich 198 Walker and Jakeman 1984, Aspect 1986). In this chapter one such meth will be discussed in which optical shutters are used to introduce detecti event triggered deadtimes in light beams (Jakeman and Walker 1985, Walk and Jakeman 1985). The presentation will not be concerned with a rigoro quantum mechanical description of the experiments, but rather with t practical impediments to the observation of the expected non-classic effects. For example, the results of an analysis of the effects of fini detector efficiencies, finite sample times, background and dark counts wi be given. Many of the ideas were developed in collaboration with my form colleague Dr J G Walker and more recently with Drs J G Rarity and Tapster. The theoretical work and computer simulation has been carried in collaboration with Dr J H Jefferson.

In the next section a brief review of the relevant existing results on detector deadtimes will be given, with some simple optical deadtime experiments described in section 3. These do not generate non-classical light fields which are free for use, but provide interesting analogues. In section 4 the use of optical deadtimes in parametric down conversion to provide freely propagating non-classical fields will be described. Preliminary experimental results will be discussed in section 5.

2. DETECTOR DEADTIMES

It is worth emphasising at the outset that in what follows it will be assumed that photon counting or photon-correlation techniques will be employed in the experimental measurements (Pike and Jakeman 1974). Time is divided into consecutive samples of duration T with the number n(t;T) of counts in each sample centred at time t entered in the appropriate channel of a frequency distribution p(n;T) (the photo-electron counting distribution) or multiplied by the number of counts in a delayed sample at time $t+\tau$ and entered in the appropriate store of correlation coefficients $\langle n(t;T)n(t+\tau;T)\rangle$. Then the photo-electron counting distribution is sub-Poissonian if

$$\mathrm{Var}\ n < \bar{n} \tag{1}$$

and the photo-detections are antibunched if

$$g^{(2)}(\tau) > n^{[2]} \tag{2}$$

for any value of $\tau>0$, where $g^{(2)}(T)=\langle n(o)n(T)\rangle/\bar{n}^2$ (for $\tau>0$) is the normalised autocorrelation function and $n^{[2]}=\langle n(n-1)\rangle/\bar{n}^2$ is the normalised second factorial moment of the photo-electron counting distribution. Neither of the above inequalities can be satisfied if the photo-electron statistics are related to a classical light intensity I(t) through the semi-classical Mandel formula (Mandel 1963)

$$p(n;T) = \int_0^\infty \frac{(\eta E)^n}{n!} e^{-\eta E} P(E)dE \tag{3}$$

where E is the integrated intensity

$$E(t;T) = \int_t^{t+T} I(t')dt'$$

and η is the detector quantum efficiency. It is worth noting that for any given light field, inequalities (1) and (2) need not both be satisfied according to figure 1.

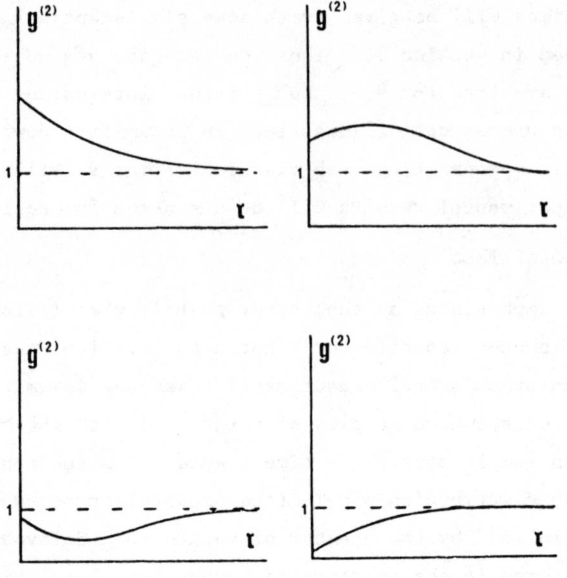

Fig 1. Possible photo-detection correlation functions

This shows some examples of the normalised autocorrelation function $g^{(2)}(\tau)$, the point τ=0 being $n^{[2]}$. The familiar situation (bunched, super Poissonian), where semi-classical detection theory can be used, is depicted in figure 1a. Figure 1d shows the opposite non-classical case (antibunched, sub-Poissonian) to be encountered later in this chapter. Other possibilities are short term antibunched but super-Poissonian (figure 1b) and short term bunched, but sub-Poissonian (figure 1c). Both of these latter cases come into the category of non-classical light.

It is well known that even if the expected photo-electron statistics are of the form shown in figure 1a, short term antibunching and even sub-Poissonian statistics may be observed due to the presence of a finite recovery or deadtime τ_D in the detector and processing electronics following the registration of each event. This is an imperfection in the measuring technique shared with other particle counting systems and leads to distortion of the event statistics. Undesirable corruption of the data of this kind is usually avoided by working in regimes where deadtime effects are small. Consequently, although a great deal of effort has been devoted to evaluating their magnitude, most results quoted in the literature on photon counting measurements are in the form of correction formulae valid in the limit $T \gg \tau_D$ (Johnson et al 1966, Teich and Vannucca 1978 and references therein). Nevertheless, some general results have been derived. In the present context the graphs shown in figure 2 are of particular interest. These show the Fano factor

$$F = \mathrm{Var}\, n_D/\bar{n}_D \tag{5}$$

plotted as a function of τ_D/T for a Poisson process corrupted by a non-paralysable deadtime following each registered event during which no further events can be detected (Muller 1974).

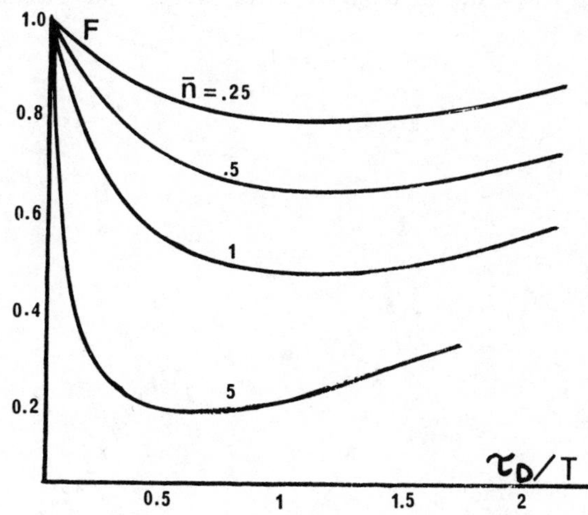

Figure 2. Fano factor versus deadtime for a Poisson process with deadtimes

The Fano factor measures the variance of the distribution of events relative to that of a Poisson distribution of the same mean. It is clear that the presence of deadtime effects can significantly reduce the noise below that expected for a Poisson process. This does not imply that measurement accuracy can be improved by introducing deadtimes into the detected event train artifically. For coherent light the reverse will be the case since photo-events have been discarded. It is conceivable, however, that the measurement of fluctuating signals might be improved under certain conditions as is found in the related post detection technique of clipping (Jakeman, Pike and Swain 1971).

The correlation properties of a Poisson sequence of events corrupted by a non-paralysable deadtime are given approximately by the formula

$$g^{(2)}_{1}(\tau) = (1+\bar{r}\tau_D)\sum_{n=1}^{m} \frac{\bar{r}(\tau-n\tau_D)^{n-1}}{(n-1)!} \exp[-\bar{r}(\tau-n\tau_D)] \qquad (6)$$

where m is the greatest integer less than τ/τ_D and $\bar{r}$ is the true mean count-rate. This formula is valid if $\bar{r}T<<1$ except for the single point $\tau = \tau_D$ (Jakeman and Jefferson 1986). In the case of a dense train of events the registered counts will be almost regularly spaced and the correlation function will have a comb-like structure. In the case of sparse events the presence of a deadtime will have almost no effect and the correlation function will retain its constant Poisson value of unity except for delays shorter than τ_D when it will be zero. An intermediate case is shown in figure 3.

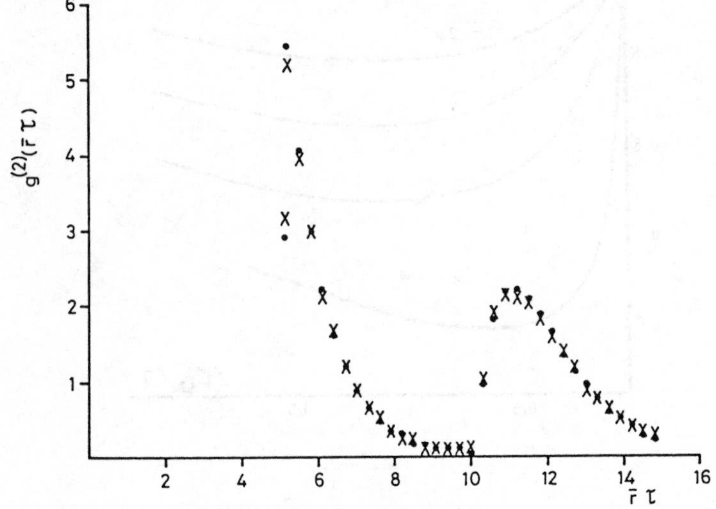

Figure 3. Autocorrelation function for a Poisson process with deadtime •exact x simulations

In all cases antibunching in the sense of inequality (2) is present since $n^{[2]} < 1$. Indeed for $T<\tau_D$, $n^{[2]}=0$.

Considerations of the above type suggest the possibility of using artificial deadtimes for the generation of sub-Poissonian and antibunched trains of events. One of the original objectives of the work described here was indeed to use this method for simulating the statistics of non-classical light for the purpose of developing new signal processing expertise. As it has turned out, the research has led to the development of a new technique for the generation of non-classical light.

3. OPTICAL DEADTIMES: SEMI-CLASSICAL EXPERIMENTS

Although it is possible to introduce deadtimes into the Poisson train of events generated by laser light using post-detection electronic techniques, it is also easy to achieve the same objective using an optical shutter in the configuration depicted in figure 4.

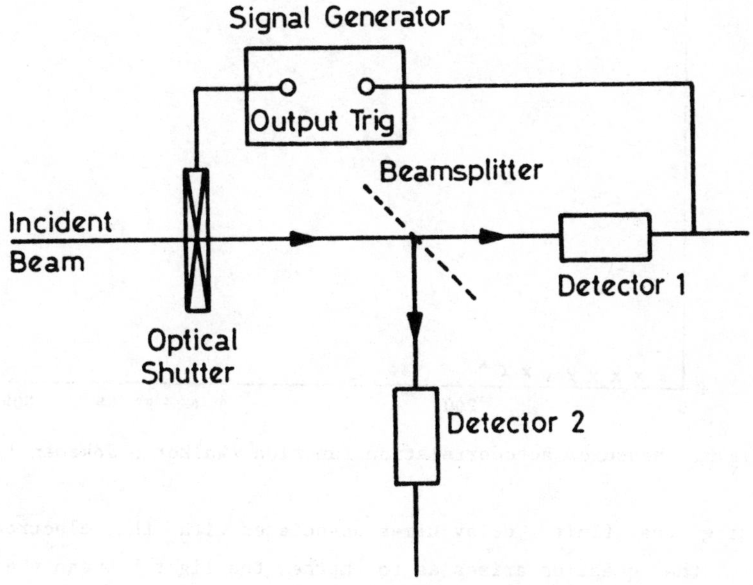

Fig 4. Optical deadtime experiment

When an event is registered in detector 1, the shutter closes for a pre-set deadtime during which no further counts will be registered. If the true event rate is $\bar{r}$ then the wait time to the next event following a deadtime will be $\bar{r}^{(-1)}$ on average and the measured event rate will be $(\bar{r}^{(-1)} + \tau_D)^{-1}$ giving

$$\bar{n}_1 = \bar{r}T/(1+\bar{r}\tau_D) \tag{7}$$

Moreover in the special case $t<\tau_D$

$$n_1^{[2]} = 0$$

with the autocorrelation function given by equation (6). Thus the events are antibunched and sub-Poissonian as expected. The results of an experiment of this kind are shown in figure 5.

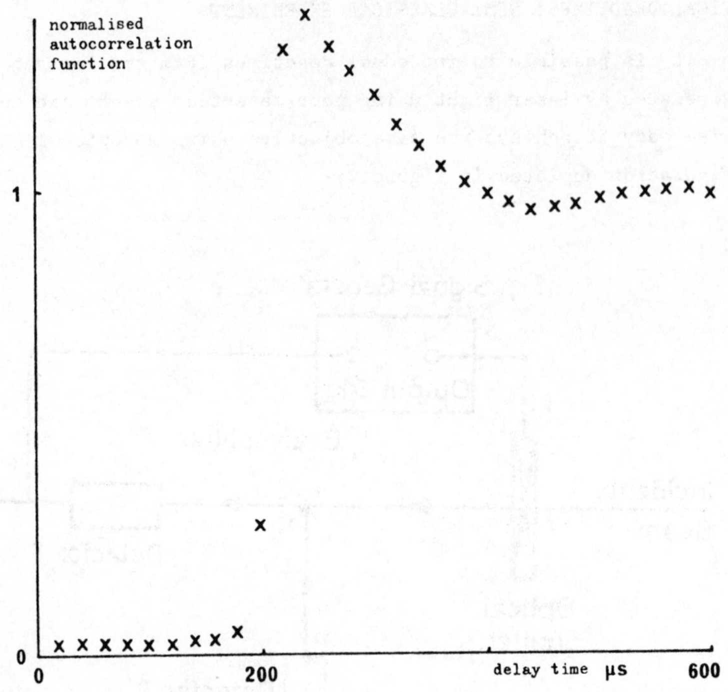

Fig 5. Measured autocorrelation function (Walker & Jakeman 1984)

Neglecting the finite delay times associated with the electronics and shutter, the question arises as to whether the light between the shutter and detector is non-classical. This cannot be tested by using a half silvered mirror or beam splitter to divert a portion of the beam into a second detector as illustrated by the dotted part of figure 4. Since this light does not participate in the triggering process the events registered in detector 2 are just those for a randomly chopped Poisson process. If the true rate in this channel is $\bar{s}$ then it is readily shown that the observed mean is given by

$$\bar{n}_2 = \bar{s}T/(1+\bar{r}\tau_D) \qquad (9)$$

$$n_2^{[2]} = 1+\bar{r}\tau_D \qquad (10)$$

and with $g_2^{(2)}(\tau) = g_1^{(2)}(\tau+\tau_D)$ being a displaced version of formula (6). This light is thus both super-Poissonian and bunched. In fact all measurable quantitites in the experiment can be derived semi-classically, by considering the post-detection event rate and the classical intensity incident on the shutter, but without invoking the quantum nature of the light at any stage. It is perhaps not surprising then that this type of arrangement does not provide a useful source of non-classical light. It is nevertheless a useful analogue configuration, and other interesting variations are possible.

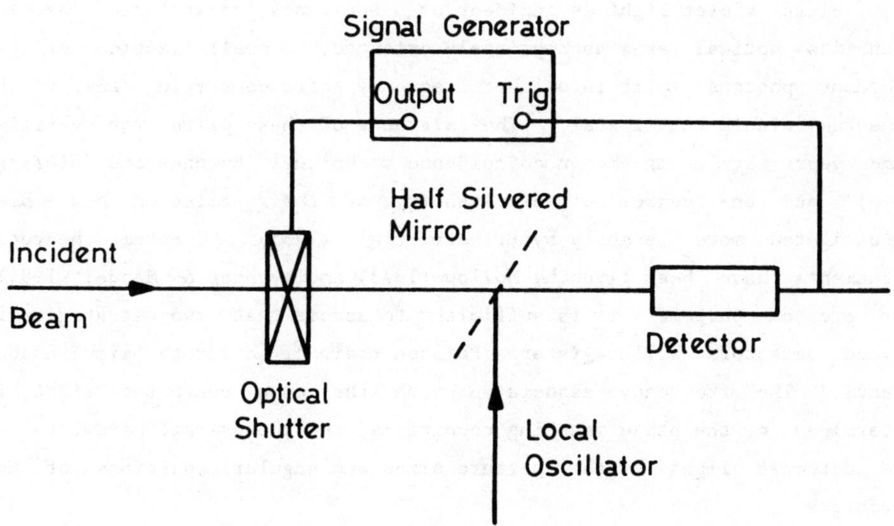

Fig 6. A homodyne experiment

For example in figure 6 a homodyne arrangement is illustrated. Theoretical analysis of this is more complicated and we do not present details of the calculation here (Jakeman and Jefferson 1986). When $\bar{r}T \ll 1$ and $T < \tau_D$ results for the second order statistics reduce to the form

$$n^{[2]} = \frac{\bar{r}_2(1 + \bar{r}\,\tau_D)}{\bar{r}\,(1 + \bar{r}_2\tau_D)} \tag{11}$$

with Fano factor

$$F = 1 - \frac{T(\bar{r} - \bar{r}^2)}{1 + \bar{r}\,\tau_D} \tag{12}$$

Here the true event rate $\bar{r}$ is given by

$$\bar{r} = \bar{r}_1 + \bar{r}_2 + 2\sqrt{\bar{r}_1\,\bar{r}_2}\cos\emptyset \tag{13}$$

The light is thus sub-Poissonian if $\bar{r} > \bar{r}_2$ when there is a higher count rate during 'open' periods, and super-Poissonian if $\bar{r} < \bar{r}_2$ when there is a lower rate during open periods. Although these properties are analogous to those expected for squeezed light, the results can again be derived without field quantization and there is no non-classical light available for use.

One remedy for this situation is to take advantage of the photon replication property inherent in parametric down-conversion.

4. NON-CLASSICAL EXPERIMENTS

When ultra violet light is incident on a non-linear crystal such as ADP with its optical axis appropriately oriented, a small fraction of the incident photons split into lower frequency pairs conserving energy and momentum within the crystal. The existence of these pairs was verified some years ago using photon coincidence techniques (Burnham and Weinberg, 1970) and the degree of simultaneity of their emission has been investigated more recently by Friberg et al (1985). Quantum theoretic treatments have been given by Mollow (1973) and by Hong & Mandel 1985). For present purposes it is sufficient to assume that two strategically placed detectors will register a Poisson train of discrete simultaneous events. The frequency associated with the down converted light is determined by the phase matching conditions, and the overall bandwidth of the detected light by the aperture sizes and angular positions of the detectors.

Consider the arrangement illustrated in figure 7.

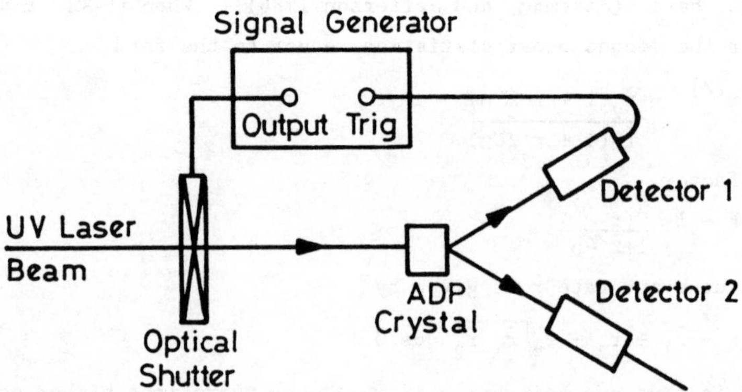

Fig 7. Optical deadtimes in parametric down-conversion

Signals from the trigger channel are now used to introduce deadtimes in the incident UV beam and as in the semi-classical experiment shown in figure 4, detector 1 will register an event train characterised by formulae (7) and (8) if $T<\tau_D$. Thus the events in this channel will be sub-Poissonian and antibunched, but as before the light is not free for use. However, in the new arrangement a replica train of events will be generated down channel 2 and can be independently observed by the monitor detector: it seems that non-classical light should be generated. Note that a quantum description of the down conversion process is necessary. Unfortunately the above description fails to take account of the practicalities of the situation. In particular the finite (and typically low) detector efficiency of the triggering detector means that only a few percent of the possible events may be registered so that only a small fraction of events in Channel 2 will be followed by the required deadtime. A finite sample time calculation of the effect of detector efficiency is presented elsewhere (Jakeman and Jefferson 1986). Here only the result obtained for the normalised second factorial moment of events in the monitor channel is given:

$$n_2^{[2]} = (1-\eta_1)(1+\eta_1\bar{r}\tau_D)\{2(1-p)/\eta_1\bar{r}T\} \quad (14)$$

where $$p = [1-\exp(-\eta_1\bar{r}T)]/\eta_1\bar{r}T \quad (15)$$

and $\bar{r}$ is the ideal rate. Although the efficiency of the monitor detector, η_2, does not figure in equation (14) it will effect the observed Fano factor F_2 which is a measure of the expected noise reduction:

$$F_2 = \eta_2 F_2 + 1-\eta_2 \quad (16)$$

Here F is the pre-detection value. It is clear that even if F is significantly reduced, the observed noise reduction will be severely limited if η_2 is very small.

Although the true detector efficiencies may be a few percent, the effective efficiency of each channel may be even lower due to the presence of background and dark counts. If these are present in the trigger channel they will lead to reduced monitor channel efficiency. Shuttered background counts, unpaired counts due to fluorescence, filter mismatch or optical misalignment lead to reductions of both monitor and trigger channel efficiencies, whilst unshuttered background counts and dark counts in the monitor channel add incoherently in the usual way (Jakeman and Jefferson 1986).

5. RESULTS AND DISCUSSION

Bearing in mind the above considerations and the instrumentation initially available, a preliminary estimate (Jakeman and Walker 1985) indicated a 1% reduction in $n^{[2]}$ might be achieved. In an autocorrelation experiment this would appear as a small reduction of the correlation coefficients below the Poisson value for delay times shorter than the deadtime. An estimate of the experiment time needed to observe such an effect can be made using the signal to noise ratio (SNR) $\sqrt{N}\,\bar{n}(1-n^{[2]})$ where N is the number of samples. For the low count rates expected the last factor on the right of equation (14) is near unity and with $\eta_1 \ll 1$

$$\mathrm{SNR} \sim \sqrt{N}\,\eta_1\ \eta_2\ \bar{r}T\ (1-\bar{r}\,\tau_D) \tag{17}$$

giving a maxiumum value of $\sqrt{N}\,\eta_1\ \eta_2 \bar{r}T/2$ when $\tau_D = 1/2\bar{r}$. For the experimental parameters given in Walker and Jakeman (1985), $n^{[2]} \sim 0.995$ with a time of a few hours to achieve SNR ~ 1. In practice, after pulsing in the monitor detector presented difficulties. These were overcome by dividing the light in this channel using a beam splitter, and cross-correlating the output from two detectors, thus increasing the experiment duration by a factor of 4. The results of this preliminary experiment have been published in full elsewhere, but one set of data is reproduced in figure 8. As expected, the observed effect is very weak!

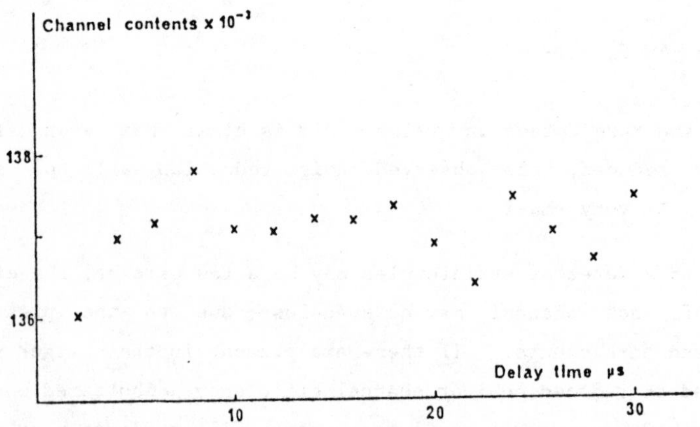

Fig 8. Measured autocorrelation function showing weak antibunching (Walker and Jakeman 1985)

The major drawback of the arrangement shown in figure 6 is the occurrence of events in the monitor channel which are not paired with events in the trigger channel due to the latter's inefficiency. This problem can be eliminated in principle by introducing a second shutter in the monitor channel which only opens when the partner of a registered event in the trigger channel is expected. In practice a fibre optic delay must be included in the monitor channel to compensate for finite response times and electrical delays so that reduced efficiency and some loss of light may occur requiring increased UV source power. Also the monitor channel shutter will be open for a finite time t during which some extra counts may be registered. If N is the number of open intervals of this type per sample time T and m is the number of extra counts registered in these intervals in addition to the expected partner event, (assuming $\eta_2 = 1$ initially) then the observed number of events per sample time will be

$$n = N(1+m) \tag{19}$$

and

$$n_2^{[2]} = \frac{\langle N(N-1)\rangle}{\bar{N}^2}\left(1+\frac{\bar{m}}{(1+\bar{m})^2}\right) + \frac{\bar{m}}{\bar{N}}\frac{(2+\bar{m})}{(1+\bar{m})^2} \tag{20}$$

independent of monitor detector efficiency. The number m is drawn from a Poisson distribution of mean $(1-\eta_1)\bar{r}t$ (neglecting background and dark counts) so that for the simple case $T < \tau_D$

$$n_2^{[2]} = \frac{(1-\eta_1)t}{\eta_1 T}\,\frac{(1+\eta_1\bar{r}\tau_D)[2+(1-\eta_1)\bar{r}t]}{[1+(1-\eta_1)\bar{r}t]^2} \tag{21}$$

This can be reduced by decreasing the ratio t/T. Although current technology still presents limitations, at the time of writing a second moment of 0.6 and a pre-detection Fano factor of 0.9 have already been achieved in an initial experiment using post-detection electronic gating as a substitute for the monitor channel optical shutter.

It is evident from the above that although the work is at an early stage, the technique of using optical shutters to introduce deadtimes in parametric down-conversion does provide a practical method for generating anti-bunched sub-Poissonian light. Even modest improvements in detector performance and shutter operation will enable much more significant effects to be generated. This will make possible both fundamental single-photon experiments (although note that the light is broadband!) and perhaps some practical applications such as the measurement of the absorption coefficients of sensitive specimens. In applications where high flux sub-

Poissonian light is required but where antibunched events are not essential, it is possible that simple feedback smoothing based on anlogue detection could be used, thus taking advantage of the high efficiencies of photo-diode detectors, for example.

REFERENCES

Aspect A, Grangier P and Roger G, 1986, Europhys Lett, 1, 173

Burnham D C & Weinberg D L, 1970 Phys Rev Lett, 25, 84.

Caves C M, 1981, Phys Rev D23, 1693.

Friberg S, Hong C K & Mandel L, 1985, Phys Rev Lett 54, 2011.

Hong C K & Mandel L, 1985, Phys Rev A31, 2409.

Jakeman E, Pike E R and Swain S, 1971, J Phys A, 4, 517.

Jakeman E & Walker J G, 1985, Opt Commun, 55, 219.

Jakeman E & Jefferson J H, 1986, Optica Acta, to be published.

Johnson F A, Jones R, McLean T P & Pike E R, 1966, Phys Rev Lett 16 589.

Kimble M J, Daganais M & Mandel L, 1977 Phys Rev Lett 39 691.

Mandel L, 1963 in Progress in Optics, ed E Wolf, Amsterdam: North Holland.

Mollow B R, 1973, Phys Rev A8, 2684.

Muller J W, 1974, Nucle Instr Meth 117, 401.

Pike E R & Jakeman E, 1974, Adv Quant Elec 2 1.

Saleh B E A & Teich M C, 1985, Opt Commun 52, 429.

Simaan H D & Loudon R, 1975, J Phys A8, 539.

Short R & Mandel L, 1983, Phys Rev Lett, 51, 384.

Slusher R E, 1986, this volume.

Stoler D, 1974, Phys Rev Lett 33, 1397.

Teich M C & Vannucci G, 1978, J Opt Soc Am, 68, 1338.

Teich M C & Saleh B E A, 1985, J Opt Soc Am, B2, 275.

Walker J G & Jakeman E, 1984, SPIE 492, ECOOSA'84, 274.

Walker J G & Jakeman E, 1985, Optica Acta, 32, 1303.

Yuen H P, 1976, Phys Rev A13, 2226.

QUANTUM FLUCTUATIONS IN THE LASER

H HAKEN

1. INTRODUCTION

In a way the year 1985 can be considered as the 25th anniversary of the laser. First indications of the achievement of laser action were reported by Maiman (1960), who found that in a ruby rod the emission linewidth decreases by a factor 4 and the emission lifetime by a factor of 16 of the normal values when the crystal is excited by an intense flash lamp. A few weeks later Collins et al.(1960) succeeded in observing important laser properties, such as intense emission, directed radiation and a high degree of coherence with pulsed ruby lasers. The first cw-laser could be operated in the same year in December in the form of the He-Ne gas laser by Javan and coworkers (1961). This latter system has become particularly suitable for comparisons of theoretical predictions with experimental observations. In this review article first I wish to stress one aspect of laser theory which is of particular importance in so far as it concerns the basic difference between the light emission from normal lamps and that of lasers. Since the talk on which this article is based was aimed at a broad audience according to the wishes of the organizers and in view of the 25th anniversary I think it will be appropriate if I present here some of the basic traits of the developent of laser theory which dates back more than 20 years ago. Because I started working on laser theory during my stay at the Bell Telephone Laboratories in 1960 even before laser action had been discovered my present survey will probably contain a personal flavour. In the second part of my article I shall present some quite recent results on the application of the maximum (information) entropy principle to laser light statistics. Here the laser proves again to be a marvellous test ground for the development or application of new concepts and methods. In the present case for the first time a way is opened for the application of the maximum (information) entropy principle to nonequilibrium phase transitions and, more generally, to selforganizing systems treated in synergetics (see e.g. Haken (1983 a,b)). At the end of my article I wish to briefly touch upon recent developments concerning gigantic fluctuations of laser light which exhibits what is nowadays called 'deterministic chaos'.

2. FROM THE MASER TO THE LASER

The laser was suggested by Schawlow and Townes (1958) who called it 'optical maser'. Indeed the laser principle can be considered as an extension of the maser principle to the optical region. The maser principle itself had been developed and applied by Gordon, Zeiger and Townes (1954) as well as Basov and Prokhorov (1954). On the other hand there are important differences between the maser and the laser, namely what mode selection is concerned. In the microwave region the electromagnetic standing waves can develop in a <u>cavity</u> (Fig.1). Because in general the spacing of mode frequencies is larger than that of the line width of molecules, the process of stimulated emission generates a single mode only (Fig. 2 and 3).

Since it seems (or at least seemed) impossible to build cavities of the size of the optical wave length, in the laser a great many modes of a cavity of the size of a few centimeters fall into the emission linewidth (Fig. 4). A first selection of modes is achieved by using a laser rod on the endfaces of which mirrors are mounted. This leads to a long life time of axial modes, whereas all other modes are, in a way, thrown away (Fig. 5, 6). Because this arrangement of mirrors acts as a Fabry-Perot interferometer, only modes with discrete frequencies can develop. Still the number of

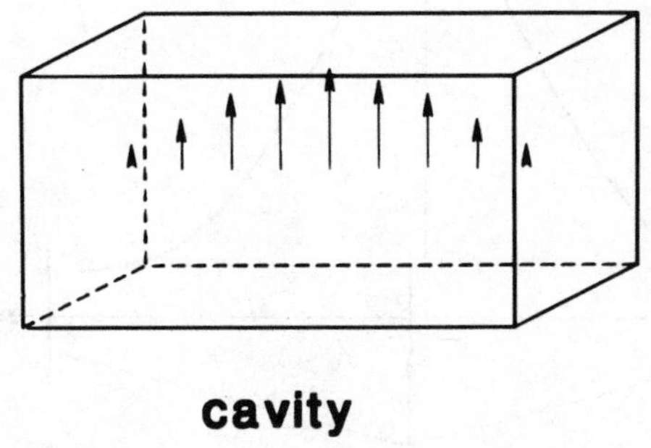

Fig. 1 Visualization of a mode of the electromagnetic field in a microwave cavity

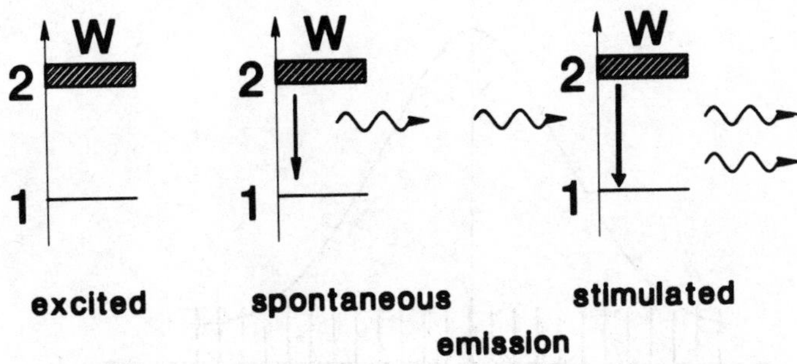

Fig. 2 The left part shows the energy diagram of a two-level system with states 1 and 2. The upper state had been excited by some pump mechanism. It may decay either spontaneously or by stimulated emission. In the spontaneously emission process a photon is emitted spontaneously, whereas in the stimulated emission process an incoming photon causes the emission of an additional photon of the same kind

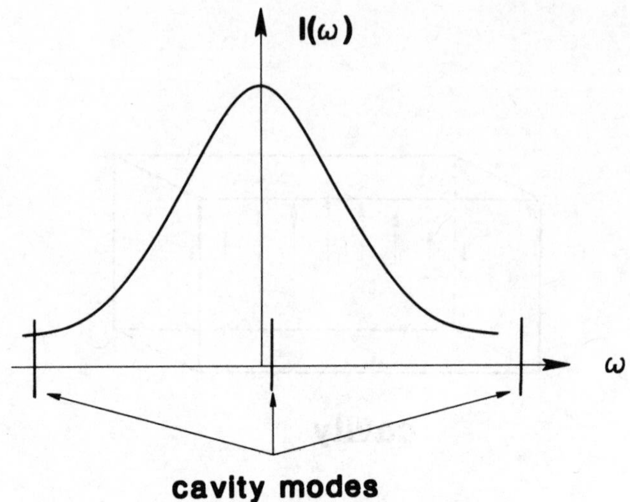

Fig. 3 In a microwave cavity only one or at most very few modes fall into the emission linewidth

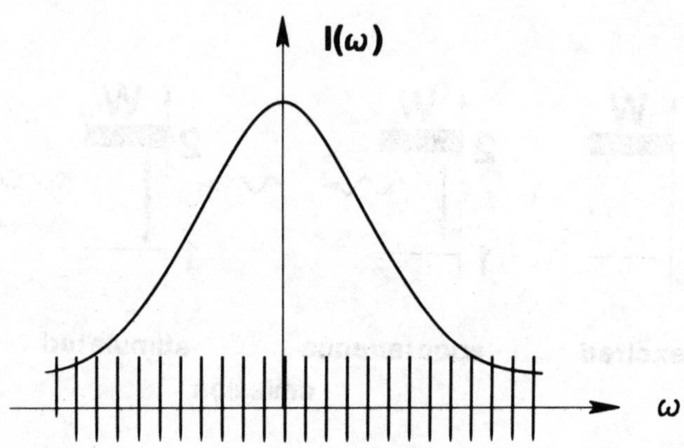

Fig. 4 In the laser very many cavity modes fall into the emission linewidth

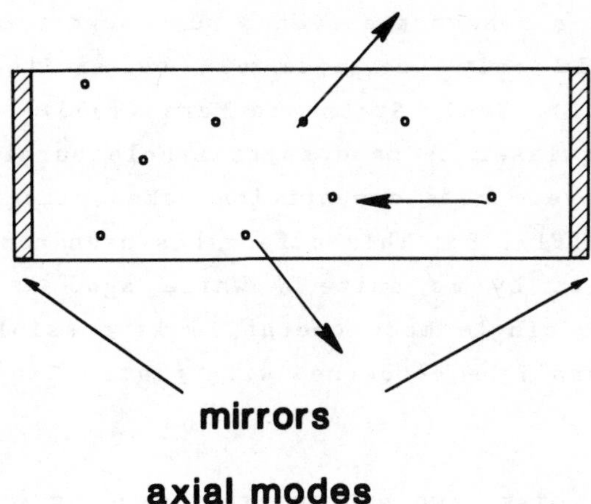

Fig. 5 Only waves running in axial direction stay for relatively long time within the laser material. All others leave the laser very quickly and do not lead to an appreciable interaction with the atoms

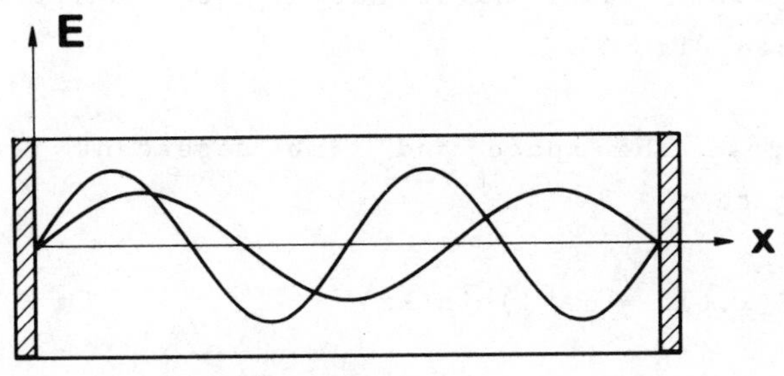

Fig. 6 The Perot-Fabry arrangement allows for standing waves with discrete wavelength only. The wavelengths in this picture are largely exaggerated

modes within the atomic optical linewidth can be numerous. Indeed, due to spatial hole burning quite a number of modes can coexist at a somewhat elevated pump power, while close to laser threshold only one mode can be excited (Haken and Sauermann (1963), Tang, Statz and Mars (1963)). On the other hand if a ring laser is used spatial hole burning is avoided (Fig. 7), a severe mode competition takes place and only one mode survives (Fig. 8). This effect has been named 'Darwinism of laser modes' by me quite a while ago. At any rate we recognize that single mode operation is possible and in the following we shall be concerned with that.

3. COHERENCE, NOISE, AND PHOTON STATISTICS OF A SINGLE MODE LASER

In order to bring out the essentials I will first present a socalled semiclassical treatment, though later on I will show that actually a fully quantum mechanical treatment is needed both what the field and the atoms are concerned. Such a quantum mechanical treatment was indeed used from the very beginning when laser noise had to be studied (Haken, 1964, 1966), (see Fig. 9).

We decompose the space and time dependent field amplitude according to

$$E(x,t) = A(t)\sin(kx), \tag{3.1}$$

where the field amplitude $A(t)$ is further decomposed in a slowly varying part $E(t)$ and a rapidly oscillating exponential function containing the frequency ω of the optical transition

$$E(t)e^{i\omega t} + E^*(t)e^{-i\omega t}. \tag{3.2}$$

We treat a solid state laser as an example. The individual atoms are distinguished by an index μ (Fig. 10). The dipole moments

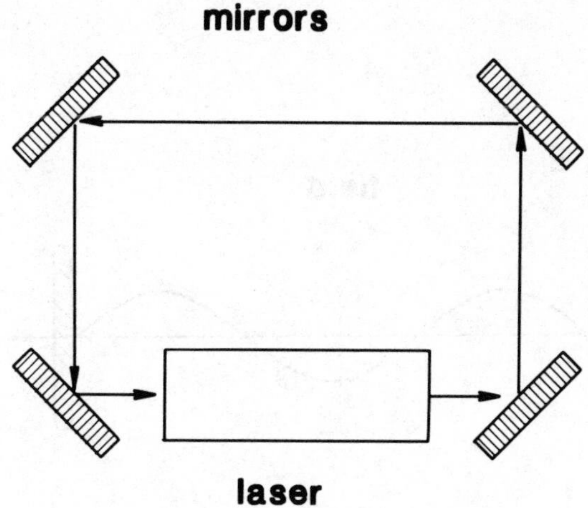

Fig. 7 A typical set-up for a ring laser which allows for mode competition

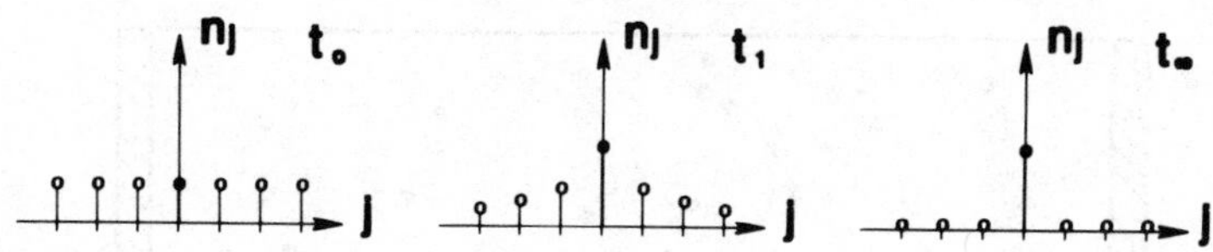

Fig. 8 Darwinism of laser modes
The number of photons belonging to different field modes distinguished by an index j is plotted versus that index for various times. At the initial time t_o all photons are given the same chance. At an intermediate time t_1 the various kinds of photons have been multiplied in different fashions. At the final time t_∞ only one kind of photons has survived, all others have died out because of mode competition in the absence of spatial or spectral hole burning

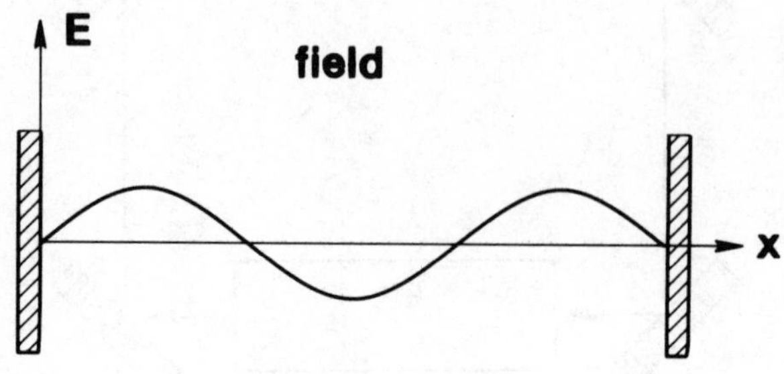

fig. 9 The field strength $E(x,t)$ in the laser

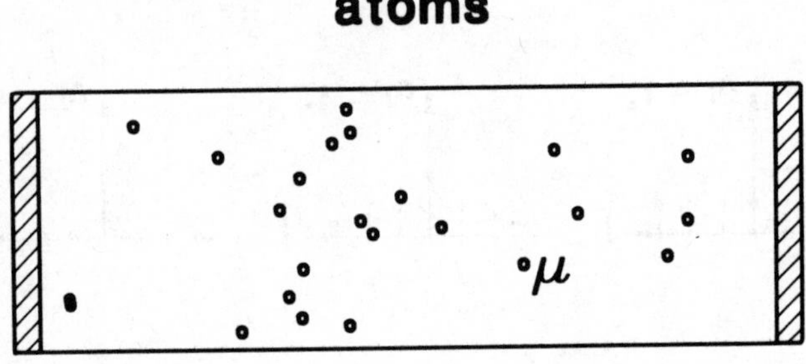

Fig. 10 The atomic variables are distinguished by an index μ

$$p_\mu(t) \tag{3.3}$$

are decomposed, in analogy to (3.2), into

$$p_\mu(t)e^{i\omega t} + p_\mu^*(t)e^{-i\omega t}. \tag{3.4}$$

The equation for the field amplitude E reads

$$\frac{dE}{dt} = -\kappa E + g\sum_\mu p_\mu + F(t). \tag{3.5}$$

The first term on the r.h.s. describes the decay of the field because of some transparency of the mirrors and other loss mechanisms. The sum over μ represents the generation of the field by the dipole moments where g is a coupling constant between field and dipole moments, and F(t) are fluctuating forces which occur whenever dissipation is present. Later we shall show how this equation as well as the following equations (3.6) and (3.7) can be derived in a quantum mechanical fashion.

The temporal change of the dipole moments is described by

$$\frac{dp_\mu}{dt} = -\gamma p_\mu + gEd_\mu + F_\mu(t), \tag{3.6}$$

where the first term describes the damping of the dipole moments. The second term describes the driving of the dipole moments by the field. This term also contains the inversion of the atoms d_μ and finally a fluctuating force $F_\mu(t)$ occurs. The occurrence of d_μ is a quantum mechanical effect which describes the direction of energy flow into or out of the field depending on the state of occupation of the atom μ.

The change of inversion d_μ is described by

$$\frac{dd_\mu}{dt} = \gamma_\parallel(d_o - d_\mu) - g(Ep_\mu^* + E^*p_\mu) + F_{d,\mu}(t). \tag{3.7}$$

The first term on the r.h.s. describes the effect of external pumping and relaxation processes of the atom μ, the second bracket the effect of its interaction with the field E. The last term is the fluctuating force.

The equations (3.5) - (3.7) are rather complicated. They are first of all numerous because of the many atoms (10^{16}-10^{18}) participating in the laser process. They are nonlinear, e.g. terms Ed, and they contain stochastic, i.e. fluctuating forces. However, as it was already shown in the case of the semiclassical laser theory which ignores fluctuations (Haken and Sauermann, 1963; Haken, 1964, 1966; Lamb, 1964) such a complicated system can be cut down to a single equation or, physically speaking, the whole dynamics of the laser system is governed by the field mode E alone. This is made possible because in most lasers the inequality

$$\gamma \gg \gamma_{\parallel} \gg \kappa \tag{3.8}$$

holds. Without going into the mathematical details we may make the following physical arguments. According to the first term on the r.h.s. of (3.5) E changes at the slow time scale $1/\kappa$. If we now consider (3.6) where E occurs as driving term we conclude that p_μ changes at the same time scale so that we immediately obtain the relations

$$\left|\frac{dp_\mu}{dt}\right| \simeq |p_\mu| \kappa \ll |p_\mu| \gamma. \tag{3.9}$$

Therefore in a good approximation we may put

$$\frac{dp_\mu}{dt} \simeq 0 \tag{3.10}$$

within (3.6). This allows us to express p_μ according to (3.6) explicitly by

$$p_\mu(t) = \frac{g}{\gamma}E(t)d_\mu(t) + \frac{1}{\gamma}F_\mu(t). \tag{3.11}$$

Actually the treatment of the fluctuating forces within the same scheme requires some additional thought. This procedure is probably one of the simplest examples of the slaving principle of synergetics (Haken, 1983a, b) according to which we may explicitly express fast relaxing variables such as p_μ by slowly varying variables such as E. Applying the same idea to d_μ and assuming a not too high field amplitude we may eliminate the atomic variables and arrive at

$$\frac{dE}{dt} = (G - \kappa)E - C\,E|E|^2 + F_{tot}, \tag{3.12}$$

where F_{tot} consists of the fluctuating forces occurring in (3.5) and (3.6). Those occurring in (3.7) have a very small effect and can be safely neglected. Eq. (3.12) allows a simple interpretation when we identify E with the coordinate of a particle q and add to (3.12) an accelleration term $m\ddot{q}$. (3.12) can then be cast into the form

$$(m\ddot{q}) + \dot{q} = - \frac{\partial V}{\partial q} + F, \tag{3.13}$$

where V is the potential of the force occurring on the r.h.s. of (3.12).

Let us first discuss the case in which the gain G is smaller than the loss κ. In such a case the potential looks as exhibited in figs. 11a, 11c for two different sizes of gain, $G_2 > G_1$. The action of the fluctuating forces can be interpreted as pushes excerted on the particle . After each push the particle relaxes towards the equilibrium position $q = 0$. Quite evidently the relaxation time increases with increasing G. When this is expressed in terms of optics we may say that the coherence time increases or, in other words, that the linewidth decreases. Because the field strength is still comparatively small we can safely neglect the nonlinear term in (3.12) and calculate the correlation function which defines the optical linewidth

$$< E^*(t_2)E(t_1) > = \text{Const.}\ e^{-\Gamma|t_2-t_1|}. \tag{3.14}$$

The result is given by

$$\Gamma = \frac{\text{Const.}}{P}, \tag{3.15}$$

where P is the output power. (3.15) is the famous Townes

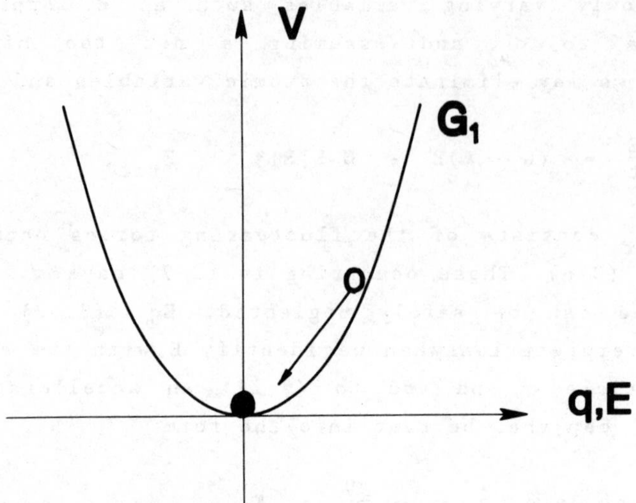

Fig. 11a The potential V versus q or E for a small gain constant G_1

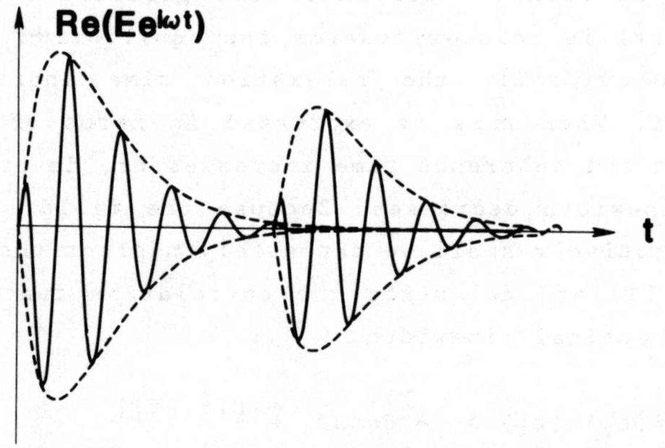

Fig. 11b The real part of the time dependent field amplitude versus time for small gain

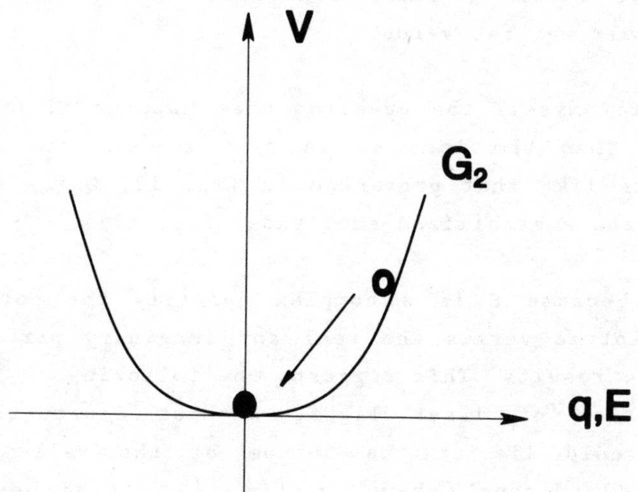

Fig. 11c Same as fig. 11a but with increased gain G_2

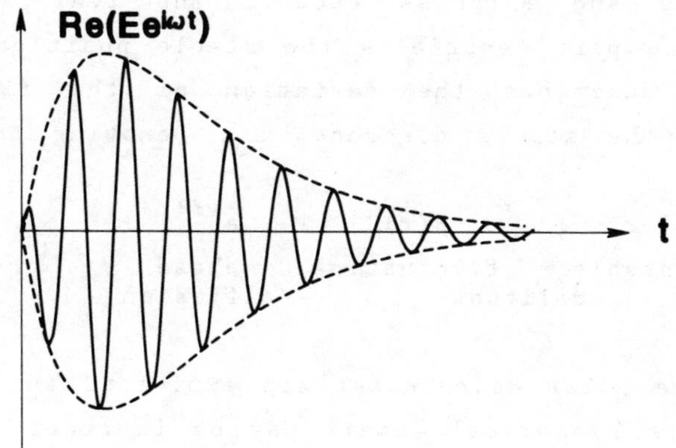

Fig. 11d Same as fig. 11b but for increased gain G_2

formula for maser or laser linewidth. So in this way the former result was recovered.

Then I asked myself the question what happens if the gain G is bigger than the loss κ. In such a case the potential curve looks like that presented in fig. 12. Quite evidently we expect now a stabilized amplitude (fig. 13).

Actually, because E is a complex quantity the potential V must be plotted versus the real and imaginary part of E so that fig.14 results. This suggests the following interpretation of laser light. Without fluctuations the particle would lie at the bottom of the valley of the potential 'landscape' shown in fig. 14. If pushes act in radial direction the stable amplitude will be quickly restored because the particle has climbed up the hill in radial direction. On the other hand, in tangential direction there are no restoring forces. The particle being subject to random forces will undergo some sort of diffusion process.

These considerations suggest to make for E the following hypothesis. We decompose the complex field amplitude into a real amplitude and a phase factor. The real amplitude consists of one part describing the stable position at r_o, whereas $\rho(t)$ describes the deviation of the fictitious particle from the stable distance, r_o. Denoting the phase by ϕ we put

$$E = \underbrace{[\, r_o}_{\text{stable amplitude}} + \underbrace{\rho(t)\,]}_{\text{fluctuating amplitude}} \cdot \underbrace{e^{i\phi(t)}}_{\text{phase diffusion}} . \qquad (3.16)$$

The prospective polar coordinates are exhibited in fig. 15. At this point a historical remark may be in order. When in 1964 I came to this kind of visualization of laser light I realized that it was in contrast to all other laser theories dealing with laser noise published till that time. Therefore I asked some of these experts whether this new situation exhibited in fig. 12 can happen and their answer was clearly 'no'. They told me that only fig. 11 is possible, the potential curve becoming flatter and flatter being

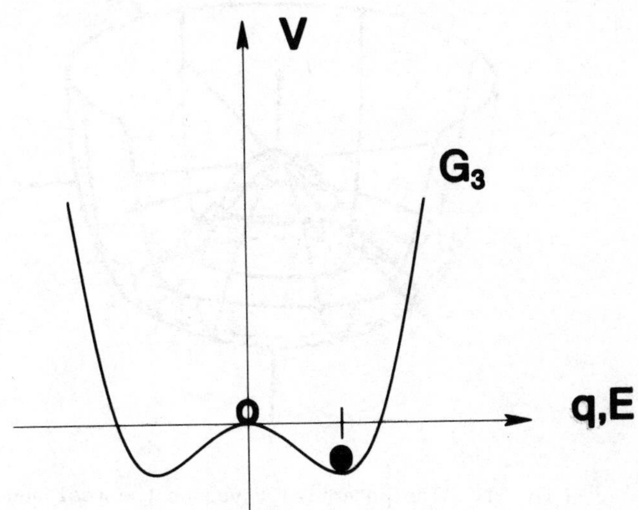

Fig. 12 The potential V versus q or E above threshold, $G - \varkappa > 0$

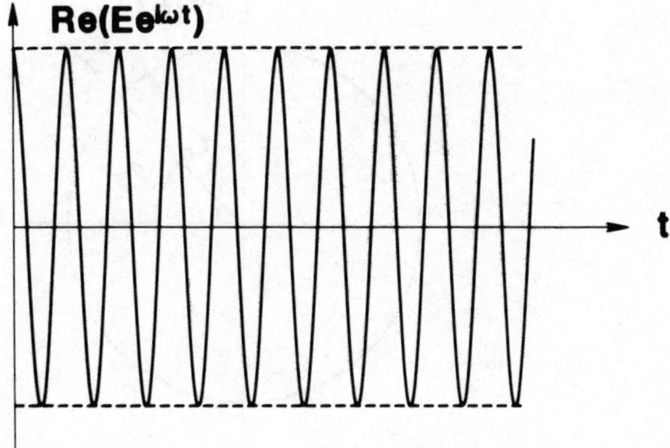

Fig. 13 The field beyond threshold but with the neglect of fluctuations

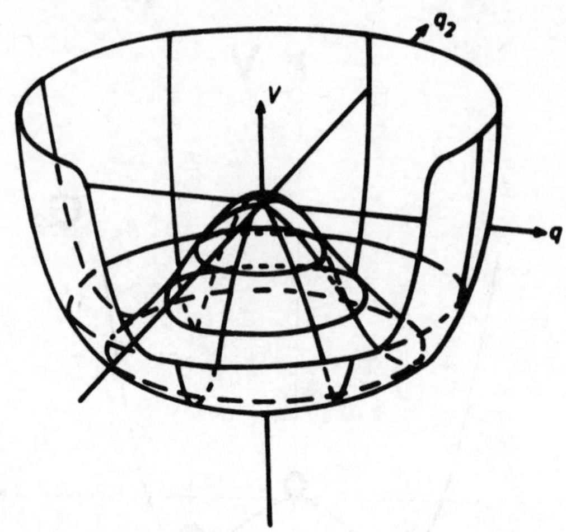

Fig. 14 The potential V versus the real and imaginary part of E above threshold. Note that the potential is rotation symmetric

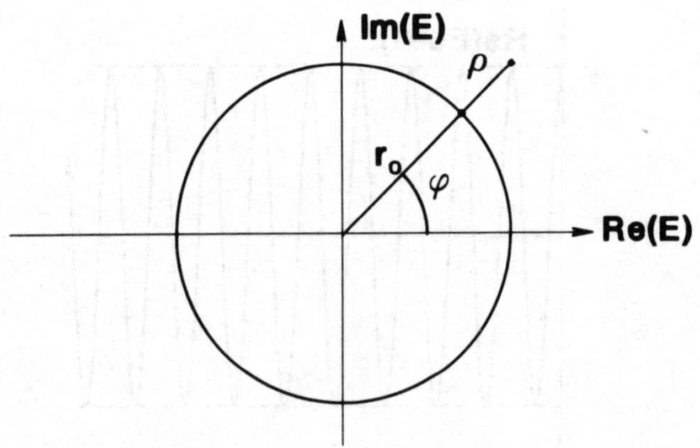

Fig. 15 The bottom of the potential of fig.14 seen from above. Note the polar coordinates

accompanied with higher and higher output. But the instability of fig. 12 can never happen in reality. But I was so much convinced that my idea was correct that I remained stubborn and published my paper (Haken, 1964). Fortunately two American experimentalists, Armstrong and Smith (1965) got interested in my theoretical predictions, and their results, which I will exhibit and discuss below, showed the correctness of my predictions. But what have these predictions been? To this end I inserted (3.16) into (3.12) and assumed that the laser is operated above threshold so that ρ can be considered as a term small compared to r_o.

Then in the leading order with no fluctuations ρ and ϕ can be put equal to 0 and we obtain

$$(G - \kappa)\, r_o \; - \; C\, |r_o|^3 \; = \; 0, \tag{3.17}$$

which allows the solutions

$$r_o = 0 \tag{3.18}$$

and

$$r_o^2 = (G - \kappa)/C. \tag{3.19}$$

Since we recognize by looking at fig. 12 that $r_o=0$ is unstable we are merely concerned with the stable solutions (3.19). In the next order we study phase fluctuations whose equation can be derived from (3.12) with (3.16)

$$\frac{d\phi}{dt} = \frac{1}{r_o}\, \mathrm{Im}\,(\, e^{-i\phi}\, F_{tot}\,). \tag{3.20}$$

$\hat{F}$

This equation can be readily solved and we find for the mean square deviation of ϕ

$$<\,(\,\phi(t_2) - \phi(t_1)\,)^2\,> \; = \; \frac{1}{r_o^2}\, Q|t_2 - t_1|, \tag{3.21}$$

i.e. the typical result of a diffusion process. In order to calculate the optical linewidth we may again use the correlation function between E at two times t_2 and t_1

which by means of (3.21) can readily be evaluated in its leading term and gives rise to

$$< E^*(t_2)E(t_1) > = r_o^2 \, e^{-\underbrace{\frac{1}{r_o^2} Q}_{\Gamma}(t_2 - t_1)} . \qquad (3.22)$$

In it quite evidently the linewidth $\Gamma = Q/r_o^2$ can be written as

$$\Gamma = \frac{\text{Const.}}{2P} , \qquad (3.23)$$

because the output power P is proportional to (3.19). As compared with the Townes formula we obtain here a correction factor of 1/2. Therefore the dependence of the linewidth on the output power P does not differ qualitatively for a laser operated below or above threshold. Therefore such measurements can not distinguish whether a laser is in its truly lasing state or still in its amplifying state below threshold. The decisive difference between laser light and light from thermal sources rests therefore in the amplitude fluctuations, or what is experimentally more easily accessible, in the intensity fluctuations.

Again making use of (3.12) and (3.16) and keeping the leading term in ρ we obtain a relaxation equation for ρ

$$\frac{d\rho}{dt} = - 2(G - \kappa)\rho + \mathrm{Re}(\hat{F}), \qquad (3.24)$$

which can be easily visualized by looking at fig. 16. The amplitude fluctuations become manifest through their correlation functions which can be easily evaluated yielding

$$< \rho(t_2)\rho(t_1) > = \text{Const.}(n_{sp} + n_{th}).e^{-2(G-\kappa)|t_2-t_1|} \qquad (3.25)$$

They tell us that with increasing gain G the relaxation becomes faster or, in other words, the relaxation time decreases with output power. What is,however, still more important is the fact that the constant in front of the exponential function in (3.25) is proportional to the number

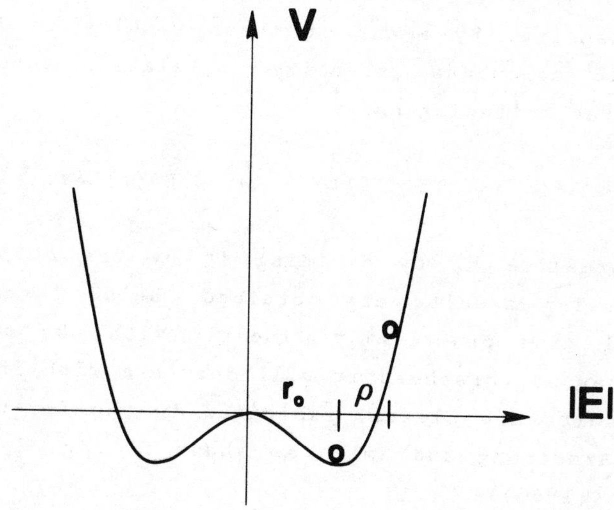

Fig. 16 How to visualize the relaxation of the coordinate ρ in the potential field V

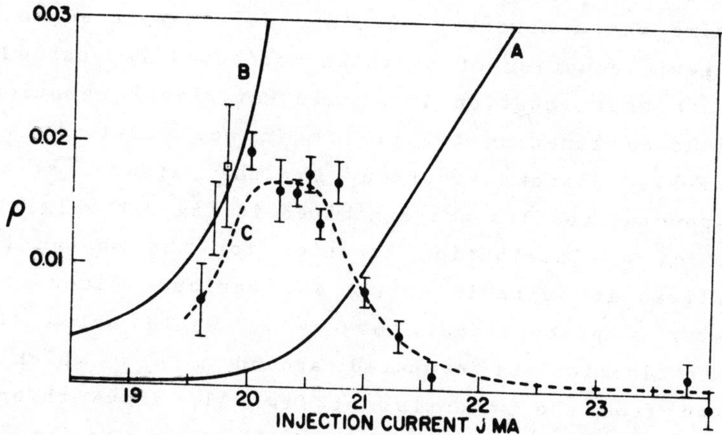

Fig. 17 Measurement of the emission intensity A (solid line) and the normalized intensity fluctuations (dashed line).(After Armstrong J.A. and Smith, A.W. Phys.Rev.Lett.14,68(1965))

of spontaneously emitted photons and thermal photons at laser threshold. In other words, a quantity independent of the laser intensity. By means of these results it is straight forward to calculate intensity correlation functions, for instance the following one

$$K_2 = < E^*(t)E^*(t)E(t)E(t) > - < E^*(t)E(t) >^2. \quad (3.26)$$

If we normalize K_2 by dividing it by the output power P the following results are obtained. Below laser threshold K_2/P will rise practically linearly with the output power whereas beyond threshold it will decrease with increasing P. These results are clearly exhibited in the first experiment done by Armstrong and Smith as shown in fig. 17 (Armstrong and Smith (1965)).

In the following years high precision measurements on these correlation functions have been made in particular by Pike and coworkers (1967), Arecchi and coworkers (1967), and others. In these experiments also the region close to laser threshold was covered which had to be excluded in my approach mentioned above. This gap was closed by Risken (1965) who at that time was a coworker of mine. He interpreted eq. (3.12) as Langevin equation of Brownian motion. This enabled him to translate that equation into a Fokker-Planck equation for a distribution function $f(E,t)$. It is not difficult to solve this Fokker-Planck equation in the steady state. The corresponding results are exhibited in fig. 18. Fig. 19 shows a plot of a distribution function for the photon number n (normalized in suitable units) for various values of a pump parameter a proportional to $G - \kappa$. While below threshold the distribution is centered around $n = 0$ which can be expected from the potential of fig. 11, above threshold it moves away to non-vanishing values again to be expected qualitatively from the potential of fig. 12. These photon distribution functions have been measured experimentally by a number of authors, in particular by Pike (1970), Arecchi (1967) and others. For sake of completeness I mention that some thoughts are required to connect the measurements of

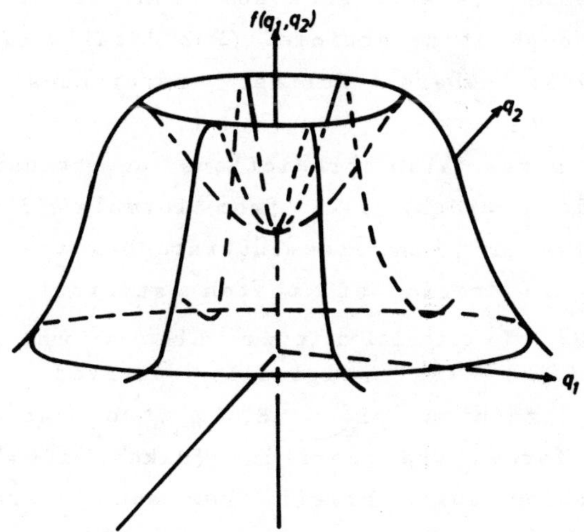

Fig. 18 The distribution function of laser light versus $q_1 \approx$ Re(E) and $q_2 \approx$ Im(E). Note the probability crater

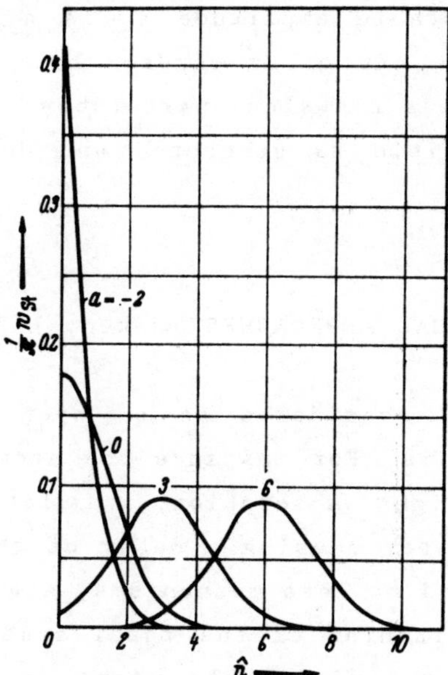

Fig. 19 The photon distribution function versus a normalized photon number for various pump parameters proportional to G - $\varkappa$ (After Risken, H. Z.Physik, 186, 85 (1965))

photons with a classical distribution function, i.e. the solution of the Fokker-Planck equation, but this question is beyond the scope of my article. (For details cf. Haken, LASER THEORY, 1984, where further references are given).

The theory makes also predictions on transients, on the change of line width, i.e. from formula (3.15) to (3.23) (fig.20). Figs. 21, 22 and 23 demonstrate how the results can be interpreted in terms of photon statistics (after Pike (1967, 69, 70)). In conculsion it should be mentioned that the equation (3.12) was originally derived in a quantum mechanical fashion, i.e. E, p_μ, d_μ as well as the fluctuating forces are operators (Haken, 1964). Therefore I wish to mention quite briefly how such operator equations result. Before doing so, let us indicate what the consequences are for the quantum mechanical properties of E as presented in (3.16). As it turns out, the phase ϕ and the amplitude ρ become operators. But what is far more interesting, the stable amplitude r_o is a c-number, i.e. a classical amplitude. In other words, the quantum mechanical operator E contains a classical part (above laser threshold), i.e. a classical field is generated and dominates over the quantum noise $\rho(t)$.

4. QUANTUM MECHANICAL APPROACHES (Haken, LASER THEORY, 1984)

The laser must be considered as an open system which is coupled to reservoirs. For instance the atoms are coupled to the pumping by light absorption, to lattice vibrations, atomic collisions etc. causing a decay of the atomic states. The field is coupled to loss mechanisms, e.g. in the mirrors. Therefore the Hamiltonian of the total system must not only comprise those of the field, the atoms and the interaction between field and atoms but also the Hamiltonians of the reservoirs and their interactions with the field and the atoms.

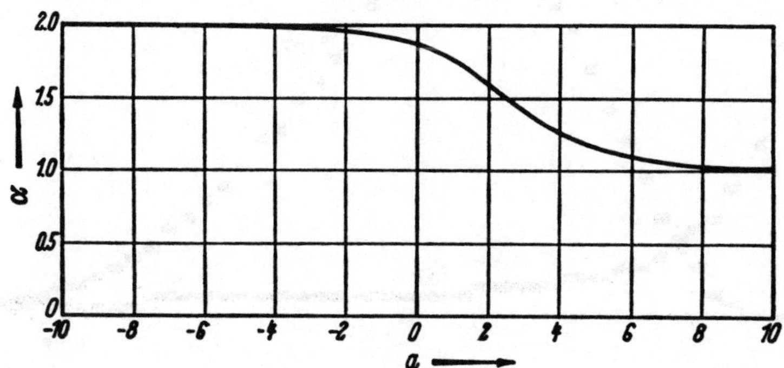

Fig. 20 The continuous change of the linewidth parameter from 2 to 1 when the laser passes through threshold (after Risken l.c.)

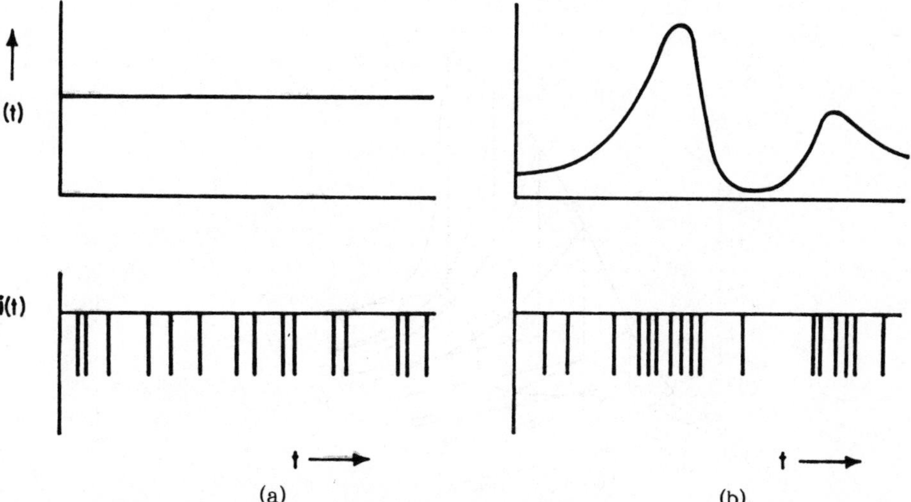

Fig 21 (a) Visualisation of the photomultiplier detector current, i(t), versus time for a constant light intensity, I(t), (laser above threshold); arrival times of photons at the counter are random.

(b) Fluctuating light intensity (laser below threshold); arrival times of photons are bunched. After Pike E R, in Quantum Optics, eds, Kay M and Maitland A, Academic Press NY, 1970.

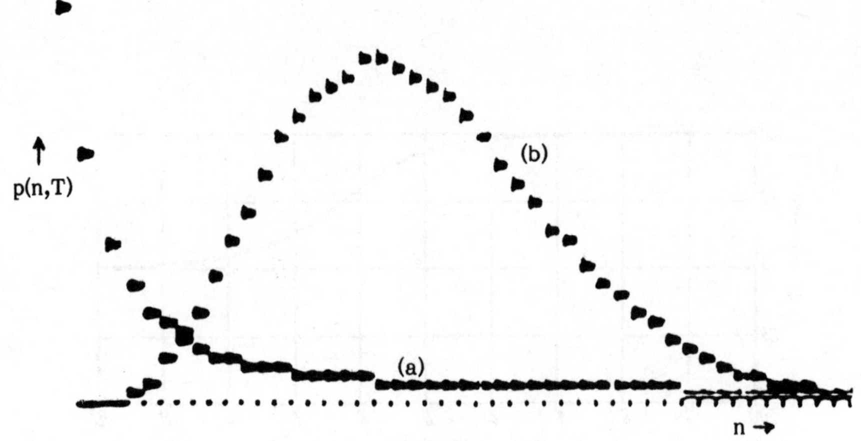

Fig 22 Photon counting distributions for (a) chaotic source (geometric distribution) (b) laser source (Poisson distribution). (After Pike E R, in Rendiconti della Scuola Internazionale di Fisica E Fermi, XLII Corso, 1967, Academic Press).

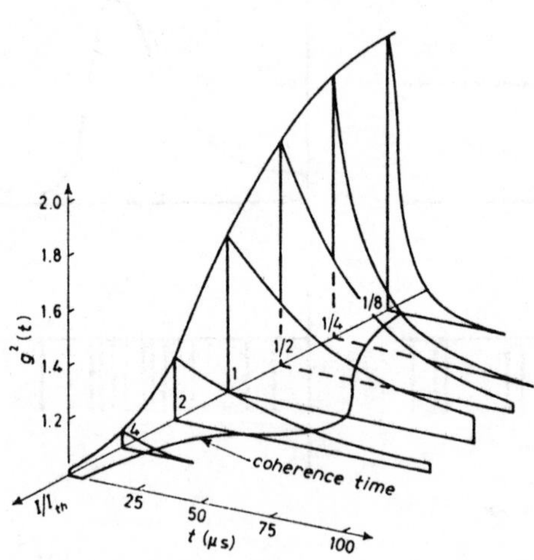

Fig 23 Delay time dependence of normalised photon correlation function $g^{(2)}(t)$ of laser near its lasing threshold (I/Ith)=1. (After Pike E R, in Rivista del Nuovo Cimento, 1, Numero Speciale, 277-314, 1969).

A solution of the corresponding Schrödinger equation seems to be a hopeless task. However, another procedure would be useful by means of Heisenberg equations of motion. Equations for the field and atomic operators can be derived and in a second step the bath variables can be eliminated. The resulting equations are operator equations. Describing the field by the creation and annihilation operators b^+, b we arrive at an equation of the form

$$\frac{db}{dt} = \frac{i}{\hbar} [H_s, b] - \kappa b + F(t). \tag{4.1}$$

In it the commutator stems from the Hamiltonian of the proper laser system (field + atoms), whereas the second and third terms stem from the elimination of the heatbath variables which gives rise to a damping term $-\kappa b$ and a fluctuating force. But this fluctuating force is an operator which does not commute with its Hermitian adjoint. For instance the following relations hold

$$< F^+(t_2)F(t_1) > = 2\kappa n_{therm} \cdot \delta(t_2 - t_1), \tag{4.2}$$

$$< F(t_1)F^+(t_2) > = 2\kappa(n_{therm}+1) \cdot \delta(t_2 - t_1). \tag{4.3}$$

When the commutator in (4.1) is evaluated we recover eq.(3.5). Similarly, eqs.(3.6) and (3.7) can be derived with the only but quite important difference that the quantities E, p_μ, d_μ as well as the fluctuating forces have become operators. A second successful approach in dealing with the quantum mechanical laser problem is based on the density matrix equation

$$\frac{d\rho}{dt} = - \frac{i}{\hbar} [H, \rho], \tag{4.4}$$

where H is the Hamiltonian of the laser including reservoirs. After elimination of the reservoir variables a density matrix equation of the proper laser system (field and atoms) is obtained (Weidlich and Haake, 1965; Scully and Lamb, 1966, 1968)

$$\frac{d\rho_s}{dt} = - \frac{i}{\hbar} [H_s, \rho_s], + \text{add. terms}, \tag{4.5}$$

where the additional terms which are not exhibited explicitly describe damping and fluctuations.

This equation has been solved directly by Scully and Lamb (1966, 1968) which thus provides another important way to find the photon distribution function. (For further work see e.g. Weidlich, Risken and Haken, (1967); and Lugiato (1979)).

Finally let us mention a third approach to the quantum mechanical laser problem which is based on quantum classical correspondence. In it according to Table 1 a correspondence is set up between operators and classical variables, as well as between the density matrix and a classical distribution function.

TABLE 1

quantum classical correspondence

operators ⟷	classical variables
$b,\ b^+$	$\beta,\ \beta^*$
density matrix	distribution function
$\rho(b,b^+;t)$	$f(\beta,\beta^*;t)$
d.m. equation	generalized Fokker-Planck equations
$\frac{d\rho_s}{dt} = L\rho_s$	$\frac{df}{dt} = \hat{L}f$

Then the density matrix equation can be translated without loss of generality into a generalized Fokker-Planck equation which after a series of elimination procedures and approximations can be transformed into the laser Fokker-Planck equation close to threshold. The method of quantum classical correspondence had been initiated by Wigner (1932) for field operators . Another kind of related representation which was also used quite successfully for field operators is that of Glauber (1963). This method was extended to atomic and field operators by Haken, Risken and Weidlich (1967), and in a still more elegant form by Lugiato et al. (1982); Haake et al. (1982, 1983).

Our above reminder of the various quantum mechanical approaches is of a rather technical nature. Therefore some general remarks on the physical meaning may be in order. As it turns out, the fluctuating forces stem from the reservoir (heatbath) variables and reflect our lack of knowledge of the microscopic states of the heatbath. The fluctuations of laser light result from this "lack of knowledge". Note that in this interpretation spontaneous emission noise stems from vacuum fluctuations of the electromagnetic field which, with its continuum of nonlasing modes, represents a reservoir.

5. SUMMARY OF THE FIRST PART OF THIS ARTICLE

In the preceding sections I have tried to present a brief review on the development of the theory on laser noise, coherence and photon statistics. As we now know laser light differs fundamentally in its physical properties from those of light from thermal sources. But these qualitative differences are not yet manifest in its linewidth but only in intensity correlation functions.

In the meantime it has turned out that the laser is a beautiful example of a large class of systems which go from a microscopically disordered state into an ordered state when more and more energy is pumped through them. The general study of these phenomena has even led to a new field of interdisciplinary research, called synergetics (Haken 1983a,b). The methods developed to cope with the properties of lasers have formed the basis of many further applications, e.g. in nonlinear optics and optical parametric oscillators as well as for further generalization, in particular what the slaving principle is concerned. In this way the laser has become a beautiful example how systems can acquire order among their parts through selforganization. As it seems the laser has become an important paradigm even for a number of biological processes such as coordination of motion. In the foregoing section I have presented a reminder of the development of the microscopic theory of laser noise. Quite recently it has become possible to derive, or to put it more precisely, to make guesses on the distribution functions of laser light for single and multimode lasers. This procedure is based on the maximum information entropy principle, to which we will turn now.

6. THE MAXIMUM INFORMATION ENTROPY PRINCIPLE FOR LASER LIGHT

The maximum information entropy principle, or in short the maximum entropy principle due to Jaynes (1957) allows one to make unbiased estimates on the probability distribution of microscopic states of systems of which otherwise only certain averages corresponding to macroscopic observations are known. This principle provides one with a very elegant access to basic relations and concepts of thermodynamics, i.e. it allows for a very nice application to systems in thermal equilibrium (Jaynes, 1957), see also Haken (1983a). On the other hand no successful attempts have been known of an application of this principle to systems far from thermal equilibrium.

In this paper I wish to show how this principle can indeed be very successfully applied to nonequilibrium systems provided they acquire macroscopic structures through selforganization. We shall in this way recover wellknown distribution functions of such systems. These functions had been previously derived from microscopic theories (cf. Haken (1983, 1984)). At the same time also generalizations of these functions now become available. In order to illustrate our procedure we focus our attention on lasers, but the general feasibility of this approach is evident.

7. THE MAXIMUM ENTROPY PRINCIPLE. A REMINDER

We define the information entropy by

$$i = -\sum_j p_j \ln p_j \tag{7.1}$$

which from a formal point of view plays a similar role as the entropy

$$S = -k \sum_i p_i \ln p_i \tag{7.2}$$

in systems in thermal equilibrium. For the following we shall adopt the notation S but shall put $k = 1$. The distribution functions p_j are considered as unknown variables still to be determined. This is done by the requirement that (7.2) acquires a maximum value under given constraints

$$\sum_i p_i f_i^{(k)} = f_k \tag{7.3}$$

and

$$\sum_i p_i = 1. \tag{7.4}$$

As we shall see below the crux of the problem consists in finding appropriate functions f_k and $f_i^{(k)}$. While in systems in thermal equilibrium this is a rather simple task new ideas are required for nonequilibrium systems as we shall see below.

The maximization of (7.2) under the constraints (7.3) and (7.4) can be performed by the use of Lagrange multipliers λ_k and λ-1. We then have

$$\delta \left[S - (\lambda-1) \sum_i p_i - \sum_{i,k} \lambda_k p_i f_i^{(k)} \right] = 0. \tag{7.5}$$

Performing the variation of (7.5) by differentiating the bracket with respect to p_i and putting the result equal to zero we obtain

$$- \ln p_i - 1 - (\lambda - 1) - \sum_k \lambda_k f_i^{(k)} = 0 \tag{7.6}$$

or equivalently

$$\ln p_i = - \lambda - \sum_k \lambda_k f_i^{(k)} \tag{7.7}$$

which after putting both sides into the exponent of an exponential function yields the required result

$$p_i = \exp \left[- \lambda - \sum_k \lambda_k f_i^{(k)} \right]. \tag{7.8}$$

Inserting (7.8) into the normalization condition (7.4) we obtain

$$e^{-\lambda} \sum_i \exp \left[- \sum_k \lambda_k f_i^{(k)} \right] = 1. \tag{7.9}$$

Introducing the abbreviation

$$\sum_i \exp \left[- \sum_k \lambda_k f_i^{(k)} \right] = Z(\lambda_1, \ldots, \lambda_n) \tag{7.10}$$

we can cast (7.9) into the form

$$\lambda = \ln Z. \tag{7.11}$$

Thus the Lagrange multiplier λ is determined provided the λ_k's have been fixed. In order to find equations for λ_k we insert (7.8) into the constraints (7.3). After some brief manipulations we arrive at the relation

$$f_k = \langle f_i^{(k)} \rangle = - \frac{\partial \ln Z}{\partial \lambda_k}. \tag{7.12}$$

Finally, by inserting (7.8) into (7.2) and using (7.4) and (7.3) we arrive at

$$S_{max} = \lambda + \sum_k \lambda_k f_k. \tag{7.13}$$

8. APPLICATION TO LASER PHYSICS

a) single mode laser

The quantities observed experimentally of a single mode laser are its intensity and the second moment of the intensity in the steady state case. Further measured quantities are intensity correlations, but because we have a time independent theory in mind we shall ignore this information here.

The space and time dependent electric field strength of a single mode laser can be written in the form

$$E(x,t) = E(t)\sin kx, \tag{8.1}$$

where the amplitude E(t) can be decomposed into its positive and negative frequency part according to

$$E(t) = \underset{b}{Be^{-i\omega t}} + \underset{b^*}{B^*e^{i\omega t}}. \tag{8.2}$$

If we measure the intensity of the lightfield over time intervals large compared to an oscillation period but small compared to the fluctuation times of B(t) the output intensity is proportional to B*B and the loss rate of the laser. For sake of simplicity we drop all other constants and put

$$I = 2\kappa B^*B. \tag{8.3}$$

Similarly the intensity squared if averaged over the same time interval turns out to be

$$I^2 = 4\kappa^2 B^{*2}B^2. \tag{8.4}$$

Because of the fluctuations of the laser B* and B are random variables which belong to a stationary process. This leads us to identify B*,B with the indices i of p_i in (7.1) but because the random variables B are no more discrete but continuous we must replace the summation over the indices i by an integration

$$i = - \int p(B,B^*) \ln p(B,B^*) \, d^2B. \tag{8.5}$$

(7.3) may be interpreted as integrals over d^2B with the probability $p(B,B^*)$ as weight function. Denoting these averages by brackets we are led to consider the following two constraints

$$f_1 = \langle 2\kappa B^*B \rangle, \tag{8.6}$$

$$f_2 = \langle 4\kappa^2 B^{*2} B^2 \rangle. \tag{8.7}$$

Furthermore by the same analogy we are led to define $f_i^{(k)}$ by means of

$$f_{B,B^*}^{(1)} = 2\kappa B^*B, \tag{8.8}$$

$$f_{B,B^*}^{(2)} = 4\kappa^2 B^{*2} B^2. \tag{8.9}$$

We are now in a position to apply formula (7.8) immediately and find

$$p(B,B^*) = \exp [- \lambda - \lambda_1 2\kappa B^*B - \lambda_2 4\kappa^2 (B^*B)^2] \tag{8.10}$$

which in a somewhat different notation reads

$$p(B,B^*) = N \exp [- \alpha|B|^2 - \beta|B|^4]. \tag{8.11}$$

This function is well known is laser physics. It had been derived by Risken (1965) solving the Fokker-Planck equation belonging to the laser Langevin equation derived previously by the present author (Haken, 1964). We note that in the laser case α must be negative. But close to threshold α can acquire values ranging from negative to positive values.

Before we go on to present some more explicit examples of more complicated cases, we may draw a number of rather far reaching conclusions. So far, the maximum (information) entropy principle had been applied to thermodynamics and irreversible thermodynamics, but not to nonequilibrium phase transitions. Among the constraints used in the former two fields is the energy. In our present case we deal with the output intensity I. But what is still more important, we have now to include the intensity correlation in form of the second moment, i.e. $\langle I^2 \rangle$. This is never done in equilibrium thermodynamics in the context of the maximum entropy principle. But now we see that the inclusion of $\langle I^2 \rangle$ is quite obvious and necessary for nonequilibrium phase transitions. At or above threshold, $\lambda_1 \geq 0$, and the integral over $\exp(\lambda_1 |b|^2)$ will diverge, reflecting the effect of critical fluctuations. Because of these, and in order to take into account their limitation due to saturation, $\langle I^2 \rangle$ must be taken into account. In other words, close to nonequilibrium phase transitions fluctuations become "observables" and must be taken into account by adequate constraints. It appears to me a safe bet that the same is true for phase transitions of systems in thermal equilibrium and thus requires an extension of that principle there also.

In our example of the single mode laser we have treated b,b* as classical (random) variables. We can immediately "translate" our approach into quantum mechanics by letting b, b* become operators, so that $I \to b^+ b$, and $I^2 \to (b^+ b)^2$ (or $b^+ b^+ bb$). At the same time, the averages (8.6) or (8.7) are translated into $\mathrm{tr}(\rho b^+ b)$, $\mathrm{tr}(\rho (b^+ b)^2)$ etc. Similar translations hold in the more complicated case to be discussed below. It is not too difficult but far beyond the scope of my article to include time-dependent phenomena, which amounts to an extension of Jaynes' maximum caliber principle (Jaynes (1985)) to nonequilibrium phase transitions.

9. MULTIMODE LASER WITHOUT PHASE RELATIONS

In this case the field strength is decomposed into its modes according to

$$E(x,t) = \sum_{\ell} E_{\ell}(t) \sin k_{\ell} x, \tag{9.1}$$

where for simplicity we consider only axial modes. Again the mode amplitudes can be decomposed into their positive and negative frequency parts

$$E_{\ell}(t) = B_{\ell}(t) e^{-i\omega_{\ell} t} + B_{\ell}^{*}(t) e^{i\omega_{\ell} t}. \tag{9.2}$$

The intensity averaged over time intervals long compared to an oscillation period and short to fluctuation periods is given by

$$I_{\ell} \approx 2\kappa_{\ell} B_{\ell}^{*} B_{\ell} \approx n_{\ell}. \tag{9.3}$$

The extension of the results of section 8 is straightforward provided we now consider either n_{ℓ} or equivalently B_{ℓ}^{*}, B_{ℓ} as stochastic variables. We obtain

$$f_{\ell} = < n_{\ell} >, \tag{9.4}$$

$$f_{\ell,\ell'} = < n_{\ell} n_{\ell'} > \tag{9.5}$$

and identifying k and i according to

$$\begin{aligned} k &\longleftrightarrow \ell \\ & \ell,\ \ell' \\ i &\longleftrightarrow (n_1, n_2, \ldots, n_M) = \underset{\sim}{n} \end{aligned} \tag{9.6}$$

we have

$$f_{\underset{\sim}{n}}^{(\ell)} = n_{\ell}, \tag{9.7}$$

$$f_{\underset{\sim}{n}}^{(\ell,\ell')} = n_{\ell} n_{\ell'}, \tag{9.8}$$

$$p_i \longrightarrow p(\underset{\sim}{n}). \tag{9.9}$$

The application of (7.8) is now straightforward yielding

$$\begin{aligned} p(\underset{\sim}{n}) &= p(n_1, \ldots, n_M) \\ &= \exp[\, -\lambda - \sum_{\ell} \lambda_{\ell} n_{\ell} - \sum_{\ell\ell'} \lambda_{\ell,\ell'} n_{\ell} n_{\ell'} \,] \end{aligned} \tag{9.10}$$

as final result. Equivalently (9.10) can be written as

$$N \exp[\, -\sum_{\ell} \alpha_{\ell} n_{\ell} - \sum_{\ell\ell'} \beta_{\ell\ell'} n_{\ell} n_{\ell'} \,]. \tag{9.11}$$

This form could be derived from a multimode Fokker-Planck equation in special cases (Haken, 1984) in which the solution could be explicitly constructed by means of the principle of detailed balance ((Graham, Haken, 1971), cf.also (Haken, 1983b)).

10. MULTIMODE LASERS WITH PHASE RELATIONS

We consider as an example the ring laser in which the electric field strength can be represented as

$$E(x,t) = \sum_{\ell} B_{\ell}(t)\, e^{ik_{\ell}x - i\omega_{\ell}t} + c.c \tag{10.1}$$

We assume that the waves are running in one direction only and the values of k and ω are taken from a certain interval small compared to k_o and ω_o, respectively

$$\begin{aligned} k_{\ell} &\in k_o \pm \Delta k, \\ \omega_{\ell} &\in \omega_o \pm \Delta\omega. \end{aligned} \tag{10.2}$$

We now consider spatial and temporal averages of E and its powers where the temporal average is defined as in the preceding sections. We readily obtain

$$\bar{E} = 0, \tag{10.3}$$

$$\overline{E_\ell(x,t)E_{\ell'}(x,t)} = 2\delta_{\ell\ell'}B_\ell B_{\ell'}, \tag{10.4}$$

$$\overline{E_\ell E_{\ell'} E_{\ell''}} = 0. \tag{10.4a}$$

By performing the integration over plane waves and using k-selection rules we readily obtain the result

$$\overline{E_\ell E_{\ell'} E_{\ell''} E_{\ell'''}} = C\sum_{\pm}\delta(\pm k_\ell \pm k_{\ell'} \pm k_{\ell''} \pm k_{\ell'''}) \cdot B_\ell^{\pm} B_{\ell'}^{\pm} B_{\ell''}^{\pm} B_{\ell'''}^{\pm}. \tag{10.5}$$

The following identifications can now be made

$$i = (B_1, B_1^*, B_2, B_2^*, \ldots, B_M, B_M^*), \tag{10.6}$$

$$f_\ell = \langle B_\ell^* B_\ell \rangle, \tag{10.7}$$

$$f_{\ell,\ell',\ell'',\ell'''} = \langle B_\ell^* B_{\ell'}^* B_{\ell''} B_{\ell'''} \rangle \delta(k_\ell + k_{\ell'} - k_{\ell''} - k_{\ell'''}). \tag{10.8}$$

Application of (7.8) yields the final result

$$p(\underset{\sim}{B}, \underset{\sim}{B}^*) = \exp\Big[-\lambda - \sum_\ell \lambda_\ell B_\ell^* B_\ell - \sum_{\ell\ell'\ell''\ell'''} \lambda_{\ell,\ell',\ell'',\ell'''} B_\ell^* B_{\ell'}^* B_{\ell''} B_{\ell'''} \delta(k_\ell + k_{\ell'} - k_{\ell''} - k_{\ell'''}) \Big]. \tag{10.9}$$

This result could be derived from a microscopic theory in the case that the principle of detailed balance holds (Haken (1983a), see also Haken (1984)).

11. SINGLE MODE LASER INCLUDING POLARIZATION AND INVERSION

In conclusion of the line of thought of how to apply the maximum entropy principle we treat a single mode laser where polarization and inversion of the atomic system are included as dynamic variables. For simplicity we shall consider running waves. Then we shall have the following variables

$$E(x,t) = B(t)e^{ikx - i\omega t} + \text{c.c.} \quad \text{field}, \tag{11.1}$$

$$P(x,t) = P(t)e^{ikx - i\omega t} + \text{c.c.} \quad \text{polarization}, \tag{11.2}$$

$$D(x,t) \quad \text{slowly varying inversion}, \tag{11.3}$$

where B(t), P(t) and D are slowly varying variables of time. When we perform spatial averages we readily obtain

$$\overline{E} = \overline{P} = 0, \quad \overline{D} \neq 0, \tag{11.4}$$

$$\overline{E^*E} = B^*B, \quad \overline{E^*E^*EE} = B^*B^*BB, \tag{11.5}$$

$$\overline{|P(x,t)|^2} = P^*(t)P(t), \quad \overline{P^*E} = P^*(t)B(t). \tag{11.6}$$

In extension of our previous considerations we shall keep all non-vanishing terms up to fourth order. Then the distribution function will acquire the form

$$p = \exp\,[\,\lambda + \lambda_1|E|^2 + \ldots + \lambda_{27}D^4\,]. \tag{11.7}$$

Unfortunately there is no microscopic theory in which a distribution function for the present problem can be calculated exactly. Therefore let us make again the adiabatic approximation in which, however, we shall keep the distribution functions for P and D explicitly. Starting again from the laser equations

$$\dot{E} = -\kappa E + gP + F, \tag{11.8}$$

$$\dot{P} = -\gamma P + gED + \Gamma, \tag{11.9}$$

$$\dot{D} = \gamma_{\|}(D_o - D) - g(EP^* + E^*P) + \Gamma_D \tag{11.10}$$

and invoking the adiabatic approximation

$$\dot{P} = \dot{D} \simeq 0 \tag{11.11}$$

we obtain from (11.9) the relation

$$\Gamma = \gamma P - gED. \tag{11.12}$$

But because we know that the fluctuating force Γ is Gaussian distributed we find that the expression

$$\gamma P - gED \qquad \text{Gaussian distributed} \tag{11.13}$$

has the same distribution function. Similarly we find from (11.10) a distribution function for the difference of the first two brackets on the r.h.s. of that equation. Finally eliminating P and D adiabatically close to threshold we obtain instead of (11.8)

$$\dot{E} = -\kappa E + GE - CE|E|^2 + F_{tot}. \tag{11.14}$$

(11.14) can easily be transformed into a Fokker-Planck equation and as we have seen before this equation can readily be solved in the steady state. The total distribution function for E, P and D is then the product of the distribution functions belonging to E, P, D which reads

$$f = e^{\lambda} \exp[\, a|E|^2 - b|E|^4]\cdot$$
$$\cdot\exp[\, -(|\gamma P - gED|^2/Q) - \{g(EP^* + E^*P) + \gamma_{\|}(D - D_o)\}^2/Q']. \tag{11.15}$$

When we compare this distribution function (11.15) with the guessed function (11.7) we find that a number of the Lagrange parameters λ vanish identically which is exhibited in table 2. Whether this vanishing of λ's is a consequence of our adiabatic elimination procedure or holds even in the case of the general equations (11.8) - (11.10) is at present an open question.

TABLE 2

	1	$\|E\|^2$	$\|P\|^2$	EP*	E*P	D	D^2
1	X	X	X	X	X	X	X
$\|E\|^2$	/	X	X				X
$\|P\|^2$	/	/					
EP*	/	/	/	X	X	X	
E*P	/	/	/	/	X	X	
D	/	/	/	/	/	X	
D^2	/	/	/	/	/	/	

12. CONCLUDING REMARKS ON SECTIONS 6 TO 11

As we have shown explicitly the maximum information entropy principle allows us to derive the general form of distribution functions of a nonequilibrium system, such as the laser, in a quite straightforward way. The results agree with distribution functions obtained from microscopic theories under certain restricting conditions.

In spite of the success of our application of that principle we must bear in mind that only little can be said about the Lagrange multipliers which can, of course, now be determined experimentally. On the other hand it has been the advantage of the microscopic theory that these constants can be determined from first principles and therefore in particular it could be predicted that α_ℓ changes sign at instability points. However, it will be tempting to deduce such properties also from a macroscopic theory.

In spite of the formal resemblance of our results to some of thermodynamics there are still basic differences. First of all we realize that the constants have a physical meaning very different from those for systems in thermal equilibrium. For instance in nonequilibrium systems, such as lasers, we have to deal with output intensities whereas in equilibrium systems we deal with e.g. energies. This is also clearly reflected by a treatment on the microscopic level. While in the microscopic treatment energies play a decisive role, in nonequilibrium systems rate constants and growth rates determine the evolving patterns.

13. GIANT FLUCTUATIONS OF LASER LIGHT: DETERMINISTIC CHAOS

In 1963 Lorenz established a by now famous model which exhibited turbulence in fluids in spite of the fact that his equations were only differential equations for three variables. His equations have the form

$$\dot{X} = \sigma X - \sigma Y, \tag{13.1}$$

$$\dot{Y} = - XZ + rX - Y, \tag{13.2}$$

$$\dot{Z} = XY - bZ. \tag{13.3}$$

Some years ago (1975) I found that the laser equations, e.g. (11.8)-(11.10) are isomorphic to the Lorenz equations. This immediately led me to the idea that laser light can exhibit deterministic chaos. Subsequently quite a number of experiments have been performed on the search of laser light chaos. Since for many lasers the instability condition for the onset of chaos is difficult, here it is assumed that the damping constant of the polarization is so large that the polarization can be eliminated adiabatically. The third variable which is needed for chaos to occur is then provided by periodic modulations of the losses, of the inversion, or by an injected signal. Among the theoretical work we mention especially that by Graham and his coworkers (see Yamada and

Graham (1980), Scholz et al (1981), for experimental works those by Arecchi et al (1982) and Klische et al (1985) to quote typical examples. Narducci et al (1985) have found that the Lorenz model allows for further instabilities applicable to laser light and quite recently Weiß et al (see Hogenboom(1985)) reported the realization of the Lorenz type of laser chaos. Therefore interest in fluctuations has been revived considerably though these fluctuations are now based on deterministic equations. Remarkable theoretical studies on the combined action of deterministic chaos and microscopic fluctuating forces were made by Graham.

References

Arecchi, F.T., Menzi, R., Pucciono, G.P., and Tredicce, J., 1982, Phys.Rev.Lett. 49 1217

Arecchi, F.T., Rodari, G.S., and Sona, A., 1967, Phys.Lett. 25A, 59

Armstrong, J.A., Smith, A.W., 1965, Phys.Rev.Lett. 14, 68

Basov, N.G., Prokhorov, A.M., 1954, Exp.Theoret.Phys.USSR 27, 431; 1955, 28 249

Collins, R.J., Nelson, D.F., Schawlow, A.L., Bond, W. Garrett, C.G.B., Kaiser, W., 1960, Phys.Rev.Lett. 5, 303

Glauber, R.J., 1963, Phys.Rev.Lett. 10, 84; Phys.Rev. 130, 2529; 131 2766

Gordon, J.P., Zeiger, H.J., and Townes, C.H., 1954, Phys.Rev. 95, 282; 99, 1264

Graham, R., and H. Haken, 1971, Z.Physik 243, 289

Haake, F. and Lewenstein, M., 1982, Z.Physik, B48 37, 1983, Phys.Rev. A27 1013

Haken, H., 1964, Z.Physik 181 96 (1964)

Haken, H., 1966, Z. Physik 190, 327

Haken, H., Z. Physik, 1969, 219, 246

Haken, H., 1983a, Synergetics. An Introduction, 3rd ed., Springer, Berlin, Heidelberg, New York

Haken, H., 1983b, Advanced Synergetics, Springer, Berlin, Heidelberg, New York

Haken, H. 1984, Laser Theory, 2nd.corrected printing Springer, Berlin,Heidelberg, New York

Haken, H., Risken, H., and Weidlich, W., 1967, Z.Physik 206 355

Haken, H., and Sauermann, H. 1963, Z.Physik 173 261

Haken, H. and Sauermann, H., 1963, Z. Physik 176, 47

Hogenboom, E.H.M., Klische, W., Weiss, C.O., and Gordone, A., 1985, Phys.Rev.Lett. 55, 2571

Javan, A., Bennett jr., W.B., Herriott, D.R., 1961 Phys.Rev.Lett. 6, 106

Jaynes, E., in 'Complex Systems - Operational Approaches in Neurobiology, Physics and Computers' Vol. 31 Springer Series in Synergetics, p. 254, Springer, Berlin, Heidelberg, New York 1985

Jaynes, E.T., 1957, Phys.Rev. 106, 4, 620; 108, 171

Klische, W., and Weiss, C.O., 1985, Phys.Rev. A31 4049

Lamb jr., W.E., 1964, 134A, 1429

Lugiato, L.A., 1979, Nuovo Cim. 50B, 89

Lugiato, L.A. Casagrande, F., and Pizzuto, L. 1982, Phys.Rev. A26 3438

Maiman, T.H., 1960, Nature 187, 493 Brit.Commun.Electron. 1960, 7, 674

Narducci, L.M., Sadiky, H., Lugiato, L.A., and Abraham, N.B., 1985, to be published

Pike, E.R., 1970, in Quantum Optics, Eds. Kay, M.S., and Maitland, A., Academic Press, New York, where further references are given

Risken, H., 1965, Z.Physik 186, 85

Schawlow, A.L., and Townes, C.H., 1958, Phys.Rev. 112, 1940

Scholz, H.J., Yamada, T., Brand, H., and Graham, R., 1981, Phys.Lett. 82A, 321

Scully, M.O., and Lamb jr., W.E., 1966, Phys.Rev.Lett. 16 853; 1968, Phys.Rev. 166 246,

Tang, C.L., Statz, H., de Mars, G.A., 1963 J.Appl.Phys. 34, 2289

Weidlich, W., and Haake, F., 1965, Z. Physik 185 30; 186, 203

Weidlich, W., Risken, H. and Haken, H., 1967, Z. Physik 201 396

Wigner, E.P., 1932, Phys.Rev. 40 749

Yamada, T., and Graham, R., 1980, Phys.Rev.Lett. 45 1322,

QUANTUM CHAOS FOR ACTIVE MEDIA

SARBEN SARKAR AND J S SATCHELL

1. INTRODUCTION

The development of ideas initiated by Poincaré (1899) has led to rapid progress in the study of systems of coupled nonlinear ordinary differential equations and the introduction of the notion of chaos (Cvitanović, P. 1984; Sarkar, S. 1986). In particular the asymptotic attracting sets of trajectories are complicated and motion on them is aperiodic. Many diverse systems have been studied and universal trends found. There is much agreement on the interpretation and description of such classical chaos. The study of the quantum analogues of chaotic classical systems is far less developed. In quantum mechanics we are dealing with probabilistic concepts and, owing to the uncertainty principle, there is smoothing out of structure at sufficiently small scales. Within quantum dissipative structures there is no concensus as to the best experimental signature for quantum chaos. In this paper we will concentrate on our recent work (Sarkar, S. et al, 1986; Sarkar, S. and Satchell, J.S., 1986; Satchell, J.S. and Sarkar, S., 1986a) on the quantisation of the Lorenz equations (or laser equations) and related Poincaré maps. We have concentrated on quantities which are the obvious generalisations of those occurring in classical chaos. The analogue of the attractor is the support of the quasi-probability distribution. The analogue of Lyapunov exponents is the separation rate of the peaks of two very sharp probability distributions which are initially close to each other.

We will introduce the concepts that we need through simple one and two dimensional examples. First we will consider systems without fluctuations:

$$\text{i.} \quad \dot{x} = -kx \tag{1}$$

where x(t) is a real variable and $k > 0$. x(o) is taken to be be non-zero. The attractor is obviously the origin. In this case it is clear that the topological dimension is zero. It will nonetheless be useful to give a general definition (Eckmann, J-P and Ruelle, D., 1985).

For every finite covering of an attractor by open sets $\{\sigma_1,\ldots,\sigma_N\}$ we can find another covering $\{\sigma_1',\ldots,\sigma_N'\}$ such that $\sigma_i' \subset \sigma_i$ for $i = 1,\ldots,N$ and any (d+2) of the σ_i will have an empty intersection.

$$\sigma_{i_o}' \cap \sigma_{i_1}' \cap \ldots \cap \sigma_{i_{d+1}}' = \emptyset \tag{2}$$

The smallest such d over all coverings is the topological dimension.

In the case of a point attractor we can have the situation in Figure 1.

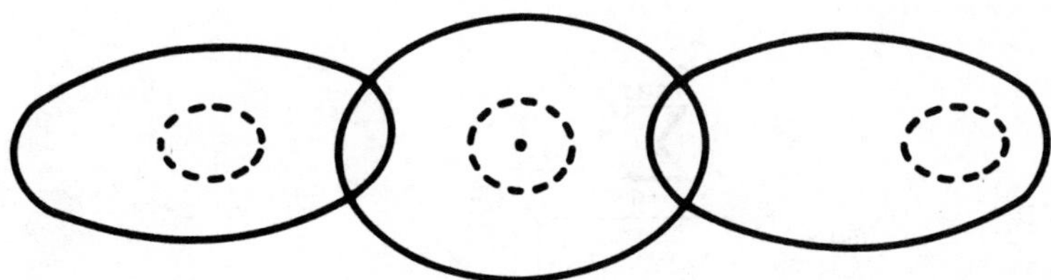

Figure 1. Coverings for a point attractor

The bold circles represent $\{\sigma_1, \sigma_2, \sigma_3\}$ whereas the broken circles represent $\{\sigma_1', \sigma_2', \sigma_3'\}$. It is clear that

$$d + 2 = 2$$

and consequently d = 0. We will now introduce the concepts of the Lyapunov exponent and dimension (Farmer, J.D., 1982; Eckmann, J-P. and Ruelle, D., 1984). The Lyapunov exponent λ determines the rate of separation or convergence of two points initially very close to each other. If x_o and x_o' are the initial points of two trajectories, then at time t their separations are given by

$$|(x_o - x_o')\, e^{-kt}|$$

$$\lambda \equiv \lim_{t \to \infty} \frac{1}{t} \log |(x_o - x_o')\, e^{-kt}| = -k \qquad (3)$$

A related quantity is the Lyapunov dimension D_L (Eckmann J-P. and Ruelle, D., 1984). If $\{\lambda_i\}$ denotes the set of Lyapunov exponents for the system and

$$\sum_{i=1}^{k} \lambda_i$$

is the sum of the k largest exponents, and k_{max} is the largest k for which this sum is non-negative, then

$$D_L = k_{max} + \frac{\sum_{i=1}^{k_{max}} \lambda_i}{|\lambda_{k_{max}} + 1|} \qquad (4)$$

In our example there is only one λ (which is negative) and so

$$D_L = 0 \qquad (5)$$

For complicated attractors, which are intricate over all scales, it is useful to introduce measures of dimension other than the topological dimension, which indicate the amount of information necessary to fix the position of a point on the attractor to a given accuracy. Moreover it is a measure of the "important" number of degrees of freedom in the problem. Although the Hausdorff-Besicovitch dimension is perhaps the best known, it is in general hard to calculate for attractors which have topological dimension three or higher. Quite recently another definition of dimension D_F, has been introduced by (Termonia, Y. and Alexandrowicz, Z., 1983; Grassberger, P. and Procaccia, I., 1983) which is practicable to use in higher dimensions. If the attractor (in the sense of a time series regularly spaced in time) is covered by N spheres containing n points each, then

$$n \sim (\bar{R}(n))^{D_F} \tag{6}$$

where $(\bar{R}(n))^d$ is the average volume of a box containing n points. $D_F = 0$ since $\bar{R}(1)$ is arbitrarily small and we require that $1 \sim (\bar{R}(1))^{D_F}$. In our example $D_F = D_L$.

ii. $$\dot{x} = y$$

$$\dot{y} = -x \tag{7}$$

with x(o) = 1, y(o) = 0 where x and y are real variables. The evolution is periodic and is a unit circle in x-y space. This circle is not an attractor in the sense of the fixed point that we had in one dimension. However we can still introduce Lyapunov exponents which are properties of dynamical systems and also dimensions which are properties of sets attracting or otherwise. If we take the two initial conditions to be (x(o) = y(o) = 1) and (x(o) = 1 + δx, y(o) = δy) where δx and δy are very small quantities, then the square of the separation of the two trajectories at time t is

$$(\cos t - ((1+\delta x)\cos t + \delta y \sin t))^2 + \\ (-\sin t - (-(1+\delta x)\sin t + \delta y \cos t))^2$$

This clearly leads to $\lambda = 0$. There are two zero Lyapunov exponents. One of these zero exponents is that which is typically associated with limit cycle attractors. The other is typical of dynamical systems which have finite attractors (which are not fixed points) and results from two initial conditions which are infinitesimal time translates of the other. The topological dimension d of a circle is one, and since there are two zero Lyapunov exponents $D_L = 2$. Moreover $D_F = 1$ since the number of points n in discs covering the circle varies linearly with the circle radius, at least for small n.

Now we will introduce fluctuations into the above examples. From systems in thermodynamic equilibrium we expect that there are fluctuations in situations when there is dissipation (Callen, H.B. and Welton, T.A., 1951). The Ornstein-Uhlenbeck process (Uhlenbeck, G.E. and Ornstein, L.S., 1930) provides a natural introduction of fluctuations into (i).

$$\text{i}'. \quad \dot{x} = -kx + D^{\frac{1}{2}}\, \xi(t) \tag{8}$$

where $\xi(t)$ is a Gaussian white noise and k and D are positive constants. This Langevin equation can be equivalently written as

$$\frac{\partial P}{\partial t} = \frac{\partial}{\partial x}(kxP) + \frac{1}{2} D \frac{\partial^2}{\partial x^2} P \tag{9}$$

where $P(x,t)$ is the conditional probability $p(x,t|x_o,o)$, ie the probability that $x(t) = x$ at time t given that at $t = 0$ $x(o) = x_o$. The solution of Eqn. (8) is

$$x(t) = x_o\, e^{-kt} + D^{\frac{1}{2}} \int_0^t e^{-k(t-t')}\, dW(t') \tag{10}$$

where W is a Wiener process. The steady state solution of Eqn. (9) is known to be

$$P(x) = \left(\frac{k}{\pi D}\right)^{\frac{1}{2}} \exp\left(-\frac{kx^2}{D}\right) \tag{11}$$

The various quantities introduced in the deterministic case will be reconsidered. Now if x(t) and x'(t) are the stochastic processes corresponding to the initial conditions x_o and x'_o then

$$x(t)-x'(t) = (x_o-x_o)e^{-kt} + D^{\frac{1}{2}}\left(\int_0^t e^{-k(t-t')}\, dW_1(t) - \int_0^t e^{-k(t-t')}\, dW_2(t')\right) \tag{12}$$

where W_1 and W_2 are independent Wiener processes. We find

$$\lambda = \lim_{t\to\infty} \frac{1}{t} \langle \log|x(t)-x'(t)|\rangle = -k \tag{13}$$

where now we have taken the expectation value of the logarithm. Hence $D_L = 0$. x(t) is a continuous curve and so its topological dimension is 1. Near the origin (which is the deterministic attractor) the probability distribution is smooth (ie there is a well-defined tangent) and so $D_F = 1$. Although in the deterministic case $D_F = D_L$, here $D_F \neq D_L$.

ii'. The stochastic 2 dimensional example is

$$\begin{pmatrix} \dot{x} \\ \dot{y} \end{pmatrix} = \begin{pmatrix} 0 & 1 \\ -1 & 0 \end{pmatrix}\begin{pmatrix} x \\ y \end{pmatrix} + \begin{pmatrix} b_1 & 0 \\ 0 & b_2 \end{pmatrix}\begin{pmatrix} dW_1 \\ dW_2 \end{pmatrix} \tag{14}$$

Writing $\underline{x} = (x,y)$ the solution of this multivariable Ornstein-Uhlenbeck process is

$$\underline{x}(t) = \left(\cos t \begin{pmatrix} 1 & 0 \\ 0 & 1 \end{pmatrix} + \sin t \begin{pmatrix} 0 & 1 \\ -1 & 0 \end{pmatrix}\right)\begin{pmatrix} x(o) \\ y(o) \end{pmatrix}$$

$$+ \int_0^t \left\{(\cos(t-t')) \begin{pmatrix} 1 & 0 \\ 0 & 1 \end{pmatrix} + \sin(t-t') \begin{pmatrix} 0 & 1 \\ -1 & 0 \end{pmatrix}\right\}$$

$$\cdot \begin{pmatrix} b_1 & 0 \\ 0 & b_2 \end{pmatrix}\begin{pmatrix} dW_1(t') \\ dW_2(t') \end{pmatrix} \qquad (15)$$

Proceeding as in (i)'

$$\lambda = \lim_{t\to\infty} \frac{1}{t} \langle \log|\underline{x}(t;\underline{x}(o)) - \underline{x}(t;\underline{x}(o)')|\rangle = 0 \qquad (16)$$

There are two independent zero Lyapunov exponents as in (ii) and so $D_L = 2$. The Fokker-Planck equation corresponding to Eqn. (14) has a Gaussian form centred around the limit cycle. This gives D_F, at least at small enough radii, to be 2. Just as in the deterministic case $D_F = D_L = 2$.

Our examples did not show chaotic behaviour. This is because we have restricted ourselves to one and two dimensional systems. Chaos is expected to appear in autonomous nonlinear dynamical systems when there are at least three variables involved. In the next section we will consider a three dimensional example first introduced by Lorenz (1963).

2. THE LORENZ MODEL

The model to be considered (Lorenz, E., 1963) is

$$\begin{aligned} \dot{x} &= \sigma(y-x) \\ \dot{y} &= zx - y \\ \dot{z} &= -xy + b(r-z) \end{aligned} \qquad (17)$$

where x, y and z are real variables and σ, b and r are real parameters. Haken (1975) has shown that these equations are equivalent to the Maxwell-Bloch equations for a single mode bad cavity laser with homogeneous broadening. A cross-section of the attractor for r = 28, b = 8/3 and σ = 10 is given in Figure 2.

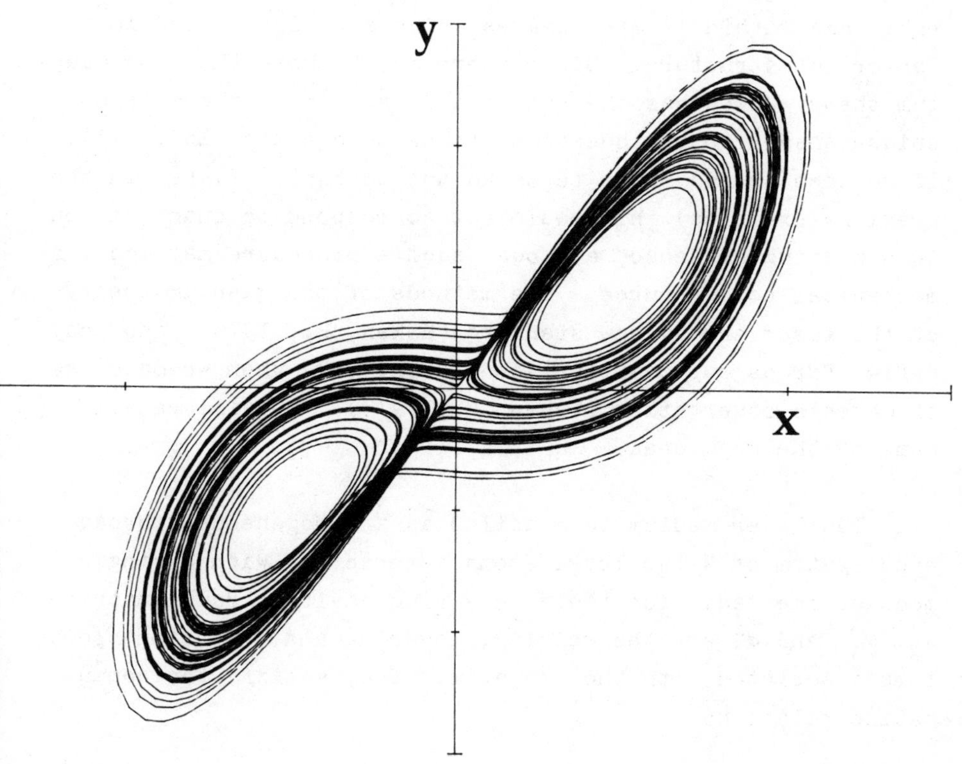

FIGURE 2. A cross-section of the Lorenz attractor

Such attractors are called strange (Guckenheimer, J. and Holmes, P., 1983). For the laser it is more realistic to consider parameters $r = 20$, $b = 1$ and $\sigma = 5$, and we will use these values from now on. There are efficient algorithms for calculating Lyapunov exponents for systems which show this complex behaviour (Eckmann, J-P. and Ruelle, D., 1985), and we find $\lambda_1 = .469$, $\lambda_2 = 0$ and $\lambda = -7.469$. Hence there is an expanding as well as contracting direction. The zero exponent is expected and arises for reasons given before. Numerically we find $D_F \sim 2.25$. The reason for this fractional dimension is the lack of regularity of the attractor, eg it has within it many leaves which are distributed in a Cantor set structure. Since there may be more than one quantum theory which has the same classical limit, there is no unique answer to the question of how to quantise Eqn. (17). If we were to add noise terms ad hoc to Eqn. (17) (as in the previous examples) this would not correspond to quantisation in a meaningful sense, although such a procedure may model a mechanical noise source. The methods of the quantum theory of the laser are by now standard (Haken, H., 1970), and they define for us our quantisation of Eqn. (17) when we make use of Haken's observation (Haken, H., 1975). We will recall some of the key ideas.

The laser medium is modelled as a homogeneously broadened system of N two level atoms interacting with a single mode of the radiation field in a ring cavity. Pauli matrices σ_j^+, σ_j^- and σ_j^z are the raising, lowering and inversion operators associated with the jth atom. They satisfy the commutation relations

$$\begin{aligned} [\sigma_j^+, \sigma_k^-] &= 2\,\sigma_j^z\,\delta_{jk} \\ [\sigma_j^z, \sigma^{\pm}_k] &= \pm\,\sigma^{\pm}_k\,\delta_{jk} \end{aligned} \tag{18}$$

For the field mode a and a^+ will denote the single-mode raising and lowering operators, and

$$[a,a^+] = 1 \tag{19}$$

The following Markovian master equation can be derived (Haken, H., 1970; Lugiato, L.A. et al, 1982).

$$\frac{d\rho}{dt} = \frac{1}{i\hbar}[H,\rho] + \kappa([a\rho,a^+] + [a,\rho\, a^+]) + L_A\,\rho \tag{20}$$

where

$$H = ig\hbar \sum_{j=1}^{N} (e^{-ik.x_j}\, a^+ \sigma_j^- - e^{ik.x_j}\, a\, \sigma_j^+) \tag{21}$$

and

$$\begin{aligned} L_A\rho = \tfrac{1}{2} \sum_{j=1}^{N} \{ & \gamma_\uparrow([\sigma_j^+,\rho\sigma_j^-] + [\sigma_j^+\rho,\sigma_j^-]) \\ & + \gamma_\downarrow([\sigma_j^-,\rho\sigma_j^+] + [\sigma_j^-\rho,\sigma_j^+]) \\ & + \gamma_0([\sigma_j^z,\rho\sigma_j^z] + [\sigma_j^z\rho,\sigma_j^z])\} \end{aligned} \tag{22}$$

Here κ is the cavity damping rate, $\gamma_\uparrow$ is the pumping rate from the ground to the excited state of an atom, $\gamma_\downarrow$ is the decay of the atomic population due to spontaneous emission, γ_0 is the rate for the collision induced phase decay of the atoms, k is the field mode wavevector, x_j is the position coordinate of the jth atom and g is the atom field coupling constant.

The master equation is an operator one and it is difficult to make progress directly with it. In many situations, when N the number of atoms is large, a c-number multivariable Fokker-Planck equation for a quasi-probability distribution can be derived as an approximation to the master equation.

Although for the moment we will not question the validity of making this approximation, we will return to this. The moments of the quasi-probability distribution give quantum mechanical averages and in order to get an unique answer an ordering of the operators has to be specified. We will choose symmetric ordering and this gives the Wigner distribution (Wigner, E.P., 1932; Lugiato, L.A., 1982). It is first necessary to construct a characteristic operator $\hat{\chi}$ for symmetric averages. We introduce

$$\chi_j(\xi, \xi^*, \eta) = \exp i(\xi^* \sigma_j^+ e^{ik.x_j} + \eta\sigma_j^z + \xi\sigma_j^- e^{-ik.x_j}) \tag{23}$$

$$\chi = \prod_{j=1}^{N} \chi_j \tag{24}$$

$$\bar{\chi} = \exp i(\zeta^* a^+ + \zeta a) \tag{25}$$

and

$$\hat{\chi} = \chi\,\bar{\chi} \tag{26}$$

The characteristic function is then given by

$$C_N(\xi, \xi^*, \eta, \zeta, \zeta^*) = \mathrm{Tr}(\hat{\chi}\rho) \tag{27}$$

C_N can be written in terms of the generalised Wigner distribution P

$$C_N(\xi,\xi^*,\eta,\eta^*) = \int\cdots\int d\bar{\zeta}_x d\bar{\zeta}_y\, d\bar{\xi}_x\, d\bar{\xi}_y d\bar{\eta}\, P(\bar{\xi},\bar{\xi}^*,\bar{\eta},\bar{\zeta},\bar{\zeta}^*) \exp i(\bar{\zeta}\zeta+\bar{\zeta}^*\zeta^*+\bar{\xi}^*\xi^*+\bar{\xi}\xi+\bar{\eta}\eta) \tag{28}$$

where

$$\bar{\zeta} = \bar{\zeta}_x + i\bar{\zeta}_y$$
$$\bar{\xi} = \bar{\xi}_x + i\bar{\xi}_y \tag{29}$$

On using the rescaled variables

$$\bar{m} = \frac{2}{N}\bar{\eta} \tag{30}$$

$$\bar{v} = \bar{v}_1 + i\bar{v}_2 = -\frac{2}{N}\left(\frac{\gamma_\perp}{\gamma_\downarrow}\right)^{\frac{1}{2}}\bar{\xi} \tag{31}$$

$$\bar{x} = \bar{x}_1 + i\bar{x}_2 = n_o^{-\frac{1}{2}}\,\bar{\zeta} \quad , \tag{32}$$

with $\gamma_\perp = 2\gamma_o + \frac{1}{2}\gamma_\downarrow$, $C = \dfrac{g^2 N}{\kappa\gamma_\downarrow}$ and $n_o = \dfrac{\gamma_\downarrow N}{8\kappa C}$,

the Fokker-Planck equation satisfied by P is

$$\begin{aligned}\frac{\partial P}{\partial t} = &\left\{\gamma_\perp \frac{\partial}{\partial\bar{v}_1}(\bar{v}_1 + \bar{m}x_1) + \gamma_\perp\frac{\partial}{\partial\bar{v}_2}(\bar{v}_2 + \bar{m}x_2)\right.\\ &+ \kappa\left(\frac{\partial}{\partial\bar{x}_1}(\bar{x}_1 + 2C\bar{v}_1) + \frac{\partial}{\partial\bar{x}_2}(\bar{x}_2 + 2C\bar{v}_2)\right)\\ &+ \gamma_\| \frac{\partial}{\partial\bar{m}}(\bar{m} - \hat{\sigma} - \bar{v}_1\bar{x}_1 - \bar{v}_2\bar{x}_2) + \frac{\gamma_\perp^2}{2C\kappa n_o}\left(\frac{1}{4}\left(\frac{\partial^2}{\partial\bar{v}_1^2} + \frac{\partial^2}{\partial\bar{v}_2^2}\right)\right.\\ &+ f^2\frac{\partial^2}{\partial m^2}(1-\sigma m) - \hat{\sigma} f^2\left.\left(\frac{\partial}{\partial\bar{m}}\frac{\partial}{\partial\bar{v}_1}\bar{v}_1 + \frac{\partial}{\partial\bar{v}_2}\bar{v}_2\right)\right)\\ &\left.+ \frac{\kappa}{4n_o}\left(\frac{\partial^2}{\partial\bar{x}_1^2} + \frac{\partial^2}{\partial\bar{x}_2^2}\right)\right\} P\end{aligned} \tag{33}$$

Here

$$\hat{\sigma} = \frac{\gamma_\uparrow - \gamma_\downarrow}{\gamma_\uparrow + \gamma_\downarrow} \tag{34}$$

$$\gamma_\| = \gamma_\uparrow + \gamma_\downarrow \tag{35}$$

$$f = \frac{\gamma_\|}{2\gamma_\perp} \tag{36}$$

When $\hat{\sigma} > 0$, we have pumping and so $\hat{\sigma}$ is a measure of pumping strength. With a further rescaling of variables

$$
\begin{aligned}
x &= -(b/2)^{\frac{1}{2}}\,\bar{x} \\
y &= (2b)^{\frac{1}{2}}\,C\,\bar{v} \\
z &= 2\,C\,\bar{m} \\
\tau &= \gamma_{\perp} t
\end{aligned}
\tag{37}
$$

the Ito-Langevin equations corresponding to Eqn. (33) is

$$
dx = \sigma\,(y-x)d\tau - \left(\frac{2}{C}\right)^{\frac{1}{2}} \sigma\varepsilon\, dW_x \tag{38}
$$

$$
dy = (xz-y)d\tau + 2\varepsilon\, dW_y \tag{39}
$$

$$
\begin{aligned}
dz = {} & -b(z-r)d\tau - (xy^{*}+x^{*}y)d\tau \\
& - \frac{br\varepsilon}{4C^2}\,(y^{*}dW_y + y\,dW_y^{*}) \\
& + 2^{3/2}\,\varepsilon\, b^{\frac{1}{2}} \left(1 - \frac{rz}{4C^2} - \frac{r^2 b}{32C^4}\,YY^{*}\right)^{\frac{1}{2}} dW_z
\end{aligned}
\tag{40}
$$

The Lorenz parameters b, σ and r within the laser context have the following expressions

$$
\begin{aligned}
b &= 2f \\
\sigma &= \frac{\kappa}{\gamma_{\perp}} \\
r &= 2C\,\hat{\sigma}
\end{aligned}
\tag{41}
$$

where

$$
\begin{aligned}
dW_x &= dW_1 + i\,dW_2 \\
dW_y &= dW_3 + i\,dW_4 \\
dW_z &= dW_5
\end{aligned}
\tag{42}
$$

and W_i ($i = 1,\ldots,5$) are independent Wiener processes. The quantisation of the Lorenz equations introduces new aspects. One is that x and y are complex, and another is the introduction of two additional parameters

$$\varepsilon = \left(\frac{Cb}{8\sigma n_o}\right)^{\frac{1}{2}} \tag{43}$$

and C. ε measures the overall scale of the noise strength. Moreover the stochastic terms are not ad hoc. They are multiplicative, and the coefficients involved in them are expressed in terms r, b, c and σ. The Ito equations (38) to (40) can be readily solved by making the infinitesimals become differences and taking the time steps very small. For $\varepsilon = 1$ (and $C = 50$) the Lorenz attractor is blurred out completely. We need to know whether for ε as large as this the calculation is legitimate, ie whether the Fokker-Planck description is valid. Now it is easy to show that

$$n_o = \frac{bN\gamma_{\downarrow}}{8\sigma C(\gamma_{\uparrow}+\gamma_{\downarrow})} \tag{44}$$

$$C = \frac{r}{2}\,\frac{(\gamma_{\downarrow} + \gamma_{\uparrow})}{(\gamma_{\uparrow} - \gamma_{\downarrow})} \tag{45}$$

and consequently

$$N = \frac{r^2}{4\varepsilon^2}\,\frac{(\gamma_{\downarrow}+\gamma_{\uparrow})^3}{\gamma_{\downarrow}(\gamma_{\uparrow}-\gamma_{\downarrow})^2} \tag{46}$$

Given r and ε fixed we need to know the minimum N which is compatible. The expression for N can be minimised, the minimum being achieved for $\gamma_{\uparrow} = 5\gamma_{\downarrow}$. We then deduce

$$N \geqslant \frac{27r^2}{8\varepsilon^2} \tag{47}$$

For $\sigma = 5$, $b = 1$ and for $\varepsilon = 1$, we have $N > 770$, while for $\varepsilon = .1$, $N > 76954$. Even for $\varepsilon = 1$ the minimum N is large enough to expect the Fokker-Planck approximation to be valid (Satchell, J.S. and Sarkar, S., 1986b). For $\varepsilon = .01$, $C = 50$, $r = 20$, $\sigma = 5$ and $b = 1$, d is 5 and we have $D_F \sim 3.27 \pm .04$ over a large range of scales. At smaller radii still D_F increases. Although our time series is not extensive enough to reveal scales sufficiently small for D_F to approach 5, the increase in D_F at the smaller radii is consistent with this picture. The attractor is expected to be noise dominated at the smallest scales (Sarkar, S. et al, 1986). When we calculate the Lyapunov exponents, we find that there are two with zero value and there are three which are non-zero. The additional zero exponent is due to the phase diffusion degree of freedom, ie there is no restoring force against phase diffusion. From Table 1 we see that the Lyapunov exponents are relatively insensitive to noise and this is also reflected in the Lyapunov dimension. Unlike

TABLE 1. Variation of Lyapunov exponents and dimensions with ε

ε	λ_1	λ_2	λ_3	λ_4	λ_5	D_F
0.0	0.469	0.0	0.0	-6.00	-7.47	3.078
10^{-3}	0.456	0.0	0.0	-6.00	-7.46	3.076
3×10^{-3}	0.461	0.0	0.0	-6.00	-7.46	3.077

D_F, even when $\varepsilon = 0$ D_L "knows" about the phase degree of freedom. D_L does not carry information about the inner scales of the attractor, ie the smearing of the attractor due to fluctuations. However it does inform us about the underlying chaotic motion due to the nonlinearities in the problem. The Lyapunov exponents are helpful in identifying the stretching and contraction inherent in the motion on the attractor. We can, at this stage, conjecture that the reason that D_F increases by about one (even for the smallest noises) is due to phase diffusion. In order to investigate further the role of phase we need to introduce phase symmetry breaking, and this is done simply by injecting a field into the laser cavity. The field can have a definite phase and, in this way, phase symmetry is broken. Formally in the master equation we can modify the cavity damping term to

$$\kappa([a-\alpha,\ \rho(a^{+}-\alpha^{*})] + [(a-\alpha)\rho,\ a^{+}-\alpha^{*}]) \qquad (48)$$

where α is a suitably rescaled injected field with phase defined so that it is real. This leads, eventually, just to a modification of the Ito stochastic differential equation for x

$$dx = \sigma(-x+y+\Omega)d\tau - \left(\frac{2}{C}\right)^{\frac{1}{2}} \sigma\varepsilon\ dW_x \qquad (49)$$

$$\text{with} \qquad \Omega = -\left(\frac{b}{2n_o}\right)^{\frac{1}{2}} \alpha$$

Writing x as $|x|\ e^{i\theta}$ the fixed points are

$$\theta = 0$$

and (50)

$$br|x| = (|x|-\Omega)(b+2|x|^2)$$

as well as

$$\theta = \pi$$

(51)

$$br|x| = (|x|+\Omega)(b+2|x|^2)$$

At the fixed point of Eqn. (50), the stability matrix in terms of the basis $\{|x|,|y|,z,\theta,\phi\}$, is

$$\begin{pmatrix} -\sigma & \sigma & & & \\ z & -1 & |x| & & \\ -2|y| & -2|x| & -b & & \\ & & & -\dfrac{\sigma}{|x|}(|y|+\Omega) & \dfrac{\sigma|y|}{|x|} \\ & & & \dfrac{|x|z}{|y|} & -\dfrac{|x|z}{|y|} \end{pmatrix} \qquad (52)$$

and ϕ = arg y. x, y and z appearing in the matrix are the fixed point values. The matrix simply factorises into the stability matrices for the phase degrees of freedom and amplitude. The eigenvalues λ for the phase part are

$$\lambda = \tfrac{1}{2}\left\{-\left(\frac{|x|z}{|y|} + \frac{\sigma}{|x|}(|y|+\Omega)\right) \pm\left(\left(\frac{|x|z}{|y|} + \frac{\sigma}{|x|}(|y|+\Omega)\right)^2 - \frac{4\sigma z\Omega}{|y|}\right)^{\frac{1}{2}}\right\} \qquad (53)$$

For positive Ω λ is by necessity negative. The expression obtained by continuing Ω to $(-\Omega)$ gives the eigenvalues for the fixed point having $\theta = \pi$. In the latter case a positive value of λ exists. The stability matrix corresponding to the amplitude gives the usual conditions for stability found in the Lorenz model.

The calculation of the Lyapunov exponent in the presence of non-zero Ω shows that one of the zero Lyapunov exponents becomes negative and increases in magnitude with increasing Ω. In the presence of Ω, D_L is again essentially independent of ε, and also decreases with increasing Ω. This indicates that it was indeed phase diffusion which was responsible for D_F in the phase symmetric model (with non-zero ε). However, when we consider D_F for the non-symmetric model, there is unexpected behaviour. Even before we reach the smallest scales at which noise dominates, D_F is not radius independent over a significant range of scales. This is illustrated in Figures 3 and 4. D_F increases as $\log_{10}R$ decreases and then saturates to a value about one greater than the deterministic value. The value of R at which there is saturation decreases when Ω is increased. In contrast it increases as ε increases. This behaviour is compatible with the interpretation in terms of phase diffusion degrees of freedom. As Ω is increased, phase diffusion is suppressed, and the scales at which phase diffusion effects appear becomes smaller. The opposite is true as ε is increased. We have also calculated $\langle(\mathrm{Im}\ x)^2\rangle^{\frac{1}{2}}$ which gives a length scale associated with phase diffusion and this is consistent with the findings in Figure 3. Since for $\varepsilon = 0$, D_F is well-defined for $\Omega \neq 0$, we see that a combination of both symmetry breaking and noise is needed to this obtain non-scaling fractal behaviour. Although non-scaling fractals can be mathematically constructed (Mandelbrot, B.B., 1982), they have not been encountered, as far as we are aware, in chaotic systems.

3. STOCHASTIC MAPS

In the theory of classical dynamical systems, the intersection of orbits with a surface of section, gives a so-called Poincaré or return map. In general it is easier to study asymptotic behaviour with maps (ie recursion relations) than with ordinary differential equations, and, of course, it is such behaviour which determines the form of attractors. Moreover geometrical ideas have been put forward to explain how strange attractors develop. In particular Smale (1967) has introduced the idea of horseshoes where a square region is mapped into a horseshoe by stretching in one direction, contracting in another and then bending. If, for a certain parameter value in a dynamical system, a horseshoe is formed then this implies the formation of a strange invariant set. The set is strange since it is not a fixed point or a limit

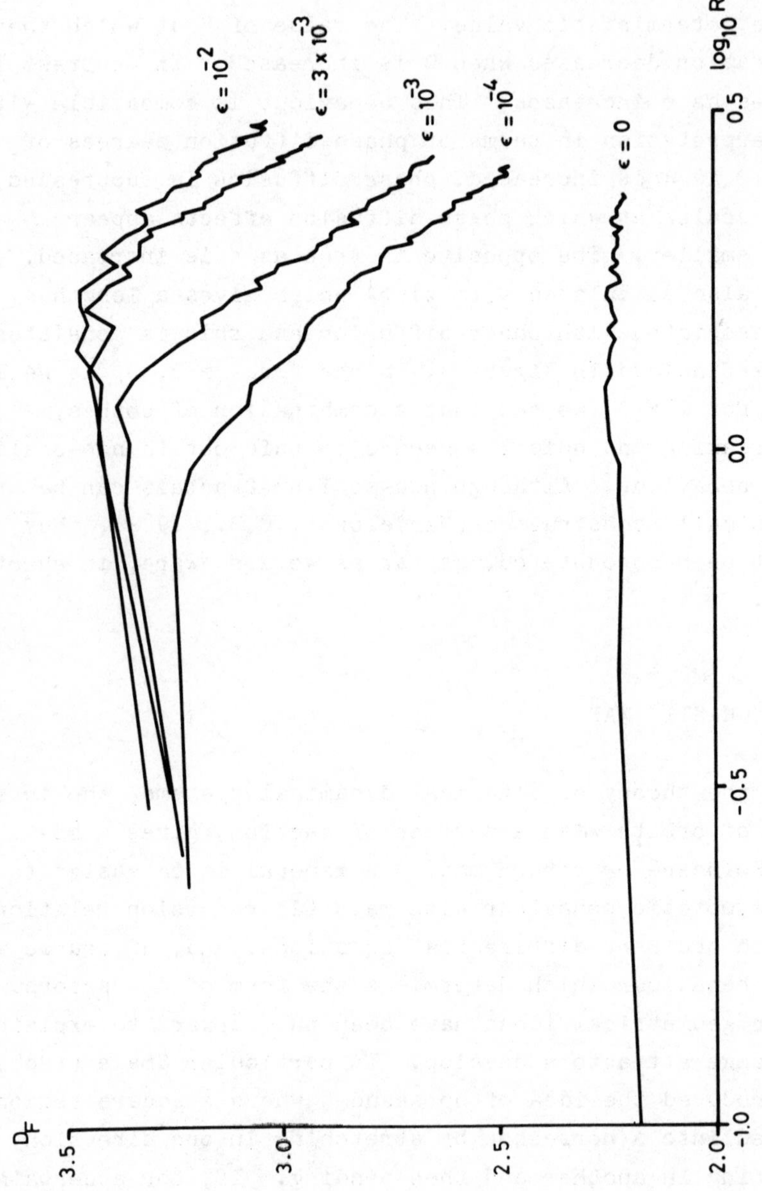

FIGURE 3. Dependence of D_F on $\log_{10} R$ with varying ε

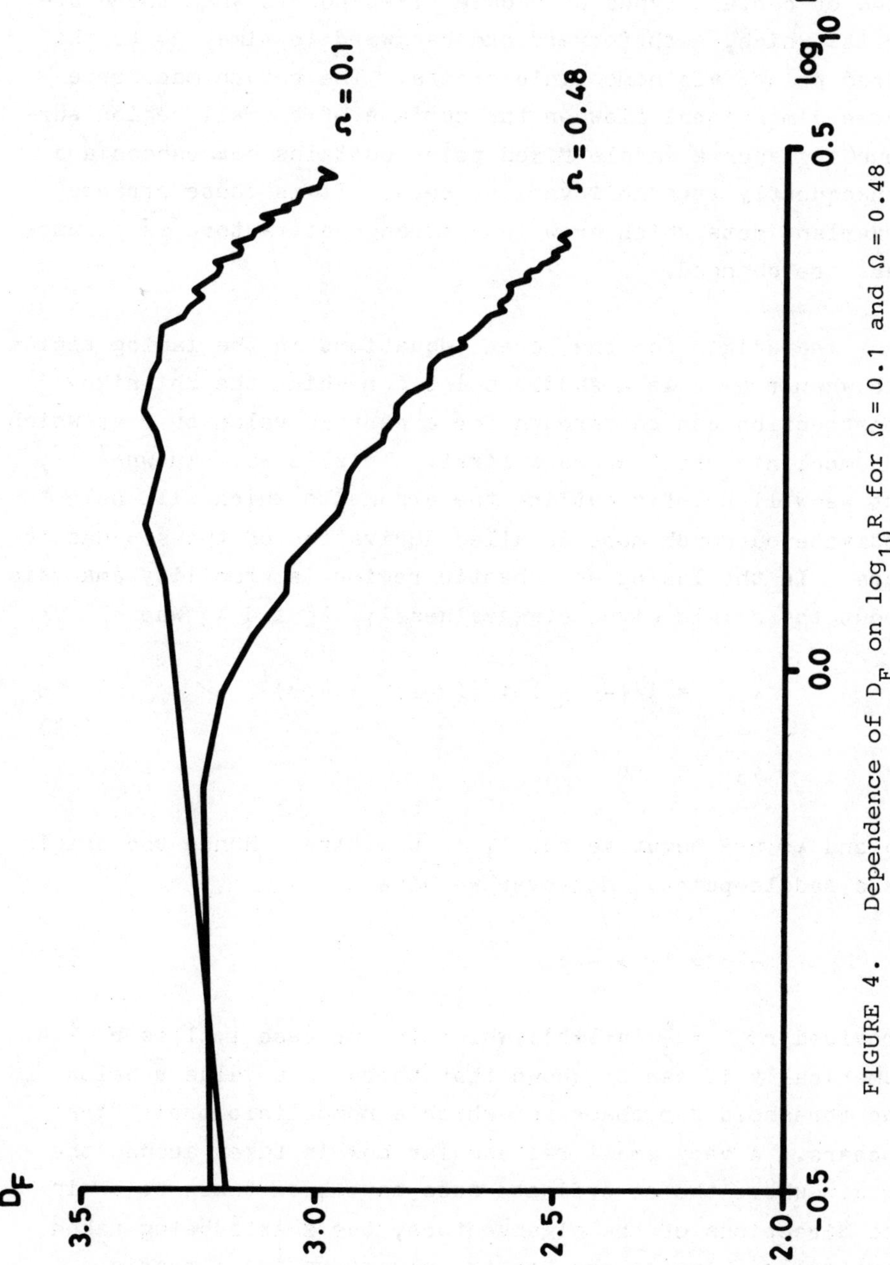

FIGURE 4. Dependence of D_F on $\log_{10}R$ for $\Omega = 0.1$ and $\Omega = 0.48$

cycle. From our calculation of Lyapunov exponents in the last section we know that in the Lorenz attractor there is stretching and contraction. In a parallel development Shilnikov (1967) has studied the behaviour in the neighbourhood of certain types of saddle fixed points when there are orbits which, both forward and backward in time, go to the fixed point, viz homoclinic orbits. The return map for a three dimensional flow on the surface of a small region surrounding such a saddle fixed point contains horseshoes and consequently strange invariant sets. It is these strange invariant sets which grow into strange attractors as parameters are changed.

The origin for the Lorenz equations in the lasing region (ie when $r > 1$) is a saddle point for which the Shilnikov construction can go through for a certain value of r at which a homoclinic orbit appears first. This is well-known[17,12], but we will briefly outline the arguments which will help to motivate our much more detailed derivation of the stochastic maps. In the lasing and chaotic regions a stability analysis about the origin gives eigenvalues λ_1, λ_2 and λ_3 where

$$\begin{aligned} \lambda_{1,2} &= \tfrac{1}{2}\left\{-\sigma - 1 \pm ((\sigma-1)^2 + 4\sigma r)^{\frac{1}{2}}\right\} \\ \lambda_3 &= -b \end{aligned} \qquad (54)$$

λ_2 and λ_3 are negative but λ_1 is positive. Hence the origin is a saddle-point. Moreover we have

$$-\lambda_2 > \lambda_1 > -\lambda_3 \qquad (55)$$

provided $r > 1 + \frac{b}{\sigma}(\sigma+1+b)$, which in our case implies $r > 2.4$. Numerically it can be shown that there is a value $\bar{r}$ below the threshold for chaos for which a homoclinic orbit first appears. A very small rectangular box is taken around the origin where the co-ordinate axes have been taken to be in the directions of the eigenvectors, the z axis being taken vertically. Inside the box the equations are linearised since the values of the variables are very small. By continuity, in the neigbourhood of the homoclinic orbit there are other orbits in a tube around it which stay close to it. The intersection of the tubes with the top of the box gives the Shilnikov return map. The contribution to the map from the part of the tubes outside the box can be parameterised generally, the only constraint being the reflection symmetry for x and y in the Lorenz equations. The more important ingredient in the map is that implied by solutions of the set of linearised equations within the box. Using this procedure, Shilnikov (1967) and others have produced maps which contain horseshoes and hence strange invariant sets.

When we quantised the Lorenz equations we found that the Lorenz equations became complex and multiplicative noise terms were introduced. For simplicity we will not examine the full five dimensional system but the ordinary Lorenz equations with possibly multiplicative noise terms. The rectangular box around the origin is shown in Figure 5.

Near the origin the stochastic Lorenz equations are given by

$$\begin{pmatrix} \dot{x} \\ \dot{y} \\ \dot{z} \end{pmatrix} = \begin{pmatrix} -\sigma & \sigma & 0 \\ r & -1 & 0 \\ 0 & 0 & -b \end{pmatrix} \begin{pmatrix} x \\ y \\ z \end{pmatrix} + \begin{pmatrix} \varepsilon_1 & \hat{\xi}_1 \\ \varepsilon_2 & \hat{\xi}_2 \\ \varepsilon_3 & \hat{\xi}_3 \end{pmatrix} \qquad (56)$$

where ε_1, ε_2 and ε_3 are small noise strengths and $\hat{\xi}_1$, $\hat{\xi}_2$ and $\hat{\xi}_3$ are independent Gaussian white noises. This linearisation is valid even in the presence of multiplicative noise terms. If we choose the variables $\bar{x}$ and $\bar{y}$ with

$$\bar{x} = \frac{-(\lambda_2+\sigma)x + \sigma y}{\sigma(\lambda_1-\lambda_2)} \qquad (57)$$

and

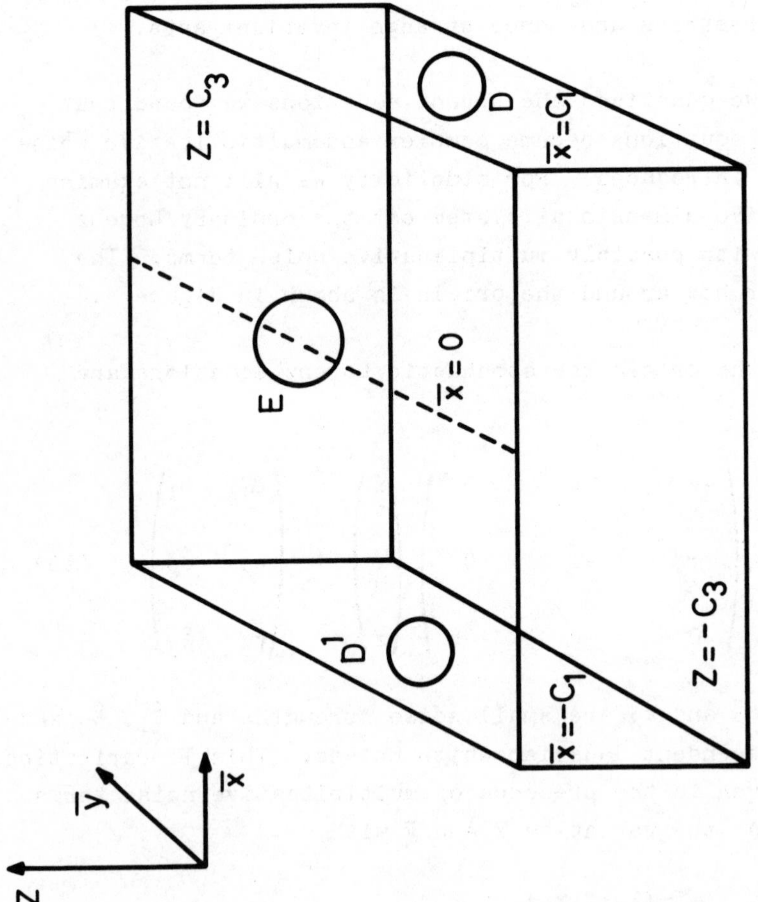

FIGURE 5. Infinitesimal box around saddle point

$$\bar{y} = \frac{(\lambda_1+\sigma)x - \sigma y}{\sigma(\lambda_1-\lambda_2)} \tag{58}$$

then

$$\begin{pmatrix} \dot{\bar{x}} \\ \dot{\bar{y}} \\ \dot{z} \end{pmatrix} = \begin{pmatrix} \lambda_1 & 0 & 0 \\ 0 & \lambda_2 & 0 \\ 0 & 0 & \lambda_3 \end{pmatrix} \begin{pmatrix} \bar{x} \\ \bar{y} \\ z \end{pmatrix} + \begin{pmatrix} \varepsilon_1' & \hat{\xi}_1' \\ \varepsilon_2' & \hat{\xi}_2' \\ \varepsilon_3 & \hat{\xi}_3 \end{pmatrix} \tag{59}$$

with new notation having the obvious meaning. The faces of the box are at $\bar{x} = \pm c_1$, $\bar{y} = \pm c_2$ and $z = \pm c_3$. E, D and D' are the intersections of the tubes around the homoclinic orbit with the faces of the box. The mapping from D to E (outside the box) is taken to have the general form

$$(c_1, \bar{y}, z) \rightarrow (X(\bar{y},z), Y(\bar{y},z), c_3) \tag{60}$$

The solution of Eqn. (59) gives

$$\bar{x}(t) = \bar{x}(o)\, e^{\lambda_1 t} + \varepsilon_1' \, e^{\lambda_1 t} \int_0^t dt' \, e^{-\lambda_1 t'} \, \hat{\xi}_1'(t') \tag{61}$$

and similarly for $\bar{y}$ and z. Now we will consider the map from E to D within the box. The time taken for the trajectory passing through a point with co-ordinate $\bar{x}$ in E to reach D is a stochastic variable and satisfies

$$e^{\lambda_1 t} = \frac{c_1}{\bar{x} + \varepsilon_1' \tilde{W}_{\lambda_1}(t)} \tag{62}$$

with

$$\tilde{W}_{\lambda_1}(t) \equiv \int_0^t dt' \, e^{-\lambda_1 t'} \, \hat{\xi}_1'(t') \tag{63}$$

It is easy to see that $\widetilde{W}_{\lambda_1}$ is a Gaussian variable and

$$\langle \widetilde{W}_{\lambda_1}(t)\, \widetilde{W}_{\lambda_1}(s)\rangle = \frac{1}{2\lambda_1}\left(1 - \theta(t-s)e^{-2\lambda_1 s} - \theta(s-t)e^{-2\lambda_1 t}\right) \quad (64)$$

where

$$\begin{aligned} \theta(x) &= 1 && \text{if} \quad x > 0 \\ &= \tfrac{1}{2} && \text{if} \quad x = 0 \\ &= 0 && \text{if} \quad x < 0 \end{aligned} \quad (65)$$

Since λ_1 is positive (and large), for long enough time t we have

$$\langle (\widetilde{W}_{\lambda_1}(t))^2 \rangle \simeq \frac{1}{2\lambda_1} \quad (66)$$

and, hence, to a good approximation

$$\widetilde{W}_{\lambda_1}(t) \backsim (2\lambda_1)^{-\frac{1}{2}}\, \xi_1(t) \quad (67)$$

The tube of trajectories around the homoclinic orbit is chosen to be very narrow, and so $(\overline{x} + \varepsilon_1'\, \widetilde{W}_{\lambda_1}(t))$ is almost surely small which implies that t is large. Moreover, depending on the sign of $(\overline{x} + \varepsilon_1'\, \widetilde{W}_{\lambda_1}(t))$ we go from E to D or D' Hence this mapping is of the form

$$(\overline{x},\overline{y},c_3) \to \left(c_1\ \mathrm{sgn}(\overline{x} + \varepsilon_1''\xi_1)\ ,\ \overline{y}\left(\frac{c_1}{|\overline{x} + \varepsilon_1''\xi_1|}\right)^{\frac{\lambda_2}{\lambda_1}} + \varepsilon_2'\xi_2\ ,\ c_3\ \frac{c_1}{|\overline{x} + \varepsilon_1''\xi_1|}^{\lambda_3/\lambda_1} + \varepsilon_3'\xi_3\right) \quad (68)$$

We shall assume that the functions X(y,z) and Y(y,z) are smooth at y = z = 0, and that a low order Taylor expansion exists. From the Lorenz symmetry the map from D' to E is

$$(-c_1,\bar{y},z) \rightarrow (-X(-\bar{y},z),\ -Y(-\bar{y},z),\ c_3) \tag{69}$$

Finally putting the maps (60), (68) and (69) together, we find owing to (55) that to a good approximation in the map for E into E, $\bar{x}$ is decoupled from the other variables and we have

$$\bar{x} \rightarrow \operatorname{sgn}(\bar{x}) \left\{ X(o,o) + \left(\frac{\partial X(o,o)}{\partial z}\right) c_3 \left(\frac{||\bar{x}| + \varepsilon_1'\xi_1|}{c_1}\right)^{-\frac{\lambda_3}{\lambda_1}} \right\} + \varepsilon_2'\xi_2 \tag{70}$$

With some straightforward redefinitions it is possible to write this map as

$$\bar{x} \rightarrow \operatorname{sgn}(\bar{x})(-A + B(||\bar{x}| + \varepsilon_1'\xi_1|)^{1-\nu}) \ + \ \varepsilon_2'\xi_2 \tag{71}$$

When we set $\varepsilon_1' = \varepsilon_2' = 0$ we recover the deterministic map for the Lorenz system of course (Sparrow, C., 1982; Lyubimov, D.V. and Zaks, M.A., 1983). The noise appears not just in a purely additive way owing to the term proportional to B. Although strictly we have derived this map just at the r value for which the origin first has a homoclinic orbit (and below the chaotic threshold) we will see that the map has bifurcations to chaos. For the Lorenz model, ν, the saddle index, is given by

$$\nu = 1 - \frac{2b}{((\sigma-1)^2 + 4\sigma r)^{\frac{1}{2}} - \sigma - 1} \tag{72}$$

and for r at the threshold value for the appearance of the homoclinic orbit this is positive. Obviously for larger r ν increases. However the fixed point can still be a saddle with ν negative although this happens not to be the case of the Lorenz model. We need $\nu < 1$ for the fixed point to be a saddle. From intersection properties of the function

$$f(x) = \mathrm{sgn}(x)(-A + B|x|^{1-\nu}) \tag{73}$$

with the line g(x) = x, it is possible to deduce a great deal of qualitative information concerning the map in Eqn. (71) in the absence of noise. In particular, if $|f'(x)| > 1$ at an intersection then the corresponding fixed point is unstable. For $\nu > 0$ it is easy to see that there is a tangent bifurcation (ie f'(x) and g'(x) coincide at the intersection) for

$$A = A_c = \frac{\nu\, B^{1/\nu}}{(1-\nu)^{1-1/\nu}} \tag{74}$$

When A is greater than A_c, there is a chaotic attractor qualitatively of the Lorenz type (Lyubimov and Zaks, 1983). Here there is no period doubling route to chaos which is qualitatively in agreement of the first threshold for chaos in the Lorenz equations (Sparrow, C., 1982; Lorenz, E.,1963). However if $\nu < 0$ it is known that there is a period doubling route to chaos (Lyubimov, D.V. and Zaks, M.A., 1983). The study of this map has already isolated an important parameter in Lorenz type systems, viz ν. The advantage of maps is that their comparative simplicity allows a detailed study of quantities such as fractal dimensions (Mandelbrot, B.B., 1982; Satchell, J.S. and Sarkar, S., 1986b). It is, of course, true that not all features of a higher dimensional system can be understood from a one-dimensional return map.

For A positive, B = 1.5 and ν between 0 and 1, three general forms of behaviour have been found in the absence of noise. They are:

i. a pair of fixed points. Depending on initial conditions the solution converges on to one or the other. This is typical of behaviour for small values of A.

ii. a pair of stable fixed points in co-existence with chaos. For initial conditions close to the origin we find chaos, but otherwise the solutions converge on to one of the fixed points. This is typical of the behaviour for ν near 1.

iii. globablly stable chaos. All solutions move in the same region (ie attractor) regardless of initial conditions. Such a case is given by A = 1 and $\nu = 0.8$, (and in fact for $A > A_c$).

More complicated behaviour is surely possible but it is not our purpose to present an exhaustive treatment of the deterministic version of Eqn. (71). The Lyapunov exponent can be regarded like an order parameter in statistical mechanics.

For ν positive the transition to chaos is found always to be of first order (ie the exponent increases discontinuously), and the possibility of scaling laws at the transition point does not exist. Unlike the case for the logistic map, the Lyapunov exponent, in the region of fully developed chaos (A = B = 1.5 and $\nu = 0.8$), is found to be insensitive to noise. Despite averaging over 10^6 iterations no change was found in the exponent for noise strengths of up to 10^{-3}. Moreover the fractal dimension D_F was 1 which is a property of one-dimensional systems with positive Lyapunov exponents. Here we do not have the analogue of the higher dimensional Lorenz attractor which has a non-integer D_F. We would like to be able to show explicitly with maps how the effects of noise increase the fractal dimension to a value equal to the integral topological dimension of the minimal space in which the system is embedded. Since for $\nu < 0$ there is a period doubling route to chaos and hence the Lyapunov exponent is zero at the onset of chaos, the change of the exponent and D_F with noise should be non-trivial. In considering $\nu < 0$ it is convenient to work with a further rescaled form of the map

$$\tilde{x} \to \mathrm{sgn}(\tilde{x})(-1 + B'(||\tilde{x}| + \sigma_1\xi_1|^{1-\nu})) + \sigma_2\xi_2 \qquad (75)$$

with B', σ_1 and σ_2 parameters. For $\sigma_1 = \sigma_2 = 0$, D_F can be calculated at $B' = B_c$, the threshold for chaos. In particular we have the results of Table 2.

TABLE 2. D_F for different ν

ν	B_c	D_F
-.25	1.209513	.34 ± .01
-.5	1.295509	.41 ± .01
-1.0	1.401155	.54 ± .005

When we make (σ_1, σ_2) non-zero, the D_F become 1 and this is further support of our earlier assertion that noise at sufficiently small scales smears out a chaotic attractor.

An interesting feature of Eqn. (75) is that alghough it produces period doubling it seems very different from the logistic map. The logistic map is smooth whereas this is discontinuous at the origin. Moreover the noisy logistic map that is usually treated (Shraiman, B. et al, 1981; Feigenbaum, M. and Hasslacher, B., 1982; Haken, H and Mayer-Kress, G., 1981) has just a simple additive term. In our case the noise is much more complicated and its form has been forced on us by the Shilnikov construction which in turn is inspired by the existence of homoclinic orbits.

CONCLUSIONS

Using simple examples we have shown how the concepts of dimensions and Lyapunov exponents are useful in describing the quantum mechanics of chaotic systems. Quantum mechanics for the Lorenz system has the effect of converting the description in terms of ordinary differential equations (valid for the deterministic system) into one in terms of stochastic differential equations. By a construction due to Shilnikov in certain circumstances we have shown how these stochastic equations give rise to stochastic maps. It has been demonstrated that the effect of this explicit stochasticity is not completely universal, but depends on the details of the route to chaos followed by the deterministic system.

REFERENCES

Callen, H.B. and Welton, T.A., 1951, Phys. Rev. 83, 34.

Cvitanović, P., Ed., 1984, Universality in Chaos, Hilger, Bristol.

Eckmann, J-P. and Ruelle, D. 1985, Rev. Mod. Phys. 57, 617.

Farmer, J.D., 1982, Z. Naturforsch. 37a, 1304.

Feigenbaum, M. and Hasslacher, B., 1982, Phys. Rev. Lett. 49, 605.

Graham, R., 1984, Phys. Rev. Lett. 53, 2020.

Grassberger, P. and Procaccia, I., 1983, Phys. Rev. Lett. 50, 346.

Guckenheimer, J. and Holmes, P., 1983, Nonlinear Oscillations, Dynamical Systems and Bifurcations of Vector Fields, Springer-Verlag, New York.

Haken, H., 1970, Laser Theory, Encycl. of Phys. XXV/2C, Springer-Verlag, Berlin.

Haken, H., 1975, Phys. Lett. A53, 77.

Haken, H. and Mayer-Kress, G., 1981, J. Stat. Phys. 26, 149.

Lorenz, E.N., 1963, J. Atmos. Sci. 20, 130.

Lugiato, L.A., Casagrande, F. and Pizzuto, L., 1982, Phys. Rev. A26, 3483.

Lyubimov, D.V. and Zaks, M.A., 1983, Physica 9D, 52.

Mandelbrot, B.B., 1982, The Fractal Geometry of Nature, Freeman.

Poincaré, H., 1899, Les Methodes Nouvelles de la Mecanique Celeste, Gauthier-Villars, Paris.

Sarkar, S., Ed., 1986, Nonlinear Phenomena and Chaos, Hilger, Bristol.

Sarkar, S., Satchell, J.S. and Carmichael, H.J., 1986, Quantum Fluctuations and the Lorenz Equations, J. Phys. A (to be published).

Sarkar, S. and Satchell, J.S., 1986, Quantum Chaos in the Lorenz Equations with Symmetry Breaking, RSRE, Malvern, preprint.

Satchell, J.S. and Sarkar, S., 1986a, Stochastic Shilnikov Maps, RSRE, Malvern, preprint.

Satchell, J.S. and Sarkar, S., 1986b, Quantum Theory of Optical Bistability for Small Systems, in this volume.

Shilnikov, L.P., 1967, Sov. Math. Dokl. 8, 54.

Shraiman, B., Wayne, C.E. and Martin, P.C., 1981, Phys. Rev. Lett. 46, 935.

Smale, S., 1967, Bull. Amer. Math. Soc. 73, 747.

Sparrow, C., 1982, The Lorenz Equations: Bifurcations, Chaos and Strange Attractors, Springer-Verlag, New York.

Termonia, Y. and Alexandrowicz, Z., 1983, Phys. Rev. Lett. 51, 1265.

Uhlenbeck, G.E. and Ornstein, L.S., 1930, Phys. Rev. 36, 823.

Wigner, E.P., 1932, Phys. Rev. 40, 749.

DYNAMIC VERSUS STATIC STABILITY

P MANDEL

1. INTRODUCTION

Quantum optics is a domain of physics which deals with the nonlinear interaction of light with matter. This nonlinear interaction is translated into nonlinear differential equations. A relevant property of these nonlinear differential equations is to have more than one solution for a given value of all parameters. Of special interest, therefore, are critical points such as bifurcations and limit points where two solutions coincide and may modify their stability. The problem can be formulated as follows. Let X(t) be a vector whose components are the dynamical variables of the problem. It verifies a differential equation

$$dX/dt \equiv X_t = F(X,\mu,p), \quad X(t=0,\mu,p) = X_0(\mu,p) \tag{1.1}$$

where F is a nonlinear function of X. The control parameter is μ whereas p stands for physical parameters which do not vary in a given experiment. A property of (1.1) is the occurence of different long-time solutions (steady or not) corresponding to different initial conditions but identical μ and p. Let x_1 and x_2 be two different long-time solutions of (1.1):

$$x_j(t,\mu,p) = X(t\to\infty,\mu,p;X_{j0}), \quad j=1,2 \tag{1.2}$$

such that x_1 and x_2 are independent of their initial conditions. Then a critical point is defined by the coincidence of x_1 and x_2:

$$x_1(t,\mu_c,p) = x_2(t,\mu_c,p) \tag{1.3}$$

This equation defines the critical control parameter μ_c. In some cases this coincidence also requires that p be either fixed at some critical value or restricted to some domain.

Examples are known where (i)x_1 and x_2 are both steady solutions, (ii)x_1 is steady but x_2 is time-dependent (periodic or chaotic) (iii)x_1 and x_2 are both time-dependent. Furthermore the number of simultaneous solutions is not restricted to two. In this contribution only steady solutions $x_{jt}=0$ will be considered.

Suppose that $F(X,\mu,p)=0$ has only two solutions x_1 and x_2. Let $F(X+\delta X,\mu,p)=F(X,\mu,p)+G(X,\mu,p)\delta X+O(\delta X^2)$.

Since F and X are vectors, $G(X,\mu,p)$ is a matrix whose determinant is $g_x(\mu,p)$. From the implicit function theorem we may prove that the coincidence equation (1.3) implies $g_x(\mu_c,p)=0$. Conversely if $g_x(\mu,p)\neq 0$ the solutions of $F(X,\mu,p)=0$ are different.

The usual approach to study the stability of the steady solutions x_1 and x_2 is to perform a linear stability analysis. This amounts to study the solutions of (1.1) which are infinitesimal perturbations of the steady state for all times, including the initial time. Let $X_j(t)=x_j+\varepsilon X_j^!\exp(\lambda_j t)\ +O(\varepsilon^2)$ be such a solution. Then to first order in ε we have from (1.1) the characteristic equation

$$\det\left(\lambda_j I-G(x_j,\mu,p)\right)=0 \qquad (1.4)$$

where I is the unit matrix. Because $g_x(\mu,p)$ vanishes at a coincidence point ($\mu=\mu_c$) there is necessarily at least one real zero root $\lambda_j(\mu_c)=0$ when $x_1=x_2$. If λ_{jn} are the roots of (1.4) we can interpret in more physical terms $(\mathrm{Re}\lambda_{jn})^{-1}$ as the relaxation times of the system when it is infinitesimally perturbed from a steady state. We then see that the coincidence of two steady solutions implies that one, at least, of the relaxation times diverges. Hence the domain surrounding $\mu=\mu_c$ will be characterized by critical slowing down.

When a system displays critical slowing down, it reacts slowly to an external perturbation. Let us assume that critical slowing down is an intrinsic property of the dynamical system (1.1) which holds even if the perturbation is not infinitesimally small. We then expect the system to relax towards its stable steady state whether a perturbation is finite in time but infinitesimal in amplitude or, conversely, when it is infinitesimal in time but fairly large in amplitude. This implies that near $\mu=\mu_c$ a more complete description of the stability properties of a steady solution requires that we specify both amplitude and duration of the perturbation.

The topic studied in this contribution is the stability properties of a nonlinear equation near a point where two steady states coincide and when the control parameter is explicitly time-dependent. The emphasis will be on analytic studies of simple model equations suggested by the laser theory and by optical bistability. Two types of time-dependences will be considered.

Swept bifurcation parameters. In this case we have

$$\mu(t)=\mu_0+vt, \quad \mu_0<\mu_c, \quad |v|<<1 \tag{1.5}$$

The motivation to study this problem is very simple. Quite often in quantum optics technical problems prevent taking data in the vicinity of μ_c with a constant bifurcation parameter. A convenient way to bypass this technical limitation is to sweep the bifurcation parameter across μ_c. It is assumed that the corresponding time-dependent state will "adiabatically" follow the exact steady state if $v<<1$. In other terms it is assumed that after transients have quickly died out, the state of the system can be described by

$$X(t,\mu(t),p)=x(\mu(t),p)+O(v)$$

which is known as the outer expansion and where $x(\mu,p)$ is the steady state. This expansion can usually be proved to be the correct expansion of the exact solution, when $v\to 0$, only far from μ_c. However the analytic expansion in powers of v diverges near μ_c as can be understood by the following argument. The sweep rate has to be compared with the decay rates $\gamma_{jn}=\mathrm{Re}\lambda_{jn}$ of the linear stability analysis. Let γ be the smallest of the γ_{jn}. The outer expansion may be expected to hold only if for a small variation $\Delta\mu=v\Delta t$ of the bifurcation parameter the system has time to relax, i.e., $\gamma\Delta t>>1$. The variation $\Delta\mu$ has to be small enough to justify the use of the linearized decay rates. Hence the inequality $1>>\Delta\mu>>v/\gamma$ is a necessary condition of the outer expansion. Because critical slowing down means $\gamma\to 0$, there is around μ_c a domain (which depends on v) where the condition $v/\gamma<<1$ is no longer verified and the outer expansion diverges. This problem will be analyzed for a steady bifurcation point and for a limit point.

Periodic modulation of the bifurcation parameter. In this case we consider a small amplitude periodic modulation of the bifurcation parameter near a limit point

$$\mu(t)=\mu_0+\varepsilon\cos\omega t \quad , \quad 0<\varepsilon<<1 \tag{1.6}$$

As we shall see this provides an alternative method to control the switching process in bistable systems based on a frequency rather than on an amplitude control. The physics of this process is fairly simple. We consider the two steady solutions shown on Fig.1:

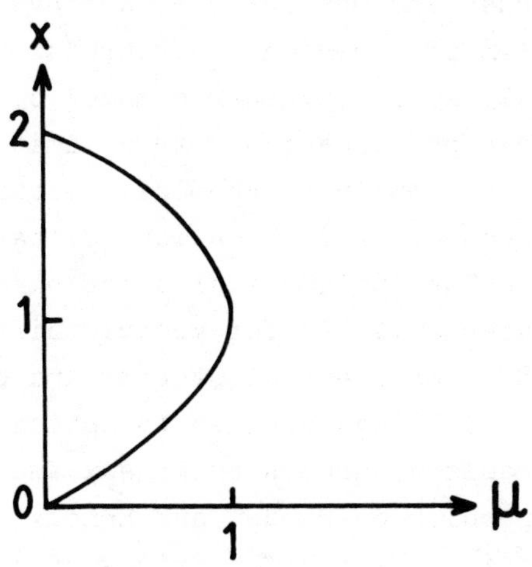

Figure 1.

Let x_2 be the unstable upper solution ($1\leq x_2\leq 2$) and x_1 the stable lower solution ($0\leq x_1\leq 1$). They coincide at $\mu=1$. The initial condition is on the stable steady state: $X(t=0,\mu_0,p)=x_1(\mu_0,p)$. If $\mu_0<<1$ the solution of (1.1) with (1.6) will be a stable limit cycle of amplitude ε: $X(t,\mu(\varepsilon,t),p)=x_1(\mu_0,p)+\varepsilon y_1(\mu_0,p,\omega,t)+O(\varepsilon^2)$ where y_1 is $2\pi/\omega$-periodic. This again is the outer solution. If $\mu_0<1$ but $\mu_0+\varepsilon>1$ the situation is somewhat different. During each period $2\pi/\omega$ the system will be dragged for some time $\Delta t(\omega,\varepsilon,\mu_0)$ in the domain $\mu>1$ where no stable solution exists at all and near x_2 which exists but is unstable. Clearly Δt is a decreasing function of ω. Let us take the simple law

$\Delta t = \omega^{-1} f(\varepsilon, \mu_0)$. In order for the instability to manifest itself it is necessary, as in the case of swept parameter, that $\gamma \Delta t >> 1$, i.e., $\omega << \gamma f$. In other terms there exists a critical frequency $\omega_c = \gamma f$ which depends on ε and μ_0 such that if $\omega > \omega_c$ there is a stable limit cycle near the lower branch even though the system spends some time in the unstable domain. On the contrary, for $\omega < \omega_c$, no such stable limit cycle exists and the solution of (1.1) diverges. As critical slowing down implies $\gamma << 1$, the critical frequency may be unexpectedly small. The existence of a stable limit cycle in these conditions results from a compensation between the destabilizing effect of the modulation and the stabilizing effect of critical slowing down.

Standard references to nonlinear analysis as used in this contribution are the books by Bender and Orszac(1978) and by Iooss and Joseph(1980). The problem of a periodic perturbation near a steady bifurcation was analyzed for a generic model by Rosenblat and Cohen(1980). Periodic perturbation near a limit point was analyzed in the context of nonlinear chemical reactions by Cohen and Matkowski(1978). It was applied to optical bistability by Mandel and Erneux(1985). The influence of swept parameter was analyzed in depth by Haberman(1979) for generic differential equations, by Kapila(1981) for chemical systems and by Matalon and Erneux for flames(1984). It was applied to optical bistability and lasers with saturable absorbers by Erneux and Mandel(1983,1984), to the laser problem by Mandel and Erneux (1984) and extended to a class of imperfect bifurcations by Erneux and Mandel(1986). Recently the influence of white noise on the laser problem with swept gain was studied by Broggi et al.(1986). In a different context the combined influence of swept parameter and fluctuations on imperfect bifurcations was analyzed by Kondepudi et al.(1983,1985,1986) in relation with symmetry breaking and chiral selection in biochemistry. Recently the problem of swept parameter was also studied for the quadratic map by Kapral and Mandel(1985). Experimental results in quantum optics are still scarce. Up to now the only reported comparisons with theoretical predictions are due to Dangoisse et al.(1985), Glorieux et al.(1985) and Arimondo et al.(1985).

This contribution is divided in three sections. Section two is devoted to the swept gain laser. We first analyze the solution of the deterministic equations. Then the influence of a white gaussian noise on the deterministic results is examined. Section three is devoted to the analysis of a simple deterministic model of a limit point. Both the sweeping at constant sweep rate and the modulation of the bifurcation parameter are studied analytically.

2. SWEPT GAIN LASER

In this section we study the semiclassical laser equations (Haken,1985) for a tuned single-mode unidirectional ring laser:

$$\begin{aligned} E_t &= -E+AP \\ P_t &= d(-P+EF) \\ F_t &= d_{\parallel}(-F+1-EP) \end{aligned} \tag{2.1}$$

In (2.1) E is the cavity field, A the pump parameter and t a dimensionless time scaled to the cavity decay rate. P is the atomic polarization whose decay rate is d. F is the atomic inversion whose decay rate is $d_{\parallel}$. F is normalized in such a way that F=1 in the absence of light-matter interaction but in the presence of pumping.

When A is time-independent, Eqs.(2.1) have two steady states: E=0 and $E^2=A-1$. The linear stability analysis of the trivial solution E=0 gives the characteristic equation

$$\lambda^2+\lambda(d+1)+d(1-A)=0 \tag{2.2}$$

Hence E=0 is stable for A<1. Likewise $E^2=A-1$ is stable for A>1, at least near A=1 (the laser second threshold will not be considered here).

Let us now assume that

$$A=A(\varepsilon t)\equiv A(t'), \quad A(0)<1, \quad 0<\varepsilon<<1 \tag{2.3}$$

We therefore consider a variation of the gain parameter which is very slow and which defines the slow time-scale t'. We note that even when $A_t\neq 0$ the trivial solution E=P=F-1=0 remains an exact steady solution of (2.1). Therefore we can linearize (2.1) around the trivial solution

$$E(t)=\eta e(t)+O(\eta^2)$$
$$P(t)=\eta p(t)+O(\eta^2) \qquad (2.4)$$
$$F(t)=1+\eta f(t)+O(\eta^2)$$

with $|\eta|<<1$ provided (2.4) also holds for the initial condition. This gives

$$e_t=-e+A(t')p$$
$$p_t=d(-p+e) \qquad (2.5)$$
$$f_t=-d_{\shortparallel}f$$

We define two auxiliary functions $g_j=g_j(\varepsilon,t')$, $j=1,2$ by

$$e(\varepsilon,t')=\sum_{j=1,2} c_j \exp\left(\frac{1}{\varepsilon}\int_0^{t'} g_j(\varepsilon,s)ds\right) \qquad (2.6)$$

They verify the nonlinear differential equation

$$g_j^2+g_j(d+1)+d(1-A)=\varepsilon\left((A_{t'}/A)(g_j+1)-g_{jt'}\right)$$

Expanding g_j in powers of ε: $g_j=h_j+\varepsilon k_j+O(\varepsilon^2)$ yields to dominant order in ε

$$h_j^2+h_j(d+1)+d(1-A)=0 \qquad (2.7)$$

Comparing (2.7) with (2.2) indicates that $h_j(t')=\lambda\{A(t')\}$. Hence $h_1(t')=\lambda_+$ is always negative whereas $h_2(t')=\lambda_-$ is negative (positive) for A<1 (A>1) and vanishes at A=1.

As in the static case ($A_t=0$) the domain of stability for e(t') given by (2.6) is defined by the requirement that the argument of the exponentials be negative. The limit of the stable domain corresponds to the vanishing of this argument:

$$\int_0^{t^*} h_2(s)ds=0 \qquad (2.9)$$

which is an implicit equation for the time t^* at which e(t') begins to diverge. A divergence of e(t') means that the linearization of (2.1) no longer holds and that the system starts the transition to reach the vicinity of the other solution $E^2=A-1$. We define an intermediate time $\bar{t}$ by the implicit equation $A(\bar{t})=1$. The stability boundary (2.9) then takes the form

$$\int_0^{\bar{t}} h_2(s)ds = \int_{\bar{t}}^{t^*} \{-h_2(s)ds\} \qquad (2.10)$$

This condition expresses the balance between the stability accumulated from 0 to $\bar{t}$ where $h_2(s)<0$ and the instability accumulated from $\bar{t}$ to t^* where $h_2(s)>0$. The balance condition (2.10) has the following consequences:

1.In opposition to the static case where A=1 is a local condition, we now have a nonlocal condition. Indeed the rate of damping below A=1 and the rate of divergence above A=1 will both contribute to the balance which determines t^*. Hence two models with a static bifurcation at A=1 may have different dynamical bifurcation points $A^*=A(t^*)$. In this respect the determination of A^* may provide a more refined test of a theory.

2.Because $t^*>\bar{t}$, the dynamical bifurcation point A^* is always delayed with respect to the static point A=1. Hence the sweep has dynamically stabilized a solution which is unstable under static conditions.

3.The balance condition is rather general since the explicit dependence of A on t' has not yet been specified.

Further insight may be gained by specifying A. Let us choose $A(t')=A_0+t'$. Then the interesting function to study is $D=(A^*-1)/(1-A_0)$ which is given from (2.10) by

$$D=\frac{1}{8\alpha}\{4\alpha-3+4\left(-3\alpha^2+19.5\alpha-11.4375+12(1-\alpha)^{3/2}\right)^{1/2}\}=\phi(\alpha) \qquad (2.11)$$

$$\alpha=4d(1-A_0)/(1+d)^2 \qquad (2.12)$$

The function D is defined, up to now, only for $\alpha\leqq1$. This restriction will be relaxed later on. Two remarks have to be made about D.

1.It is a function of a single variable which combines two parameters of the problem, d and A_0. Furthermore α is invariant for a change of d into 1/d: the delay with respect to the static bifurcation is not a property restricted to either a bad or a good cavity.

2.It is independent of ε. This property corresponds to a discontinuous behavior of D in the limit of small sweep rate since (2.11) implies

$$\lim_{\varepsilon\to 0} D(\varepsilon)=D+O(\varepsilon), \quad D=O(1) \tag{2.13}$$

whereas D does not exist when $\varepsilon=0$. The origin of the discontinuity lies in the fact that the dynamical instability condition is (2.9) iff $\varepsilon\neq 0$ as seen from (2.6).

An alternative and simple way to see this discontinuity is to analyze the good cavity limit $d>>1$ and $d_{\shortparallel}>>1$ (Lugiato et al., 1984). Then the field equation becomes

$$E_t=E(-1+\frac{A}{1+E^2}) \tag{2.14}$$

This equation belongs to a class of differential equations

$$X_t=X\{-1+Af(X)\}, \; f(0)=1, \quad X\geq 0 \tag{2.15}$$

where f(X) is analytic in X around X=0. The steady solutions of (2.15) are X=0 which is stable below A=1 and Af(X)=1 which coincides with X=0 at A=1. The linearization of (2.15) around X=0 yields $x_t=x(A-1)$. When $A=A(t)=A_0+vt$ with $A_0<1$ and $v>0$ the solution of the linearized equation is

$$x(t,v,A_0)=x_0\exp\{t(A_0+\frac{v}{2}t-1)\} \tag{2.16}$$

The instability condition becomes

$$A_0+\frac{v}{2}t^*-1=0 \tag{2.17}$$

When v=0 we recover the static instability condition A=1. If v is non-zero but otherwise arbitrary we find

$$A^*-1=1-A_0, \quad t^*=2\bar{t} \tag{2.18}$$

and therefore D=1. Thus in the good cavity limit, Eq.(2.14) predicts that A^* is always independent of the sweep rate. This is no longer true for the more general equations (2.1). However owing to the invariance property of α, we can state without further calculation that (2.18) also holds in the limit of a bad cavity. In addition the property D=1 for all non-zero sweep rates which holds for a class of equations such as (2.15) is independent of the slope f'(X) at A=1: the bifurcation can be supercritical as in a laser or subcritical as can happen in the laser with saturable absorber (Erneux et al,1984).

The limitation $\alpha \leqq 1$ can be bypassed by solving (2.5) exactly and taking the limit of small sweep rate on the exact solution (Mandel et al.,1984). The exact solution of (2.5) with $A(t)=A_0+vt$, $v>0$, is

$$p(t)=\pi e^{-(d+1)t/2}\{\alpha_1 Ai(\xi)-\alpha_2 Bi(\xi)\} \tag{2.18}$$

$$\alpha_1=p(0)Bi'(\xi_0)-(vd)^{-1/3}(p_t(0)+p(0)\frac{d+1}{2})Bi(\xi_0) \tag{2.19.a}$$

$$\alpha_2=p(0)Ai'(\xi_0)-(vd)^{-1/3}(p_t(0)+p(0)\frac{d+1}{2})Ai(\xi_0) \tag{2.19.b}$$

with $\xi=(vd)^{-2/3}\{dA(t)+(\frac{d-1}{2})^2\}$, $\xi_0\equiv\xi(t=0)$. The functions Ai and Bi are the Airy functions (Abramowitz et al.,1965). Using the asymptotic properties of the Airy functions, it is easy to verify that for all $\alpha>0$ we have

$$\lim_{v\to 0} D(v)=D+O(v)$$

$$D=D(\alpha)=O(1)$$

The discontinuous behavior of D(v) at v=0 is therefore confirmed with the exact solution of (2.5).

The next two pages show the results of a numerical integration of Eqs.(2.1). Fig.2 shows the influence of A_0 which is equal to 0.5,0 and -0.5 starting from the top figure, respectively. The other parameters are v=0.01, $d=d_{\shortparallel}=10$. Fig.3 shows the influence of $d_{\shortparallel}$ which is equal to 5,1 and 0.5 starting from the top figure, respectively. The other parameters are $A_0=0$, v=0.01 and d=10.

To assess the influence of noise on a delayed bifurcation, Broggi et al.(1986) have analyzed Eq.(2.14), which is the good cavity laser equation, with an additive gaussian white noise. The corresponding Fokker-Planck equation is

$$\frac{\partial}{\partial t}P(x,t)=\frac{\partial}{\partial x}\{x(1-\frac{A(t)}{1+x^2})+q\frac{\partial}{\partial x}\}P(x,t) \tag{2.20}$$

This equation has been solved numerically for $A(t)=A_0+\varepsilon t$ with q=0.001. The following observations have been made:

1.The addition of noise completely washes out the dependence of the delay on A_0. This is not surprizing since noise randomly

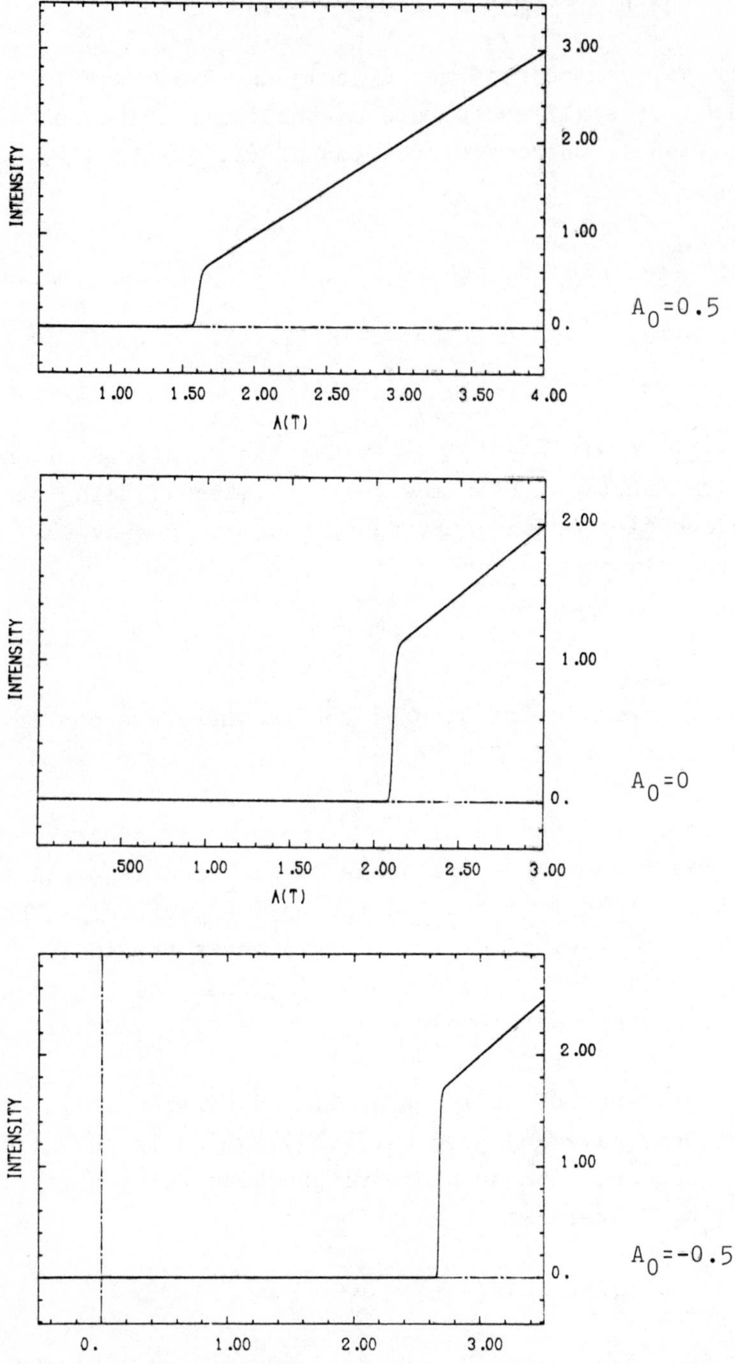

Figure 2. Numerical integration of Eqs.(2.1) with $A(t)=A_0+10^{-2}t$ and $d=d_{\parallel}=10$. Initial conditions: $E(0)=0.01$, $P(0)=0$, $F(0)=1$.

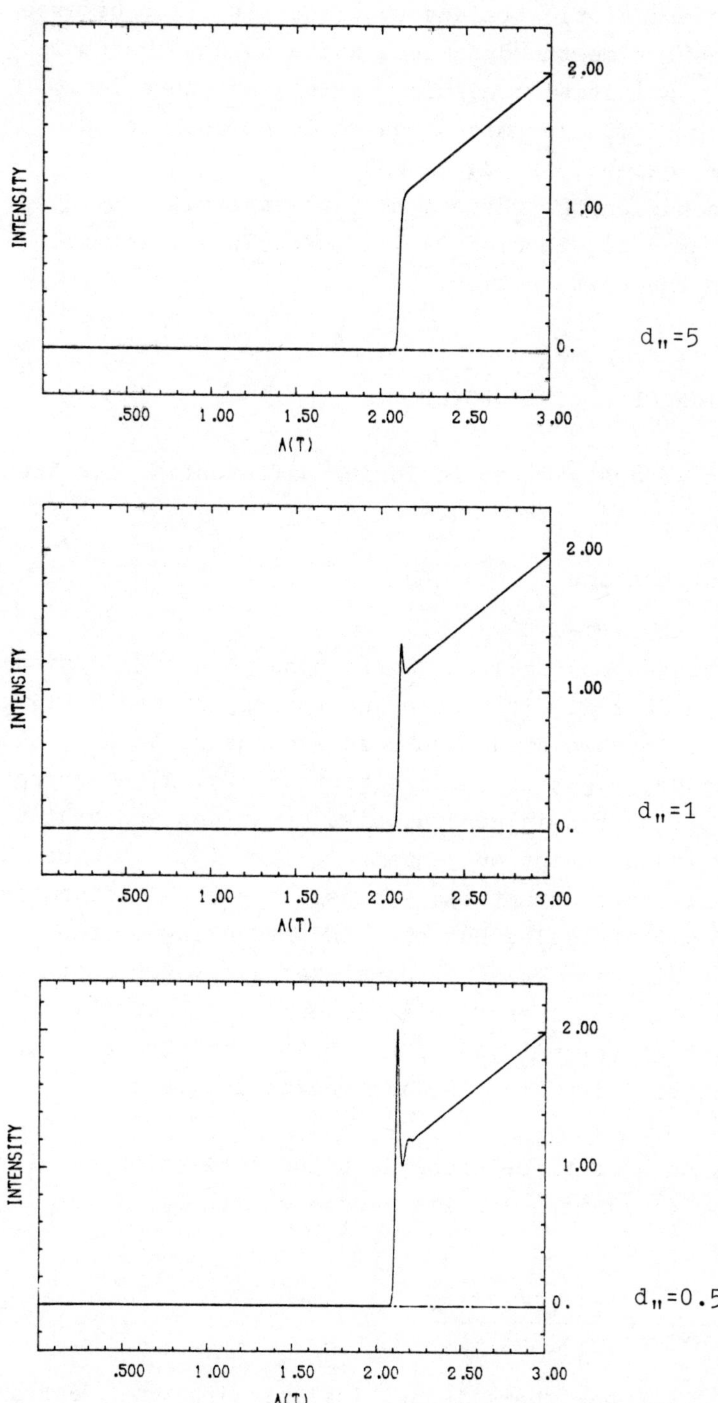

Figure 3. Numerical integration of Eqs.(2.1) with $A(t)=10^{-2}t$ and d=10. Same initial conditions as in Fig.2 .

perturbs the deterministic trajectory. Hence it is to be expected that after a short time the dynamical state of the system becomes independent of its initial condition as long as there is a single attractor. This also means that colored noise should restore, to some extend, the dependence on A_0.
2.The delay depends on the ratio q/ε. Qualitatively, the delay remains $O(1)$ if $\varepsilon>q$ but becomes $O(\varepsilon)$ if $\varepsilon<q$. This point will be commented on in the next section.

3. PERTURBATION OF A LIMIT POINT

In this section we analyze the following differential equation

$$x_t=x^2-2x+\mu \tag{3.1}$$

Its steady solutions are

$$x_1=1-(1-\mu)^{1/2}, \quad x_2=1+(1-\mu)^{1/2} \tag{3.2}$$

They are represented on Fig.1 and correspond to a limit point. The lower branch ($0\leqq x_1\leqq 1$) is stable and the upper branch ($1\leqq x_2 \leqq 2$) is unstable. The two steady solutions coincide at $\mu=1$. Such limit points occur, e.g., in optical bistability. Although optical bistability requires in general five equations and boundary conditions for its description, equations like (3.1) can be derived by an adequate asymptotic expansion (such as adiabatic elimination of all variables but one) and rescaling of the limit point coordinates. This was done explicitly for instance in Erneux et al.(1983). Therefore it is expected that (3.1) will display generic properties of a limit point when a local analysis in the vicinity of this limit point is justified.

We note that when μ is time-independent and smaller than unity the solution of (3.1) with initial condition $x(t=0)=x_0$ is

$$x(t)=\frac{x_1(x_0-x_2)e^{\lambda t}-x_2(x_0-x_1)e^{-\lambda t}}{(x_0-x_2)e^{\lambda t}-(x_0-x_1)e^{-\lambda t}} \tag{3.3}$$

where $\lambda=(1-\mu)^{1/2}$. Hence there is critical slowing down near $\mu=1$ as expected from the general argument developed in the introduction.

3.1 Linear sweep across the limit point.

Let us consider Eq.(3.1) with

$$\mu(t)=\mu_0+vt, \quad \mu_0<1, \quad v>0 \tag{3.4}$$

The system is initially on the stable lower branch $x_0=1-(1-\mu_0)^{1/2}$ and the control parameter μ is increased linearly in time. The sweep rate v is yet arbitrary. It turns out that this problem can be integrated exactly in terms of the Airy functions (Abramowitz et al.,1965):

$$x(t)=1+v^{1/3}\{Ai'(\xi)+\gamma Bi'(\xi)\}/\{Ai(\xi)+\gamma Bi(\xi)\} \tag{3.5}$$

where the prime means the derivative of the Airy function with respect to its argument

$$\xi=v^{-2/3}(1-\mu_0-vt) \tag{3.6}$$

whereas γ is related to the initial condition

$$\gamma=-\{(x_0-1)Ai(\zeta)-v^{1/3}Ai'(\zeta)\}/\{(x_0-1)Bi(\zeta)-v^{1/3}Bi'(\zeta)\}, \tag{3.7}$$
$$\zeta=\xi(t=0)$$

These expressions significantly simplify if we consider the limiting case of a very small sweep rate. Using the asymptotic expansions of the Airy functions for $v<<1$ but $1-\mu_0=O(1)$ we find

$$x(t)=1+v^{1/3}Ai'(\xi)/Ai(\xi) \quad +O(v^{4/3}) \tag{3.8}$$

With this approximate solution we can follow the evolution of x(t) using the properties of the Airy function Ai. When $\mu(t)<<1$, ξ is positive and we find from (3.8)

$$x(t)\cong 1-v^{1/3}\xi^{1/2}+O(v^{4/3})=1-\{1-\mu(t)\}^{1/2}+O(v^{4/3}) \tag{3.9}$$

Hence the system initially follows adiabatically the stable steady state. The expansion (3.9) diverges when $\mu(t)\cong 1$ because critical slowing down will induce a deviation from the adiabatic solution. Beyond $\mu(t)=1$ there is no longer a stable attractor and the solution diverges in a finite time. In optical bistability this divergence corresponds to the initial phase of the jump transition from the lower to the upper branch. This later part of the dynamical evolution can be characterized by two values of $\mu(t)$:

(i)When $Ai'(\xi)=0$

$$x(t)=1 \text{ and } \quad \mu_1=1+v^{2/3}1.01879 \tag{3.10}$$

(ii)When $Ai(\xi)=0$

$$x(t)=\infty \text{ and } \mu_2=1+v^{2/3}2.3381 \tag{3.11}$$

This evolution is shown schematically on Fig.4.

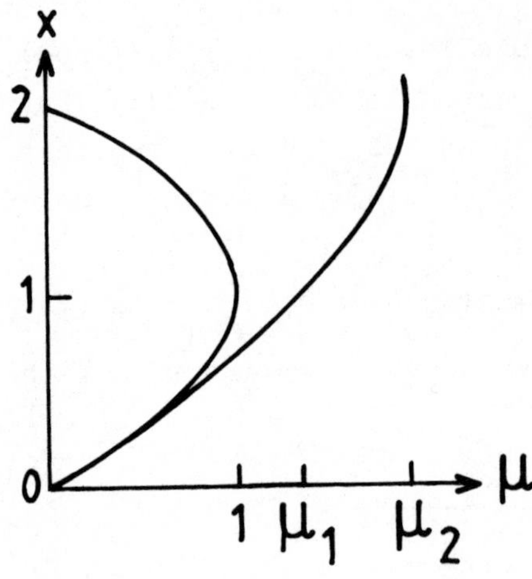

Figure 4. Steady and time-dependent solutions of (3.1) and (3.4).

Contrary to the situation analyzed in section 2 where the sweep along a constant solution produced an O(1) delay, we have here a vanishing delay in the limit of vanishing sweep rate

$$\lim_{v\to 0} D(v)=(\mu_1-1)/(1-\mu_0)=O(v^{2/3}) \tag{3.12}$$

We can now draw some conclusions on the sweeping process and its associated delay in the limit of vanishing sweep rate. We have studied analytically two models. In the first model (section 2) the steady solution along which the sweep began was a constant and the delay was O(1). In this section a sweep along a μ-dependent solution produced a vanishing delay. These are the two extreme cases of a general problem which can be described as follows. Let us consider a family of stable steady solutions of (1.1) $X=X(\mu,\delta)$ where μ is the control parameter. When $\delta=0$, $X(\mu,0)$ has a branch which is a constant solution bifurcating at $\mu=\mu_c$. Without loss of generality this constant is set equal to zero. The bifurcation can be either supercritical or subcritical. The family of solution is parametrized by the magnitude of the so-called imperfection δ (Matkowski et al.,1977). It is a measure of the deviation of X from zero at μ_c. Figs.5a and 5b display two such imperfect bifurcations for supercritical and

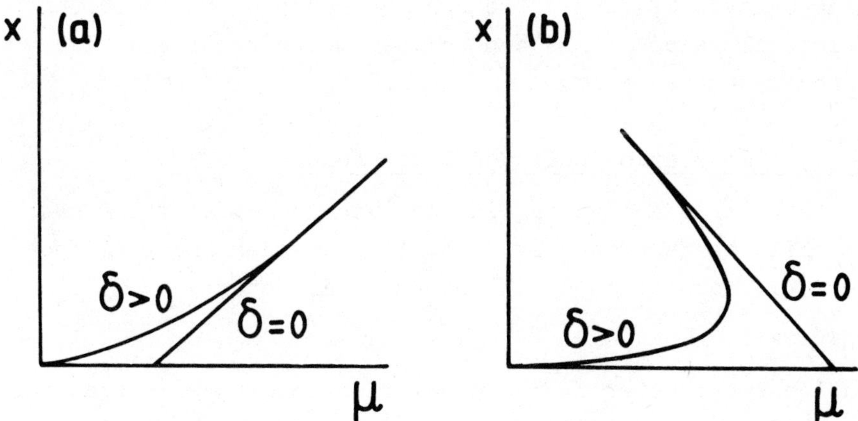

Figure 5. Supercritical (a) and subcritical (b) imperfect bifurcations.

subcritical bifurcation diagrams, respectively. The limit point of Fig.1 which has been studied in this section appears as an extreme case of imperfect subcritical bifurcation with $\delta=O(1)$. In the limit $0<v<<1$ the delay $D(v)$ results from an asymptotic expansion of $X(t,v,\delta)$. The presence of a second parameter, δ, may therefore modify this expansion. For quadratic nonlinearities as in (3.1) and for cubic nonlinearities, Erneux et al. (1986) have shown analytically that $D=O(1)$ when $0<\delta<v<<1$; conversely $D=O(v)$ when $0<v<<1$ but $\delta>>v$. The delicate point is to determine the transition region, in parameter space, where the delay is no longer $O(1)$. This domain depends critically on the nonlinearity. For a quadratic nonlinearity, it is given by $\delta=O(v)$ whereas for a cubic nonlinearity it is given by $\delta=O(v^{3/4})$. Hence the delay D continually decreases, at constant small sweep rate, when δ increases.

Although all these results have been derived from a purely deterministic theory, it is interesting to notice that they still provide a relevant guide to estimate the delay in the presence of small fluctuations. Indeed the mean intensity as determined by the Fokker-Planck Eq.(2.20) behaves somewhat like the imperfect supercritical bifurcation of Fig.5a. In this case q plays the role of the imperfection and the relation $q=O(v)$ does give the transition between the domains of large and vanishing delays. Of course the disappearance of the dependence on the initial value of the control parameter remains a clear signature of stochasticity.

3.2 Periodic modulation near the limit point.

In this section we analyze again the response of a system near a limit point as described by Eq.(3.1) but under the influence of a periodic modulation of the control parameter:

$$\mu(t)=\mu_0+\varepsilon\cos\omega t \tag{3.13}$$

The smallness parameter is the modulation amplitude whereas the modulation frequency is O(1). If $\mu_0<<1$ Eqs.(3.1) and (3.13) have a convergent outer expansion in powers of ε if the initial condition is on the stable lower steady branch:

$$x(t,\mu,\varepsilon)=1-(1-\mu_0)^{1/2}+\sum_{n=1}^{\infty}\varepsilon^n x_n(t,\mu_0)$$

A more interesting situation is when $\mu_0<1$ but $\mu_0+\varepsilon>1$. Then the control parameter periodically goes into the domain where there is no attractor. We analyze this case with a multiple time-scale method as follows. We express the fact that μ_0 lies near the limit point by expanding μ_0 in a power series in ε

$$\mu_0=1-\gamma_1\varepsilon-\gamma_2\varepsilon^2+O(\varepsilon^3)<1 \tag{3.14}$$

We seek a solution of (3.1) also in a power series in ε

$$x(t,\varepsilon)=1+\varepsilon\alpha_1(t,\tau)+\varepsilon^2\alpha_2(t,\tau)+O(\varepsilon^3) \tag{3.15}$$

where we have introduced an auxiliary slow time $\tau=\varepsilon t$. The two times τ and t are treated as independent variables:

$$dx/dt=\partial x/\partial t+\varepsilon\partial x/\partial\tau$$

Inserting these series into (3.1) and (3.13) leads to a sequence of linear differential equations. To first order in ε we have

$$\alpha_{1t}=-\gamma_1+\cos\omega t$$

whose solution is

$$\alpha_1(t,\tau)=a(\tau)-\gamma_1 t+\frac{1}{\omega}\sin\omega t \tag{3.16}$$

Therefore we must impose $\gamma_1=0$ in order to ensure the that α_1

remains bounded. To second order in ε we have

$$\alpha_{2t}=-\alpha_{1\tau}+\alpha_1^2-\gamma_2$$

whose solution is

$$\alpha_2(t,\tau)=b(\tau)+(-\gamma_2-a_\tau+a^2+\frac{1}{2\omega^2})t+\frac{1}{4\omega^3}\sin 2\omega t-\frac{2a}{\omega^2}\cos\omega t \qquad (3.17)$$

Requiring that α_2 be a bounded function yields the differential equation which $a(\tau)$ must verify

$$a_\tau=a^2+\frac{1}{2\omega^2}-\gamma_2 \qquad (3.18)$$

With this equation $\alpha_1(t,\tau)$ is fully determined.

The steady solutions of (3.18) are

$$a_\pm=\pm(\gamma_2-\gamma_2^*)^{1/2},\quad \gamma_2^*=1/2\omega^2 \qquad (3.19)$$

A linear stability analysis of these solutions indicates that a_+ is unstable and a_- is stable. Hence in the long time limit defined by $2a_+\varepsilon t>>1$ the solution of (3.1) and (3.13) is

$$x(t)=1-\varepsilon(\gamma_2-\gamma_2^*)^{1/2}+\frac{\varepsilon}{\omega}\sin\omega t+O(\varepsilon^2) \qquad (3.20.a)$$

corresponding to

$$\mu(t)=1-\gamma_2\varepsilon^2+\varepsilon\cos\omega t+O(\varepsilon^3) \qquad (3.20.b)$$

The stability boundary is reached when $\gamma_2=\gamma_2^*$. Then

$$\mu_0^*=1-\frac{1}{2}(\frac{\varepsilon}{\omega})^2+O(\varepsilon^3) \qquad (3.21)$$

On the boundary

$$x^*(t)=1+\frac{\varepsilon}{\omega}\sin\omega t+O(\varepsilon^2) \qquad (3.22)$$

which also expresses a balance between accumulated stability and accumulated instability. Indeed the expression (3.22) implies that $x^*(t)$ spends exactly half a period in the domain $x<1$ and half a period in the domain $x>1$. When $\gamma_2>\gamma_2^*$, $x(t)$ as given by (3.20.a) spends more time per period in the domain $x<1$ than in the domain $x>1$. Finally when $\gamma_2<\gamma_2^*$ the function $a(\tau)$ diverges and so does $x(t)$. In the case of optical bistability this will correspond to the initial stage of a jump transition to the upper branch. The instability condition can be interpreted in terms of a critical frequency:

$$\gamma_2<\gamma_2^* \rightarrow \omega<\omega_c=1/(2\gamma_2)^2 \qquad (3.23)$$

This justifies the argument presented in the introduction. As $\gamma_2=(1-\mu_0)/\varepsilon^2$ the critical frequency depends on the initial value of the control parameter and on the modulation amplitude. It is physically relevant to notice that on the stability boundary μ_0^* and $x^*(t)$ depend on the modulation amplitude only via the ratio ε/ω. Although this property holds only to dominant order in ε as shown below, it provides a useful approximate scaling law of the stability boundary. In optical bistability, e.g., x and μ are the output and input intensities, respectively. We can then use (3.21) to express the maximum allowed modulation amplitude, A*, as a function of the modulation frequency in the form

$$A^*=\omega\{2(1-I_0/I_\uparrow)\}^{1/2} \qquad (3.24)$$

where I_0 is the holding intensity and $I_\uparrow$ is the input intensity coordinate of the limit point, both at zero modulation amplitude.

Under the influence of the modulation, the dynamical limit point is shifted with respect to its static value $(x^*,\mu^*)=(1,1)$. In terms of time-averages over a period we have

$$(\langle x^*\rangle,\langle\mu^*\rangle)=(1,1-\frac{\varepsilon^2}{2\omega^2})$$

The symmetry of Fig.1 is not broken in the sense that $\langle x^*\rangle$ remains equal to its static value 1, but the abcissa of the dynamical point is reduced by an amount which, again, depends only on the ratio of the modulation amplitude to the modulation frequency.

We can relate the stability boundary condition (3.21) to known properties of the Mathieu equation. The transformation

$$x(t)=1-d(\ln f)/dt$$

applied to (3.1) and (3.13) leads to the canonical form of the Mathieu function

$$f_{zz}+(a+2q\cos 2z)f=0 \qquad (3.25)$$

with the definitions

$z=\omega t/2, \quad a=-4(1-\mu_0)/\omega^2, \quad q=2\varepsilon/\omega^2$

For an initial condition x(0) on the lower branch, $a(0)=-\sqrt{\gamma_2}$. The solution of (3.18) is therefore

$$a(\tau)=\Delta\tanh(-\tau\Delta+\text{arctanh}\frac{a(0)}{\Delta}), \quad \Delta=(\gamma_2-\gamma_2^*)^{1/2}$$

The corresponding solution of the Mathieu equation (3.25) is

$$f(z)=C_1e^{\tilde{\mu}z}\phi(z)+C_2e^{-\tilde{\mu}z}\phi(-z)$$

where $\tilde{\mu}=2\varepsilon\Delta/\omega$, $\phi(z)=\exp(\varepsilon\omega^{-2}\cos 2z)$ and $2C_1=1/2C_2=\{\frac{\Delta+a(0)}{\Delta-a(0)}\}^{1/2}$.

Hence we see that the instability condition (3.21) precisely corresponds to the stability boundary ($\tilde{\mu}=0$) of the periodic Mathieu functions. In the limit of small amplitude of modulation ($\varepsilon<<1$) the next terms in the power expansion of μ^* versus ε can be found in standard tables (Abramowitz et al.,1965):

$$\mu_0^*=1-\frac{\varepsilon^2}{2\omega^2}+\frac{7\varepsilon^4}{2^5\omega^6}-\frac{29\varepsilon^6}{3^2 2^4\omega^{10}}+\frac{68687\varepsilon^8}{3^2 2^{15}\omega^{14}}+\cdots \qquad (3.27)$$

$$=1-\omega^2F(\varepsilon^2/\omega^4)$$

From this series we can easily derive an expansion for γ_2^* and ω_c:

$$\gamma_2^*=\frac{1}{2\omega^2}\{1-\frac{7\varepsilon^2}{2^4\omega^4}+\frac{29\varepsilon^4}{2^3 3^2\omega^8}-\frac{68687\varepsilon^6}{2^{14}3^2\omega^{12}}+\ldots\}$$

$$\omega_c=\frac{1}{(2\gamma_2)^{1/2}}\{1-\frac{7}{2^3}(\varepsilon\gamma_2)^2+\frac{29}{3^2 2^2}(\varepsilon\gamma_2)^4-\frac{3\times 79}{2^8}(\varepsilon\gamma_2)^6+\ldots\}$$

4. CONCLUSION.

We have tried to show on two simple examples that the introduction of a time-dependent bifurcation parameter may drastically alter a static bifurcation diagram. The origin of this investigation is an attempt to understand seemingly irreducible differences between experimental and theoretical results. In search for an explanation it was realized that quite often a time-dependence is added to the control parameter, usually for technical reasons (see,e.g., Arimondo et al.,1985 for lasers with saturable absorbers and Kimble et al.,1984 for optical bistability). In the course of this analysis it was discovered that O(1) effects can be induced by O(ε) causes and that the subject of time-dependent perturbations in dissipative structures is of interest per se.

A general feature which has been analyzed in this contribution is that with dynamical perturbations unstable states can be stabilized (as in section 2) and dynamically stable states may appear which do not exist under static conditions (as in section 3.2). Apart from the obvious interest of such dynamically stable states, the method of swept parameter has a drawback. It is sometimes used as a convenient way to obtain at once the whole bifurcation diagram in a given range of the control parameter. Because the delay is in the direction of increasing control parameter, a pair of forward and backward sweeps over the same range of parameters will not produce the same diagrams (Bandy et al.,1985; Brun et al.,1985). For instance, in the case of the swept gain laser discussed in section 2 a backward sweep starting on the stable upper branch E^2=A-1 will follow this steady branch down to A=1 with a small deviation of O(ε) near A=1 if the sweep rate is ε<<1. Such an effect should not be called optical bistability because the two dynamical states obtained in the forward and backward sweeps are not available at the same time. However if the static bifurcation diagram displays a domain of bistability, a forward sweep followed by a backward sweep will give an artificially enhanced hysteresis.

The methods used to analyze the response to time-varying parameters are simplest for one-dimensional problems but are by no means restricted to this one-dimensional case. This is obvious for the swept-gain laser where three- and one-dimensional equations were studied. It is also true for the multiple time-scale analysis of a periodic modulation used in section 3.2 which has been applied, with similar conclusions, to dispersive optical bistability in two dimensions (Mandel et al.,1985).

Although the coincidence of two steady state solutions implies a domain of critical slowing down, the converse is not true. In the theory of an optical transistor (Mandel, 1985) the steady state equation $x^3-3x^2+3x-\mu=0$ appears. It has a single real root for $\mu>0$. The point $x=\mu=1$ is characterized by an infinite slope $dx/d\mu$ and at this point the linear stability analysis gives a zero root. Hence in the vicinity of $x=\mu=1$ there will be critical slowing down. This is especially important since it is precisely this vicinity which is exploited when the optical transistor is used to amplify a small amplitude periodic modulation of the input beam.

In the swept gain laser and in the periodic modulation near the limit point it has been shown that the instability condition could be described with a balance condition. It expresses the exact compensation between the accumulated stability during the first part of the sweep or of the cycle and the accumulated instability during the second part of the sweep or of the cycle. On the basis of these two results and other yet unpublished results it seems that the notion of stability and instability accumulated during the variation of the control parameter provides a physical key to the determination of dynamical stability criteria. The important fact which we have stressed in this contribution is that there is no reason why a dynamical bifurcation should appear as a small deviation from the corresponding static bifurcation. In some cases it has even been proved that the bifurcation point from the trivial solution can be shifted to $+\infty$ under the influence of a sweep: in such a case the jump to the finite solution never occurs (Mandel et al., 1984; Erneux et al.,1984).

ACKNOWLEDGMENTS.

I wish to express my gratitude to T. Erneux for his unflagging patience and longstanding friendship. This work was supported in part by the FNRS (Belgium), the Stimulation Action program of the European Community and the Association Euratom-Etat Belge.

REFERENCES.

Abramowitz, M. and Stegun, I. 1965, Handbook of mathematical functions, Dover Publications, New York

Arimondo, E., Gabbanini, C., Menchi, E., Dangoisse, D. and Glorieux, P. 1985, Transient bistability in a laser containing a saturable absorber, preprint

Bandy, D. K., Narducci, L. M. and Lugiato, L.A. 1985, J.Opt. Soc.Am. 2B,148

Bender, C. M. and Orszag, S.A. 1978, Advanced mathematical methods for scientists and engineers, McGraw-Hill, New York

Broggi, G., Colombo, A., Lugiato, L. A. and Mandel, P. 1986, Influence of white noise on delayed bifurcations, preprint

Brun, E., Derighetti,B., Meier, D., Holzner, R. and Ravani, M. 1985, J.Opt.Soc.Am. 2B,156

Cohen, D.S. and Matkowski, B. J. 1978, SIAM J.Appl.Math.35,307

Dangoisse, D., Glorieux, P. and Midavaine, T 1985, Observation of chaos in a frequency modulated CO_2 laser, preprint

Erneux, T. and Mandel, P. 1983, Phys.Rev.28A,896

Erneux, T. and Mandel, P. 1984, Phys.Rev.30A,1902

Erneux, T. and Mandel, P. 1986,SIAM J.Appl.Math.46,1

Glorieux, P. and Dangoisse, D.1985,Dynamical behavior of a laser containing a saturable absorber, preprint

Haberman, R. 1979,SIAM J.Appl.Math.37,69

Haken, H. 1985, Light, vol.2, North-Holland,Amsterdam

Iooss, G. and Joseph, D.D. 1980, Elementary stability and bifurcation theory, Springer Verlag, Heidelberg

Kapila, A.K. 1981, SIAM J.Appl.Math.41,29

Kapral, R. and Mandel, P. 1985,Phys.Rev.32A,1076

Kimble, H. J., Rosenberger, A. T. and Drummond, P. D. 1984, in Optical bistability 2 edited by Bowden, C. M., Gibbs, H. M. and McCall, S. L., Plenum Press, New York

Kondepudi, D. K. and Nelson, G. W. 1983, Phys.Rev.Lett.50,1023

Kondepudi, D. K. and Nelson, G. W. 1985, Nature 314,438

Kondepudi, D. K., Moss, F. and McClintock, P. V. E. 1986, Observation of symmetry breaking, state selection and sensitivity in a noisy electronic system, preprint

Lugiato, L. A., Mandel, P. and Narducci, L.M. 1984,Phys.Rev. 29A,1438

Mandel, P. 1985,Dynamical theory of an optical transistor, preprint

Mandel, P. and Erneux, T. 1984, Phys.Rev.Lett.53,1818

Mandel, P. and Erneux, T. 1985, Nonlinear control in optical bistability, preprint

Matalon, M. and Erneux, T. 1984, SIAM J.Appl.Math.44,734

Matkowski, B. J. and Reiss, E. 1977, SIAM J.Appl.Math.33,230

RYDBERG ATOMS AND QUANTUM OPTICS

P L KNIGHT

1 INTRODUCTION

Rydberg atoms with principal quantum numbers n > 30 have extremely large dipole moments but at the same time have long spontaneous emission lifetimes. Their large dipole moments make them exceedingly sensitive to weak radiation fields. We review the properties of these Rydberg atom systems and discuss their use as a testbed for fundamental problems in quantum optics. The detection of weak radiation fields is discussed. We then examine the spontaneous decay properties of highly excited atoms in resonant cavities and show how the Einstein A-coefficients, Lamb shifts and other radiative corrections are modified by the confining cavity, and by collective multi-atom interactions. Finally we discuss the problem of the strongly-driven Rydberg atom which executes an oscillatory Rabi time-evolution which is damped by environmental, dissipative influences at the same time as responding to the detailed photon statistics of the cavity field.

2 RYDBERG ATOMIC STATES

Rydberg atomic states are those in which the valence electron is promoted to an orbit with a very large principal quantum number. In this review, the principal quantum numbers we have in mind are typically n = 30 to 40. In such orbits, the electron spends most of its time a large distance from the ionic core. The energies of these states are hydrogenic and given by the Bohr formula (Bohr, 1913)

$$E_n = -R/n^{*2} \qquad (1)$$

where R is the Rydberg constant (= 13.6 eV or 3.289 828 x 10^9 MHz) and the effective quantum number $n^* = n - \delta_\ell$ where n is the principal quantum number and δ_ℓ the quantum defect. The quantum defect depends on the electron-core interaction and is sensitive to the penetration of the core and depends therefore on the orbital angular momentum quantum number ℓ. The atomic physics of Rydberg states is well-reviewed in Stebbings and Dunning (1983), Latimer (1979) and Kleppner, Littmann and Zimmerman (1981). The correspondence principle suggests that for extremely large n, the transition frequency $E(n \rightarrow n-1)/\hbar$ approaches the Bohr orbit frequency itself.

The radius of such a Rydberg atom is approximately $r_n \sim n^{*2} a_o$ where a_o is the Bohr radius, and for $n \sim 100$ is approximately the radius of a virus. Such huge, fragile atoms can be sorted by size by their transmission (or blocking) through an atomic "sieve" made up of μm-size slots (Fabre et al 1983). Although these states are fragile (they are so close in energy to the photoionization threshold) they can have long spontaneous decay lifetimes because their wave functions overlaps so little with the lower-lying final states. This overlap is small even for low angular-momentum states, where the spontaneous decay lifetime $\tau \sim n^{*3}$ but is even smaller for large ℓ Rydberg states where $\tau \sim n^{*5}$. Hulet and Kleppner (1983) have produced 'circular' Rydberg atoms by an ingenious series of adiabatic rapid passage transitions.

In the laboratory, principal quantum numbers n ~100 have been produced, but the champions in the search for really highly excited atoms are radio-astronomers. The emission lines from transitions n = 301 to n = 300, in the radio frequency region have been observed from hydrogen atoms in interstellar space and in gaseous nebulae. For example, the n = 95 to n = 94 transition in atomic hydrogen is called H94α. The H 94α and H 110α lines have been observed by astronomers at Kitt Peak National Observatory (Chaisson, 1976) and their Doppler shifts used to find the distribution of radial velocities in, for example, the Triffid Nebula. More recently, astronomers have seen Rydberg transitions from states with principal quantum numbers n ~733 (Dalgarno 1985).

The spontaneous decay rate A for a transition from state n to n' is (Haroche 1984)

$$A_{n,n'} = 16\pi^2\nu^3 |\langle n|r|n'\rangle|^2 / 3\epsilon_0 hc^3 \tag{2}$$

where the dipole matrix element $\langle n|r|n'\rangle$ is very small if n' << n (small overlap). For n $\sim$ 30, the spontaneous decay lifetime $\tau = A^{-1}$ is about 30 μs for decay back to low-lying states. If, on the other hand, n' $\sim$ n, then the dipole matrix element is large (and proportional to n^2), the transition frequencies $E_n - E_{n'} = h\nu$ scale as n^{-3} and have wavelength between 20 to 50μm for n between 30 and 50, then the lifetime for transitions to nearby states (n = 30 to n = 29, say) is 10^{-2} s, $A_{n,n-1} \sim 10^2 s^{-1}$. This A may seem small (it is n^{-5} times a conventional optical Einstein A coefficient for the spontaneous decay) but is huge in comparison with normal spontaneous emission rates in the mm-wave region where for example the ammonia maser spontaneous rate A is approximately 10^{-6} s^{-1}. This enhancement of 10^{8} is entirely due to the large size of Rydberg atoms, and offers the opportunity (discussed later) of observing mm-wave spontaneous decay in the laboratory.

Stimulated emission rates are similarly enhanced by the large dipole matrix elements for transitions to nearby states. A Rydberg atom is therefore an extremely good detector of very weak mm-wave fields. Experiments by Haroche and his group (Haroche 1984) have demonstrated that one-photon ($n \rightarrow n \pm 1$) and two-photon ($n \rightarrow n \pm 2$) transitions can be observed in Rydberg atoms with picowatt radiation fields. The spontaneous emission lifetimes are so long for these mm-wave transitions that for $n, \ell \sim 30$, a <u>single photon</u> field saturates the transition and stimulates transitions at a rate equal to the spontaneous rate.

The strong coupling of Rydberg atoms to radiation fields makes them ideal testbeds for fundamental experiments in atom-field interactions, (Haroche 1984, Rempe and Walther 1985). An additional motivation for such studies is derived from the field of quantum chaos (eg, Schuster 1984; Zaslav sky 1985). Classical calculations of highly excited bound electrons driven by intense radiation fields indicate that the motion becomes chaotic (eg Delone et al 1983). What happens in a quantum system is therefore of great interest, and a "diffusive" photon absorption has been suggested as a signature for the onset of quantum stochasticity. Bayfield and Pinnaduwage (1985) have produced an atomic beam of highly excited hydrogen with principal quantum numbers $n \sim 60$, and induced up to 16-photon microwave multiphoton transitions between Rydberg states. They report evidence for a diffusive absorption.

3 RYDBERG ATOMS AND BLACK-BODY RADIATION FIELDS

A typical experiment to produce Rydberg atoms employs either single-step UV laser or two-step visible laser excitation from a low-lying state. Population is produced in a high-lying state $|n_1 \ell_2\rangle$. These high-lying states are easily detected by field ionization in a modest dc electric field (Fig 1). The atoms are excited at t = o in

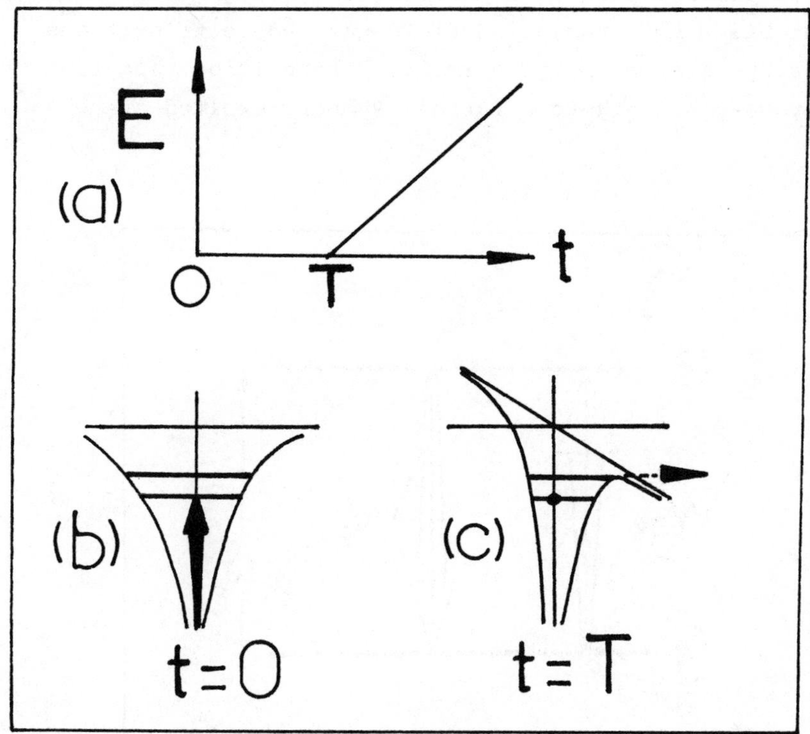

Fig 1 : Field ionization detection of Rydberg atoms by a ramped dc electric field (a). In (b) the Rydberg states are populated by a laser in field free region and (c) selectively ionized at time T.

a field free region. At time T a ramp dc electric field is applied. Each Rydberg state, with principal quantum number n, can ionize when the applied electric field equals the electric field which the electron feels from the nucleus plus core, ie when

$$E \sim E_c = \frac{e}{4\pi\epsilon_0 (n^2 a_0)^2} \tag{3}$$

The field ionization rate is proportional to n^{-4} and so selective detection of one Rydberg state compared with one a little lower in energy is possible.

When the laser populates a particular Rydberg state, the ambient black-body radiation field can induce transitions to nearby states. The stimulated transition rate from this Rydberg state g to a further Rydberg excited state e (Fig 2)

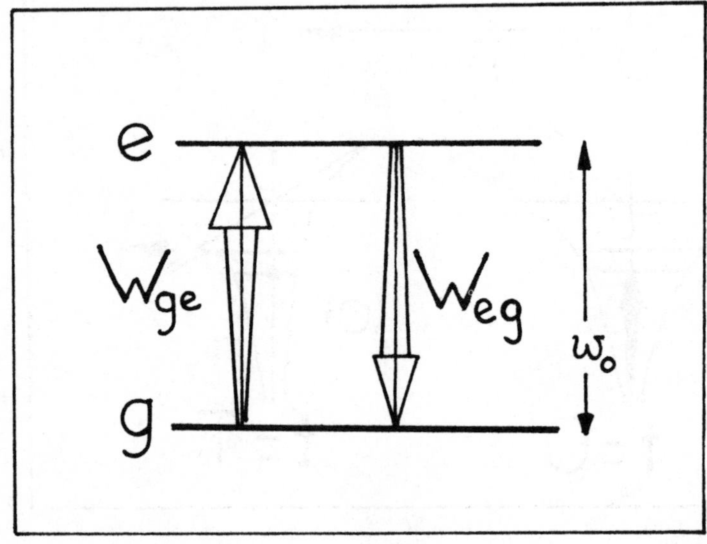

Fig 2 : Stimulated transitions between two Rydberg atomic states.

is given by

$$W_{ge} = B\,\bar{n} \tag{4}$$

where $B = d^2 w_o^{\,3}/3\pi\epsilon_o \hbar c^3$ is the Einstein B-coefficient and $\bar{n}$ is the number of resonant ($w = w_o$) black-body photons interacting with the Rydberg atoms. For a Rydberg mm-wave transition $\hbar w_o \ll kT$ even at room temperatures, so we can use the Rayleigh-Jeans limit to the Planck law for $\bar{n}$,

$$\bar{n} \sim kT / \hbar w_o \gg 1 \tag{5}$$

so

$$W_{ge} = \frac{d^2 \omega_o^2 kT}{3\pi\epsilon_o \hbar^2 c^3} \tag{6}$$

At 300^oK, $kT/\hbar \sim 6 \times 10^{12}$ Hz. The black-body radiation induced transition rate will exceed the spontaneous emission rate between the same states at room temperatures for a principal quantum number $n \sim 25$ ($d^2 \sim n^4$, $w_o^2 \sim n^{-6}$ so $W_{eg} \sim n^{-2}$, whereas $A \sim n^{-3}$).

So although we initially populate only a certain n, black-body radiation will induce transitions into neighbouring levels and reduce the effective lifetime of Rydberg states (Gallagher and Cooke 1979; Beiting et al 1979; Figger et al, 1980; Koch et al, 1980). The flow of population into more highly excited states is easily seen using the ramped field ionization detection scheme (eg Gallas et al 1985). Different states ionize at different parts of the ramped electric field, with more highly excited states ionizing at lower electric fields. If the ramp (Fig 1) is applied at a short time T after the laser excitation, only the initial laser-excited states are seen in the field ionization signal. But if T is increased, the black-body radiation field induces transitions to higher-lying states which are then seen in the ion signal. At delays T of 12μs, a whole series of Rydberg states are populated above the initially-pumped state.

The strong coupling of Rydberg atoms to weak mm-wave black-body radiation has been employed in ultra-high sensitivity thermal radiation detectors (Ducas et al 1979, Figger et al 1980, Moi et al 1983). Figger et al (1980) excited a sodium atomic beam with two tunable dye lasers in a cold box (14°K) to the $22^{2}D$ state. A field ionization detector was employed which produced a field which does not ionize the $22^{2}D$ state (but would ionize higher states). In the cold box containing the Rydberg atomic beam a small flap could be opened to admit radiation from an infrared source. This thermal radiation induces upwards transitions to higher levels and produces a field ionization signal. Figger et al (1980) deduced a noise equivalent power (NEP) of 10^{-17} W/$(Hz)^{1/2}$ with an output bandwidth of 1Hz and a detector quantum efficiency of ~3×10^{-3}. Moi et al (1983) used a liquid helium cryostat to reduce background noise and a stabilized laser excitation to produce an N E P of 10^{-19} W/$(Hz)^{1/2}$.

4 SHIFTS OF ATOMIC LEVELS BY BLACK-BODY RADIATION

Atomic energy levels can be shifted by radiation fields. The archetype radiative shift is of course the Lamb shift, but external fields can also "dress" atomic states (Barrat and Cohen-Tannoudji 1961, Pancharatnam 1966, Series 1970). Induced level shifts variously called ac Stark shifts, light shifts and "lamp" shifts have been observed in many atom-laser experiments and discussed using a variety of semiclassical and quantum methods (eg Knight and Milonni 1980). Rydberg states have, as we have seen, very large dipole moments and can have their energies shifted by large amounts by intense laser fields. Liberman et al (1983) observed shifts of order 5 x 10^8 Hz in transitions induced by a tunable UV laser from the $5^2S_{1/2}$ ground state of rubidium to the $22^2P_{3/2}$ Rydberg state when the transition states were dressed by an intense non-resonant Nd : YAG laser operating at 1.06 μm and at 5 to 10 MW/cm^2. Such large shifts in intense fields are relatively straightforward to observe. Harder to observe are the changes in atomic energy levels caused by background black-body radiation, but again Rydberg atoms provide the sensitivity to see such small effects (Hollberg and Hall, 1984). Thermal corrections to energy levels were first correctly calculated by Knight (1972) and Barton (1972) and have recently been reviewed by Barton (1985). The energy level shift of a state 0 can be calculated using second order perturbation theory to be (Barton 1985)

$$\Delta(T) = \frac{-4\alpha^3}{3\pi\beta^2}\sum_i E_{io}|\underline{r}_{io}|^2\int_0^\infty dy \cdot \frac{y^3}{(e^y-1)} \cdot \frac{1}{(\beta^2 E_{io}^2 - y^2)} \quad (7)$$

where $\beta = 1/kT$ and we use units so that $\hbar = m_e = 1$. The dipole moment matrix element $r_{io} = \langle i|r|0\rangle$ and the transition energies $E_{io} = E_i - E_o$. At high temperatures $(\beta E_{io})^2 \ll 1$. Note that the apellations "high" or "low" depend upon E_{io}, and 300^oK would be "low" for ground states but "high" for Rydberg states.

At high temperatures $(\beta E_{io})^2$ can be neglected in the denominator of eq (7), and the integral performed to give $(\pi^2/6)$, independent of i (or O). The atomic state summation can be performed using the TRK sum rule

$$\sum_i E_{io} |\underline{r}_{io}|^2 = 3/2 \tag{8}$$

to give a result independent of O, so all Rydberg states are shifted equally. For these frequencies the Rydberg atom interacts as if the electron were unbound. The final high-temperature result is

$$\Delta(T) = \frac{\alpha^3 \pi}{3} (kT)^2 = 2{\cdot}42 \times 10^3 (T/300)^2 \text{ Hz} \quad (\beta E_{io} \ll 1) \tag{9}$$

At low temperatures, or for low-lying atomic states, we have $(\beta E_{io})^2 >> 1$ so may neglect y^2 in the denominator of eq (7). Again the integral may be performed to give $(\pi^4/15)(\beta E_{io})^{-2}$, so that

$$\Delta(\text{low} T) = -\frac{4\pi^3}{45} \alpha^3 (kT)^4 \sum_i |\underline{r}_{io}|^2 / E_{io} \tag{10}$$

or in terms of the static polarisability $\Pi(o)$ and the mean square electric field of the black-body radiation,

$$\Delta(\text{low} T) = -\tfrac{1}{2} \Pi(0) \langle \mathcal{E}^2 \rangle_T \tag{11}$$

For a 1s hydrogenic ground state $\Pi(o) = 9/2Z^4$ (Barton 1985) and

$$\Delta_{1s}(\text{low} T) = 0.0388 (T/300)^4 Z^{-4} \text{ Hz} \tag{12}$$

For intermediate temperatures for hydrogenic states, Farley and Wing (1981) have numerically calculated $\Delta(T)$.

Such shifts are extremely small, but Hollberg and Hall (1984) observed them in an ultra-high resolution

Doppler-free two-photon Ramsey excitation of an atomic beam of sodium atoms from their ground state to 36s state. The T^4 ground state shift is entirely negligible, so the thermally-induced frequency shift is dominated by the 2.42 kHz shift at 300°K. The thermal radiation in this experiment is derived from an external source and imaged on to the atoms, with temperatures up to 10^3 °K. They saw frequency shifts in their Ramsey fringe pattern of from 200 to 1400 Hz as their source increased from 300°K to 1000°K. Their experiment indicates that atoms always have properties determined in part by their environment. Itano et al (1981) have investigated the sensitivity of the caesium atomic clock primary frequency standard to background thermal radiation-induced level shifts. They calculate as above, a T^4 energy level shift of the $^2S_{1/2}$ hyperfine fractional frequency shift of this international time standard of -1.7×10^{-14} at 300 °K which is a significant shift for these primary frequency standards. The shifts due to magnetic interactions are believed to be insignificant. Barton (1985) has calculated the temperature-dependence of the electron spin magnetic moment (see also Donoghue and Holstein 1983, Donoghue et al 1984) and finds

$$\delta\mu(T)/\mu = -(5\pi\alpha/9)(kT/m)^2 \qquad (13)$$

and has also investigated magnetic interactions at finite temperatures for atomic systems. Again Rydberg atoms offer a possible test of these important problems.

5 SPONTANEOUS DECAY OF RYDBERG STATES IN A CAVITY

So far we have discussed the radiative properties of Rydberg atoms in free space. Transitions between Rydberg states have transition wavelengths in the mm-wave range and very high Q cavities can be constructed in this wavelength range to enhance local field strengths. In a high Q cavity, the Rydberg atom couples to the resonant cavity mode very strongly and both stimulated and spontaneous rates are affected by the cavity mode structure. The modification of spontaneous emission lifetimes by a cavity Q was considered many years ago by Purcell (1946). In a very high Q cavity, the spontaneous emission ceased to be an irreversible loss ; instead the atom and field mode periodically exchange the energy of a single photon through self-induced, spontaneous Rabi oscillations. Under such conditions, the atom-single-photon coupling is larger than any other interaction and the dynamics reflect this in Rabi oscillations, the spectrum in AC Stark splitting. At more modest Q values, the spontaneously-radiated field is absorbed by the cavity before it can re-excite the atom, and such Q's enhance the decay.

We use a Schrodinger amplitude method to describe an atom coupled to a single mode cavity radiation field which in turn is coupled to a continuum of cavity wall oscillator states which act as a "bath" or damping reservoir responsible for the cavity Q .Our method is precisely that used for caculations of resonant multiphoton excitation (Knight, 1977; 1978; 1979; 1980). The level scheme is indicated in Fig 3. The Hamiltonian for the combined atom-field mode-cavity "bath" system

$$H = H_a + wa^+a - \underline{d}\cdot\underline{\hat{E}} + H_b + V \qquad (14)$$

where Ha is the atomic unperturbed Hamiltonian, H_b the bath Hamiltonian, wa^+a the Hamiltonian for the cavity mode of frequency w described in terms of annihilation

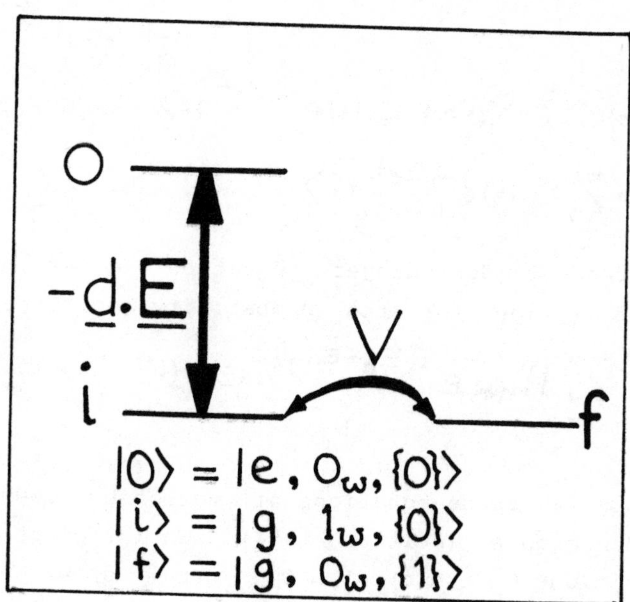

Fig 3 : Energy level scheme for excited atom spontaneously emitting a photon into a cavity mode, frequency w. This cavity mode excitation is damped by a resistive coupling to the walls V, depositing the excitation in one of the cavity "bath" states.

and creation operators a, a^+. The atom interacts with the quantized cavity mode electric field through a dipole coupling $-\underline{d}\cdot\hat{\underline{E}}$, whereas the field mode interacts with the continuum of cavity "bath" states by an interaction V. The cavity mode electric field is given by

$$\hat{\underline{E}} = i(\omega/2\pi \mathcal{V})^{1/2}\hat{\underline{\epsilon}}\,(a-a^{+}) \tag{15}$$

where $\hat{\underline{\epsilon}}$ is the polarisation unit vector and $\mathcal{V}$ is the cavity volume. We use units such that $\hbar = c = 1$.

We can expand the wavefunction for the total system in terms of the states shown in Fig 3:

$$\Psi(t) = C_o(t)e^{-iE_ot}|o\rangle + C_i(t)e^{-iE_it}|i\rangle$$
$$+ \sum_f C_f(t)e^{-iE_ft}|f\rangle. \tag{16}$$

The time-dependent Schrodinger equation gives the equations of motion for the probability amplititudes

$$i\dot{C}_n = \sum_m H_{nm}\, e^{i(E_n-E_m)t}\, C_m(t). \tag{17}$$

We choose to solve these equations of motion by Fourier transform techniques (Heitler 1954). We introduce the initial conditions $C_o(t = o) = 1$, $C_{n\neq o}(t = o) = o$ directly into the equations of motion:

$$i\dot{C}_n = \sum_m H_{nm}\, e^{i(E_n-E_m)t}\, C_m(t) + i\delta_{no}\,\delta(t) \tag{18}$$

and imagine all interactions to be abruptly turned on at $t = o$. We define Fourier transforms

$$C_n(t) = \frac{-1}{2\pi i}\int_{-\infty}^{\infty} dE.\, g_n(E)\, e^{i(E_n-E)t} \tag{19}$$

$$i\delta(t) = \frac{-1}{2\pi i}\int_{-\infty}^{\infty} dE.\, e^{i(E_o-E)t} \tag{20}$$

The Fourier probability amplitudes $g_n(E)$ satisfy linear algebraic equations

$$(E-E_n)g_n(E) = \sum_m H_{nm}\, g_m(E) + \delta_{no} \tag{21}$$

which for our problem are explicitly

$$(E-E_o)g_o(E) = H_{oi}g_i(E) + 1 \tag{22}$$

$$(E-E_i)g_i(E) = H_{io}g_o(E) + \sum_f H_{if}\, g_f(E) \tag{23}$$

$$(E-E_f)g_f(E) = H_{fi}\, g_i(E) \tag{24}$$

The matrix elements H_{oi}, H_{io} will produce spontaneous Rabi oscillations between the two discrete atom-field states i and o. The matrix elements H_{if} will produce a cavity decay rate (ω/Q) of the field mode energy from i into the infinitely many states f in which one of the cavity states is excited.

We use eq (24) to eliminate the continuum amplitude $g_f(E)$ from eqs (22), (23), to obtain

$$(E-E_i+\tfrac{i}{2}\gamma)g_i(E) = H_{io}\, g_o(E) \tag{25}$$

where the complex γ is given by

$$-\frac{i}{2}\gamma = \sum_f \frac{|H_{if}|^2}{(E-E_f)}\,. \tag{26}$$

The real part of γ will describe cavity damping, and the imaginary part a small cavity dissipation frequency shift. At this point we make the usual Weisskopf-Wigner pole approximation (Knight and Milonni 1976, Cohen-Tannoudji et al 1977) (or equivalently a Born-Markov approximation) in which we assume the cavity bath states are infinite in number and distributed with a flat spectrum representing a reservoir with little or no "memory", and set the energy E in the expression for equal to its weak-coupling pole value $E = E_i$. Then

γ is now a constant with real and imaginary parts

$$-\frac{i}{2}\gamma = \sum_f \frac{|H_{if}|^2}{(E_i - E_f)}$$
$$= \sum_f |H_{if}|^2 \left[\mathbb{P}\frac{1}{(E_i - E_f)} - i\pi\delta(E_i - E_f) \right] \tag{27}$$
$$= \Delta E_i - \frac{i}{2}\Gamma$$

The principal value is the cavity-induced shift of the transition frequency, and $\Gamma = (\omega/Q)$ is the cavity decay rate. From now on we will assume the shift ΔE_i is incorporated into E_i and forget it, so we can write Γ as γ.

For the initial state probability amplitude, we find from eq (22) and (25)

$$g_o(E) = \frac{1}{(E - E_o + \frac{i}{2}\Gamma_o(E))} \tag{28}$$

where $\Gamma_o(E)$ is the "off-shell" excitation complex self-energy of state o,

$$-\frac{i}{2}\Gamma_o(E) = \frac{|H_{oi}|^2}{(E - E_i + \frac{i}{2}\gamma)} . \tag{29}$$

If we treated $\Gamma_o(E)$ in pole approximation as the constant $\Gamma_o(E_o)$, then $g_o(E)$ has a simple pole and the Fourier inversion gives the initial state probability amplitude

$$C_o(t) = \exp(-\Gamma_o(E_o)t) . \tag{30}$$

The initial state o decays at a rate Γ_o, which has a real part γ_o, a shift ΔE_o,

$$\gamma_o = \gamma |H_{oi}|^2 / (\Delta^2 + \tfrac{1}{4}\gamma^2) \tag{31}$$

$$\Delta E_o = 2\Delta |H_{oi}|^2 / (\Delta^2 + \tfrac{1}{4}\gamma^2) \tag{32}$$

where $\Delta = E_o - E_i$ is the detuning of the atomic transition $(E_e - E_g)$ from the cavity mode frequency ω. On resonance, the dissipative shift of state O is zero, but the width $\gamma_o = 4|H_{oi}|^2/\gamma$. In terms of the resonant Rabi frequency Ω (see later) and the cavity Q, this "weak-coupling" decay rate of the initially excited state is

$$\gamma_o = \Omega^2 Q / \omega \tag{33}$$

and we see that the cavity Q enhances the decay (Purcell, 1946). If $|H_{oi}|$ is not small, this weak coupling approximation will be inadequate, and we turn next to the strong coupling problem.

If we retain the energy dependence of $\Gamma_o(E)$, we find from eqs (22 - 24, 28, 29)

$$g_o(E) = \frac{(E - E_i + \frac{i}{2}\gamma)}{(E - E_o)(E - E_i + \frac{i}{2}\gamma) - |H_{oi}|^2} \tag{34}$$

$$g_i(E) = \frac{H_{io}}{(E - E_o)(E - E_i + \frac{i}{2}\gamma) - |H_{oi}|^2} \tag{35}$$

$$g_f(E) = \frac{H_{fi} H_{io}}{(E - E_f)[(E - E_i + \frac{i}{2}\gamma)(E - E_o) - |H_{oi}|^2]} \tag{36}$$

These probability amplitudes are dominated by the roots $E(\pm)$ of the quadratic in the denominator of eq (34)

$$E(\pm)=\tfrac{1}{2}(E_i+E_o-\tfrac{i}{2}\gamma)\pm\tfrac{1}{2}\left[(E_i-E_o-\tfrac{i}{2}\gamma)^2+4|H_{oi}|^2\right]^{1/2}. \quad (37)$$

The probability amplitudes as a function of time are found by inverting the Fourier amplitudes using contour integration. We find

$$C_o(t)=(E(+)-E(-))^{-1}\Big\{(E(+)-E_i+\tfrac{i}{2}\gamma)e^{i(E_o-E(+))t} - (E(-)-E_i+\tfrac{i}{2}\gamma)e^{i(E_o-E(-))t}\Big\} \quad (38)$$

$$C_i(t)=H_{io}(E(+)-E(-))^{-1}\Big\{e^{i(E_i-E(+))t} - e^{i(E_i-E(-))t}\Big\}. \quad (39)$$

The probability of decay into the cavity bath states is

$$P = 1 - P_o(t) - P_i(t) \quad (40)$$

where the state probabilities $P_o(t)$, $P_i(t)$ from eqs (38), (39) are

$$P_o(t)=|\Omega|^{-2}\,|a\sin(\Omega t/2)-i\Omega\cos(\Omega t/2)|^2 e^{-\gamma t/2}, \quad (41)$$

$$P_i(t)=\frac{4|H_{io}|^2}{|\Omega|^2}\,|\sin(\Omega t/2)|^2 e^{-\gamma t/2}, \quad (42)$$

$$\Omega = E(+)-E(-),$$

where $a = \Delta - i\gamma/2$. In Fig 4 we plot $P_o(t)$, $P_i(t)$, P as a function of time. For this choice of strong coupling we note that the atom and field periodically exchange energy, which is slowly dissipated. If $|H_{oi}| \ll \gamma$, then of course no oscillations are completed.

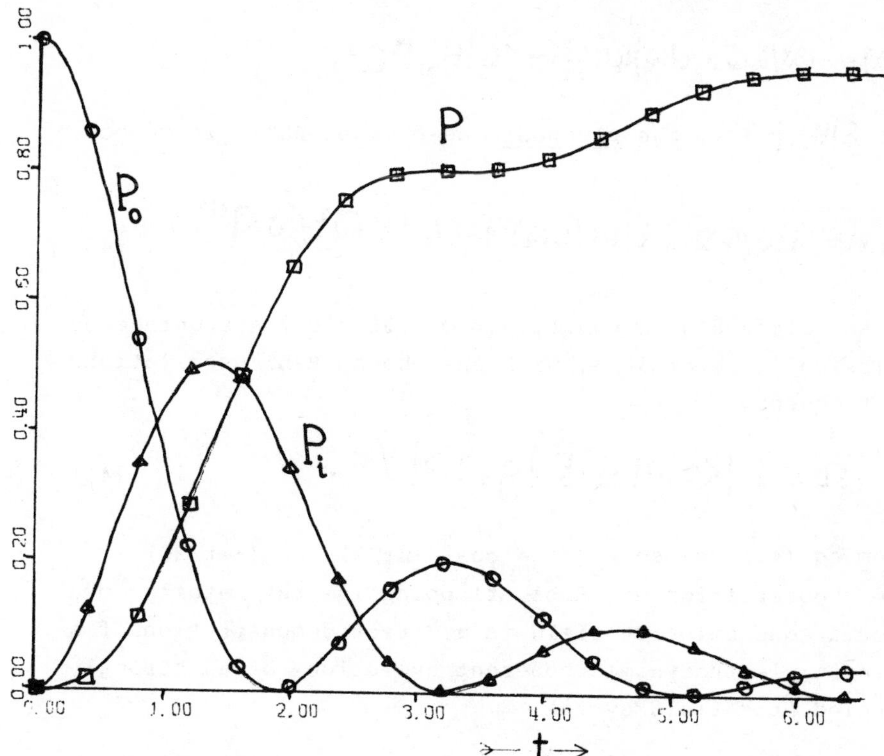

Fig. 4 : Time evolution of state populations for an atom spontaneously decaying in a very high Q cavity.

We can also calculate the spectrum of both states produced by the decay from eq (36). We invert eq (36) to find $C_f(t)$, and $|C_f(t)|^2 = P_f(t)$, the probability that a bath state f is produced:

$$P_f(t) = \frac{b|H_{oi}|^2}{|\Omega|^2}\left|\frac{\left(e^{i(E_f - E(+))t} - 1\right)}{(E_f - E(+))} - \frac{\left(e^{i(E_f - E(-))t} - 1\right)}{(E_f - E(-))}\right|^2 \tag{43}$$

where b is a collection of constants involving $|H_{if}|^2$ If the cavity mode frequency is exactly resonant with the Rydberg atomic transition frequency, $\Delta = 0$, the roots $E(\pm)$ can be written as

$$E(\pm) = -i\omega/4Q \pm i(\omega/4Q)[1 - 16|H_{oi}|^2 Q^2/\omega^2]^{1/2} \tag{44}$$

but $2|H_{oi}| = \Omega$, the <u>resonant</u> one-photon Rabi frequency, so

$$E(\pm) = -i\omega/4Q \pm i(\omega/4Q)[1 - 4\Omega^2 Q^2/\omega^2]^{1/2}. \tag{45}$$

In an ultra-high Q cavity, $(\omega/Q \ll \Omega$ and the spontaneous emission is reversible, with one-photon Rabi oscillations at frequency

$$\Omega = 2|\langle e,0|\underline{d}.\hat{\underline{E}}|g,1\rangle|/\hbar. \tag{46}$$

From eq (41) we see $P_0(t) = \cos^2(\Omega t/2)\exp(-\omega t/2Q)$. The observation of Rabi flopping in the atom's own spontaneous emission field is a direct demonstration of a quantum electrodynamic coherent evolution of a strongly coupled atom-field system.

In a more modest Q cavity, $(\omega/Q) > 2\Omega$ and the roots $E(\pm)$ become imaginary. The root E (+) is the smallest and governs the evolution of the amplitudes

$$E(+) \cong -i(\omega/4Q) + i(\omega/4Q)[1 - 2\Omega^2 Q^2/\omega^2]$$

$$= -i\Omega^2(Q/2\omega) = -i\gamma_0/2. \tag{47}$$

in agreement with eq (33). The decay rate in the cavity in terms of the cavity density of states $N_c(\omega)$ (Haroche 1984)

$$\gamma_0 = \frac{2d_{eg}^2 Q}{\hbar\epsilon_0 V} = \frac{\pi d_{eg}^2 N_c(\omega)\hbar\omega}{\epsilon_0\hbar^2} \tag{48}$$

where the density of modes/unit frequency/unit volume is

$$N_c(\omega) = \frac{2}{\pi}\left(\frac{Q}{\omega}\right)\left(\frac{1}{V}\right). \tag{49}$$

In free space, $N(\omega) = \omega^2/\pi^2 c^3$, and the ratio of the spontaneous decay rate of state o in the cavity γ_o, to that in free space, γ_s is

$$\frac{\gamma_o}{\gamma_s} = \frac{3}{4\pi^2} Q\left(\frac{\lambda_o^3}{V}\right) \tag{50}$$

a result first obtained by Purcell (1946). In such a cavity, the Q enhanced the strength of the vacuum fluctuations responsible for the spontaneous emission. In a waveguide, the cavity can completely cut off the resonant modes so that $N_c(\omega) = o$ and spontaneous emission cannot occur (Kleppner, 1981). This termination of spontaneous decay has been observed indirectly in the changes in black-body radiation absorption by Rydberg atoms in a confining cavity (Vaidyanathan et al 1981). Cavity Q-enhanced spontaneous decay has been observed in Rydberg transitions by Goy et al (1983). Hulet et al (1985) have succeeded in observing completely turned off, or inhibited, decay of Rydberg states in agreement with theoretical results of Knight and Milonni (1973).

In a remarkable experiment by Gabrielse and Dehmelt (1985), the cyclotron spontaneous decay rate of a free electron in a trap was decreased by a factor of 10 in a cavity so small that it excluded the vacuum fluctuations at the cyclotron resonance frequency. In the experiment of Goy et al (1983), sodium atoms were excited to the 23s state by tunable dye lasers in a niobium superconducting cavity cooled so that Q is high and the background thermal radiation $\bar{n} \ll 1$ at the Rydberg resonance 23s $\rightarrow$ 22p at 340 GHz. The free-space decay rate γ_s for this transition is 150 s^{-1}. In the cavity the spontaneous emission increased to $8 \times 10^4\ s^{-1}$ ($\omega/Q = 2.8 \times 10^6\ s^{-1}$) as the photons get absorbed in the cavity walls faster than the atoms decay. Recent experiments by Walther and colleagues, to be described briefly in section 7 on spontaneous decay of Rubidium in an ultrahigh Q cavity see completely reversible one-photon Rabi oscillations (Meschede et al 1985). In this high Q (= 8×10^8 at 2^oK) cavity Rabi oscillations at the one-photon level are observed through their power broadening of resonance curves.

The final state spectrum P_f also exhibits the Rabi one-photon oscillations. If $|H_{oi}| \ll \gamma/4$ and the atom is driven weakly on resonance ($\Delta = 0$), we find from eq (43)

$$P_f(t\to\infty) \cong \frac{b\gamma|H_{io}|^2}{E_f^2+\gamma^2/4}\left[\frac{1}{E_f^2+4|H_{oi}|^4/\gamma^2}\right], \quad (51)$$

which is a single line with a power-broadened width $2|H_{oi}|^2/\gamma = \Omega^2 Q/2\omega$. If the Rabi frequency is instead much greater than ω/Q so the atom is strongly driven on re sonance, we dress the atom by the radiative interaction into doublets of atom-field states (eg Knight and Milonni 1980) each member of which decays separately at rate $\gamma/2$:

$$P_f(t\to\infty) \cong b\frac{\gamma}{4}\left[\frac{1}{(E_f-|H_{oi}|)^2+\frac{1}{4}\gamma^2}+\frac{1}{(E_f+|H_{oi}|)^2+\frac{1}{4}\gamma^2}\right] \quad (52)$$

which describes an AC-Stark split doublet with lines of width $\gamma/2$ (Sanchez-Mondragon et al 1983). Again the atom-field coupling is so strong that the spectrum is self-split by its radiative interaction. Sanchez-Mondragon et al (1983) also calculate the influence of coherent cavity radiation on the cavity emission and describe a cavity spectrum analogous to the free-space Mollow triplet spectrum (see eg Knight and Milonni 1980).

If there are many atoms in the cavity, collective effects modify the vacuum emission spectrum (Agarwal 1984), essentially by increasing the one-photon Rabi frequency by a factor $\sqrt{N}$ where N is the number of radiating atoms.

The modification of spontaneous emission rates has also been studied by depositing Langmuir-Blodgett mono-molecular layers on conducting surfaces and studying the fluorescence from dye layers a fixed distance from the surface (Drexhage 1974). The related problem of spontaneous decay of an excited atom in a Fabry-Perot cavity has been considered by Barton (1970), Philpott (1973) and Milonni and Knight (1973).

6 CAVITY MODES AND RADIATIVE CORRECTIONS OF RYDBERG STATES

As we have seen from the previous section, atoms in cavities see quite different cavity field mode structure to atoms in free space. Power (1966) proposed that radiative corrections, and especially self-masses are changed if the electron is confined in a small cavity. Such mass shifts are better regarded as energy shifts and are altered by atomic binding. If the mass is changed by cavity confinement, the hyperfine structure would alter through the dependence of the Fermi contact interaction on the electron magnetic moment and its mass-dependence. Guttrich and Billman (1967) failed to find such alterations in a Cs atomic clock which had such a cavity (actually two long plates) inserted in its C-magnet resonance region. The binding was shown by Golub and Guttrich (1967) to cancel the energy shift calculated by Power. Extensive calculations of energy shifts (and lifetime changes) between conducting plates of free electrons and atoms were performed by Barton (1970) who calculated in addition a cavity-dependent correction to the Lamb shift. Dobiasch and Walther (1985) have the calculated the modifications to the Bethe part of the Lamb shift and discuss the possibility of observation in Rydberg atom systems.

Cavity-dependent changes to the electron anomalous magnetic moment (g - 2) and free electron cyclotron orbit frequency shifts have been discussed by Dehmelt (1984), and by Brown et al (1985). The present precision in the measurement of $a = \frac{1}{2}(g-2)$ in the University of Washington electron trap experiments ($a \sim 1.2 \times 10^{-3}$) is

$$\frac{\Delta a}{a} \sim 4 \times 10^{-9} \qquad (53)$$

(Van Dyck et al, 1984 Ekstrom and Wineland 1980). The anomaly is $a = \omega_s - \omega_c)/\omega_c$ where ω_s is the spin precession frequency and ω_c the cyclotron frequency of the electron in a magnetic field. If the shift in the cyclotron frequency in the cavity is $\bar{\omega}_c - \omega_c = \Delta\omega_c$, where $\bar{\omega}_c$ is the observed frequency in a cavity and ω_c is the frequncy in free space and this shift is not taken into account, an error in the anomaly

$$\frac{\delta a}{a} = \frac{1}{a}(\Delta\omega_c / \omega_c) \qquad (54)$$

and current precision in recent measurements will be upset if $\Delta\omega_c/\omega_c \sim 5 \times 10^{-12}$. Brown and coworkers (1985) calculate a fractional frequency shift $\Delta\omega_c/\omega_c$ $\sim 90 \times 10^{-12}$ in the high Q cavity ($Q \sim 10^3$) used in the Washington experiments. It seems that cavity confinement effects profoundly alter radiative corrections at a measurable level. Calculations for bound electrons (and especially for Rydberg states) are under way (Boulware and Brown 1985; Boulware et al 1985).

7 COOPERATIVE RADIATIVE INTERACTIONS OF RYDBERG ATOMS

An ensemble of many Rydberg atoms in a cavity behaves as a single quantum system, whose collective interaction with radiation fields has been extensively studied experimentally and theoretically by Haroche and his co-workers (Haroche 1984, Haroche and Raimond 1985). The Rydberg atom maser in a high Q single-mode cavity is described by single mode superradiance theory (Bonifacio et al 1971). Raimond et al (1982a) observed the evolution of the photon statistics of the Rydberg atom maser from a Bose-Einstein distribution below threshold to a bell-shaped curve appropriate to coherent emission above threshold in qualitative agreement with single-mode superradiance theory (De Giorgio and Ghielmetti 1971). The same system absorbs black body radiation as if it were a single quantum system described as a Bose gas (Raimond et al 1982b), which again stresses the strong coupling between Rydberg atoms and radiation.

We saw in section 5 how difficult it is to observe single-photon spontaneous Rabi oscillations unless a very high Q cavity is used, when $\Omega \gg \omega/Q$. Collective, cooperative coupling between many atoms enhances the effective one-photon Rabi frequency of the superrariant interaction of N atoms with the single-mode field. The cooperative Rabi frequency is

$$\Omega_N = 2\,(d/\hbar)\mathcal{E}_0\sqrt{N} = \Omega\sqrt{N} \tag{55}$$

where $\mathcal{E}_0$ is the "electric field per photon" (Knight and Allen 1983). Kaluzny et al (1983) observed such collective Rabi oscillations in a Rydberg maser as N was increased from 2,000 atoms to 40,000 atoms in the high Q decay from $36_{1/2}S$ to $35_{1/2}P$ states in a cavity of volume 1.74 cm^3 with a resonant field mode of frequency $\omega/2\pi$ = 82 058 MHz. The cavity damping rate ω/Q = 5 x 10^6 s^{-1}. For the Rydberg transition, the atomic dipole is 592 atomic units, so the single-photon single-atom Rabi frequency $\Omega/2\pi$ = 20kHz. For $N \sim 2000$ atoms, the decay begins to deviate from an exponential ($\sqrt{N}\Omega > \omega/Q$ requires $N \gtrsim 14000$) and for N = 19000 to 40000, collective Rabi oscillations, with a period of fractions of microseconds are observed. Haroche (1984) has described this collective maser evolution using semiclassical pendulum models triggered by vacuum fluctuations and by black-body radiation fields. With a higher Q cavity, Walther and colleagues have succeeded in obtaining Rydberg maser C W oscillation with only a single atom present in the cavity at any time (Meschede et al 1985). Heidmann et al (1985a, b) have suggested that such Rydberg masers may produce sub-poissonian photon statistics and squeezed light fluctuations. No experimental work has yet been reported on this interesting problem of squeezing in Rydberg systems.

8 PHOTON STATISTICS AND RABI OSCILLATIONS

In an ultra-high Q cavity, an atom can execute many Rabi oscillations in a cavity decay time. If the field in the cavity were prepared in a number state $|n\rangle$, and an atom injected in its ground state, the probability of being excited at time t by the resonant field mode is

$$P_e(t) = \sin^2 \sqrt{n}\lambda t \tag{56}$$

where λ is a coupling constant such that $2\sqrt{n}t$ is the n-photon Rabi oscillation frequency (Jaynes and Cummings 1963). The inversion $W(t) = P_e(t) - P_g(t) = 1 - 2P_e(t)$ is

$$W_n(t) = -\cos 2\sqrt{n}\,\lambda t. \tag{57}$$

In practice, the field mode will have a distribution of photon numbers p(n). For example, if it is prepared in a coherent state $|\alpha\rangle$, p(n) is a Poisson distribution

$$p(n) = \frac{|\alpha|^{2n}}{n!} \cdot e^{-|\alpha|^2} \tag{58}$$

with a mean $\langle n \rangle = |\alpha|^2$ and a width $\Delta n = |\alpha| = \langle n \rangle^{1/2}$. The inversion produced by a coherent state, from eqs (57) and (58) is

$$W_\alpha(t) = -\sum_{n=0}^{\infty} e^{-|\alpha|^2} \frac{|\alpha|^{2n}}{n!} \cos 2\lambda\sqrt{n}\,t. \tag{59}$$

The Poisson spread in n gives a dephasing of the Rabi oscillations (a collapse) studied by Cummings (1965) and others (see eg Knight and Milonni 1980 for earlier references, and Knight and Radmore 1982). Eberly and coworkers have shown that the Rabi oscillations will revive (Eberly et al 1980, Narozhny et al 1981, Yoo et al 1981, Yoo and Eberly 1985). Both collapses and revivals in the coherent state Jaynes-Cummings model are purely quantum features and have no classical counterpart. The collapse time is λ^{-1} and the revival times is $2\pi\sqrt{\langle n \rangle}/\lambda$ for coherent state initial field distributions.

In a chaotic, Bose-Einstein field, the probability distribution is broader and the average inversion reflects this broader distribution p(n)

$$p(n) = \frac{1}{1+\langle n \rangle}\left(\frac{\langle n \rangle}{1+\langle n \rangle}\right)^n. \tag{60}$$

The average inversion for an initially unexcited atom interaction with a single-mode chaotic field is

$$W_{ch}(t) = -\sum_{n=0}^{\infty} p(n) \cos 2\lambda\sqrt{n}\,t. \qquad (61)$$

The inversion again collapses and revives (Knight and Radmore 1982b). Although the revival is a clear quantum feature, the collapse of Rabi oscillations is less clear-cut as a quantum effect because the thermal field can be represented as a classical Gaussian distribution of mode electric field strengths which would dephase the oscillations without invoking quantum effects. The detailed analysis of the Jaynes-Cummings model with full references is given elsewhere in this volume.

The experimental investigation of collapses and revivals in the Jaynes-Cummings model requires an ultra-high Q cavity so that the cavity damping time (Q/ω) is longer than a collapse time λ^{-1} (which is equal to a one-photon rabi period) or a revival time $2\pi \langle n \rangle^{1/2}/\lambda$. Meschede et al (1985) have constructed a superconducting cavity at a frequency ω = 21506.51 (5) MHz with a Q at 2° K of 8×10^8. At such temperatures the mean number of thermal photons $\langle n \rangle = 1.5$. The Rydberg atom system employed in Walther's experiments is a rubidium beam excited by a frequency-doubled ring dye laser to the 63 $P_{3/2}$ state. Such a highly-excited state has a very large one-photon Rabi frequency $\Omega \gg \omega/Q$. The 63 $P_{3/2}$ state is detected by field ionization. The interaction time is varied by employing a Fizeau slotted wheel velocity selector, and Rabi oscillations, collapses and revivals are observed (Walther 1985). The problem of accounting for cavity dissipation whilst retaining all the quantum features of the Jaynes-Cummings model has been examined by Barnett and Knight (1985). Their analysis shows that the revivals, due essentially to the "graininess" of the photon numbers in the coherent state Poisson sum, are much more sensitive to cavity dissipation than the collapses. This final example demonstrates the versatility of Rydberg atom experiments in testing our most fundamental ideas concerning the interaction of radiation with atoms.

Rydberg experiments have been successful in demonstrating directly quantum features of atom-radiation interactions, which until recently were considered to be unrealizable in the laboratory. Spontaneous decays and other quantum electrodynamic radiative corrections have been shown not to be immutable but sensitive to the local environment (especially the local geometry). Black-body radiation in which all atoms are bathed has been shown to clothe atoms in energy-changing and lifetime-shortening interactions. Collective atom-atom interactions modelled by the early primitive single-mode theories of the late 1960's have been dusted off and shown to be applicable to the cooperative interaction of Rydberg atoms with cavity radiation. All of these fundamental properties have been explored in the past five years at the same time as Rydberg masers have been developed as ultra sensitive detectors of mm-wave radiation, a practical application of basic quantum optics.

ACKNOWLEDGEMENTS

I would like to thank S M Barnett, G Barton, J H Eberly, P Meystre, P M Radmore and H Walther for discussions. This work was supported in part by the Science and Engineering Research Council.

REFERENCES

Agarwal G S 1984, Phys Rev Lett 53, 1732.
Barnett S M and Knight P L 1985 to be published.
Barrat J P and Cohen-Tannoudji C 1961, J Phys Radium Paris 22, 239, 443.
Barton G 1970, Proc Roy Soc A320, 251.
Barton G 1972, Phys Rev A5, 468.
Barton G 1985, in "The Spectrum of Atomic Hydrogen", ed G W Series (World Scientific, Singapore), to be published.
Barton G 1986, to be published.
Bayfield J E and Pinnaduwage L A 1985, Phys Rev Lett 54, 313 and J Phys B 18, L49.
Beiting E J, Hildebrandt G R, Kellert F G, Foltz G W, Smith K A, Dunning F B and Stebbings R F, 1979, J Chem Phys 70, 3551.
Bohr N 1913, Phil Mag 26, 1.
Bonifacio R, Schwendimann P and Haake F 1971, Phys Rev A4 302.
Boulware D G and Brown L S 1985, Phys Rev Lett 55, 133 (C).
Boulware D G, Brown L S and Lee T 1985, to be published.
Brown L S, Gabrielse G, Helmerson K and Tan J 1985, Phys Lett 55, 44.
Chaisson E J 1976, in "Frontiers of Astrophysics" ed E H Avrett, (Harvard University Press, Boston).
Cohen-Tannoudji C and Haroche S 1966, Comptes Rend hebd Seanc Acad Sci Paris 262, 268.
Cohen-Tannoudji C, Diu B and Laloe F, 1977, "Quantum Mechanics" (Wiley, New York).
Cummings F W 1965, Phys Rev 140, A1051.
Dalgarno A, 1985 in "Proc NATO ASI on Atoms in Unusual Circumstances, Cargese, July 1985", to be published.
De Giorgio V and Ghielmetti F 1971, Phys Rev A4, 2415.
Dehmelt H G 1984, Proc Nat Acad Sci (USA) 81, 8037.
Delone N B, Krainov B P and Shepelyansky D L 1983, Usp Fiz Nauk 140, 355 (Sov Phys Usp 26 (1) 551).
Dobiasch P and Walther H 1985, to be published.
Donoghue J F and Holstein B R 1983, Phys Rev D28, 340.
Donoghue J F, Holstein B R and Robinett R W 1984, Phys Rev D30, 2561.
Drexhage K 1974, Progress in Optics 12, 165 (North-Holland, Amsterdam) ed E Wolf.
Ducas T W, Spencer W P, Vaidyanathan A G, Hamilton W H and Kleppner D 1979, Appl Phys Lett 35, 382.
Eberly J H, Narozhny N B and Sanchez-Mondragon J J 1980 Phys Rev Lett 44, 1323.
Ekstrom P and Wineland D J 1980, Sci Amer 243 105.
Fabre C, Gross M, Raimond J M and Haroche S, 1983, J Phys B 16, L671.
Farley J W and Wing W H 1981, Phys Rev A 23, 2397.
Figger H, Leuchs G, Straubinger R, and Walther H 1980, Opt Commun 33, 37.
Gabrielse G and Dehmelt H G 1985, Phys Rev Lett 55, 67.
Gallagher T F and Cooke W E 1979, Phys Rev Lett 42, 835.

Gallas J A C, Leuchs G, Walther H and Figger H 1985, in Adv in At Mol Phys 20, eds Sir David Bates and Benjamin Bederson (Academic Press, Orlando).
Golub R and Guttrich G L 1967, Phys Lett 24A, 224.
Goy P, Raimond J M, Gross M, and Haroche S 1983, Phys Rev Lett 50, 1903.
Guttrich G L and Billman K W 1967, Phys Lett 24A, 53.
Haroche S 1984, in "New Trends in Atomic Physics," Proceedings of the Les Houches Summer School, Session XXXVIII, edited by G Grynberg and R Stora (North-Holland, Amsterdam).
Haroche S, and Raimond J M 1985, in Adv in At Mol Phys 20, 347, eds Sir David Bates and Benjamin Bederson, (Academic Press, Orlando).
Heidmann A, Raimond J M and Reynaud S 1985, Phys Rev Lett 54, 326.
Heidmann A, Raimond J M, Reynaud S and Zagury N 1985, Opt Commun 54, 54 and 189.
Heitler W 1954, "The Quantum Theory of Radiation" (Clarendon Press, Oxford).
Hollberg L and Hall J L 1984, Phys Rev Lett 53, 230.
Hulet R G and Kleppner D 1983, Phys Rev Lett 51, 1430.
Hulet R G, Hilfer E S and Kleppner D 1985, Phys Rev Lett 55, 2137.
Jaynes E T and Cummings F W 1963, Proc I E E 51 89.
Itano W, Lewis L L and Wineland D J 1981, J de Phys Colloque C8, suppl 12, Tome 42, p C8 - 283.
Kaluzny Y, Goy P, Gross M, Raimond J M and Haroche S 1983, Phys Rev Lett 51, 1175.
Kleppner D 1981, Phys Rev Lett 47, 233.
Kleppner D, Littmann M G and Zimmerman M L 1981, Sci Amer, May Issue, 108.
Knight P L 1972, J Phys A5, 417.
Knight P L 1977, Opt Comm 22, 173.
Knight P L 1978, J Phys B11, L511.
Knight P L 1979, J Phys B 12 3297.
Knight P L 1980, in "Laser Physics" eds D F Walls and J D Harvey (Academic Press, Sydney).
Knight P L and Allen L 1983, "Concepts of Quantum Optics" (Pergamon Press, Oxford).
Knight P L and Milonni P W 1976, Phys Lett 56A, 275.
Knight P L and Milonni P W 1980, Physics Reports 66C, 21.
Knight P L and Radmore P M 1982b, Phys Lett 90A, 342.
Koch P R Hieronymous H, Van Raan A F J and Raith W 1980, Phys Lett 75A, 273.
Latimer C J 1979, Contemp Phys 20, 631.
Liberman S, Pinard J and Taleb A 1983, Phys Rev Lett 50, 888.
Meschede D, Walther H and Muller G 1985, Phys Rev Lett 54, 551.
Milonni P W and Knight P L 1973, Opt Comm 9, 119.
Moi L, Goy P, Gross M, Raimond J M, Fabre C and Haroche S 1983, Phys Rev A27, 2043 and 2065.
Narozhny N B, Sanchez-Mondragon J J and Eberly J H 1981 Phys Rev A23, 236.
Pancharatnam S 1966, J Opt soc Am 56, 1636.

Philpott M R 1973, Chem Phys Lett 19, 435.
Power E A 1966, Proc Roy Soc A292, 424.
Purcell E M 1946, Phys Rev 69, 681.
Raimond J M, Goy P, Gross M, Fabre C and Haroche S 1982a Phys Rev Lett 49, 1924.
Raimond J M, Goy P, Gross M, Fabre C and Haroche S 1982b, Phys Rev Lett 49, 117.
Rempe G and Walther H, 1985, in "Fundamentals of Laser Interactions", ed F Ehlotzky, (Springer-Verlag, Berlin).
Sanchez-Mondragon J J, Narozhny N B and Eberly J H 1983, Phys Rev Lett 51, 550.
Schuster H G, 1984, "Deterministic Chaos", (Physik-Verlag, Weinheim, F D R).
Series G W 1970, in "Quantum Optics" eds S M Kay and A Maitland (Academic Press, London), 395.
Stebbings R F and Dunning F B (eds) 1983, "Rydberg States of Atoms and Molecules," (Cambridge University Press, Cambridge).
Vaidyanathan A, Spencer W P and Kleppner D 1981, Phys Rev Lett 47, 1592.
Van Dyck Jr, R S, Schwinberg P B and Dehmelt H G 1984, "Proc 9th International Conference on Atomic Physics", Washington, July 1984.
Walther H 1985, private communication.
Yoo H-I and Eberly J H 1985, Physics Reports 118, 239.
Yoo H-I, Sanchez-Mondragon J J and Eberly J H 1981, J Phys A14, 1383.
Zaslavsky G M 1985, "Chaos in Dynamical Systems", (Harwood Academic Publishers, London).

THE JAYNES–CUMMINGS MODEL AND BEYOND

S M BARNETT, P FILIPOWICZ, J JAVANAINEN, P L KNIGHT AND P MEYSTRE

1. INTRODUCTION

Recent experiments by Walther and coworkers (Meschede et al 1985, Rempe and Walther 1985) have demonstrated that the Rydberg atom maser is capable of detecting the atom-single-photon coupling. In this, it approaches the idealization of fundamental two-level atom-single mode radiation field interaction of the Jaynes-Cummings model (Jaynes and Cummings 1963). In the JCM, the familiar Rabi oscillations in the evolution of the atomic inversion as the atom and field exchange energy are affected by the distribution of photon numbers, which causes a dephasing or 'collapse' as the range of possible Rabi frequencies interfere (Cummings 1965, von Foerster 1975, Meystre et al 1975, Knight and Milonni 1980). In a coherent state field, the Poisson distribution of photon numbers is responsible for this collapse and is a purely quantum effect. The discrete nature of the photon number distribution leads to a further purely quantum mechanical effect as the dephased Rabi oscillations partially rephase or 'revive'. The revivals are governed by the single-photon Rabi frequency or effectively the 'granularity' of the field (Eberly et al 1980, Narozhny et al 1981, Yoo et al 1981, Hioe et al 1983, Yoo and Eberly 1985).

In Rydberg maser experiments, the finite cavity Q introduces field dissipation into the system (Haroche 1984, Sachdev 1984). The cavity losses introduce Langevin quantum noise sources which react back on the JCM and affect the collapses and revivals in a nontrivial way (Barnett and Knight 1985). Further, the injection of a low density beam of atoms leads to modifications of the conventional JCM, the intracavity field being 'kicked' coherently by a succession of two-level atoms (Filipowicz et al 1985a,b). This paper discusses these novel developments 'beyond' the JCM. Section 2 reviews the standard JCM, both in its semiclassical and fully quantized versions and Section 3 generalizes these results to the case of a thermal field. In Section 4, we analyze, again both semiclassically and quantum mechanically, the injection of a low density stream of atoms inside a single mode, lossless cavity (Filipowicz et al 1985a). The last two Sections address the problem of dissipation. In Section 5, we generalize the treatment of Section 4 to the case of a lossy cavity, but under the assumption that the losses are so weak that the Jaynes-Cummings and cavity damping dynamics can be decoupled (Filipowicz et al 1985b, 1986a). This is a theoretical model of the micromaser (Meschede et al 1985). We show that in general, a truly microscopic maser does not produce coherent radiation. The assumptions about the damping are further relaxed in Section 6 which describes the theory of JCM with damping, with no decorrelation between atom and field operators required or employed (Barnett and Knight 1985). These results demonstrate that dissipation profoundly influences revivals by smearing out the discrete nature of the sum over Rabi frequencies. Such effects appear well before cavity field lifetimes damp out the Rabi oscillations.

2. THE JAYNES-CUMMINGS MODEL

The Jaynes-Cummings model is the simplest model of quantum optical resonance (Allen and Eberly 1975), yet it displays non-perturbative evolution, is sensitive to photon statistics, and most importantly can be realized in the laboratory (Meschede et al 1985, Rempe et al 1985). The model consists of a single two-level atom with ground state g and excited state e, interacting with a quantized single mode cavity radiation field of frequency ω close to the transition frequency ω_0 of the atom (Fig. 1). In the rotating wave approximation (RWA) the Hamiltonian for the atom, field mode and interaction is

$$H_Q = 1/2\, \hbar\omega_0\sigma_3 + \hbar\omega\, a^\dagger a + \hbar\lambda\, (a^\dagger\sigma_- + \sigma_+ a) \quad . \tag{1}$$

In the RWA, the excitation number in field and atom is conserved. The coupling constant λ depends on cavity mode volumes, dipole strengths and mode frequencies (Allen and Eberly 1975, Knight and Milonni 1980).

We first examine the classical field version of the JCM in which the field annihilation and creation operators a and $a^\dagger$ are replaced by c-numbers α and α^* which are the normalized positive and negative frequency parts of the cavity mode electric field amplitudes. The semiclassical Hamiltonian is

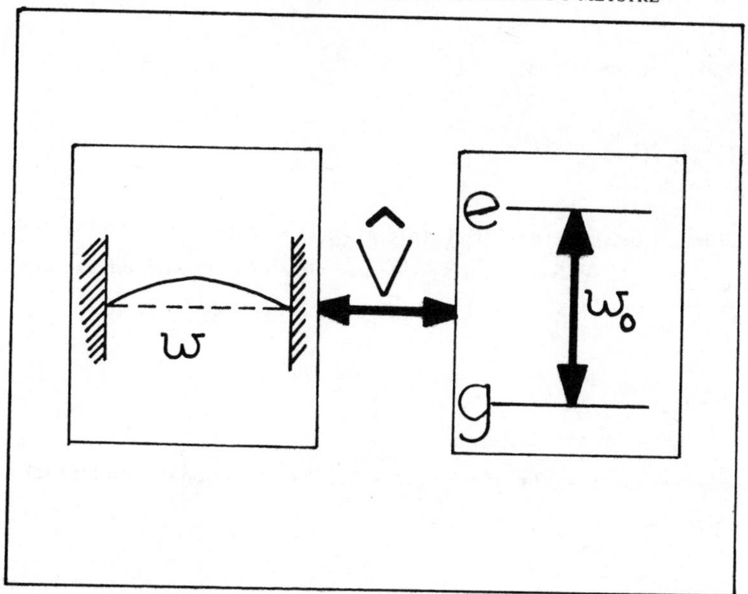

Fig.1: The Jaynes-Cummings model of a single two-level atom interacting with a single quantized field mode in a cavity.

$$H_{SC} = 1/2\, \hbar\omega_0\sigma_3 + \hbar\lambda\, (\alpha^* \sigma_- e^{i\omega t} + \sigma_+ \alpha\, e^{-i\omega t}) \quad . \tag{2}$$

The interaction induces Rabi oscillations between the ground state g and excited state e. We expand the wave function as

$$\Psi(t) = c_g(t)\, e^{-iE_g t/\hbar}\, |g\rangle + c_e(t) e^{-iE_e t/\hbar}\, |e\rangle \ , \tag{3}$$

where the probability amplitudes $c_g(t)$ and $c_e(t)$ obey the evolution equations in the RWA

$$\dot{c}_g = -i\lambda\alpha\, e^{i\Delta t/2}\, c_e \ , \tag{4}$$

$$\dot{c}_e = -i\lambda\alpha^*\, e^{-i\Delta t/2}\, c_g \ , \tag{5}$$

where the detuning $\Delta \equiv \omega - \omega_0$. The equations (4) and (5) have the solutions for the initial condition $c_g(t{=}0) = 1$,

$$c_e(t) = i(2\lambda\alpha/\Omega)\, e^{-i\Delta t/2} \sin \Omega t/2 \tag{6}$$

$$c_g(t) = e^{-i\Delta t/2}\, [\cos(\Omega t/2) + i(\Delta/\Omega) \sin(\Omega t/2)] \tag{7}$$

where the Rabi frequency Ω is

$$\Omega = \sqrt{\Delta^2 + 4|\lambda\alpha|^2} \ . \tag{8}$$

The probability of being in the initial state g, $P_g(t) = |c_g(t)|^2$ and the excited state e, $P_e(t) = |c_e(t)|^2$ oscillate at Ω as a function of time (Fig.2) with a maximum level of excitation

$$P_e(\max) = \frac{4|\lambda\alpha|^2}{\Delta^2 + 4|\lambda\alpha|^2} \tag{9}$$

which approaches unity as the detuning $\Delta \to 0$. The inversion driven by such a monochromatic field is (for $\Delta = 0$)

$$W_c(t) = -\cos 2\lambda|\alpha|t \ . \tag{10}$$

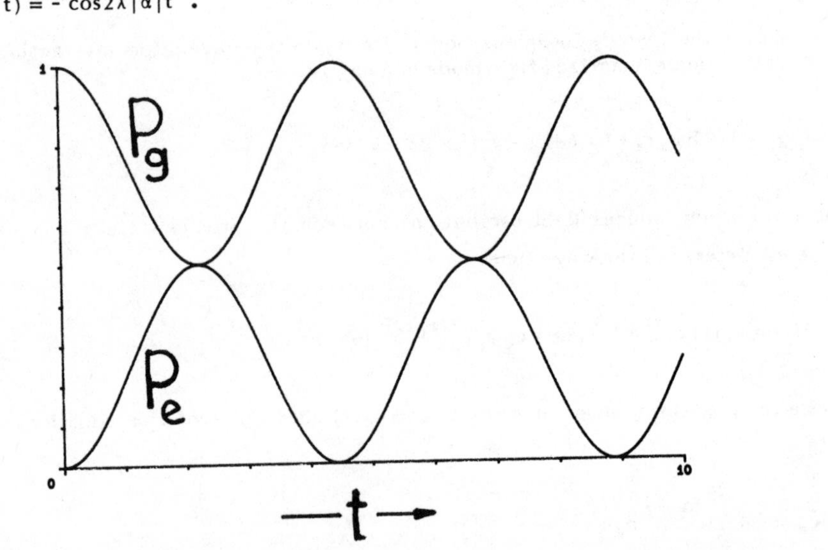

Fig. 2: Rabi oscillations excited in two-level atom by a classical field.

The quantum electrodynamic calculation of the Rabi problem proceeds from the fully quantized Hamiltonian H_Q. We take as the initial state an atom in its ground state g and the cavity in a number state n,

$$|i\rangle = |n\rangle|g\rangle$$

with energy $E_i = E_g + n\hbar\omega$. In RWA H_O couples i only to the state f

$$|f\rangle = |n-1\rangle|e\rangle$$

with energy $E_f = E_e + (n-1)\hbar\omega$. We expand the wavefunction into atom-field states i and f

$$\Psi(t) = C_i(t)\, e^{-iE_it/\hbar}\, |i\rangle + C_f(t)\, e^{-E_ft/\hbar}\, |f\rangle \quad . \tag{11}$$

The equations of motion for $C_i(t)$ and $C_f(t)$ are solved to give the probability of excitation to the state f

$$P_f(t) = \sin^2 \sqrt{n}\, \lambda t \tag{12}$$

and quantum inversion

$$W_Q(n) = -\cos 2\lambda\sqrt{n}\, t \quad , \tag{13}$$

where we have assumed resonance $\Delta = 0$. Unexpectedly we find the evolution of the inversion excited by a coherent classical field of magnitude $|\alpha|$ to be precisely that induced by a number state field of effective magnitude $\sqrt{n}$. Had we initiated the evolution in the excited state i

$$|i\rangle = |n\rangle|e\rangle \tag{14}$$

which couples to f

$$|f\rangle = |n+1\rangle|g\rangle \quad , \tag{15}$$

then

$$P_i(t) = \cos^2 \sqrt{n+1}\, \lambda t \tag{16}$$

which if $n = 0$ exhibits the purely quantum feature of periodic reversible spontaneous emission. ($P_i(t) = 0$ for the classical version of this initial condition.)

We might have expected the Jaynes-Cummings model excited not by a number state field but rather by a coherent field to have the closest correspondence to the semi-classical problem. That this is not so reflects new light on the role of photon statistics in atom-field strong coupling physics. The electric field of the number state is zero

$$\langle n|\,\hat{E}\,|n\rangle = 0 \quad . \tag{17}$$

To obtain a non-zero electric field we need a coherent superposition, for example a coherent state (e.g. Loudon 1983)

$$|\alpha\rangle = e^{-|\alpha|^2/2} \sum_{n=0}^{\infty} \frac{\alpha^n}{\sqrt{n!}}\,|n\rangle \tag{18}$$

where $|\alpha|^2 = \bar{n}$ is the mean number of photons in the mode. If we expand the electric field operator as

$$\hat{E} = ie_0\,(a\,e^{-i\omega t + i\mathbf{k\cdot r}} - a^\dagger\,e^{i\omega t - i\mathbf{k\cdot r}}) \tag{19}$$

then, using $\alpha = |\alpha|\exp(i\phi)$,

$$\langle\alpha|\,\hat{E}\,|\alpha\rangle = -\,2e_0\,|\alpha|\,\sin(\mathbf{k\cdot r} - \omega t + \phi) \tag{20}$$

$$\langle\alpha|\,\hat{E}^2\,|\alpha\rangle = e_0^2\,[4\,|\alpha|^2\sin^2(\mathbf{k\cdot r} - \omega t + \phi) + 1] \tag{21}$$

so that the uncertainty in electric field strengths is

$$\Delta E = e_0 = \sqrt{\hbar\omega/2\varepsilon_0 V} \tag{22}$$

where e_0 is the 'electric field per photon'. An atom interacting with a coherent state field sees a spread e_0 in field strengths and a spread in Rabi interaction frequencies. The photon number distribution p_n

$$p_n = \frac{|\alpha|^2}{n!}\,e^{-|\alpha|^2} \tag{23}$$

for n > 1 is a bell-shaped function with a width $\Delta n = |\alpha| = \sqrt{\bar{n}}$. A coherent state quantized field induces an inversion

$$W_Q(\alpha) = \sum_{n=0}^{\infty} p_n W_Q(n) = -\sum_{n=0}^{\infty} e^{-\bar{n}} \frac{\bar{n}^n}{n!} \cos 2\lambda\sqrt{n}\, t \ . \tag{24}$$

This infinite sum has no known finite expression but can be evaluated numerically, or dealt with using asymptotic expansions. The most interesting property of $W_Q(\alpha)$ is that it does not tend to $W_C(\alpha)$ as $\bar{n} >> 1$, contrary to naive semiclassical limit suppositions. The spread of Rabi frequencies leads to a dephasing of the Rabi oscillations (Cummings 1965, Stenholm 1973, von Foerster 1975, Meystre et al 1975). The range of dominant Rabi frequencies are from $\Omega = 2\lambda\sqrt{\bar{n}+\Delta n}$ to $\Omega = 2\lambda\sqrt{\bar{n}-\Delta n}$ and dephase in a collapse time t_c

$$t_c^{-1} = \left[2\lambda\sqrt{\bar{n}+\Delta n} - 2\lambda\sqrt{\bar{n}-\Delta n}\right] \simeq \lambda \tag{25}$$

which is independent of $\bar{n}$. The one-photon Rabi frequency is λ, which emphasises that it is the quantum uncertainty $\Delta E = e_0$ in electric field strengths in coherent state which dephases the Rabi oscillations. Cummings (1965) showed that the envelope of the collapse is approximately Gaussian. To observe a Jaynes-Cummings collapse requires a one-photon Rabi frequency to be observable. This cannot be achieved with optical transitions, but as we shall see is attained in transitions between highly-excited Rydberg states in ultra-high Q cavities.

That the collapse in the coherent-state Jaynes-Cummings oscillations is due to the vacuum fluctuations can be shown directly using a transformed representation of the quantum Hamiltonian H_Q (Eq.(1)) (Knight and Radmore 1982). We write the Hamiltonian as

$$H'_Q = 1/2\, \hbar\omega_0 \sigma_3 + \hbar\lambda \left(a^\dagger e^{i\omega t} \sigma_- + \sigma_+ a\, e^{-i\omega t}\right) \ . \tag{26}$$

We transform H'_Q using the inverse of the Glauber coherent state generator (Glauber 1963) $D(\alpha)$

$$D(\alpha) = \exp(\alpha a^\dagger - \alpha^* a) \tag{27}$$

where

$$D^{-1}(\alpha)\ |\alpha\rangle = |0\rangle \tag{28}$$

to find our new Hamiltonian

$$H''_Q = D^{-1}(\alpha) H'_Q D(\alpha) \tag{29}$$

as

$$H''_Q = 1/2\, \hbar\omega_0\, \sigma_3 + \hbar\lambda\ [\ (a^\dagger + \alpha^*) e^{i\omega t}\, \sigma_- + \sigma_+ (a+\alpha)\, e^{-i\omega t}\ ]\ . \tag{30}$$

In this representation, the initial state is a ground state atom and the vacuum cavity field state, and transitions are induced by both the quantum interaction and a semi-classical interaction with a coherent classical field of strength α (Mollow 1975, Samson and Ben-Reuwen 1975, Pegg 1980, Knight and Allen 1983). In this transformed representation the classical field α induces sinusoidal Rabi oscillations between g and e without changing the number of photons in the field mode (and this number is initially zero). The quantised field interaction, through a, $a^\dagger$, can change the number of photons in the mode. For example it can take the two-state system from $|0, e\rangle$, the atom excited and no photon present, to $|1,g\rangle$ with one photon in the mode and the atom in the ground state. In the RWA, the 'essential states' coupled to H''_Q are represented in Fig.3 as a ladder of states connected by couplings

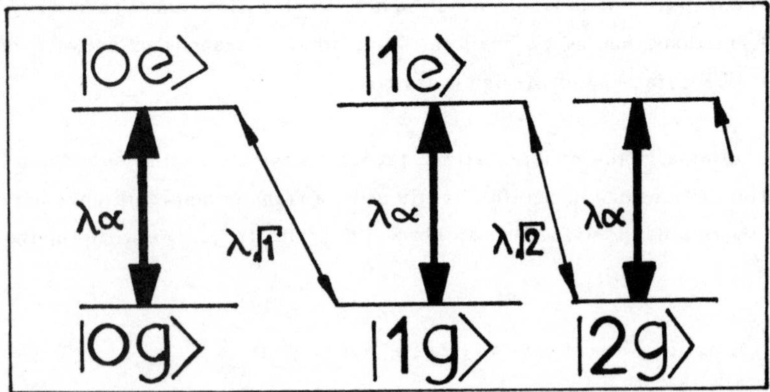

Fig.3: Energy levels and couplings of the transformed JCM in which the classical field Rabi couplings $\lambda\alpha$ compete with quantum couplings $\lambda\sqrt{n+1}$.

$$\langle n,e|\, H''_Q \,|n+1,g\rangle = \hbar\lambda\sqrt{n+1}\, e^{-i\omega t} \tag{31}$$

$$\langle n,e|\, H''_Q \,|n,g\rangle = \hbar\lambda\alpha\, e^{-i\omega t} \tag{32}$$

and the distribution of Rabi frequencies responsible for the JCM collapse arises as a consequence of quantum dynamics. As successive states $|n,e\rangle$, $|n,g\rangle$ are coupled into the evolution, quantum Rabi frequencies $\lambda\sqrt{n}$ are generated after a succession of semiclassical oscillations. If the field mode is initially highly excited ($\alpha \gg 1$) then the initial time evolution is just that of the first rung on the ladder of couplings shown in Fig.3. A small proportion of population (reversibly) leaks into the ground state plus a photon through the quantum electrodynamic interaction. The evolution therefore takes place through an alternating sequence of strong ($\lambda\alpha$) and quantum ($\lambda\sqrt{n}$) couplings, which start out as very weak ($\lambda\sqrt{1}$) but strengthen as n increases. The role of the one-photon Rabi frequency (the first quantum transition) is to initiate a cascade of quantum eigenfrequencies. It is clear that no simple semiclassical result is obtained when $\alpha \gg 1$, because eventually large quantum Rabi frequencies $\lambda\sqrt{n}$ are generated which compete with $\lambda\alpha$ for the atomic population.

A further quantum feature of the coherent JCM manifests itself after the Rabi oscillations have collapsed. Because the photon numbers n are discrete in the quantum sum $W_Q(\alpha)$, the oscillations will partially rephase in a revival time t_R (Eberly et al 1980, Narozhny et al 1981, Yoo et al 1981, Hioe et al 1983, Yoo and Eberly 1985),

$$t_R \simeq 2\pi\alpha/\lambda = 2\pi\sqrt{\bar{n}}\, t_c \;, \tag{33}$$

as illustrated in Fig.4. This revival property is a much more unambiguous signature of quantum electrodynamics: any spread in field strengths will dephase Rabi oscillations, but the revivals are entirely due to the grainy nature of the field so that the atomic evolution is determined by the individual field quanta. Eventually, the revivals (which are never complete and get broader and broader) overlap to give a quasi-random time evolution. The Jaynes-Cummings model clearly exhibits interesting quantum features of atom-field interaction and has proven popular with theorists as a test-bed for theoretical ideas. Stenholm (1981) used a Bargman representation to derive $W_Q(\alpha)$. Multilevel and multimode generalisations of the two-state JCM have been reported (Buck and Sukumar 1981, 1984; Sukumar and Buck 1981, 1984; Li and Bei 1984; Walls and Barakat 1970), some of which have the virtue of being exactly

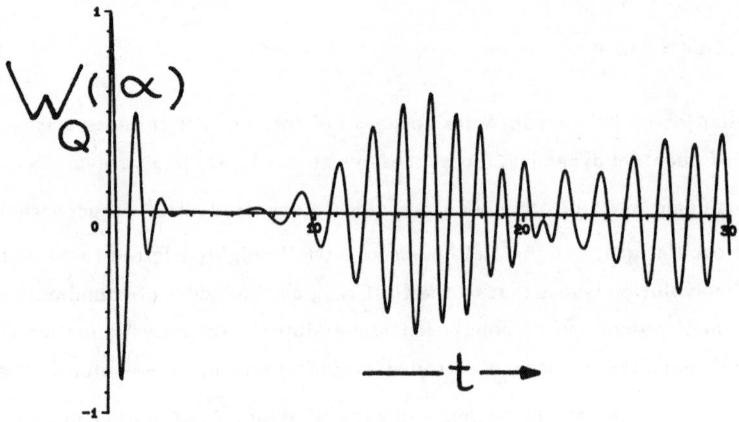

Fig. 4: Evolution of inversion for an initially excited atom in a cavity with a resonant field mode initially in a coherent state with $\bar{n} = |\alpha|^2 = 5$ photons. Note collapse and revivals of Rabi oscillations.

summable in closed form with remarkably simple dynamics. Filipowicz and Mostowski (1981) have demonstrated that the motion of a relativistic electron in a static magnetic field exhibits quantum collapses and revivals remarkably similar to those of the JCM. Singh (1982) discusses the field statistics of generalised (particularly multiphoton) JCM.

3. THERMAL FIELD JAYNES-CUMMINGS MODEL

So far we have concentrated on the problem of an atom interacting with a quantized coherent state field. Experimental work on very highly excited Rydberg atoms with large dipole moments indicate that single atom-mm-wave photon interactions can be detected (Haroche et al 1982, Haroche 1984; Rempe and Walther 1985). At such large wavelengths, thermal occupation of the resonant cavity mode will be important and the strong coupling of two-level atoms with Bose-Einstein fields needs to be considered. Cummings (1965) has given graphs of the evolution of the excited state probability of an atom interacting with an initially chaotic, or Bose-Einstein thermal field. Meystre (1974) and von Foerster (1975) have examined numerically the collapses and revivals of excitation for this chaotic initial condition. Riti and Vetri

(1982) have also reported on the short-time evolution of an atom excited by a chaotic field. Knight and Radmore (1982b) have studied both the quantum and semiclassical JCM driven by chaotic radiation, and again find clear differences between quantum and classical predictions.

The probability distribution p_n for chaotic, Bose-Einstein radiation is given by

$$p_n = \frac{1}{1+\bar{n}}\left[\frac{\bar{n}}{1+\bar{n}}\right]^n \tag{34}$$

with a most probable value of n = 0 and a photon uncertainty

$$\Delta n = \sqrt{\bar{n}^2 + \bar{n}} \ . \tag{35}$$

The large spread in photon numbers will lead to a large spread in participating Rabi frequencies in the JCM sum for the inversion evolution from an initially unexcited state (Knight and Radmore 1982b)

$$W_Q(ch) = -\sum_{n=0}^{\infty} p_n \cos(2\lambda\sqrt{n}\, t) \ . \tag{36}$$

The inversion will rapidly collapse, with a spread of Rabi frequencies $\Omega(\bar{n}+\Delta n)$ to $\Omega(\bar{n}-\Delta n)$ giving a collapse time for $\bar{n} >> 1$

$$t_c^{-1} \simeq \lambda\sqrt{\bar{n}} \tag{37}$$

or for $\bar{n} << 1$

$$t_c^{-1} \simeq \lambda \ . \tag{38}$$

For $\bar{n} >> 1$, the collapse is now dependent on the initial field strength. For an initially unexcited atom

$$W_Q(ch,g) = -1 + \frac{2}{1+\bar{n}} \sum_{n=0}^{\infty} \left[\frac{\bar{n}}{1+\bar{n}}\right]^n \sin^2(\lambda\sqrt{n}\, t) \ . \tag{39}$$

This is evaluated and plotted numerically in Fig.5 for $\bar{n} = 1$.

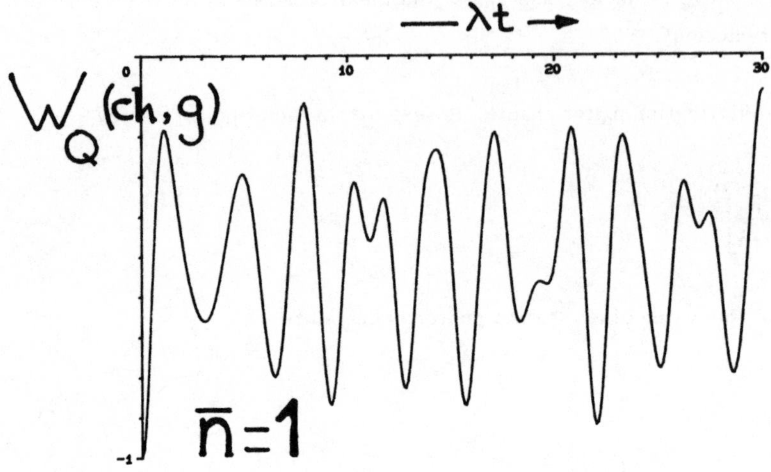

Fig. 5: Evolution of inversion for an initially unexcited atom driven by a chaotic field (from Knight and Radmore 1982b)

To order t^4,

$$W_Q(ch,g) \simeq -1 + 2(\lambda t\sqrt{\bar{n}})^2 - (4/3)\,(\lambda t\sqrt{\bar{n}})^4(1+1/2\bar{n}) \ldots \qquad (40)$$

In the semiclassical approach to the JCM, the probability of an atom remaining in the ground state when driven by a coherent field of intensity I is

$$P_g(t) = 1/2\;(1 + \cos 2\sqrt{I}\;\lambda' t) \qquad (41)$$

where λ' is the semiclassical equivalent of the quantum coupling constant λ. If the field is driven instead by a classical chaotic field of mean intensity <I> and an intensity distribution

$$p(I) = \langle I\rangle^{-1} \exp(-I/\langle I\rangle) \qquad (42)$$

then the ground state population is

$$P_g(t) = \frac{1}{2\langle I\rangle} \int_0^\infty \exp(-I/\langle I\rangle)\,(1 + \cos 2\sqrt{I}\,\lambda' t)\,dI \tag{43}$$

and the inversion driven by such a classical field is

$$W_c(ch,g) = 1 - 2P_g(t) \ . \tag{44}$$

The intensity integration in Eq.(43) can be performed to give (Knight and Radmore 1982b)

$$W_c(ch,g) = 2\theta\, D(\theta) - 1 \tag{45}$$

where $\theta = \langle I\rangle^{1/2}\lambda' t$ and $D(\theta)$ is Dawson's integral

$$D(\theta) = \exp(-\theta^2) \int_0^\theta \exp(x^2)\, dx \ . \tag{46}$$

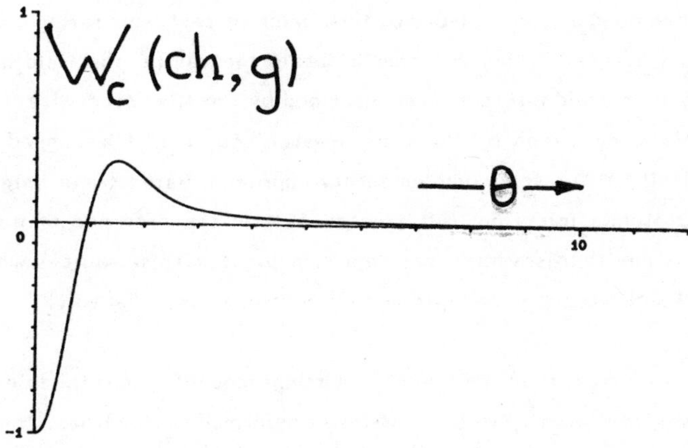

Fig. 6: Time evolution of the inversion of a two-level system driven by a classical chaotic field vs. $\theta = \langle I\rangle^{1/2}\lambda' t$ (from Knight and Radmore 1982b).

The inversion driven by a classical chaotic field is shown in Fig.6 as a function of the variable $\theta = \langle I\rangle^{1/2}\lambda' t$. There is a single collapse, and of course no quantum revivals. To order t^4,

$$W_c(ch,g) = -1 + 2\langle I\rangle(\lambda't)^2 - 4/3\,\langle I\rangle^2(\lambda't)^4 \,\ldots. \qquad (47)$$

in agreement with Eq.(40) for large $\bar{n}$.

4. COHERENTLY KICKED RESONATOR FIELD

Until now, we have considered atoms at rest in a single-mode, lossless cavity. The Rydberg atom maser (Meschede et al 1985; Rempe et al 1985; Walther 1985), in contrast, employs a low-density atomic beam to ensure collision-free conditions for the highly excited Rydberg atoms. Also, there is no coherent initial field. Rather, the field builds up from noise as further atoms are injected one by one into the resonator. Finally, the resonator has a finite Q, so that eventually a steady state is reached, in contrast to the lossless JCM. In order to determine to which extent this system can be used to test the predictions of the JCM, it is necessary to include both the effects of atomic injection and of resonator losses. Recently, Filipowicz et al (1985b) have developed a simple theory of this "micromaser", separating the problem into two distinct stages. When an atom is inside the cavity, the field damping is ignored and the atom-field interaction is described by the JCM; otherwise, the evolution of the field is described by the usual master equation of a damped harmonic oscillator (Louisell 1973). In a complementary approach, Barnett and Knight (1985) have neglected atomic injection, but treated JCM in the presence of dissipation. They showed that cavity losses introduce Langevin quantum noise sources which react back on the JCM and affect the collapse and revivals in a nontrivial way.

In this Section, we first extend the Jaynes-Cummings model to treat the injection of a sequence of two-level atoms inside a lossless single mode resonator. This problem was originally considered by Jaynes and Cummings (1963) for the case of absorbing atoms. Cavity damping will be introduced for the maser in Section 5, and the damped JCM will be the object of Section 6.

When interacting with a low-density stream of two-level atoms, a single mode cavity field may be seen as a "coherently kicked harmonic oscillator", the kicking being due to the consecutive dipole interactions. "Kicked systems" have received considerable attention in recent years, mostly in connection with the problem of chaos vs. quantum chaos. A widely studied model in this context is the "δ-kicked rotator" (Casati et al

1979) which has been shown to lead to quantum-mechanical localization in regions where the classical version of the model is chaotic. Similar results characterize the dynamics of highly excited atoms driven by radio-frequency fields (Casati el al 1984; Bayfield and Pinnaduwage 1985). Recently, Graham and coworkers (Graham 1985; Graham and Tel 1985) have extended these studies to periodically kicked dissipative systems.

The coherently kicked harmonic oscillator differs from such models in that it does not exhibit chaos. Nonetheless, its semiclassical version may display a sensitive dependence on initial conditions and it operates just at the threshold of type-I intermittency (see e.g. Schuster 1984). Specifically, as successive atoms are injected the system evolves towards one of infinitely many fixed points. These fixed points correspond to values of the field such that each atom undergoes a $2q\pi$ coherent interaction (q integer) during its passage through the resonator (Allen and Eberly 1975). They are only marginally stable, i.e. stable when approached from one side but unstable when approached from the other, depending on the initial atomic inversion.

In contrast, the quantum version of the system usually does not reach a steady state if the injected atoms are in a state of inverted population. Rather, the intracavity intensity grows ad infinitum. This is because in general the intrinsic intensity fluctuations of quantized fields prohibit the existence of exact $2q\pi$ pulses. For particular parameters of the system the field may, however, evolve towards a pure number state. Then the fluctuations are not effective, and a steady state is possible. We discuss this fully quantized version first.

At time t_i, an atom in a state given by the density matrix ρ_A enters the cavity containing the field in state ρ_F, so that the density operator of the combined system is $\rho(t_i) = \rho_A \rho_F$. After the interaction, the state of the system is $\rho(t_i+t_{int}) = U(t_{int})\rho(t_i)$, where $U(t_{int}) \equiv \exp(-iH_Q t_{int}/\hbar)$ and H_Q is the Jaynes-Cummings Hamiltonian (1). If the density matrices of both the field and the atom are initially diagonal they remain so for all times. When the atom leaves the resonator, the field is left in the state

$$\rho_F(t_i+t_{int}) = \mathrm{Tr}_A[U(t_{int})\rho(t_i)] \tag{48}$$

where Tr_A denotes trace over the atomic states. The probability distribution $p_n = \langle n|\rho_F|n\rangle$ is explicitly given by (Jaynes and Cummings 1963)

$$p_n(t_i+t_{int}) = \frac{1}{e^x+1}\,[(\alpha_{n+1} + \alpha_n e^x)p_n(t_i) + \beta_{n+1}e^x p_{n+1}(t_i) + \beta_n p_{n-1}(t_i)] \quad . \tag{49}$$

Here

$$\beta_n = 1 - \alpha_n = (4\lambda^2 n/\Omega_n^2)\,\sin^2(\Omega_n t_{int}/2)\,, \tag{50}$$

and $\Omega_n = \sqrt{\Delta^2 + 4\lambda^2 n}$ is the Rabi frequency for the state $|n\rangle$ of the cavity field mode. The parameter x is proportional to the temperature of the two-level atoms entering the cavity; the populations ρ_{ee} and ρ_{gg} of the upper and lower levels e and g are

$$\rho_{ee} = e^{-x}\,\rho_{gg} \tag{51}$$

with $\rho_{ee} + \rho_{gg} = 1$.

Jaynes and Cummings (1963) considered absorbing atoms ($\rho_{ee} < \rho_{gg}$) and showed that <u>if</u> the system reaches steady state, the condition $p_n(t_i+t_{int}) = p_n(t_i)$ yields

$$B_n \equiv \beta_n\,(\rho_{n-1} - e^x \rho_n) = 0 \tag{52}$$

for all $n \geq 1$. Hence, at least one of the two factors on the RHS of Eq.(52) must be zero. If all coefficients β_n are different from zero the steady state condition (52) becomes

$$p_n = e^{-x}\,p_{n-1} \tag{53}$$

If the injected atoms have a positive temperature ($\rho_{ee} < \rho_{gg}$), a steady state can always be found. It has the statistics of a thermal field, as readily seen by comparison of Eqs.(53) and (34) (Jaynes and Cummings 1963). In contrast, when the atoms are at least partially inverted, $e^{-x} \geq 1$, the steady state density matrix for the field is not normalizable, indicating that no steady state is possible.

The situation changes drastically if $\beta_k = 0$ for at least one k, as is seen by inspecting the relevant part of the infinite tridiagonal matrix on the RHS of Eq.(49):

$$\frac{1}{1+e^x}\begin{bmatrix} \alpha_{k-1}+e^x\alpha_{k-2} & \beta_{k-1}e^x & 0 & 0 \\ \beta_{k-1} & \alpha_k+e^x\alpha_{k-1} & \text{'0'} & 0 \\ 0 & \text{'0'} & \alpha_{k+1}+e^x\alpha_k & \beta_{k+1}e^x \\ 0 & 0 & \beta_{k+1} & \alpha_{k+2}+e^x\alpha_{k+1} \end{bmatrix} .$$

Now there is no coupling between the diagonal elements of the field density matrix (photon numbers) below and above $n = k$. At least at exact resonance $\Delta = 0$, it is readily seen that if one k satisfies $\beta_k = 0$, an infinite sequence of integers that fulfil the same condition can be found and the photon number space breaks up into an infinite number of disconnected blocks. A truncated thermal-type steady state can then be reached within each block separately. It has positive or negative 'temperature', depending upon the sign of x. This result can be applied to the generation of a number state $|k-1\rangle$ of the electromagnetic field (Filipowicz et al 1985a).

In the resonant case $\Delta = 0$, β_k is zero if

$$\lambda\sqrt{k}\, t_{int} = q\pi \ , \tag{54}$$

with q integer. Fig.7 shows the average photon number $\langle n\rangle$ and the rms spread $\Delta n = (\langle n-\langle n\rangle\rangle^2)^{1/2}$ as a function of the number of atoms having passed through the cavity. In this example all atoms are injected in the upper state, $x \to -\infty$. The initial field is taken to have a Poisson distribution with $\langle n\rangle = 8$, and $\lambda t_{int} = 2\pi/\sqrt{199}$, whence $\beta_k = 0$ for $k = 199$ (and for 796, 1791, ...). Except for the extremely small contribution due to its initial tail above $n = 198$, the photon statistics collapses towards that of the number state $|198\rangle$.

We now turn to the semiclassical description of this problem. As usual, it is obtained by factorizing the expectation values in the Heisenberg equations of motion for the atomic and field operators (Allen and Eberly 1973). This gives

$$\dot{\sigma}_- = i\Delta\sigma_- + i\lambda\sigma_3 E$$

$$\dot{\sigma}_3 = -2i\lambda(\sigma_+ E - E^*\sigma_-) \tag{55}$$

$$\dot{E} = -i\lambda\sigma_-$$

and $\sigma_+ = \sigma_-^*$, where

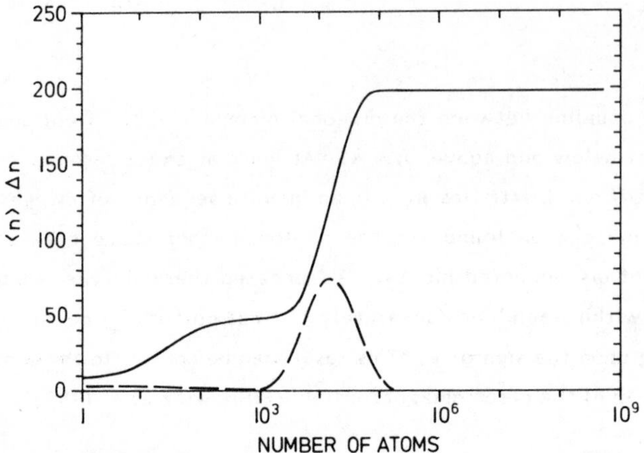

Fig.7: Average (solid line) and rms spread (dashed line) of the photon distribution evolving from an initial Poisson distribution with $\langle n\rangle = 8$, as a function of the number of atoms that have traversed the cavity for the interaction time $\lambda t_{int} = 2\pi\sqrt{199}$.

$$E = \langle a\rangle \exp(i\omega t)$$

$$\sigma_- = \langle S_-\rangle \exp(i\omega t) \tag{56}$$

$$\sigma_3 = \langle S_3\rangle .$$

We focus on the resonant case, and for convenience choose the phase of the classical field such that it only couples to the out-of-phase component of the polarization. In terms of the length R of the Bloch vector and the tipping angle θ defined through the equations

$$i(\sigma_+ - \sigma_-) = R\sin\theta \ ,$$

$$\sigma_3 = R\cos\theta \ , \tag{57}$$

and with the dimensionless time $\xi = 2\lambda t$, Eqs.(55) become

$$\frac{\partial R}{\partial \xi} = 0$$

$$\frac{\partial \theta}{\partial \xi} = E \tag{58}$$

$$\frac{\partial E}{\partial \xi} = (R/4) \sin\theta \ .$$

As in the quantum mechanical case, the atoms are assumed to be injected inside the resonator without initial dipole moment, $\theta = 0$. We take advantage of the conservation of the magnitude R of the Bloch vector, $R = \sigma_3(\tau{=}0) = (\rho_{ee}-\rho_{gg})(0)$, to carry out the further transformation

$$\tau = \xi\sqrt{|R|}/2 \ ,$$

$$\varepsilon = \frac{2E}{\sqrt{|R|}} \ . \tag{59}$$

The semiclassical model then reduces to the pendulum equations (Allen and Eberly 1973)

$$\frac{\partial \theta}{\partial \tau} = \varepsilon \tag{60a}$$

$$\frac{\partial \varepsilon}{\partial \tau} = \eta \sin\theta \tag{60b}$$

where $\eta \equiv \mathrm{sign}(R)$. The sign of the inversion alone determines the qualitative behaviour of the model.

Let us denote by ε_n the dimensionless intracavity field before injection of atom n+1. Depending upon whether the atom is initially inverted or not, the Bloch vector starts in upright or downright position with angular velocity ε_n. The field ε_{n+1} at the instant of escape of this atom is obtained by integrating Eqs. (60) over the dimensionless interaction time τ_{int}. This yields the return map

$$\varepsilon_{n+1} = f(\varepsilon_n). \tag{61}$$

The function $f(\varepsilon)$, which also depends on the control parameters η and τ_{int}, can be implicitly obtained by taking advantage of the energy conservation relation

$$\frac{\dot{\theta}^2}{2} + \eta\cos\theta = \frac{\varepsilon_n{}^2}{2} + \eta \tag{62}$$

which yields when inserted into Eq.(60b) the elliptic integral (Filipowicz et al 1985a)

$$\int_{\varepsilon_n}^{f(\varepsilon_n)} d\varepsilon\left[1-\left(\frac{\varepsilon_n{}^2-\varepsilon^2}{2}+\eta\right)^2\right]^{-1/2} = \eta\tau_{int} \ . \tag{63}$$

If the intracavity field ε_n before injection of atom n+1 is large enough and the field consequently varies only little under the influence of one atom, Eq.(60) gives approximately

$$\varepsilon_{n+1} \simeq \varepsilon_n + (2\eta/\varepsilon_n)\sin^2(\varepsilon_n\tau_{int}/2) \tag{64}$$

but in general the easiest way to obtain the map f is to integrate numerically the pendulum Eqs.(60) directly.

Fig.8 shows the return map $\varepsilon_{n+1} = f(\varepsilon_n)$ for the case of initially inverted atoms ($\eta > 0$) and $\tau_{int} = 5$. The evolution of the field as successive atoms are injected inside the resonator is obtained by successive iterates of the map. Since for $\eta > 0$ the function f satisfies $f(\varepsilon) \geq \varepsilon$ and its oscillations decrease in amplitude for increasing ε (see Eq.(61)), the iteration always converges towards one of the fixed points $0 = \varepsilon_0^* < \varepsilon_1^* < \ldots < \varepsilon_j^* < \ldots$ given by the solutions of $\varepsilon^* = f(\varepsilon^*)$.

This is proven by noticing the existence of a "critical" fixed point $\varepsilon_{\ell_c}^*$ that cannot be passed for any initial condition $\varepsilon_0 < \varepsilon_{\ell_c}^*$. It is the smallest nontrivial ($\neq 0$) fixed point such that $\mathrm{Max}(f(\varepsilon)) \leq \varepsilon_{\ell_c}^*$ for all $\varepsilon \leq \varepsilon_{\ell_c}^*$. Furthermore, no fixed point ε_ℓ^* above $\varepsilon_{\ell_c}^*$ can be passed if the system starts at a field value $\varepsilon_0 < \varepsilon_\ell^*$. All fixed points are only marginally stable, i.e. stable when approached from below but unstable for $\varepsilon > \varepsilon^*$. They are just at the threshold of type I intermittency (Schuster 1984).

The fixed point ε_q^* clearly corresponds to a $2q\pi$ field-atom interaction: for this value of the intracavity field, the interaction time τ_{int} is precisely such that the Bloch vector performs a $2q\pi$ rotation (Allen and Eberly 1973), leaving the total system unchanged when the atoms leave the cavity. The coherent, nonlinear nature of the semiclassical atom-field interaction dynamically prevents the unlimited growth of

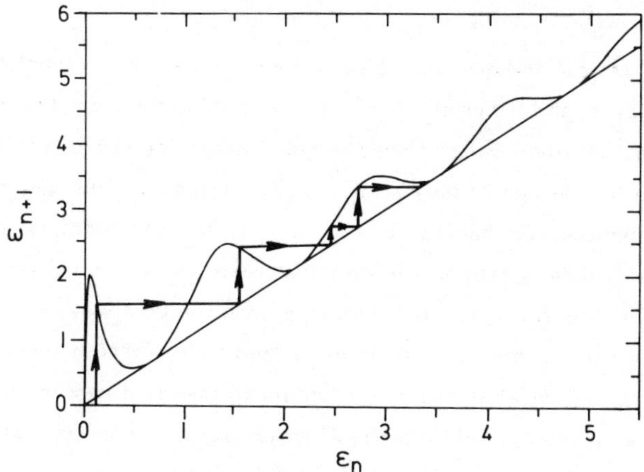

Fig.8: Classical return map $\varepsilon_{n+1} = f(\varepsilon_n)$ for initially inverted atoms (η = 1) for the dimensionless interaction time $\tau_{int} = 5$. Also shown are the straight line $\varepsilon_{n+1} = \varepsilon_n$, and the iteration of the field towards $\varepsilon_3^* \simeq 3.5$ from the initial value $\varepsilon_0 \simeq 0.2$ (after Filipowicz et al 1985a).

the field even in the absence of damping.

It is in principle possible to determine the basin of attraction of any fixed point ε_ℓ^* by repeatedly applying the inverse mapping f^{-1} to the connected interval of the field values ε such that $\varepsilon_{\ell-1}^* < \varepsilon \leq \varepsilon_\ell^*$ and $f(\varepsilon) \leq \varepsilon_\ell^*$. This construction leads to qualitatively different results depending on the interaction time τ_{int}.

Consider first the limit when τ_{int} is short. The first nontrivial fixed point is roughly given by the correspondingly strong intracavity field $\varepsilon_1^* \simeq 2\pi/\tau_{int}$, as seen from Eq.(64). This shows that the field grows indefinitely in the limit $\tau_{int} \to 0$. For $\tau_{int} \lesssim 2.5$, ε_1^* itself becomes the critical fixed point $\varepsilon_{\ell_c}^*$. The basins of attraction of all fixed points are then neatly separated, with all $\varepsilon_{j-1}^* < \varepsilon_0 \leq \varepsilon_j^*$ leading to the fixed point ε_j^*.

The situation is much more intricate for large τ_{int}. The successive fixed points are now close to one another, and there can be a substantial number of them below $\varepsilon_{\ell_c}^*$. Because the map $f(\varepsilon)$ is not reversible, this leads to the construction of tightly interwoven basins of attraction of the various fixed points, with a concomitant sensitive dependence on initial conditions. The iterates of the mapping $\varepsilon_{n+1} = f(\varepsilon_n)$ are 'almost chaotic': Although any initial point always leads asymptotically to one of the fixed points ε_j^*, which one it will be depends sensitively on the initial value ε_0 of the field (Filipowicz et al 1985a).

To compare the semiclassical and quantum descriptions we assume for simplicity that the initial field ε_0 is large, and hence the field changes only little under the influence of a single atom. Eq.(64) shows that then the fixed points of the return map are approximatively given by the condition $\varepsilon_q^*\tau_{int} = 2q\pi$, q integer. This is the obvious counterpart of the quantum-mechanical condition (54) for the blocking of phase space. The major difference is that in the classical case, there is no restriction on the possible values of the field, so that blocking eventually always occurs. In contrast, the intensity fluctuations present in quantized fields force it eventually to grow past these points. A notable exception occurs if the field evolves towards a number state, a field with suppressed intensity fluctuations. Of course, classical intensity fluctuations would lead to a situation similar to that of a (fluctuating) quantum field, so that this system does not provide a good experimental test of QED, in contrast to e.g. the Jaynes-Cummings revivals. In this sense, this situation is analogous to the Cummings collapse, which as we have seen Section 3 can be produced by fluctuating classical chaotic fields as well as by quantized fields.

5. THE MICROMASER

A perfectly lossless cavity can of course not be built in practice. But if cavity damping is slow enough to be negligible during the transit time of a single atom through the resonator, as is expected to be the case in the experiments of Walther and coworkers (Meschede et al 1985), there is a straightforward approximate way to handle dissipation: if an atom is present inside the cavity, the dynamics is described by the JCM, and if no atom is present the evolution of the field is given by the master equation describing the coupling of a single harmonic oscillator to a thermal bath (Louisell 1973). This is of course only approximately correct, as the Langevin forces associated with dissipation are expected to react back on the JCM dynamics and influence the collapse and revivals. In the next Section, we discuss quantitatively the conditions under which such a separation is possible, but for now assume that it is a good approximation in the case of very high-Q resonators. This allows us to build a quantum theory of a truly microscopic maser (Filipowicz 1985b). We show that this system displays features alien to usual masers and lasers (Sargent et al 1975). In particular, the intracavity field is in general sub-Poissonian, and the onset of maser oscillations is followed by a succession of abrupt transitions in the state of the field. The familiar coherent output of masers and lasers is recovered only after introducing a sufficient amount of stochasticity in the system.

As we have seen in Section 4, Eq.(48), the state of the field after exit of atom i is given by

$$\rho_F(t_i+t_{int}) = Tr_A[U(t_{int})\rho(t_i)] \equiv F(t_{int})\rho_F(t_i) \quad . \tag{48}$$

But in the interval between $t_i+ t_{int}$ and the time t_{i+1} at which the next atom is injected, the field now evolves at the damping rate $\gamma = \omega/Q$ towards a thermal steady state with a temperature-dependent mean photon number n_b. This evolution is described by the standard master equation (Louisell 1973)

$$\dot{\rho}_F \equiv L\rho_F = (\gamma/2)(n_b+1)[2a\rho_F a^\dagger - a^\dagger a\rho_F - \rho_F a^\dagger a] + \\ + (\gamma/2) n_b [2a^\dagger\rho_F a - aa^\dagger\rho_F - \rho_F aa^\dagger] \tag{65}$$

of a damped harmonic oscillator. Hence, at time t_{i+1} the state of the field is

$$\rho_F(t_{i+1}) = \exp(Lt_p)\, F(t_{int})\rho_F(t_i) \, , \tag{66}$$

where $t_p = t_{i+1}-t_i-t_{int} \simeq t_{i+1}-t_i$ is the time interval between atom i leaving the resonator and atom i+1 entering it. Iteration of Eq.(66) eventually leads to a steady-state photon density matrix $\rho_{F,st}$, which is the solution of this equation with $\rho_F(t_{i+1}) = \rho_F(t_i)$.

We only treat the case when many atoms traverse the cavity during the lifetime $1/\gamma$ of the field, $N_{ex} \equiv R/\gamma >>1$, where R is the atomic flux, and invoke the technical assumption that the arrivals of the atoms obey a Poisson process with mean spacing 1/R between events. Its effect is to replace the operator $\exp(Lt_p)$ by its average $(1-L/R)^{-1}$ over the exponential distribution $P(t_p) = R\exp(-Rt_p)$ of the intervals t_p. The steady-state field occupation probabilities can then be solved analytically up to a normalization constant:

$$p_n = C\cdot\left(\frac{n_b}{1+n_b}\right)^n \prod_{k=1}^{n} \left(1+\frac{N_{ex}}{n_b}\frac{\beta_k}{k}\right) \, . \tag{67}$$

From now on we focus on exact resonance, $\Delta = \omega - \omega_0 = 0$.

Fig.9 shows the mean photon number <n> inside the resonator as a function of the scaled interaction time $\tau_{int} \equiv \sqrt{N_{ex}}\, \lambda t_{int}$, with $N_{ex}=200$ and the number of thermal photons $n_b = 0.1$. Physically, τ_{int} plays the role of the pump parameter of the microscopic maser. For small τ_{int}, <n> is nearly zero, but a finite <n> emerges at $\tau_{int} = 1$. For τ_{int} increasing past this value, <n> first grows rapidly, but then starts decreasing to reach a minimum at about $\tau_{int} \simeq 2\pi$. At this point, the field abruptly jumps to a higher intensity. This general behaviour recurs roughly at integer multiples of 2π, but becomes less pronounced for increasing τ_{int}. Finally, a stationary regime with <n> nearly independent of τ_{int} is reached. However, outside the time scale of Fig.9 there is still additional structure reminiscent of the Jaynes-Cummings revivals.

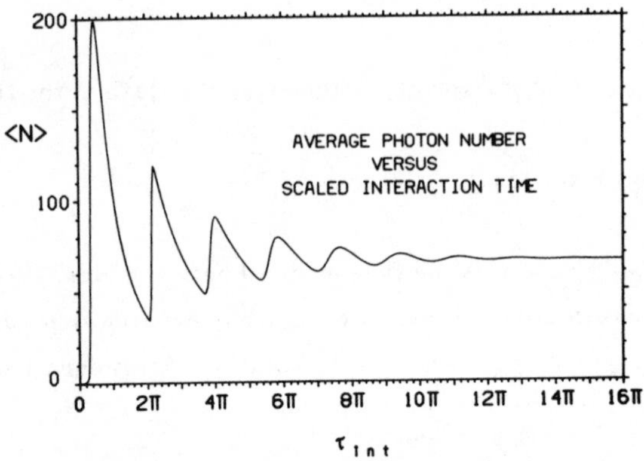

Fig.9: Intracavity mean photon number <n> as a function of the dimensionless interaction time τ_{int} for $N_{ex} = 200$, $n_b = 0.1$ and $\Delta = 0$, in the case of a monoenergetic atomic beam. These parameters are consistent with the experimental values of Meschede et al (1985).

For N_{ex} not too small, at the very first transition the function $\langle n(\tau_{int})\rangle / N_{ex}$ essentially does not depend on N_{ex}. In contrast, the subsequent transitions become sharper for increasing N_{ex}. We conclude that in the limit $N_{ex} \to \infty$ the first transition is analogous to a continuous phase transition, while the others are similar to first-order phase transitions.

The origin of the transitions can be understood intuitively via the stochastic coherent evolution equation for the photon number n

$$\dot{n} = \gamma N_{ex} \sin^2(\lambda\sqrt{n+1}\, t_{int}) - \gamma n + \xi(n,N_{ex},t) \; . \tag{68}$$

The first term in Eq.(68) is the gain due to the change in atomic inversion as deduced from the Jaynes-Cummings model (compare with Eq.(16)), and the second term describes cavity losses (here $n_b = 0$). The third term is a stochastic force reflecting both the classical random character of the arrival times of the atoms and the intrinsic quantum fluctuations. An involved analysis is needed (Filipowicz et al 1986b) to obtain $\xi(n,N_{ex},t)$, but for $N_{ex} >> 1$ the noise force clearly can be regarded as small since the standard deviation and the mean of the photon number are of the order of $\sqrt{N_{ex}}$ and N_{ex}, respectively.

The possible mean photon numbers $\langle n\rangle$ are approximately given by the stable stationary solutions of Eq.(68) with $\xi = 0$. For $\tau_{int} < 1$ the only solution for the field is $\langle n\rangle \propto \tau_{int}^2 << 1$. The maser threshold $\tau_{int} = 1$ occurs when the linearized gain for $n \simeq 0$ compensates the cavity losses:

$$R \frac{d}{dn} \sin^2(\lambda\sqrt{n+1}\, t_{int})\Big|_{n=0} \simeq R\,(\lambda t_{int})^2 = \gamma . \tag{69}$$

The threshold condition is thus recovered from a simple gain-loss point of view (Sargent et al 1975).

When τ_{int} is further increased, Eq.(68) exhibits an increasing number of metastable steady states, and the fluctuating force ξ causes transitions between them. We assume for simplicity that ξ is not a function of n. Then the system predominantly tends to reside in the <u>global</u> minimum of the potential

$$V(n) = -\int dn\, \gamma\{N_{ex} \sin^2(\lambda\sqrt{n+1}\, t_{int}) - n\}. \tag{70}$$

An inspection of Eq.(70) shows that the value of n giving the global minimum is a discontinuous function of τ_{int}, which explains the successive transitions of Fig.9.

Fig.10 shows the normalized standard deviation $\sigma \equiv \sqrt{\langle n^2\rangle - \langle n\rangle^2}/\sqrt{\langle n\rangle}$ of the photon distribution as a function of τ_{int} for the parameters of Fig.9. Above the threshold $\tau_{int} = 1$ the photon statistics is first strongly super-Poissonian, with $\sigma \simeq 4$. Further super-Poissonian peaks occur at the positions of the subsequent transitions.

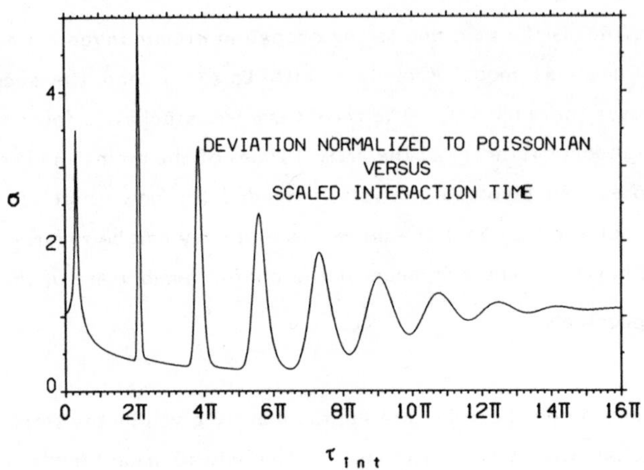

Fig.10: Normalized second moment σ of the photon distribution for the conditions of Fig.9. For a sub-Poissonian distribution, $\sigma < 1$.

In the remaining intervals of τ_{int}, σ is typically of the order of 0.5, a signature of the sub-Poissonian nature of the field. Note that the value $\sigma \simeq 1$ at large interaction times does not signal a Poissonian photon statistics, because the photon number distribution p_n then typically turns out to have multiple peaks, very much like those illustrated in Fig.11. In general, the microscopic maser does not produce coherent radiation.

The difference between our conclusions and the traditional laser theories arises because our microscopic maser possesses less stochasticity and noise than macroscopic masers and lasers. Of course, we have introduced some noise with the averaging over the intervals t_p between atoms. But because many atoms traverse the cavity during the lifetime of the field, it has no practical influence on the results, as was checked both analytically and numerically. This average is primarily a technical trick to obtain the analytic form (67) of the photon statistics. Another source of noise is associated with cavity damping, as implied by the fluctuation-dissipation theorem. The major effect of this damping is to force the evolution of the system towards a steady state. We described in the preceding Section a situation where the

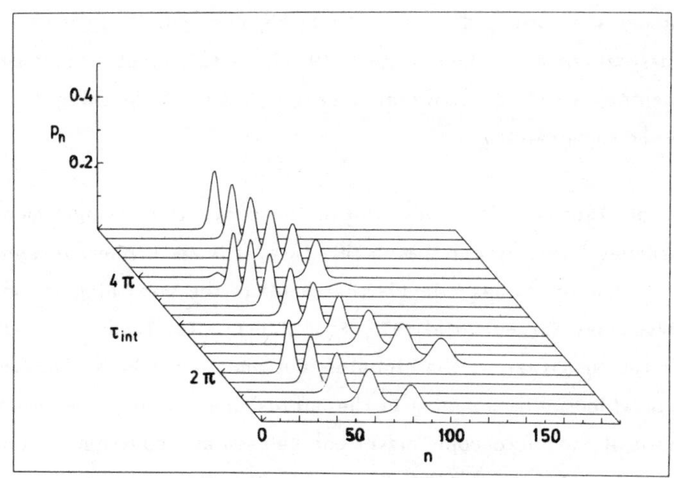

Fig.11: Photon statistics of the microscopic maser for relatively short interaction times.

maser can reach a steady state also in the absence of losses, but such cases are non-generic and characterized by a field either with a truncated thermal distribution or in a pure number state, i.e. by a field of even less coherence than when cavity losses are introduced. A coherent output is reached only if a sufficient amount of stochasticity is introduced in the model.

To illustrate this point, we introduce a velocity distribution by averaging $\beta_k(t_{int})$ in Eq.(67) over interaction times, a correct procedure when the interaction times of different atoms are statistically independent. The first few transitions survive a velocity spread as large as $\Delta v/v \simeq 10\%$, but all features except the maser threshold are washed out when the full Maxwellian width is kept. In this case, the factors β_k are replaced by their average 1/2 over a broad distribution of t_{int}, and we essentially regain the results of conventional laser theory (Sargent et al 1975). Eq.(67) becomes

$$p_n = C\cdot\left(\frac{n_b}{1+n_b}\right)^n \prod_{k=1}^{n}\left(1+\frac{N_{ex}}{2n_b k}\right) , \tag{71}$$

so that far above threshold, $N_{ex} \to \infty$, a field with Poissonian photon statistics is created inside the resonator.

Paradoxically, it is an incoherent average that is responsible for the coherent field (or more precisely, Poissonian photon statistics) of the maser. The reason is that the granulated character of the quantum-mechanical phases involved in the coherent atom-field dynamics is averaged over, so that the electrons in the atoms act much more like a classical current (Glauber 1963, 1965). A sufficient amount of inhomogeneous broadening, with a concomitant averaging over the detuning Δ in Eq.(67), would produce the same result.

The same general argument also holds true in the conventional single mode, homogeneously broadened laser, except that in this case, it is an incoherent average over the radiative lifetime of the laser levels that leads to the smoothing of the quantum mechanical phases (see Sargent et al 1975, Eq.(17.11)). The fundamental difference between these two situations is that spontaneous emission, which is unavoidable in optical resonators, looses its meaning in the microscopic maser. The (usually) sub-Poissonian output of the microscopic maser can be seen as a consequence of inhibited spontaneous emission (Drexhage 1974; Goy et al 1982; Gabrielse and Demhelt 1985; Hulet et al 1985; Meschede et al 1985).

6. DISSIPATION AND THE JAYNES-CUMMINGS MODEL

In the preceding Section, we described an approximate way to handle dissipation in a case where the interaction of a single atom with the cavity field mode is interrupted after a time short compared to the resonator damping time. This is quite appropriate to describe a number of features of the micromaser of Walther and coworkers (Meschede et al 1985), but clearly does not allow to discuss the back reaction of the dissipative forces responsible for damping of the JCM collapse and revivals. Also, when exactly this approximation looses its validity remains to be determined.

The finite cavity Q introduces cavity losses and bath Langevin fluctuations into the JCM and will modify the quantum collapses and revivals. Of course, if the field damping rate $\gamma = \omega/Q$ is very large, the field energy will be dissipated before a Rabi oscillation is completed. For larger Q cavities, many Rabi oscillations can occur in a cavity damping time and a quantum collapse can be observed if $\gamma < \lambda$. Finally, in very high Q cavities, the revival time $t_R \simeq 2\pi\alpha/\lambda$ lies within the cavity damping time γ^{-1}. Haroche (1984) has used dressed atom methods to discuss the dissipation effects on Rabi oscillations. Sachdev (1984) has obtained an analytic solution for spontaneous emission in an unoccupied ($\bar{n} = 0$) damped field mode.

The effects of dissipation on collapse have been examined by Barnett and Knight (1985). Their method is based on an inductive solution of the equations of motion of the atom-field reduced density operator (Agarwal 1974; Louisell 1973). The density matrix for the atom-field-cavity bath system is obtained by standard master equation techniques. The resistive cavity dissipation is described as in the preceding Section by coupling to a bath of harmonic oscillators sufficiently broad and slowly-varying with frequency to allow the use of the Born-Markov approximation (Louisell 1973) and obtain a cavity damping $\gamma = \omega/Q$ of the field energy. At 0° K and for a mode exactly resonant with the two-level transition, we have

$$\frac{d\rho}{dt} = -i\lambda[(a^{\dagger}\sigma_{-} + \sigma_{+}a), \rho] + \gamma a\rho a^{\dagger} - (\gamma/2)\, a^{\dagger}a\rho - (\gamma/2)\, \rho a^{\dagger}a \quad . \tag{72}$$

This master equation could be analyzed directly (e.g. Puri and Agarwal 1985). Instead, we transform the density matrix to a new 'dissipation' picture using the operators J and L defined by their action to the right of the density matrix (Davies 1976)

$$J\rho = -\gamma a\rho a^{\dagger} \tag{73}$$

$$L\rho = -(\gamma/2)\, a^{\dagger}a\rho - (\gamma/2)\, \rho a^{\dagger}a \tag{74}$$

so that the density matrix in the dissipation picture is

$$\chi = \exp(-(J+L))\, \rho \quad . \tag{75}$$

The evolution of this dissipation picture density matrix is given by the transformed master equation

$$\begin{aligned}\frac{d\chi}{dt} = -i\lambda[&e^{\gamma t/2}\, a^{\dagger}\sigma_{-}\chi - 2\sinh(\gamma t/2)\, \sigma_{-}\chi a^{\dagger} + e^{-\gamma t/2}\, \sigma_{+}a\chi \\ &- e^{\gamma t/2}\, \chi\sigma_{+}a + 2\sinh(\gamma t/2)\, a\chi\sigma_{+} - e^{-\gamma t/2}\, \chi a^{\dagger}\sigma_{-}] \quad .\end{aligned} \tag{76}$$

This equation resembles the undamped Jaynes-Cummings evolution equation, except for the presence of terms in which χ is sandwiched between operators. These terms, together with those involving $\exp(+\gamma t/2)$, account for the change in loss rate when the atom makes a transition. In the form of the semiclassical limit in which the field is taken as a classical entity uneffected by the atomic evolution, Eq.(76) becomes

$$\frac{d\rho}{dt} = -i\Omega e^{-\gamma t/2}\,[(\sigma_+ + \sigma_-),\, \rho]$$

where Ω is the semiclassical Rabi frequency and $\rho = \chi$ in this approximation. In this limit we note that the resistive coupling merely damps the atom-field perturbation in an exponential form expected from a phenomenological viewpoint. Departures from this form are a consequence of the quantised nature of the atom-field interaction.

In the terms of atom-field components, Eq.(76) becomes

$$\begin{aligned}\frac{d}{dt}\chi(n,m;e,e) &= -i\lambda(n+1)^{1/2}e^{-\gamma t/2}\chi(n+1,m;g,e)\\ &\quad + i\lambda(m+1)^{1/2}e^{-\gamma t/2}\chi(n,m+1;e,g)\end{aligned} \tag{77}$$

$$\begin{aligned}\frac{d}{dt}\chi(n+1,m;g,e) &= -i\lambda(n+1)^{1/2}e^{\gamma t/2}\chi(n,m;e,e)\\ &\quad + i\lambda(m+1)^{1/2}e^{-\gamma t/2}\chi(n+1,m+1;g,g)\\ &\quad + 2i\lambda(m+1)^{1/2}\sinh(\gamma t/2)\,\chi(n+1,m+1;e,e)\end{aligned} \tag{78}$$

$$\begin{aligned}\frac{d}{dt}\chi(n,m+1;e,g) &= -i\lambda(n+1)^{1/2}e^{-\gamma t/2}\chi(n+1,m+1;g,g)\\ &\quad + i\lambda(m+1)^{1/2}e^{\gamma t/2}\chi(n,m;e,e)\\ &\quad - 2i\lambda(n+1)^{1/2}\sinh(\gamma t/2)\,\chi(n+1,m+1;e,e)\end{aligned} \tag{79}$$

$$\begin{aligned}\frac{d}{dt}\chi(n+1,m+1;g,g) &= -i\lambda(n+1)^{1/2}e^{\gamma t/2}\chi(n,m+1;e,g)\\ &\quad + i\lambda(m+1)^{1/2}e^{\gamma t/2}\chi(n+1,m;g,e)\\ &\quad + (e^{\gamma t}-1)\frac{d}{dt}\chi(n+1,m+1;e,e)\ .\end{aligned} \tag{80}$$

For instance, the coefficient of $|n\rangle\langle m+1|\otimes|e\rangle\langle g|$ is $\chi(n,m+1;e,g)$. We note that the dissipation picture density matrix elements are coupled only to elements with either the same total excitation number, or to elements containing one more quantum in the bra and ket sides of the density matrix. It is this feature of the representation that allows us to solve for the atomic inversion. If we were to take the atom to be initially excited with the field in the vacuum state as did Sachdev (1984) the number of participating atom-field states would shrink enormously and the inhomogeneous terms (terms corresponding to different number of quanta in this representation) in Eqs.(77)-(80) would be absent.

The general solution of Eqs.(77)-(80) is complicated. If we are interested only in the atomic inversion, it is sufficient to trace over the photon numbers after calculating

$\chi(n,n;e,e)$. We proceed by eliminating $\chi(n,n+1;e,g)$, $\chi(n+1,n;g,e)$ and $\chi(n+1,m+1;g,g)$ to generate a third-order inhomogeneous ordinary differential equation for $\chi(n,n;e,e)$. Writing χ_n as short-hand for $\chi(n,n;e,e)$,

$$\frac{d^3\chi_n}{dt^3} + \frac{3\gamma}{2}\frac{d^2\chi_n}{dt^2} + \left[4\lambda^2(n+1) + \frac{\gamma^2}{2}\right]\frac{d\chi_n}{dt} + 2\gamma\lambda^2(n+1)\,\chi_n$$

$$= 2\lambda^2(n+1)\left[\gamma\,\chi_{n+1} + 2(1 - e^{-\gamma t})\,\frac{d\chi_{n+1}}{dt}\right]. \tag{81}$$

We define

$$x_n = e^{\gamma t/2}\,\chi_n \tag{82}$$

which obeys the third order differential equation

$$\frac{d^3x_n}{dt^3} + [4\lambda^2(n+1) - \gamma^2/4]\,\frac{dx_n}{dt} =$$

$$2\lambda^2(n+1)\left[\gamma e^{-\gamma t}\,x_{n+1} + 2(1 - e^{-\gamma t})\,\frac{dx_{n+1}}{dt}\right]. \tag{83}$$

Consider the case of an initially excited atom in a field containing precisely ℓ photons. Within the secular approximation,

$$\ell^2\gamma \ll \lambda\sqrt{\ell+1} \tag{84}$$

x_ℓ satisfies the equation

$$\frac{d^3x_\ell}{dt^3} + 4\lambda^2(\ell+1)\,\frac{dx_\ell}{dt} = 0 \tag{85}$$

with solution

$$x_\ell = \frac{1}{2}\,(1 + \cos 2\lambda\sqrt{\ell+1}\;t)\ . \tag{86}$$

This solution can be used in Eq.(83) to derive inductively the form of $x_{\ell-1}$. This process is repeated to give all the x_n. The x_n for $n > \ell$ are all zero from the initial conditions. The equations of motion are linear in the x_n. Therefore, for an initial field with a spread of photon numbers the x_n are given by photon number weighted sums over the solutions for an initial field with a precise number of photons. Inverting the transformations to obtain density matrix elements of ρ, we find after extensive algebra for the excited state probability

$$P_e(t) = \frac{1}{2} e^{-\gamma t/2} [\sum_{\ell=0}^{\infty} e^{-\ell\gamma t} p_\ell \cos 2\lambda\sqrt{\ell+1}t$$

$$+ \sum_{n=0}^{\infty} \sum_{\ell=0}^{\infty} \frac{(\ell-n-1/2)!}{(\ell-n)!} (1-e^{-\gamma t})^{\ell-n} p_\ell] \ . \qquad (87)$$

Note that there is no damping modification of the Rabi frequency in this under-damped limit. The simplicity of this result is a consequence of the very weak cavity damping: many Rabi oscillations occur in a cavity lifetime. In an untransformed density matrix treatment in a dressed-state basis, the secular approximation enables us to neglect the couplings between coherences belonging to different manifolds (that is, off-diagonal density matrix elements that do not correspond to the same number of quanta in the atom-field system). This much simplifies the dynamics of the 'cascade' through dressed states as the cavity Q resistively damps out the atom-field excitations.

The atomic inversion W_Q is given in terms of P_e by

$$W_Q(t) = 2P_e(t) - 1 \ . \qquad (88)$$

In Fig.12, the atomic inversion is plotted as a function of time for an initially excited atom in interaction with a coherent field of mean photon number equal to 5 for a small cavity damping $\gamma = 0.03\lambda$. As the damping increases the revivals rapidly diminish in amplitude. The revivals depend upon the discrete nature of the photon number distribution; this 'granularity' is translated into a discrete spectrum of Rabi frequencies. The cavity damping broadens these spectral components (or 'dressed atom' eigenenergies); when this broadening becomes comparable with the spacing between the Rabi frequencies the spectrum becomes continuous, the inversion collapses and never revives. The revivals may thus be significantly attenuated, even though they occur at times $t < \gamma^{-1}$, the cavity field lifetime. Note, however, that our simple solution cannot be used to demonstrate the <u>complete</u> destruction of revivals as this requires a stronger damping than that permitted by our secular or under-damped approximation and would require a less restrictive approach to the solution of Eq.(81).

The collapse depends on the spread of the Rabi frequencies and is affected much less

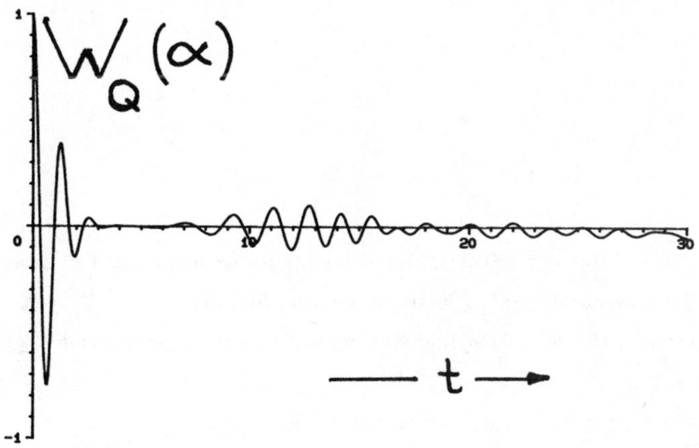

Fig.12: The evolution of the atomic inversion $W_Q(\alpha)$ for an initially excited atom in a coherent field with a mean photon number of 5 and $\gamma = \omega/Q = 0.03\lambda$.

by the broadening of the dressed eigenstates. Correspondigly, the collapse in Fig. 12 is practically the same as in the absence of cavity losses.

Cavity damping will significantly attenuate the revivals unless the cavity Q is such that the field damping rate γ is very much less than the one-photon Rabi frequency λ. An elementary calculation shows that for a Rydberg atom mm-wave transition, $\gamma/\lambda \simeq 10^{15} Q^{-1} (\Delta n V/n^7)$, where the cavity volume V is in m^3, n is the initial state principal quantum number and Δn the change in n in the transition. For the experiments of Haroche and coworkers (Haroche 1984), $\gamma/\lambda \simeq 2$, whereas those of Meschede et al (1985) have $\gamma/\lambda \simeq 2.10^{-3}$, and are much less affected by dissipation. The recent experimental observation of collapse and revivals by the Munich group (Walther 1985) has a wide significance as an example of a quantum coherence phenomenon competing against an environmental damping.

ACKNOWLEDGMENTS

We would like to thank G. S. Agarwal, J. H. Eberly, M. A. Lauder, L. Lugiato, M. Sargent III, M. O. Scully, S. Stenholm, P. M. Radmore, G. Rempe and H. Walther for discussions. The Imperial College part of this work was supported in part by the Science and Engineering Research Council.

REFERENCES

Agarwal G. S. 1974, "Quantum Statistical Theories of Spontaneous Emission and their Relation to Other Approaches", (Springer Verlag, Berlin)

Allen L. and Eberly J. H. 1975, "Optical Resonance and Two-Level Atoms" (John Wiley and Sons, New York)

Barnett S. M. and Knight P. L. 1985, to be published

Bayfield J. E. and Pinnaduwage A. 1985, Phys. Rev. Lett. **54**, 313

Buck B. and Sukumar C. V. 1981, Phys. Lett. **81A**, 132

Buck B. and Sukumar C. V. 1984, J. Phys. **A17**, 877

Casati G., Chirikov B. V., Israilev F. M. and Ford J. 1979, in Stochastic Behaviour in Classical and Quantum Systems", Eds. G. Casati and J. Ford, Lecture Notes in Physics **93** (Springer, New York)

Casati G., Chirikov B. V. and Shepelyanski D. L. 1984, Phys. Rev. Lett. **53**, 2525

Cummings F. W. 1965, Phys. Rev. **140**, A1051

Davies E. B. 1976, "Quantum Theory of Open Systems", (Academic Press, London)

Drexhage K. H. 1974, in "Progress in Optics" Vol. 12, ed. E. Wolf (North-Holland, Amsterdam)

Eberly J. H., Narozhny N. B. and Sanchez-Mondragon J. J. 1980, Phys. Rev. Lett. **44**, 1323

Filipowicz P., Javanainen J. and Meystre P. 1985a, J. Opt. Soc. Amer. B. (in press)

Filipowicz P., Javanainen J. and Meystre P. 1985b, submitted to Phys. Rev. Lett

Filipowicz P., Javanainen J. and Meystre P. 1986a, in "Coherence, Cooperation and Fluctuations", eds. F. Haake, L. M. Narducci and D. F. Walls (Cambridge University Press, Cambridge)

Filipowicz P., Javanainen J. and Meystre P. 1986b, to be published

Filipowicz P. and Mostowski J. 1981, Phys. Lett. **86A**, 356

Gabrielse G. and Dehmelt H. 1985, Phys. Rev. Lett. **55**, 67

Glauber R. J. 1963, Phys. Rev. **131**, 2766

Glauber R. J. 1965, in "Quantum Optics and Electronics", eds. C. DeWitt, A. Blandin, and C. Cohen-Tannoudji (Gordon and Breach, New York)

Goy P., Raimond J. M., Gross M. and Haroche S. 1982, Phys. Rev. Lett. **50**, 1903

Graham R. 1985, to be published in the Proceedings of the Rochester International

Meeting on Instabilities and Dynamics of Lasers and Nonlinear Optical Systems, eds. R. Boyd, L. Narducci, and M. Raymer (Cambridge University Press, Cambridge)

Graham R. and Tel T. 1985, Z. Physik **B**, in press

Haroche S., Goy P., Raymond J. M., Fabre C. and Gross M. 1982, Phil. Trans. Roy. Soc. Lond. **A307**, 659

Haroche S. 1984, in "New Trends in Atomic Physics", Proc. Les Houches Summer School Session XXXVIII, eds. G. Grynberg and R. Stora (North-Holland, Amsterdam)

Hioe F.T, Yoo H. I. and Eberly J. H. 1983, in "Coupled Nonlinear Oscillators" eds. J. Chandra and A. C. Scott, (North-Holland, Amsterdam)

Hulet R. G., Hilfer E. S. and Kleppner D. 1985, Phys. Rev. Lett. **55**, 2137

Jaynes E. T. and Cummings F. W. 1963, Proc. IEEE **51**, 89

Knight P. L. and Allen L., 1983, "Concepts of Quantum Optics" (Pergamon Press, Oxford)

Knight P. L. and Milonni P. W. 1980, Physics Reports **66C**, 21

Knight P. L. and Radmore P. M. 1982a, Phys. Rev. **A26**, 676

Knight P. L. and Radmore P. M. 1982b, Phys. Lett. **90A**, 342

Li Xiao-shen and Bei Nian-yu 1984, Phys. Lett. **101A**, 169

Loudon R. 1983, "The Quantum Theory of Light", 2nd edition, (Oxford University Press, Oxford)

Louisell W. H. 1973, "Quantum Statistical Properties of Radiation" (Wiley, New York)

Meschede D., Walther H. and Müller G. 1985, Phys. Rev. Lett. **54**, 551(1985)

Meystre P. 1974, PhD Thesis, Swiss Federal Institute of Technology, Lausanne (unpublished)

Meystre P., Geneux E., Faist A. and Quattropani A. 1975, Nuovo Cimento **25B**, 521

Mollow B. R. 1975, Phys. Rev. **A12**, 1919

Narozhny N. B., Sanchez-Mondragon J. J. and Eberly J. H. 1981, Phys. Rev. **A23**, 236

Pegg D. T. 1980, in "Laser Physics", eds. D. F. Walls and J. D. Harvey (Academic Press, Sidney)

Puri R. R. and Agarwal G. S. 1985, submitted to Phys. Rev. **A**

Rempe G. and Walther H. 1985, in "Fundamentals of Laser Interactions", ed. F. Ehlotzky (Springer Verlag, Berlin)

Riti C. and Vetri G. 1982, Opt. Comm. **44**, 105

Sachdev S. 1984, Phys. Rev. **A29**, 2627

Samson R. and Ben-Reuwen A. 1975, Chem. Phys. Lett. **36**, 523

Sargent III M., Scully M. O. and Lamb, Jr. W. E. 1974, "Laser Physics" (Addison-Wesley, Reading). See in particular Chapter 17.

Schuster H. G. 1984, "Deterministic Chaos: An Introduction" (Physik-Verlag,

Weinheim)

Singh S. 1982, Phys. Rev. **A25**, 3206

Stenholm S. 1973, Phys. Rep. **6C**, 1

Stenholm S. 1981, Opt. Comm. **36**, 75

Sukumar C. V. and Buck B. 1981, Phys. Lett. **83A**, 211

Sukumar C. V. and Buck B. 1984, J. Phys. **A17**, 885

von Foerster T 1975, J. Phys. A **8**, 95

Walls D. F. and Barakat R, Phys. Rev. A **1**, 446 (1970)

Walther H. 1985, private communication

Yoo, H. I., Sanchez-Mondragon J. J. and Eberly J. H. 1981, J. Phys. **B14**, 1383

Yoo H. I. and Eberly J. H. 1985, Phys. Rep. **118**, 239

THE DYNAMIC CHARACTER OF QUANTUM ELECTRODYNAMICS IN QUANTUM OPTICS

H J KIMBLE

1. INTRODUCTION

A deceptively simple optical experiment that has played an important role in the development of the quantum theory is the Young's two-slit interference experiment. A straightforward calculation in either classical or quantum theory yields the field in the detection plane as the sum of the amplitudes of the electric fields propagated from the two apertures. There is thus a sense in which the quantum interference of amplitudes is trivial and already embodied in classical physics. This point was made quite clearly by Born in his correspondence with Einstein (1).

> "Let me begin with an example. A beam of light falls on to a plate of doubly refracting crystal, and is split into two beams. The direction of polarisation of one of the beams is determined by measurement: it is then possible to deduce that that of the second beam is perpendicular to the first. In this way one has been able to make a statement about a system in a certain part of space as a result of a measurement carried out on a system in another part of space. That this is possible depends on the knowledge that both beams have originated from one beam which has passed through a crystal; in the language of optics, that they are coherent. It seemes to me that this case is closely related to your abstract example, which is apparently connected with collision theory. But it is simpler and shows that such things happen within the framework of ordinary optics. All quantum mechanics has done is to generalise it."

Of course in certain circumstances there is a profound difference between the correlations that arise in classical theory and quantum theory, as is illustrated by the consideration of experiments of the type described by Einstein, Podolsky and Rosen (2). This distinction has been sharply drawn by Bell (3), with the confirmation of the quantum theory being made in a series of beautiful experiments (4-6). Indeed it is in the investigation of phenomena for which the rules of classical probability theory are violated that one is able to address perhaps the single most important aspect of the quantum theory, as Dirac has pointed out:

> "I believe that the concept of the probability amplitude is perhaps the most fundamental concept of the quantum theory."(7)

Following this point of view, I wish to examine in these notes several examples involving the interaction of simple atomic systems with the electromagnetic field. The emphasis will be on the distinction between quantum theory as a simple extension of classical theory as in the quote from Born and quantum theory as a revolutionary description of nature as expressed by the concept of the probability amplitude. In contrast to many of the canonical illustrations of the quantum theory, I will attempt to stress situations that reveal the quantum nature of dynamics as opposed to circumstances in which, roughly speaking, the quantum theory serves as a kind of constraint or boundary condition. My goal is to identify in quantum optics examples for which the paradoxes of quantum theory are embedded inextricably within the structure of the system's dynamic evolution. The interest here is not so much to test QED as it is to explore in search of qualitatively new phenomena which have their origin in the nonclassical nature of QED.

Fortunately there is in quantum optics a means to delineate clearly between classical and nonclassical behaviour. The Optical Equivalence Theorem assures us that if the Glauber-Sudarshan phase-space functional P is well-behaved, then the quantum description reduces to that of classical statistics (8). Hence P serves to demarcate the boundary of nonclassical phenomena in quantum optics (9).

2. PHOTON CORRELATIONS IN AN ATOMIC CASCADE

Since the work of Kocher and Cummins (4) almost 20 years ago, the study of polarization correlations in an atomic cascade has provided an extremely important test of the quantum theory relative to classical statistical theories (local realistic theories). As Mandel points out (10), the quantum predictions rely on the constraints imposed by the angular momenta of the initial and final states and, most importantly, on the quantum superposition of probability amplitudes for a two-particle state. However, predictions of the quantum theory in this experiment do not depend in any detailed way upon the dynamics of the photon emissions.

One might attempt to obtain dynamic information about the emitted field by asking for the joint probability distribution of photon frequencies or for the joint distribution of emission times independent of frequency. Weisskopf and Wigner long ago considered the first problem and showed a strong correlation between the frequencies of photons emitted in a cascade. The statistical coupling of emission frequencies ensures that energy is conserved in each microscopic event. Kimble, Mezzacappa and Milonni (11) considered the complementary question of temporal correlation and found that the photon cascade is sequential but otherwise uncorrelated with respect to its time evolution. There is, however, a slight exception that occurs over an initial interval of order $\Delta t \leqslant$ (optical period) during which there is an intrinsic indistinguishability as to which of two photons is emitted first. One thus obtains an interference between probability amplitudes that decays away as time evolves and the frequencies become (in principle) distinguishable. While the intervals involved in optical transitions are generally much too short to make this effect observable, it nonetheless serves to illustrate the type of behaviour discussed in the introduction. The concept of the quantum amplitude is interwoven with the dynamical evolution of the system.

One should perhaps stress that the statement that the photons in the cascade are sequential but otherwise uncorrelated except for some initial interval does nonetheless imply correlations of nonclassical degree in analogy with parametric down conversion, which is discussed in two other contributions to this volume.

3. QUANTUM JUMPS IN A THREE-LEVEL ATOM

As a second example consider the atomic level scheme shown in Figure 1 in which a "strong" transition, $0 \leftrightarrow 1$, and a "weak" transition, $0 \leftrightarrow 2$, share a common lower level. The designations "strong" and "weak" refer to the relationship of the Einstein A-coefficients for the two transitions, $A_1 \gg A_2$. The transition frequencies are denoted by ω_{10} and ω_{20} with the difference $|\omega_{20}-\omega_{10}|$ itself of the order of an optical frequency. If the strong transition is illuminated by light of frequency $\omega_1=\omega_{10}$ and of Rabi frequency $\Omega_1 \sim A_1$, then the nature of the fluorescent field is well understood (12–14). In particular the fluorescence exhibits antibunching with deviations from classical behaviour occurring over intervals of duration of order A_1^{-1}. A completely analogous result is found if the weak transition alone is illuminated, except that the time scale for nonclassical fluctuations in the field radiated from the $0 \leftrightarrow 2$ transition is $A_1^{-1} \gg A_2^{-1}$.

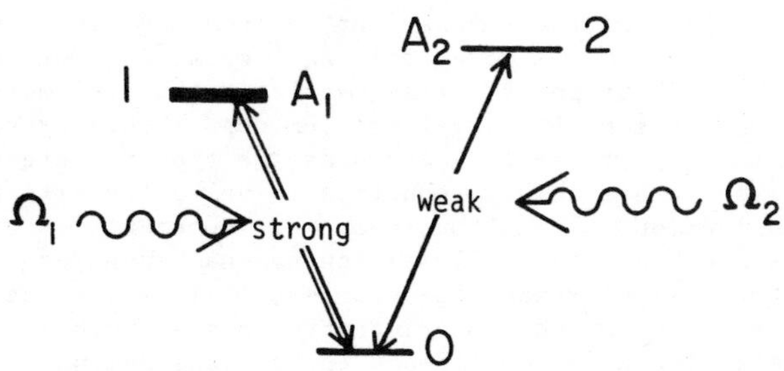

Figure 1.- Three-level atom with "Vee" configuration. The Einstein A-coefficients for the transitions $i \to 0$ are denoted by A_i. Each transition is illuminated by a field in a coherent state corresponding to Rabi frequency Ω_i.

Imagine now that both transitions are illuminated simultaneously. Within the dipole and rotating wave approximations, it is straightforward to calculate the various field correlation functions following the formalism of Ref (11). In particular if $P_{ij}(t,\tau)$ denotes the joint probability of photoelectric detection of a photon from the transition $i\to0$ at time t followed by a photon from the transition $j\to0$ at time $t+\tau$, we find (neglecting retardation in the notation)

$$P_{ij}(t,\tau) \ \alpha \ \langle\hat{\sigma}_{ii}(t)\rangle\langle\hat{\sigma}_{jj}(\tau)\rangle_0 \quad . \tag{1}$$

Here $\langle\hat{\sigma}_{ii}(t)\rangle$ is the expectation value of the Heisenberg operator that represents the probability of occupation of the state i at time t, and the subscript "0" denotes the same probability given that the atom is in the state 0 at time $\tau=0$. In other words, the detection of the first photon occurs with a probability proportional to the probability of occupation of the level i. However, given this first event, the atom is projected into the ground state from which it must be subsequently excited to either of the states 1 or 2.

It is this subsequent re-excitation that is of interest in determining the influence of one transition on the other. For example, if the emission of a photon from the state 1 is followed by an excitation from the ground state to the state 2, then the fluorescence from the $1 \leftrightarrow 0$ transition will terminate for the duration of the interval that the atomic electron spends in level 2, a time of order A_2^{-1}. Since the fluorescence from the $1 \leftrightarrow 0$ transition produces roughly $A_1\sim 10^8$ photons/sec, this termination results in a rather large amplification that led Dehmelt to propose this scheme of "electron shelving" as a means to produce an " atom amplifier" (15).

From the perspective of the current discussion it is of interest to examine the telegraphic atomic fluorescence from the strong transition somewhat more carefully (16,17). If we examine the limit $(A_1\sim\Omega_1)\gg(A_2\sim\Omega_2)$, then we can neglect the detailed nature of processes occurring on a time scale A_1^{-1} since the interruptions of the strong fluorescence due to transitions to the weak level occur on a time scale $A_2^{-1} \gg A_1^{-1}$. One can show quite generally that for such disparate time scales, the three-level problem reduces to an equivalent two-level problem describable by rate equations involving a combination of levels 1 and 0 as the ground state and the level 2 as the excited state. This reduction is valid for either coherent or incoherent illumination. For the case of coherent excitation the equivalent two-level system includes phenomena such as Autler-Townes splittings and three-level Raman-type processes. In either case, one can identify regimes in which the interruptions of the strong signal fluorescence due to the presence of weak transitions dominate the character of this signal. For example, for resonant excitation of the strong transition and for a detuning of the weak field $\Delta_2=\pm1/2\Omega_1$ $(\Delta_2,\Omega_1 \gg A_1)$ to one component of the Autler-Townes doublet, a population $\langle\sigma_{22}\rangle=1/3$ is achieved for Ω_2 of order $(A_1A_2)^{1/2}$. The strong fluorescence is thus off one third of the time. Ω_2 can be made much smaller while still maintaining a population in level 2 of order unity by exciting both transitions on resonance $(\Delta_1=0=\Delta_2)$ with $\Omega_1\ll A_1$. In this case $\langle\sigma_{22}\rangle = 1/3$ for

$\Omega_2 = \Omega_1(A_2/A_1)^{1/2}$. Of course there are regimes in which the interrruptions become extremely rare, such as for $\Omega_1 \gg A_1$, $\Delta_1 = 0 = \Delta_2$. In this case $\langle\sigma_{22}\rangle \sim \Omega_2^2 A_1^3/A_2\Omega_1^4 \ll 1$ unless Ω_2 is of order $\sqrt{A_1 A_2}\,\Omega_1^2/A_1^2$, which is a field much larger than the previous two limits. Due to the Autler–Townes splitting, the weak transition is essentially driven far from resonance.

Subject only to the conditions $\{(A_1 \sim \Omega_1) \gg (A_2 \sim \Omega_2)\}$, it is in fact possible to obtain a complete statistical description of the fluorescence from the strong transition (16–18). This is of course an exceptional circumstance not only in quantum optics but also in statistical physics. Since the analysis outlined above relies upon a coarse graining of time with respect to the lifetime of the strong transition, we necessarily lose such features as photon antibunching of the light from the $1 \leftrightarrow 0$ transition and obtain a classical stochastic process. From the point of view of the overall qualitative nature of the "telegraphic" fluorescence, such features are not of importance since their extent in time A_1^{-1} is extremely small as compared to the time scale A_2^{-1} of the interruptions for the situation experimentally envisioned $(A_1/A_2 \sim 10^{-6})$. However, from the point of view of the classical versus nonclassical nature of the fluorescence, these features are essential. Indeed it is of interest to inquire about the fashion in which one moves from a nonclassical to a classical statistical process by time averaging in the general case of arbitrary $(A_1, \Omega_1, A_2, \Omega_2)$. The three-level problem is ideally suited for such an investigation since at either the quantum or classical extreme the model is quite tractable and as well realistic with regard to experiment.

4. CAVITY QUANTUM ELECTRODYNAMICS

Since our understanding of many radiative problems is based to a large extent upon continuum limits of discrete cavity-mode interactions, I would next like to consider the interaction of a collection of N two-level atoms with a single cavity mode. The Hamiltonian $\hat{H}$ is taken to be of the form (19,20)

$$\hat{H} = \hbar\omega_a/2\ \hat{J}_z + \hbar\omega_c \hat{a}^+\hat{a} + \hbar g[i\hat{J}_-\hat{a}^+ + \text{H.C.}]$$

$$+ [\text{Decay: } \kappa, \gamma] + [\text{Excitation}]. \qquad (2)$$

Here $\{\hat{J}_z, \hat{J}_\pm\}$ are collective atomic operators for the N atoms and obey the usual Pauli algebra. $\{\hat{a}, \hat{a}^+\}$ are boson annihilation and creation operators for the singe mode of the cavity. The atoms and the field mode are coupled through an assumed dipole-interaction with coupling coefficient g given by

$$g=\left(\frac{\omega_c\mu^2}{2\hbar\varepsilon_o V}\right)^{1/2} \tag{3}$$

ω_a and ω_c are the atomic transition frequency and cavity resonance frequency, respectively. Dissipation in the field mode due to (for example) resistive losses at the cavity boundaries is expresed as a field decay rate κ. Atomic decay is assumed to be purely radiative to a set of continuum modes and proceeds at an energy decay rate γ. Both field and atomic decay can be treated by well-known reservoir techniques. The possibility of excitation other than that expressed by a particular initial state is also included.

This Hamiltonian has of course been extensively studied in quantum optics. A partial listing of limiting cases and associated phenomena follows (21-25).

A. $N=1$, $g >> (\gamma,\kappa)$; Jaynes-Cummings.
Quantum revivals
"Vacuum-field" Rabi splittings

B. $N>>1$, $g << (\gamma,\kappa)$; Optical Bistability
Nonequilibrium critical phenomena
Dynamical instabilities
Photon antibunching, squeezing

C. $N=1,(g,\gamma,\kappa)$ arbitrary;
Modified Jaynes-Cummings
Single-atom Optical Bistability

D. $N>>1$, $\kappa>> g >> \gamma$; Single-Mode Superradiance

E. $N>>1$, $g >> \kappa,\gamma$; Tavis-Cummings Model

While there is truly an incredible range of distinct physical processes described by (A-E), it seems that many of the common features inherited from the parent Hamiltonian (2) are often overlooked. One important example can be drawn from the weak field behaviour. The weak field limit is defined by the requirement $x << 1$, with $x=(\langle\hat{a}^+\hat{a}\rangle/N_s)^{1/2}$, $N_s=\gamma^2/8g^2$, and exists for all of the cases (A-E). As Carmichael has shown by employing the Schwinger representation (24), in this limit the atomic system can be represented as an atomic oscillator obeying a boson algebra $[\hat{b},\hat{b}^+]=1$. In the absence of dissipation, the Hamiltonian of Eq (2) becomes

$$\hat{H} = \hbar\omega_a\hat{b}^+\hat{b} + \hbar\omega_c\hat{a}^+\hat{a} + \hbar g\sqrt{N}[\hat{a}^+\hat{b}+\hat{a}\hat{b}^+] \tag{4}$$

describing an atomic oscillator $\hat{b}$ coupled to a field oscillator $\hat{a}$ (25). By introducing the normal-mode operators

$$\hat{A} = \frac{1}{\sqrt{2}}(\hat{a}+i\hat{b})$$

$$\hat{B} = \frac{1}{\sqrt{2}}(\hat{a}-i\hat{b}) \qquad (5)$$

and by considering the case of resonance $\omega_a = \omega_c \equiv \omega$, one can rewrite $\hat{H}$ as a sum of two uncoupled harmonic oscillations,

$$\hat{H} = \hbar(\omega+\sqrt{N}g)\hat{A}^+\hat{A} + \hbar(\omega-\sqrt{N}g)\hat{B}^+\hat{B} \quad . \qquad (6)$$

The normal-mode frequencies are

$$\lambda_\pm = i(\omega \pm \sqrt{N}g) \quad , \qquad (7)$$

with the lifting of the degeneracy of the atom-field system $(\pm\sqrt{N}g)$ having been termed "vacuum-field" Rabi splittings. We see that this nomenclature is somewhat inappropriate since it is not the vacuum field which is split or which induces a splitting. Rather the splitting is a result of the mutual coupling of two systems to produce a spectrum of energies that is no longer doubly degenerate. Note that there remains a single nondegenerate ground state.

Again following Carmichael (24), we inquire as to the role of dissipation in the weak-field limit. For (κ,γ) nonzero, one finds more generally that

$$\lambda_\pm = \frac{1}{2}\left(\frac{\gamma}{2}+\kappa\right) \pm \left[1/4\left(\frac{\gamma}{2}-\kappa\right)^2 - Ng^2\right]^{1/2}$$

$$\equiv \frac{1}{2}\left(\frac{\gamma}{2}+\kappa\right) \pm i\Omega \quad . \qquad (8)$$

We see that $\lambda_\pm$ contains an imaginary part only for

$$\frac{1}{2}\left|\frac{\gamma}{2} - \kappa\right| < \sqrt{N}g \quad . \tag{9}$$

That is, since $\hat{a}$ and $\hat{b}$ are independently coupled to separate reservoirs, a periodic exchange of energy occurs only if the decay rates of the two oscillators are not too dissimilar. For $\gamma/2=\kappa$,

$$\lambda_{\pm} = \frac{\gamma}{2} \pm i\sqrt{N}g \quad , \tag{10}$$

showing not only the previously described splitting but also finite width in the eigenvalue structure of the coupled system.

While many of the above results have appeared in the literature recently in connection with the study of the Jaynes–Cummings problem, one should note that precisely the same results exist in the literature on optical bistability. In that language, Eq.(8) is rewritten in terms of the cooperativity parameter $C=(Ng^2)/(\gamma\kappa)$ as

$$\lambda\pm = \frac{1}{2}\left(\frac{\gamma}{2} + \kappa\right) \pm \left[\frac{1}{4}\left(\frac{\gamma}{2} - \kappa\right) - \gamma\kappa C\right]^{1/2} \quad . \tag{11}$$

As I have stressed above, "vacuum-field" Rabi splitting is a quite general feature of the Hamiltonian (2). The splitting would be manifest in any one of a number of observations including the transient response of the system, the spectral densities, and the intensity correlation functions. In experiments with Rydberg atoms in the microwave region, the conditions $N\sim1$, $g\gg(\kappa,\gamma)$ have been achieved only in superconducting cavities (26). In the visible, the condition $|\gamma/2-\kappa|\ll\sqrt{N}g$ is usually not met, due either to the operating condition $\gamma\gg\kappa$ or to the small values of g. It is, however, possible to observe "vacuum-field" Rabi splitting in the visible as a rather large effect in the weak-field behaviour of a collection of atoms within a high finesse resonator. We are currently pursuing such an investigation in our atomic beam apparatus (27).

With this rather brief introduction to the study of QED in a cavity, let me now return to the central point of these notes and attempt to draw a distinction between classical and quantum features in this problem. Please recall that quantum in the current context refers to nonclassical in the sense of the violation of the classical laws of probability. Furthermore note that nonclassical behaviour is not necessarily implied by what I shall term as violations of one's "Bloch" intuition, which is based upon the behaviour of free-space atoms with a cavity interaction added after the fact. Such violations of "Bloch" intuition are connected with the relationship of time scales $\gamma\leftrightarrow g\leftrightarrow\kappa$. For example, the ratio g/γ expresses the ratio of rates of coherent cavity processes to incoherent decay and can be written rather concisely as

$$\frac{g}{\gamma} = \frac{1}{2}\left(\frac{V_o}{V_{cavity}}\right)^{1/2} \quad . \tag{12}$$

Here V_{cavity} is the cavity volume while V_o is a volume of size (absorption cross-section)×(atomic lifetime) × c . In order to have the coherent processes associated with the interchange of energy between atom and cavity field compete with those of the atomic decay we require $g/\gamma \sim 1$. For example, the enhancement of spontaneous emission into the cavity mode by a single atom is given by an altered decay rate $\gamma' = \gamma\ (1 + 2g^2/\gamma\kappa)$ (23,30,31). In the microwave experiments with Rydberg atoms (26,28,29), $g/\gamma \sim 10^2 - 10^4 >> 1$, so that the coupling to the cavity dominates free-space emission. On the other hand, in the visible, one typically finds that $(g/\gamma) << 1$ and single atom radiative processes are largely unaffected by the cavity. However, my students and I are investigating certain optical systems for which it should be possible to obtain $g/\gamma \sim 1$. With the recent availability of mirrors of very low loss and using allowed dipole transitions in the visible, we hope to investigate a system for which $\gamma \sim g \sim \kappa$. That is, the coherent processes of the atom-cavity field interaction will proceed on the same time scale as the damping processes. While such a system will certainly produce violations of one's "Bloch" intuition, it should go much further to produce a variety of nonclassical effects. In very general terms this is a result of the nonclassical nature of the fluctuation-relaxation processes for atom-field interactons that are purely radiative. Any attempt at a classical statistical analysis will fail due to the singularity of the Glauber-Sudarshan P-function.

One example of nonclassical behaviour is photon antibunching. For a weak field Carmichael (24) has shown that the normalized second-order intensity correlation function $\lambda(\tau) = g^{(2)}(\tau) - 1$ is given from a linearized analysis by

$$\lambda(\tau) = -\frac{8g^2}{\gamma^2}\left\{\frac{\gamma}{(\gamma+2\kappa)}\frac{2C}{(1+2C)}e^{-1/2(\kappa+\gamma/2)\tau}\left[\cos\Omega\tau + \frac{(2\kappa+\gamma)}{\Omega}\sin\Omega\tau\right]\right\} \tag{13}$$

with Ω defined in Eq. (8) and $C = Ng^2/\gamma\kappa$. Note that $N_s \equiv \gamma^2/8g^2$ sets the scale for deviations from classical behaviour ($\lambda(\tau) < 0$). N_s is the saturation photon number. From the values quoted above, one sees that for transitions in Rydberg atoms $N_s << 1$, while typically in the visible $N_s >> 1$. Certainly an exciting prospect would be the investigation of optical systems for which $N_s \sim 1$. For such systems not only would the nonclassical features predicted by the linearized theory become observable, but that theory would itself become inadequate as the quantum fluctuations become increasingly large (note that for N_s sufficiently small, $\lambda(0)$ as calculated from the linearized theory becomes less than -1, which is impossible).

Carmichael, Sakar , and Satchell investigate these issues in their contribution to this volume.

Indeed one might imagine a cirumstance in which the whole of one's intuitive picture of the nature of the quantum dynamical processes becomes unreliable in a regime for which $\gamma \sim g \sim \kappa$. The phenomenon of quantum switching in optical bistability perhaps serves as an illustration (23,32). The existence of two operating states for a given set of external control parameters can at least figuratively be associated with a generalized potential in the multidimensional space of the variables (atomic inversion, atomic polarization, intracavity field). Steady states in this space correspond to local minima in the potential. In the absence of fluctuations, these states are both locally and globally stable. However, the inherent quantum noise associated with the atom-field interaction causes some states to lose their stability as the system explores the potential surface in the fashion of a random walk. Explicit results for the decay or switching from one state to another have been presented only in the "good-cavity" limit, for which $\gamma \gg \kappa \gg g$, $N \gg 1$, and only for the case of resonant atomic and cavity detunings. Furthermore, atomic correlations have been neglected. From the analyses one can examine the spectrum of eigenvalues to obtain the dominant (smallest) eigenvalue λ_1, which governs the rate of global decay. This eigenvalue of course depends upon the external control parameters, but I wish to focus on the point of bistable operation corresponding to the minimum in λ_1. If this value is denoted by $\bar{\lambda}_1$ then a plot of $\bar{\lambda}_1$ versus system size N_s is as shown in Figure 2 for a value of atomic cooperativity parameter C=8, which is twice that for the critical onset of bistability. Note that the decay rate $\bar{\lambda}_1$ has an exponential dependence on N_s, so that for example N_s=100 gives a decay time of 10^{60} cavity lifetimes while N_s=10 results in a time of 10^6 cavity lifetimes.

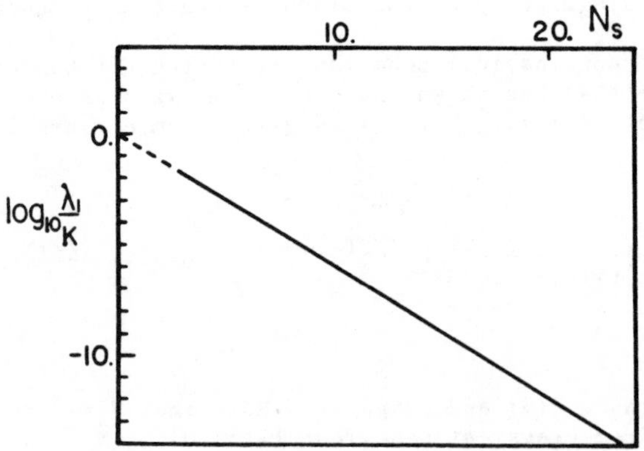

Figure 2.- Switching eigenvalue $\bar{\lambda}_1$ evaluated at the most stable point of operation versus system size N_s. The case considered is absorptive bistability in the good cavity limit, C=8. The figure is constructed from the results of Ref. (32).

Unfortunately these results from the existing literature on quantum switching provide us with only qualitative information for the regime of interest here ($\gamma \sim g \sim \kappa$). One difficulty is of course that all are one dimensional calculations. More importantly, with the exception of the work by Lugiato (23), none are capable of addressing the issue of global stability when the atom-field interaction is purely radiative and nonclassical processes (such as photon antibunching) are involved. The quantum character of the processes involved in optical bistability is demonstrated by noting that in the general case both the Glauber-Sudarshan and the Wigner distributions become singular and nonpositive. This situation seems to be unique to the problem of optical bistability since quantum switching in previous problems in quantum optics has referred to a classical diffusive process with quantum aspects entering to specify the noise strength. However, when nonclassical processes are involved, this classical correspondence can no longer be made and the standard methods of computation fail. The switching or decay problem cannot be understood as a classical statistical process and should perhaps be termed quantum tunneling to denote the role of probability amplitudes. Since the two possible states of the intracavity field between which the system makes transitions are macroscopically distinguishable, the potential exists for an investigation of quantum tunneling in a macroscopic system (33). Furthermore, with reference to Figure 2, one notes that as N_s becomes smaller and smaller the eigenvalue $\bar{\lambda}_1$ approaches unity. This implies that the rate determining global stability approaches the rates governing local stability so that coherent mixing of these macroscopic states might occur on a time scale comparable to the dissipative time scale (γ^{-1}, κ^{-1}). The experimental challenges to the observation of such phenomena are formidable, as are the theoretical difficulties in obtaining an adequate description. For example Drummond has discussed the question of the representation dependence or nonuniqueness of the tunneling eigenvalues (34).

5. CONCLUSION

I believe that it is most often the case in quantum optics that the quantum nature of the atom-field interaction leads to phenomena which are completely classical in their dynamical character. The quantum aspects of the problem usually enter in rather simple ways, such as in fixing noise strengths and decay rates, or in randomizing initial conditions. However, in the spirit of this Symposium on Frontiers in Quantum Optics, I wish to venture that a frontier does exist in quantum optics capable of exciting the imagination with a wealth of new phenomena. That frontier lies in the investigation of systems for which the paradoxes of the quantum theory expressed by the concept of probability amplitude are embedded within the structure of the dynamics. I have attempted in these notes to present examples that will hopefully confirm this optimism.

This work was supported in part by the National Science Foundation (Grants PHY8211194 and PHY8351074), the Venture Research Unit of British Petroleum, Int., and the Joint Services Electronics Program.

REFERENCES

1. M. Born, The Born-Einstein Letters (Walker and Company, New York,1971), 173-174.
2. A. Einstein, B. Podolsky, and N. Rosen, Phys. Rev. 47, 777 (1935).
3. J. S. Bell, Rev. Mod. Phys. 38, 447 (1966).
4. C. A. Kocher and E. D. Cummins, Phys. Rev. Lett. 18, 575 (1967); E. S. Fry and R.C. Thompson, Phys. Rev. Lett. 37, 465 (1976).
5. J. F. Clauser and A. Shimony, Rep. Prog. Phys. 41, 1881 (1978).
6. A. Aspect, J. Dalibard, and Gérard Roger, Phys. Rev. Lett. 49, 1804 (1982).
7. P. A. M. Dirac, Fields Quanta 3, 154 (1972).
8. E. C. G. Sudarshan, Phys. Rev. Lett. 10, 277 (1963); J. R. Klauder, Phys. Rev. Lett. 16, 534 (1966).
9. M. D. Reid and D. F. Walls, Phys. Rev. A (to be published) 1986.
10. L. Mandel, Progress in Optics, Vol. XIII, ed. E. Wolf (North-Holland, New York, 1976), 29.
11. H. J. Kimble, A. Mezzacappa , and P. W. Milonni, Phys. Rev. A 31, 3686 (1985).
12. H. J. Kimble and L. Mandel, Phys. Rev. A 13, 2123 (1976); H. J. Kimble, M. Dagenais and L. Mandel, Phys. Rev. A 18, 201 (1978).
13. B.R.Mollow, in Progress in Optics Vol XIX, ed. E. Wolf (North-Holland, New York, 1981) 1; D. F. Walls, Nature 250, 451 (1979).
14. J. D. Cresser, J. Häger, G. Leuchs, M. Rateike and H. Walther, in Dissipative Systems in Quantum Optics ed. R. Bonifaccio, vol. 27 of Topics in Current Physics (Springer, New York, 1982), 49.
15. H. Dehmelt, in Advances in Laser Spectroscopy, eds F. T. Arecchi, F. Strumia, and H. Walther, NATO Advanced Study Institute Series. Vol. 95 (Plenum, New York, 1983).
16. R. J. Cook and H. J. Kimble, Phys. Rev. Lett. 54, 1023 (1985).
17. R. J. Cook and H. J. Kimble, Phys. Rev. A (to be published).
18. J. Javanainen, Phys. Rev. A (to be published, 1986).
19. H. Haken, Laser Theory (Springer-Verlag, New York, 1984).
20. H. J. Carmichael, Quantum-Statistical Methods in Quantum Optics, Lecture Notes in Physics (Springer, to be published).
21. E.T. Jaynes and F. W. Cummings, Proc IEEE 51, 89 (1963).
22. H. I. Yoo and J. H. Eberly, Phys. Reports 118, 240 (1985).
23. L. A. Lugiato, Progress in Optics Vol XXI, ed. E. Wolf (North Holland, New York, 1984), 71.
24. H.J. Carmichael, Phys. Rev. A (to be published, 1986).
25. G. S. Agarwal, J. Opt. Soc. Am. B2, 480 (1985).
26. D. Meschede, H. Walther and G. Muller, Phys. Rev. Lett. 54, 551 (1985).
27. A. T. Rosenberger, L. A. Orozco and H. J. Kimble, Phys. Rev. A 28, 2569 (1983); in Fluctuations and Sensitivity in Nonequilibrium Systems, eds. W. Horsthemke and D. Kondepudi (Springer-Verlag, New York, 1984), p 62.
28. R. S. Hulet, E. S. Hilfer, and D. Kleppner, Phys. Rev. Lett. 55, 2137 (1985).
29. Y. Kaluzny, P. Goy, M. Gross, J. M. Raimond, and S. Haroche, Phys. Rev. Lett. 51, 1175 (1983).

30. P. Stehle, Phys. Rev. A 2, 102 (1970).
31. P. Goy, J. M. Raimond, M. Gross and S. Haroche, Phys. Rev. Lett. 50, 1903 (1983).
32. J. C. Englund, R. R. Snapp, and W. C. Schieve, Progress in Optics, Vol XXI, ed. E. Wolf, (North-Holland, New York, 1984), p. 357.
33. A. J. Leggett, Supl. Prog. Th. Phys. 69, 80 (1980); A. O. Caldeira and A. J. Leggett, Annals of Phys. 149, 374 (1983).
34. P. D. Drummond, preprint.

AMPLIFIERS, ATTENUATORS AND THE QUANTUM THEORY OF MEASUREMENT

R J GLAUBER

I. Introduction: Two Paradoxes

Although quantum mechanics has been a mature field for many years now, it has one area, concerned mainly with the theory of measurement, that seems to be permanently adolescent. It is an unruly area, in which opinions remain fragmented and philosophical contention thrives.

One of the traditional bones of contention involves the connection of the quantum mechanical micro-world with our classical everyday world. Quantum mechanical measurements made on elements of the micro-world are usually imagined to be recorded by large-scale classical instruments. Where, in the account of those measurements, is the transition to be made from the quantum mechanical to the classical description -- and how can any such transition be carried out in a mathematically consistent way?

Following some work on the theory of quantum mechanical amplifying devices, it has occurred to me that at least part of that question can be answered in largely physical terms. Arbitrary quantum states of the electromagnetic field, for example, can be used as inputs for the amplification process and the amplified output fields can be so intense that they qualify as fully classical -- that is to say, accurately measurable without any significant disturbance of the field. Such amplification processes are in fact actually used in many experiments to render quantum phenomena observable but,

---

* Prepared from a tape recording of the talk, December 18, 1985.

even where they are not used in practice, it is interesting to imagine the effect their use would have. They suggest, in other words, an interesting class of gedanken experiments and I shall try presently to illustrate a few of them.

There was time, well over a century ago, when clever schemes to construct perpetual motion machines were all the rage. The effort spent on them was not all wasted; they did help teach us two important principles of thermodynamics. We're not so wrapped up in that venerable subject any more, but we do seem still to be learning about two others: relativity theory and quantum mechanics. So it should not be too surprising that the same infernal ingenuity that once went into perpetual motion machines is now suggesting means for communicating faster than light, and for confounding the principle of complementarity. Some of these are interesting schemes; they too might just be capable of teaching us something.

One interesting proposal amounts to using the Einstein, Podolsky, Rosen paradox[1)] as a means of communication. To be able to do that would mean, in fact, being able to communicate faster than light, and so that's not a possibility to be dismissed lightly. Such a thought must indeed have occured to many of us as soon as we began reading of the EPR paradox and of the species of non-locality it imposes on quantum mechanics.

Let me remind you, therefore, of the most elementary scheme for superluminal communication that you might try to base on the EPR paradox. We begin with a light source which gives out photons in pairs. An ordinary atom that does that will give them off in random directions, but if we don't mind using the gamma rays of annihilation radiation, those photons can, in fact, be given off back to back - and with zero total angular momentum. That, of course, is an essential point. The two photons must have closely correlated angular momenta so that any statement you make about the angular momentum of one of them, or any observation, conveys a unique implication about the other.

Now, what must you do to communicate by this strange medium? Well, you set up your own equipment at a certain distance from the source, not too far because you don't want to lose sight of the source, but let's just say, one light-week away or perhaps a bit less. The friend with whom you want to communicate may be a comparable distance from the source, let's say down in the foreground of our Figure 1. Now all you

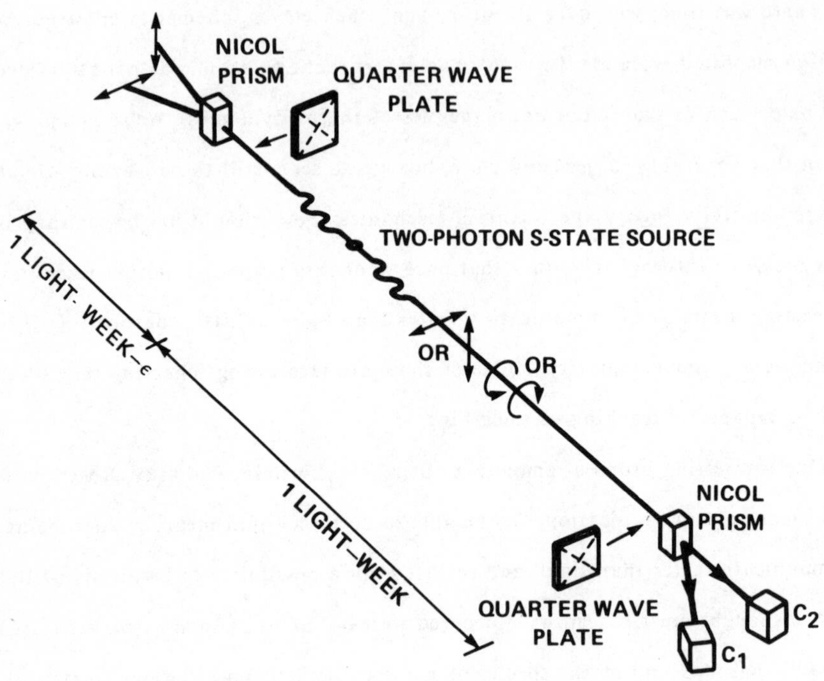

Fig. 1

have to do is to make polarization measurements upon the photons which come your way. It's a bit wasteful in practice if you do that with polarizing filters; they just absorb half the photons and there is no point in letting that happen. It's a much better idea to use a Nicol prism or, better still, a Wollaston prism, which will send vertically plane polarized photons in one direction, and the others, with horizontal polarization, in another direction, so that none at all are lost. Then, of course, you can add to the equipment a quarter-wave plate, if you like, with its two axes at 45 degree angles with respect to the polarization directions separated by the prism. Putting the quarter-wave plate in front of the prism makes of it, in effect, a circular polarization detector for the two possible circular polarizations.

How do we send the message? We have no control whatever over the polarization of the photons being produced by our source. But what we can do, of course, is to exercise our choice either to leave things as they stand in Fig. 1, and observe one or the other of two plane polarizations, or alternatively to push the quarter-wave plate into the beam we are detecting and then we will inevitably observe one or the other of two circular polarizations.

The correlation that exists between the polarizations of the photons going off in opposite directions then assures us that we, off at this great distance, have a certain control over the photons that are appearing, virtually instantaneously at our friend's position. We are able to determine, simply by pushing the quarter wave plate in or out, whether the individual photons our friend receives are going to be plane polarized or circularly polarized. If they're plane polarized, we can't control which of the two plane polarizations they have. But we can send him circularly polarized photons whenever we please and they are presumably distinguishable, in principle, from plane polarized ones. We can let photons polarized in those ways represent the dots and dashes, respectively, of Morse code, or the binary digits, and repeat our message as often as our friend needs in order to overcome any statistical problems he may have.

The drawback of this communication scheme, of course, is that it doesn't work at all. Its failing lies in the very essence of what is necessary to send a message. Our friend has no idea, to begin with, whether any given photon he receives is polarized circularly or linearly. His ability to detect the message depends on his ability to determine the polarizations of those individual photons, and unfortunately there is only a limited amount of measurement that he can make on any one photon before it is absorbed, or he otherwise loses track of its initial state. He can, for example, pass it through a plane-polarization filter. But the fact that it has passed throught the filter does not tell him whether its initial polarization was plane or one of the two circular varieties; that remains really quite undetermined. Let's say, for example, that a photon has passed through our own prism arrangement and has been revealed to have plane polarization. Our distant friend then receives a plane polarized photon but he still can not say with better than 50/50 likelihood that the arriving photon is either plane polarized or circularly polarized. So there is no information transmitted at all - and the scheme collapses. That much has probably occurred to most of us as students.

A new and interesting suggestion has been added to this picture, however, by N. Herbert[2)], who proposes to improve our friend's polarization measurements greatly by adding, in effect, an amplifier to his system. He calls it a "laser gain tube", which you'll recognize as a characteristically innocent and passive sounding name, and he uses another well chosen term for its action; it "clones" the photons. Once the "gain tube", which is shown in place in Fig. 2, has "cloned" for our friend lots of photons identical to the original one, he should no longer find it difficult to determine their state of polarization. What he can do, for example, is to install a beam splitter to send, say 90 percent of those cloned photons into another laboratory where he has all sorts of equipment available and can make measurements on as many photons as he likes; so in that way he can get an excellent idea of their polarization. All he needs to determine

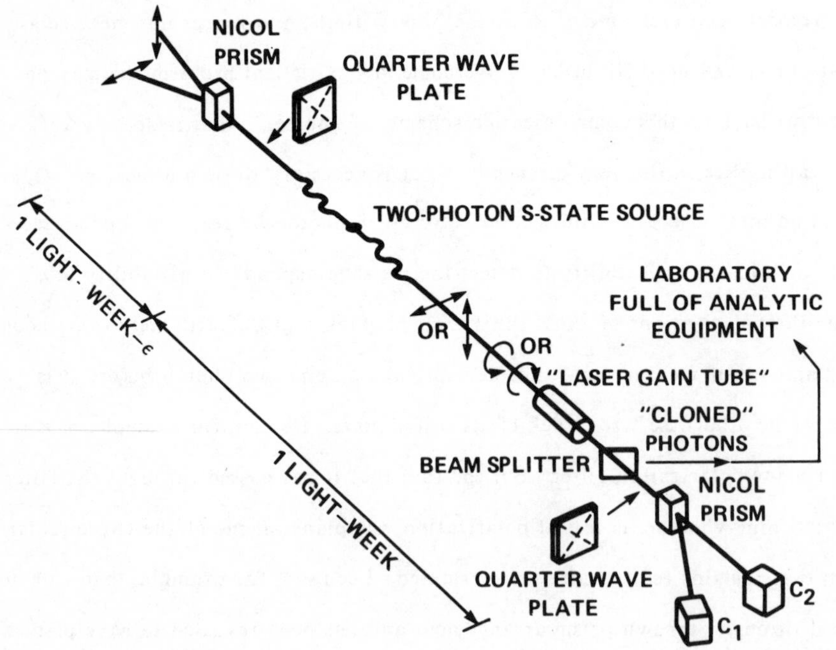

Fig. 2

is what the polarization was, whether plane or circular, of the photon that first entered his laser gain tube.

If we've found, at last, the means for him to do that - as the argument indicates - then we can indeed communicate outside the light cone and create miracles of all sorts. Well, does the scheme work? Let me postpone telling you the answer until we've developed the means of describing a bit better the role played by the quantum amplifier. Then we'll be able to say what the insertion of an amplifier really does in that scheme, and in several other sorts of experiments as well.

In the meantime, I should like to tell you another story. This one is much older and much better known. It deals with one of the fundamental dilemmas of the quantum theory of measurement. Schrödinger[3)], responding to questions about completeness of quantum mechanics in 1935, indicated quite some dissatisfaction with the field he had pioneered. He was troubled by the fact that one never sees in our everyday world anything that looks like a superposition of quantum mechanical states. By that he meant a superposition of vectors in an infinite-dimensional Hilbert space, to be technical about it. The microscopic world, on the other hand, is full of them.

You can easily imagine experimental arrangements in which events in the microscopic world would seem to imply the creation of such superpositions in our everyday world. Let us suppose that the quantum state $|a\rangle$ of something in the micro-world implies, by some kind of rigorous dynamics, that the state of something else that we can actually observe in the laboratory goes to a particular state $|A\rangle$, and suppose too that the alternative micro-state $|b\rangle$, whatever it may be, leads to a large-scale state that we can label $|B\rangle$. Now, let us imagine a third microstate $|c\rangle$ which is physically quite distinguishable from both $|a\rangle$ and $|b\rangle$ but can be expressed as a superposition of those two states with non-vanishing coefficients α and β,

$$|c\rangle = \alpha\,|a\rangle + \beta\,|b\rangle$$

That state should obviously lead to a similar linear combination

$$|C\rangle = \alpha\,|A\rangle + \beta\,|B\rangle$$

as a final state of whatever it is we observe. That looks like no more than the postulate of the linearity of quantum mechanics. Now, how could you disagree with that?

But Schrödinger did indeed have some trouble with that idea, and described a gedanken experiment which has become the definitive illustration of his dilemma. He imagined a diabolic arrangement in which a cat is confined to a box which contains a lethal device that may be either triggered or left passive according to whether a radioactive nucleus decays or fails to decay. If the radioactive decay takes place, which he assumed would happen half the time, a hammer would strike a vial of cyanide, and the cat would be dispatched. It's not immediately clear, of course, how a radioactive nuclear decay process trips a hammer. Schrödinger evidently didn't see that as a problem, but we may. The hammer won't budge unless the quantum signal is amplified greatly and that amplification is what we want to discuss.

We can simplify Schrödinger's example a bit by thinking, not of radioactive decay processes, but of photons that can be polarized either horizontally or vertically. (This next transparency looks, I'm afraid, rather like a transcription from an Egyptian papyrus of the 2nd dynasty.) We can let the state of horizontal polarization for a single photon $|h\rangle$ be the state $|a\rangle$ and let the state of vertical polarization $|v\rangle$ be the state $|b\rangle$,

$$|a\rangle = |h\rangle \tag{1}$$

$$|b\rangle = |v\rangle \tag{2}$$

And for our two states in the macroscopic world, let us take as state $|A\rangle$, the cat, alive and healthy,

$$|A\rangle = | \quad \rangle , \tag{3}$$

and as the state $|B\rangle$, the cat, not so alive and not so healthy,

$$|B\rangle = | \quad \rangle , \tag{4}$$

Now, let's take our two coefficients, alpha and beta both to be $1/\sqrt{2}$. The state

$$|c\rangle = \frac{1}{\sqrt{2}} (|h\rangle + |v\rangle) , \tag{5}$$

the superposition of the two polarizations, is a state which is quite meaningful to us in the micro-world. Such a superposition of the horizontal and the vertical polarizations is just another linear polarization inclined at 45 degrees to the two axes.

That gives us a simple way of carrying out Schrödinger's experiment. We can send a single photon polarized at 45 degrees, relative to the axes of a Wollaston prism through the prism and let it go off in either of the two emergent polarized beams. In one of those two beams we can place a photo-detector that, when it registers, does something awful to the cat, and we can leave the other beam free so that if the photon goes there nothing happens to the cat. According to Schrödinger's analysis then, and according to our earlier argument the cat is left finally in a linear superposition of two states,

$$|c\rangle = \frac{1}{\sqrt{2}}\left\{ |\quad\rangle + |\quad\rangle \right\} . \qquad (6)$$

It has, to be precise, an amplitude .7071 for being alive and an amplitude .7071 for being dead. That is a very strange state of affairs because we don't recognize the existence of any such state in the everyday world. We do in the microscopic world, but in our everyday world, we certainly don't. (Some of us, arriving here in jet-lagged shape last Sunday may have had the impression we were in something rather like that state, but we'd have no right to be quite that specific about the coefficients.) So let's agree, there is really nothing we can identify with such precise superpositions in our macroscopic world.

Well, the cat paradox may not seem too closely related to superluminal communication but if, in fact, the polarization of a single photon is to induce the tripping of a hammer then it can do that only if some process of amplification intervenes. So we are back once more to talking about amplifiers, and once more I'd like to postpone further consideration of the problem until we've managed to discuss how they work.

II. A Quantum Mechanical Attenuator: The Damped Oscillator

It may be a good idea, before discussing amplification, to say a few words about dissipation or attenuation, which is a much more natural state of affairs.

Amplification, we'll then show, can be described by making only a few small -- but highly significant -- changes in the basic mechanism. Here, I must admit, I am in the position of a merchant with a particular model to sell; the truth is it has been on the market for a long time, though not well advertised [4], and has registered only a few sales to date. Please forgive me then if I carry on a bit like a vacuum cleaner salesman. The model of the damped harmonic oscillator I want to describe to you gives about the simplest description you can have of the attenuation process.

The damped oscillator is something that you can easily construct from an assembly of simple harmonic oscillators in the following way: you couple one oscillator, the central oscillator, which is described by the amplitude operators a and $a^\dagger$, to a whole heat bath full of oscillators, described by the amplitude operators b_k and $b_k^\dagger$ for $k = 1, \ldots N$ where N is quite large, or perhaps even infinite. The Hamiltonian for the system need only couple the central oscillator to the heat bath oscillators through the familiar "rotating wave" terms. We write

$$H = \hbar\omega a^\dagger a + \sum_k \hbar\omega_k b^\dagger_k b_k + \hbar\sum_k (\lambda_k a^\dagger b_k + \lambda_k^* b_k^\dagger a), \tag{7}$$

in which the frequencies ω_k cover a range in the neighborhood of ω fairly densely, and the λ_k are a set of coupling parameters. This Hamiltonian remains invariant when we make the phase transformation

$$a \rightarrow a e^{i\theta}$$
$$b \rightarrow b e^{i\theta}$$

on all of the operators it contains. That invariance means that there is a corresponding conservation law, and in the present case it's just staring at you. This coupling obviously conserves the total number of quanta in the system. And that property helps to make of it a highly soluble model.

The equations of motion, of course, are linear;

$$\dot{a} = -i\omega a - i\sum_k \lambda_k b_k$$
$$\dot{b}_k = -i\omega_k b_k - i\,\lambda_k^* a\,. \tag{8}$$

Furthermore, for this simple coupling the Schrödinger states have the wonderful property that an initially coherent state remains coherent at all later times.[5] So, in fact, the equations of motion could every bit as well have been written as equations for a set of c-number amplitudes $\alpha(t)$ and $\beta_k(t)$ for the coherent states of all the oscillators. The Schrödinger state of the oscillator, in other words, apart from a time dependent phase factor of no interest, is just given by

$$|t\rangle = |\alpha(t)\rangle \prod_k |\beta_k(t)\rangle , \tag{9}$$

a product of coherent states in which the amplitudes $\alpha(t)$ and $\beta_k(t)$ obey the equations of motion (8). There is a sense, then, in which this fully quantum mechanical system behaves as if it were classical. The fact that it is quantum mechanical even becomes a bit of a joke. The evolution of the system just carries its zero-point uncertainty cloud around the classical trajectory without altering it in any way. That's all that ever happens.

If the entire system is initially in a pure coherent state its density operator in the Schrodinger picture at time t will be just

$$\rho(t) = |\alpha(t), \{\beta_k(t)\}\rangle\langle\alpha(t), \{\beta_k(t)\}| , \tag{10}$$

which is still a pure coherent state for all of the individual oscillators. Note that the phase factor left undetermined in the state vector has now cancelled out.

It will be useful now to write the solutions to the equations of motion for the time-dependent amplitudes in an abbreviated form. The functions $\alpha(t)$ and $\beta_k(t)$ can be expressed always as linear combinations of their initial values, which we can write as α and β_k respectively. The coefficients will be time-dependent functions which we can write as u(t) and $v_k(t)$, so that we have for example,

$$\alpha(t) = \alpha\, u(t) + \sum_k \beta_k v_k(t) . \tag{11}$$

and the initial conditions

$$u(0) = 1 \quad , \quad v_k(0) = 0 . \tag{12}$$

We shall presently determine the functions u(t) and $v_k(t)$ explicitly, but there is a good deal we can say before doing that.

The only one of the oscillators that really interests us is the central one. We can find the reduced density operator for that one by taking the trace over all of the variables for the heat bath oscillators. What we have left then is just

$$\rho_A = \mathrm{Trace}_B\, \rho(t) = |\alpha(t)\rangle\langle\alpha(t)|, \tag{13}$$

which still represents a pure state for the central oscillator.

Of course, if the heat bath begins in a mixed state rather than a pure coherent state, then what we have to do is to mix together coherent states of this sort, corresponding to different values of the initial amplitudes β_k. The amplitudes $\alpha(t)$, according to Eq.(11), depend linearly on the β_k, and so the appropriate reduced density operator should take the form

$$\rho_A(t) = \int |\alpha(t)\rangle\langle\alpha(t)|\, \mathcal{P}(\{\beta_k\}) \prod_k d^2\beta_k \ . \tag{14}$$

That's the way the whole notion of mixing enters the problem; it is unavoidable, of course, when the heat bath is not at zero temperature.

Now, what is it that results from all of this? Again, let me spare you the calculations; I'm sure you'll believe their results. It's most convenient to express the reduced density operator for the central oscillator in terms of what I have called the P-representation, that is to say a mixture of pure coherent states $|\gamma\rangle\langle\gamma|$ with a weight function $P(\gamma)$, which is, strictly speaking, a quasi-probability distribution.

If we assume that the central oscillator is indeed initially in a coherent state with amplitude α, then its reduced density operator can be written in the general form

$$\rho_A(t) = \int P(\alpha,0|\gamma,t)\, |\gamma\rangle\langle\gamma|\, d^2\gamma \ . \tag{15}$$

The weight function $P(\alpha,0|\gamma,t)$ in this expression can be thought of as a conditioned quasi-probability density for the occurrence of a coherent state amplitude γ at time t, given the amplitude α at time zero. If we then take the heat bath oscillators to be initially in chaotic states, i.e. Gaussian mixture states, with mean occupation numbers $\langle n_k\rangle$, finding the function P is simply a matter of carrying out some Gaussian integrations. The result, we find, is always Gaussian in form,

$$P(\alpha,0|\gamma,t) = \frac{1}{\pi D(t)} \exp\left\{-\frac{|\gamma - \alpha\mu(t)|^2}{D(t)}\right\} \ . \tag{16}$$

The dispersion of this Gaussian function is given by

$$D(t) = \sum_k \langle n_k\rangle\, |v_k(t)|^2 \ . \tag{17}$$

and its mean value by $\alpha u(t)$. To evaluate the dispersion, or the mean value of the Gaussian function, we must of course solve the equations of motion to find the functions $u(t)$ and $v_k(t)$.

According to the initial condition, Eq.(12), the functions $v_k(t)$ must all vanish at $t = 0$, so the dispersion $D(t)$ is zero initially and the function P begins life as a delta-function. That, of course, is no surprise since the central oscillator starts out in the pure coherent state $|\alpha\rangle$. What is more interesting is that if we take the heat bath to be at zero temperature, that is take all of the $\langle n_k \rangle$ to vanish, the dispersion $D(t)$ remains zero at all times and the function P always remains a two-dimensional delta function,

$$P(\alpha, 0 \mid \gamma, t) = \delta^{(2)}(\gamma - \alpha u(t)). \tag{18}$$

In this case, in other words, the central oscillator always remains in a pure coherent state. It will typically lose its initial excitation to the heat bath oscillators, or to express that somewhat differently, we should expect the function $u(t)$ to decrease in modulus as the time t increases. That, at least, is the behavior we should anticipate in a dissipative system.

It's worth stressing two points, therefore, at this stage of the calculation. Firstly, the results we have reached at this point are exact; we have made no approximations. Secondly, because the result given by Eq.(18) is exact, it still reflects the intrinsic reversibility of the equations of motion. It is entirely possible for the heat-bath oscillators to re-excite the central oscillator and such Poincaré recurrences, though they may be enormously delayed, are in fact inevitable. The function $u(t)$ then, when solved for exactly, will not in general decrease monotonically in modulus forever. But the Poincaré recurrence times go rapidly to infinity as the number N of heat bath oscillators increases. In practice N needn't be very large before it becomes an excellent approximation - over reasonable lengths of time like the age of the universe - to ignore the Poincaré recurrences altogether and use approximations in which the modulus of $u(t)$ does decrease steadily. Those are the approximate ways of approaching the equations of motion which introduce the notion of irreversibility.

The equations of motion (8) have essentially the same structure as a set of coupled equations derived after a number of approximations by Weisskopf and Wigner[7]), in order to describe the process of radiation damping. If the coupling constants λ_k are not too large in modulus and there are enough heat bath oscillators with frequencies ω_k near ω, then the function u(t) can be approximated quite well by a complex exponential function,

$$u(t) = e^{-[K+i(\omega+\delta\omega)]t} = e^{-(K+i\omega')t} \quad . \tag{19}$$

This function has a certain frequency shift $\delta\omega$ in it, and it has a certain damping constant K as well. Those constants are given by the relation

$$\delta\omega - iK = \lim_{\epsilon\to 0} \sum_k \frac{|\lambda_k|^2}{\omega - \omega_k + i\epsilon} \quad . \tag{20}$$

which is characteristic of second-order perturbation theory, although the overall approximation retains terms of all orders in the coupling strengths.

With the heat bath at temperature zero then, the conditional quasi-probability density is given, according to eqs. (18) and (19), by

$$P(\alpha,0\,|\,\gamma,t) = \delta^{(2)}(\gamma - \alpha e^{-(K+i\omega')t}) \quad . \tag{21}$$

Since the state of the central oscillator, in this case, remains at all times a pure coherent state, it always has minimal uncertainty. This model for the dissipation process, in other words, is completely noise-free.

The randomness we refer to as noise only enters the dissipation process when the heat bath possesses initial excitations that exert random driving forces on the central oscillator. For those cases, which correspond to $\langle n_k\rangle = 0$ in our model, the noise is described by the dispersion D(t). We can evaluate that expression approximately by noting that the functions $v_k(t)$ have a resonant character; they are only large in modulus for modes with frequencies ω_k within an interval of width K about the frequency $\omega' = \omega + \delta\omega$. If the mean occupation numbers $\langle n_k\rangle$ take on the constant value $\langle n_{\omega'}\rangle$ within this narrow band, we can write Eq.(17) as

$$D(t) = \langle n_{\omega'}\rangle \sum_k |v_k(t)|^2 \,. \tag{22}$$

To evaluate the sum in this expresssion we can note that the functions u and v_k obey an identity equivalent to the conservation law,

$$|u(t)|^2 + \sum_k |v_k(t)|^2 = 1 \quad , \tag{23}$$

which follows from the equations of motion (8) and the initial conditions (12). The dispersion can then be written as

$$D(t) = \langle n_{\omega'} \rangle (1 - |u(t)|^2) \,, \tag{24}$$

which in the Weisskopf-Wigner approximation reduces to

$$D(t) = \langle n_{\omega'} \rangle (1 - e^{-2\kappa t}) \,. \tag{25}$$

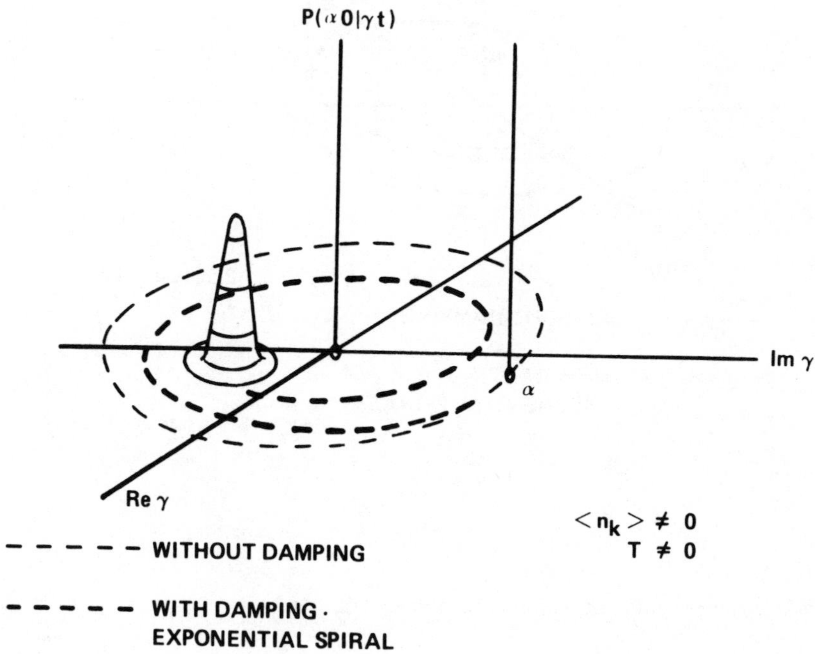

Fig. 3

The dispersion D(t) then increases from the initial value zero, while the center of the Gaussian distribution given by Eq.(16) circles about on the exponential spiral given by Eq.(19). A three-dimensional portrait of the quasi-probability density as a function of the complex amplitude γ is shown in Fig. 3. For times t much greater than the damping time κ^{-1} the central oscillator comes to equilibrium with the heat bath oscillators. The dispersion then reaches the limiting value $\langle n_{\omega'} \rangle$ and the Gaussian distribution settles down into the stationary form shown at the center of Fig. 4.

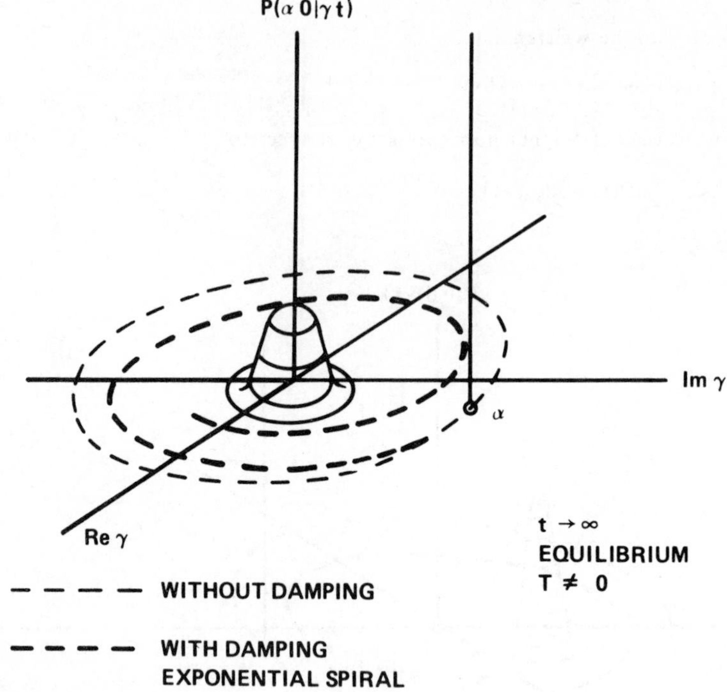

Fig. 4

III. More General Models; Coupled Damped Oscillators

I'd like to mention at this point some simple generalizations of the damped oscillator model we have just discussed. We can begin by asking what happens if the central oscillator is coupled not just to one, but to two different heat baths? The two heat baths, shown schematically in Fig. 5, can be at different temperatures.

That question can be answered in large measure without altering at all the calculation we have just gone through. The two heat baths are just equivalent to a single compound heat bath. The central oscillator doesn't come to equilibrium with either of the two heat baths individually; instead it comes to an equilibrium characterized by an average occupation number $\langle a^{\dagger} a \rangle$ equal to the mean of the occupation numbers $\langle n_1 \rangle$ and $\langle n_2 \rangle$ of the two heat baths.

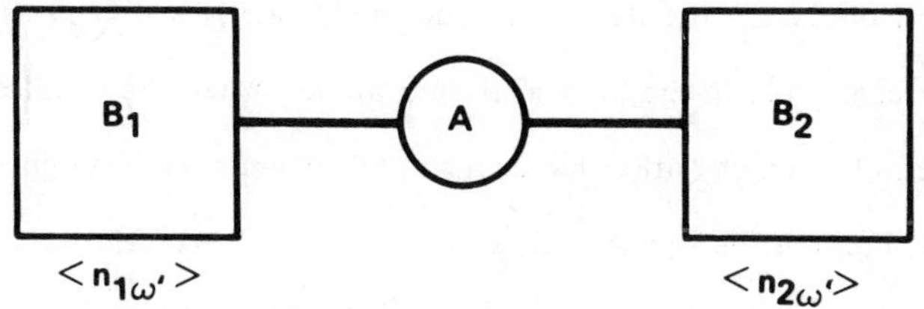

Fig. 5

V.I. Man'ko and I have recently looked at the behavior of dissipative systems[8) somewhat more complicated than this one, and considerably more varied in their behavior. You might ask what happens, for example, when the central oscillator of the foregoing example is replaced by a pair of coupled oscillators? Suppose we let A_1 and A_2 be two identical oscillators coupled by means of the "rotating wave" terms -- with coupling strength λ . We may now imagine each of these oscillators to be coupled to its own heat bath. That coupling scheme is shown in Fig. 6. Let us assume, for simplicity, that the two heat baths and their couplings to the A-oscillators are identical in structure, and that each of the A oscillators, in the absence of the other, would have a damping constant κ .

It is not difficult to show, for this model, that the A oscillators relax to equilibrium excitations which are intermediate between those of the two heat baths, but different from one another's. For the A_1 oscillator, for example, we find the equilibrium excitation

$$
\begin{aligned}
\langle a_1^\dagger a_1 \rangle = \tfrac{1}{2} (\langle n_1 \rangle + \langle n_2 \rangle) \\
+ \tfrac{1}{2} (\langle n_1 \rangle - \langle n_2 \rangle) \frac{\kappa^2}{\kappa^2 + \lambda^2} \; .
\end{aligned}
\tag{26}
$$

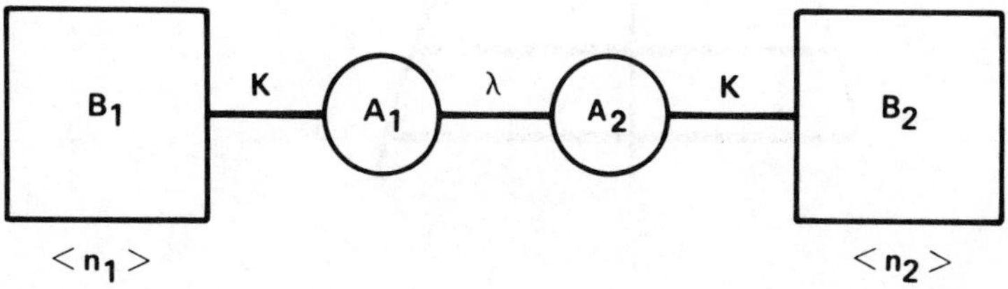

Fig. 6

When $\lambda = 0$ and the two A oscillators are uncoupled this is just $<n_1>$ as it should be for an oscillator connected to the first heat bath alone. When the coupling parameter λ greatly exceeds κ, on the other hand, the equilibrium excitation goes to $\frac{1}{2}(<n_1> + <n_2>)$, because the two A oscillators are then, in effect, directly coupled to both heat baths. For intermediate coupling strengths we find a range of equilibrium excitations. For $\lambda = \kappa$, for example, we have

$$<a_1^\dagger a_1> = \frac{3}{4}<n_1> + \frac{1}{4}<n_2>$$
$$<a_2^\dagger a_2> = \frac{1}{4}<n_1> + \frac{3}{4}<n_2> . \qquad (27)$$

It is interesting to see that we can vary the nature of the equilibria reached by these systems by adjusting the strengths of their couplings.

IV. A Quantum Mechanical Amplifier

How can we build a device that amplifies signals at the quantum level? In fact, we can do that too with harmonic oscillators, but at least one of them must be rather special[9]. I'm sure there is very little I need to tell you about harmonic oscillators, but let me just introduce the special one to you. There it is in Fig. 7. You see, there is something a bit funny about that harmonic oscillator. Its Hamiltonian is just minus

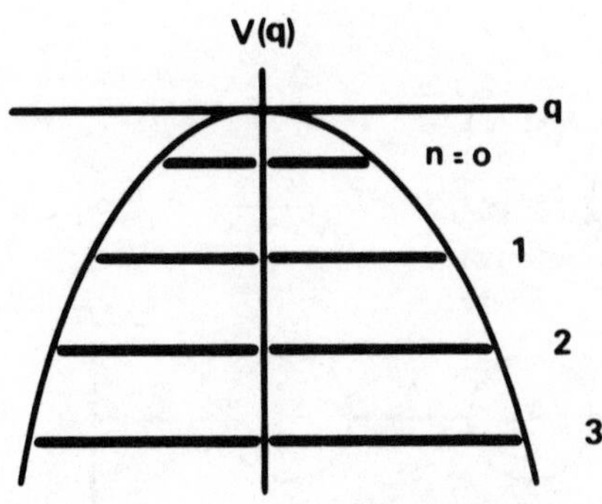

Fig. 7

the usual one; both the potential and kinetic energies go down instead of up. That simple change of sign doesn't change the algebraic properties of the amplitude operators a and $a^{\dagger}$, however. They obey the same commutation relation they did before. In fact only one thing changes; reversing the sign of the Hamiltonian implies reversing the sign of the frequency ω in the equation of motion. And so a(t) varies as $a(0)e^{i\omega t}$ instead of $a(0)e^{-i\omega t}$.

Our inverted oscillator, strictly speaking, no longer has a ground state; it has a state of maximum energy which we must inevitably associate with quantum number n=0. That leads to a certain dilemma of terminology. I talked about that in Poland a few weeks ago and suggested a name alternative to the "ground state" for this highest-lying of states. But word later came back to me explaining that in Poland the name "ground state" remains most appropriate because everything interesting is underground. Things are no doubt a bit different here in England, but I'm afraid I've gotten into the habit of calling that state the "ground state" and hope you'll bear with me.

The inverted oscillator still has a discrete succession of eigenstates which can be generated by applying powers of $a^{\dagger}$ to the "ground state",

$$|n\rangle = \frac{1}{\sqrt{n!}}(a^{\dagger})^{n}|0\rangle\ . \tag{28}$$

You have to bear in mind, however, that the energies of these states have the minus sign in them, $E_n = -(n+\frac{1}{2})\hbar\omega$, so the creation operator $a^{\dagger}$ really creates de-excitations, rather than positive energy quanta. And the annihilation operator a actually raises the energy of the oscillator by decreasing the quantum number n by one unit.

Once we've absorbed those trifling changes we're prepared to let the inverted oscillator take over the role of the central oscillator in the scheme we used to discuss dissipation. The heat bath can remain precisely the same as we had it before, but when we couple the inverted oscillator to it, we must interchange the roles of a and $a^{\dagger}$ in the coupling terms in order to retain the "rotating wave" form. The coupling terms which tend to conserve energy, in other words, take the forms ab_k and $b_k^{\dagger}a^{\dagger}$ rather than $ab_k^{\dagger}$ and $b_k a^{\dagger}$. When we make this interchange the Hamiltonian for the coupled system becomes

$$H = -\hbar\omega a^{\dagger} a + \sum_k \hbar\omega_k b_k^{\dagger} b_k + \hbar\sum_k (\lambda_k a b_k + \lambda_k^{*} b_k^{\dagger} a^{\dagger}) \quad . \tag{29}$$

Now this Hamiltonian has a certain invariance in it, which is a bit different from the one we saw earlier. If we alter the phase of a , then the b_k's must all undergo the complex conjugate transformation,

$$\begin{aligned} a &\longrightarrow a e^{i\vartheta} \\ b_k &\longrightarrow b_k\, e^{-i\vartheta} \quad . \end{aligned} \tag{30}$$

That gives us a different sort of conservation law. The quantity conserved is the difference of the number of quanta in the central oscillator and the number in the heat bath,

$$a^{\dagger} a - \sum_k b_k^{\dagger} b_k = \text{constant} \quad . \tag{31}$$

The equations of motion, of course, are still linear,

$$\begin{aligned} \dot{a} &= i\omega a - i\sum_k \lambda_k^{*} b_k^{\dagger} \\ \dot{b}_k &= -i\omega_k b_k - i\,\lambda_k^{*} a^{\dagger} \quad , \end{aligned} \tag{32}$$

but as you can see, they mix the adjoint operators in with the non-adjoints and that does change the character of the solutions. Before undertaking the explicit solution of those equations, however, it is useful once more to go as far as we can with the generic forms the solutions must have. So let us write the solution to the equations of motion for a(t) in the form

$$a(t) = a(0)\, U(t) + \sum_k b_k^{\dagger}(0)\, V_k(t) \tag{33}$$

and define thereby a set of functions U(t) and $V_k(t)$ that obey the initial conditions

$$U(0) = 1 \quad , \quad V_k(0) = 0 \ . \tag{34}$$

We shall eventually have to solve for those functions but, as before, we can gain a number of insights before having to do that.

Now, we can no longer use the trick of saying a coherent state remains always a coherent state. In fact for this model a coherent state does <u>not</u> remain a coherent state; its intrinsic uncertainty cloud grows explosively. That's just in the nature of this device; quantum mechanical linear amplifiers do that. So, if we are to evaluate the reduced density operator for the inverted oscillator, we must fall back on some more general analytic technique than we have used earlier.

It is sufficient for this purpose to introduce the characteristic function for the unknown density operator, in fact, to use the specific form in which the a and $a^\dagger$ operators are normally ordered,

$$\chi_N(\mu,t) = \text{Trace}\,\{\rho(0)\, e^{\mu a^\dagger(t)}\, e^{-\mu a(t)}\}$$
$$= \text{Trace}\,\{\rho(t)\, e^{\mu a^\dagger(0)}\, e^{-\mu a(0)}\}\,. \tag{35}$$

The trace which defines this characteristic function is written on the upper line in the Heisenberg picture and on the lower line in the Schrödinger picture. Let us again take the initial state of the system to be one in which the central (inverted) oscillator is in the coherent state $|\alpha\rangle$, while the heat bath oscillators are in chaotic states with occupation numbers $\langle n_k\rangle$;

$$\rho(0) = |\alpha\rangle\langle\alpha| \prod_k \int e^{-\frac{|\beta_k|^2}{\langle n_k\rangle}} |\beta_k\rangle\langle\beta_k| \frac{d^2\beta_k}{\pi\langle n_k\rangle}\,. \tag{36}$$

We can use this form for the density operator then and the expression for a(t) given by Eq. (33) to evaluate the Heisenberg form for the characteristic function.

If we now write the reduced density operator in the Schrödinger picture in precisely the same form as we used in Eq.(15), that is to say the P-representation form, we find that the characteristic function $\chi_N(\mu,t)$ is just a two-dimensional Fourier transform[10] of the unknown weight function $P(\alpha,0\,|\,\gamma,t)$. That transform is easily inverted and we then find a Gaussian form for the conditioned quasi-probability density,

$$P(\alpha,0\,|\,\gamma,t) = \frac{1}{\pi\,\eta(t)} \exp\Big\{-\frac{|\gamma - \alpha U(t)|^2}{\eta(t)}\Big\}\,, \tag{37}$$

in which the dispersion $\eta(t)$ is given by

$$\eta(t) = |U(t)|^2 - 1 + \sum_k \langle n_k\rangle\, |V_k(t)|^2\,. \tag{38}$$

The functions U(t) and $V_k(t)$ obey an identity in this case too, one which is equivalent to the conservation law of Eq.(31) and states

$$|U(t)|^2 - \sum_k |V_k(t)|^2 = 1\,. \tag{39}$$

We can use this relation to write the dispersion as

$$\eta(t) = \sum_k (1 + \langle n_k\rangle)\, |V_k(t)|^2\,. \tag{40}$$

This dispersion vanishes initially, since the central oscillator begins in a pure coherent state. But at later times the dispersion takes on positive values in general, and it does that even when the heat bath has no initial excitation; i.e., all $<n_k>$ are zero.

The dispersion $\eta(t)$ is a measure of the noise present in the excitation of the central oscillator. There is no avoiding the occurrence of such noise. It results, for example, from the spontaneous emission of quanta from the inverted oscillator into the heat bath, even when the latter is at zero temperature, and when initial amplitude vanishes as well.

The functions $U(t)$ and $V_k(t)$ can be approximated by the same sort of approximation we used earlier. But this time it should perhaps be called the anti-Weisskopf-Wigner approximation since the function $U(t)$ now blows up exponentially instead of decreasing. We find we can write $U(t)$ as

$$U(t) = e^{[\kappa + i(\omega - \delta\omega)]t} = e^{[\kappa + i\omega'']t}, \tag{41}$$

where the amplification constant κ and the line shift $\delta\omega$ are given in terms of the coupling strengths and the frequencies ω_k once again by Eq.(20).

The functions $V_k(t)$ have a resonant character for the case of amplification as well as dissipation. They are largest in modulus for frequencies ω_k close to ω. We can use that fact once more to approximate the dispersion function given by Eq.(39) as

$$\begin{aligned} \eta(t) &= (1 + <n_{\omega''}>) \sum_k |V_k(t)|^2 \\ &= (1 + <n_{\omega''}>)(|U(t)|^2 - 1) \\ &= (1 + <n_{\omega''}>)(e^{2\kappa t} - 1) \quad . \end{aligned} \tag{42}$$

The initial amplitude α may be regarded, in a sense, as an input signal for our amplifier. Then it is clear from the general form of the quasi-probability function (36) that the mean value of the distribution circles about the complex γ-plane in an exponentially increasing spiral. That is to say the amplified signal $\alpha U(t)$ blows up exponentially in modulus. And at the same time the dispersion $\eta(t)$, or the noise present blows up just as dramatically. A somewhat subdued picture of the way the function $P(\alpha,0|\gamma,t)$ changes with time is given in Fig. 8. The dispersion tends to increase more rapidly than the figure indicates -- but flattened-out Gaussian functions are very difficult to portray in three dimensions.

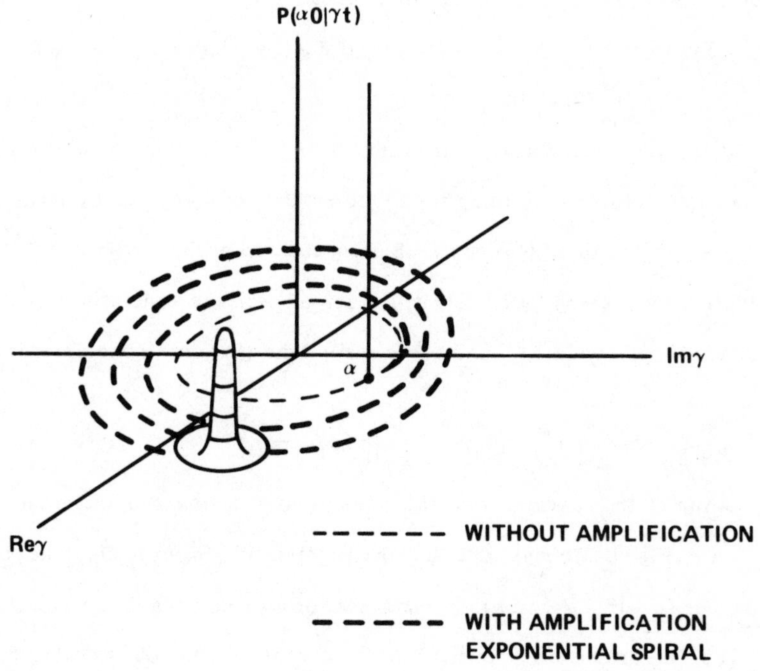

Fig. 8

Now that we know what happens when the amplifier begins in a pure coherent state $|\alpha>$, it is easy to find out what happens for a great variety of other initial states. To see what the amplifier does when it begins in the n = 0 state, for example, we just substitute $\alpha = 0$ in Eq.(37) to find the quasi-probability distribution

$$P_0(\gamma,t) = \frac{1}{\pi\, n(t)} \exp\{-\frac{|\gamma|^2}{n(t)}\}. \tag{43}$$

This simple Gaussian function, we can say, represents the purest sort of noise. A large part of it comes from the amplification of zero-point fluctuations -- or equivalently, amplified spontaneous emission. The remainder comes, when the $<n_k>$ are not equal to zero, from random forcing of the central oscillator by the heat bath.

It's not much more work to find the P-distribution that corresponds to any initial n-quantum state. To do that we simply observe that the density operator for the pure coherent state $|\alpha>$ can be written as

$$|\alpha><\alpha| = e^{-|\alpha|^2} \sum_{n,m} \frac{\alpha^n \alpha^{*m}}{\sqrt{n!\, m!}} |n><m|. \tag{44}$$

It is, in other words, a species of generating function for the operators $|n\rangle\langle m|$, Included among these are the alternative initial density operators

$$\rho_A(0) = |n\rangle\langle n| \ . \tag{45}$$

We can find the quasi-probability distributions which evolve from the entire set of these initial states just by expanding the function $P(\alpha,0|\gamma,t)$ given by Eq.(37) in a power series analogous to Eq.(44) and evaluating the appropriate coefficients. The function $P(\alpha,0|\gamma,t)$ given by Eq.(37) is, in fact, a species of generating function[11)] for the Laguerre polynomials L_n and so we easily find that an n-quantum initial state leads to

$$P_n(\gamma,t) = \frac{(-1)^n}{\pi\,\eta^{n+1}(t)} L_n\left(\left(1+\frac{1}{\eta(t)}\right)|\gamma|^2\right) e^{-\frac{|\gamma|^2}{\eta(t)}} \ . \tag{46}$$

Some inkling of the appearance of these functions at a time t of the order of κ^{-1} is given in Fig. 9. We may note that the function P_1, for example, takes on negative values for $|\gamma| < 1$, and then the higher order functions P_n do as well for values of $|\gamma|$ that are of order unity. The mean value of $|\gamma|$, on the other hand, increases exponentially as $e^{\kappa t}$ and for values of $|\gamma|$ large compared to unity all of the functions P_n assume positive values and vary quite smoothly. While the quasi-probability densities P_n don't behave like ordinary probability densities within the quantum domain, for which $|\gamma|$ is of order unity, they do indeed behave like probability densities in the classical domain $|\gamma| \gg 1$, to which the amplification process inevitably brings the excitation. When the amplification has taken place over a period several times κ^{-1} in duration, almost the entire normalization

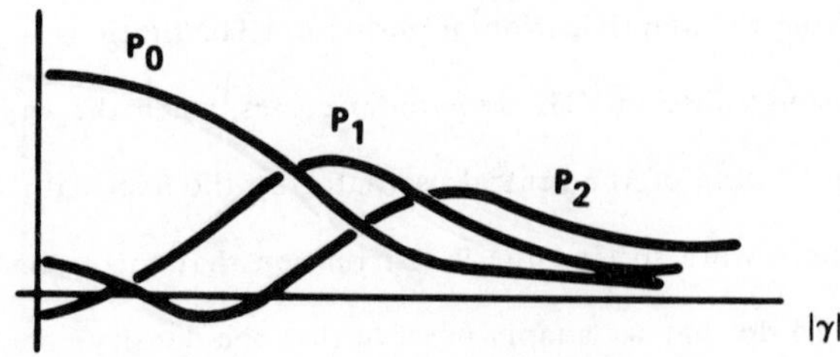

Fig. 9

integral of P_n comes from the classical domain $|\gamma| \gg 1$, and it becomes quite correct asymptotically to interpret P_n as a classical probability distribution for the oscillator amplitude. When the heat bath is initially at zero temperature, for example, the functions $P_n(\gamma,t)$ become asymptotically

$$P_n(\gamma,t) = \frac{1}{\pi n!} \frac{|\gamma|^{2n}}{(|U(t)|^2-1)^{n+1}} e^{-\frac{|\gamma|^2}{|U(t)|^2-1}} . \tag{47}$$

Sometimes we may be interested in finding the behavior of the amplifier for initial states which are superpositions of different n-quantum states. In fact, the examples we shall discuss involve some uncertainty in whether the amplifier begins in an $n = 0$ or an $n = 1$ state, so we need to know what is the effect of adding non-diagonal terms like $|0\rangle\langle 1|$ and $|1\rangle\langle 0|$ to the initial density operator. The density operator (44) for a pure coherent state is evidently a generating function for these operators too, and so their contribution to the time dependent density operator can also be found from the appropriate terms of the same expansion of the function P.

Sometimes we are also interested in knowing the matrix elements of the density operator in the n-quantum state basis. We can find those as well by using the generating function device. For an initially pure coherent state $|\alpha\rangle$, for example, we thus find the matrix element for $\rho_A(t)$:

$$\langle n|\rho_A(t)|m\rangle = \sqrt{\frac{n!}{m!}} \frac{\bar{n}^n}{(1+\bar{n})^{n+1}} \left(\frac{\alpha U(t)}{1+\bar{n}}\right)^{m-n} L_n^{m-n}\left(\frac{-|\alpha|^2|U|^2}{\bar{n}(1+\bar{n})}\right) \exp\left(-\frac{|\alpha|^2|U|^2}{1+\bar{n}}\right) . \tag{48}$$

It is interesting to examine the ratio of the off-diagonal terms of this matrix to the diagonal ones. For $m > n$, and $t \gg 1$ we find

$$\frac{\langle n|\rho_A(t)|m\rangle}{\langle n|\rho_A(t)|n\rangle} \longrightarrow \frac{1}{(m-n)!}\sqrt{\frac{m!}{n!}} \left(\frac{\alpha}{U^*(t)}\right)^{m-n} , \tag{49}$$

but this ratio goes to zero as $\exp[-(m-n)\kappa t]$. The off-diagonal matrix elements tend therefore to become quite small in comparison to the diagonal ones. That is not quite the same thing as saying that the matrix assumes a precisely diagonal or stationary form, since there are so very many non-vanishing off-diagonal matrix elements present.

It is a kind of asymptotic near-diagonality that the density matrix possesses, and it assumes that property eventually, no matter what the initial state of the central oscillator may have been.

This asymptotic descent of the density matrix into quasi-diagonality is even more dramatic in the coordinate-space or momentum-space representations. The coordinate-space representation of the general expression (15) for the density operator is

$$\langle q' | \rho_A(t) | q'' \rangle = \int P(\alpha,0|\gamma,t) \langle q' | \gamma \rangle \langle \gamma | q'' \rangle d^2\gamma . \qquad (50)$$

The coherent state wave functions $\langle q' | \gamma \rangle$ and $\langle \gamma | q'' \rangle$ are Gaussian in form[6] and vanish quite rapidly except for values of q' and q'' which lie close to $(2\hbar/\omega)^{1/2}(\mathrm{Re}\,\gamma)$. The coordinate values for which the product $\langle q' | \gamma \rangle \langle \gamma | q'' \rangle$ is significantly different from zero only allow $|q' - q''|$ to take on values comparable to the range of zero-point fluctuations, $(2\hbar/\omega)^{1/2}$. When this product of wave functions is multiplied by a smooth P-function and integrated over γ, as in Eq.(50), the constraint becomes even tighter. With the function P given by the pure noise distribution of Eq.(43), for example, we find that the matrix element (50) vanishes for $|q - q''|$ larger than $\left\{\frac{2\hbar(2n+1)}{\omega n(n+1)}\right\}^{1/2}$, a distance which shrinks exponentially to zero.

This interesting property of asymptotic or macro-diagonality that the density matrix possesses will have an important consequence when we begin looking at experiments. It will imply that we cannot use our amplifier to generate any of those weird large-scale superposition states that quantum mechanics seems in principle to admit but that we have never in our lives seen. Those are the states that Schrödinger maintained were intrinsically absurd consequences of quantum mechanics. They are not so much absurd as unrealizable.

How do we define the gain and noise figures for our amplifier? We can define the amplitude gain in an obvious way. It's just the exponential $e^{\kappa t}$. The power gain then is just the square of that,

$$G = |U(t)|^2 = e^{2\kappa t} . \qquad (51)$$

The noise figure requires a bit more subtlety. A conventional way of defining the noise figure of an amplifier is to imagine a noiseless amplifier with the same gain and ask how much noise you would have had to put in initially in order for the noiseless amplification process to produce the same final noise that the real amplifier produces. That figure then offers a fair measure of how strong your input signal ought to be if it is to be detectable above the final noise.

The amount of Gaussian noise you must add to an initially coherent state $|\alpha\rangle$ in order for this hypothetically noiseless amplification process to lead to the P-distribution of Eq.(37) is

$$\begin{aligned} N_{noise} &= \frac{n(t)}{|U(t)|^2} \\ &= (1 + \langle n_{\omega}"\rangle)(1 - 1/G) \ . \end{aligned} \tag{52}$$

In a recent paper on amplifiers Caves[12)] has established a lower bound for this noise figure by invoking general principles rather than any specific model. His result for the noise figure was , $N_{noise} > 1 - 1/G$, which is entirely consistent with Eq.(52).

There is an interesting game that you can play with this sort of amplifier and with the attenuator that you can construct by turning the inverted oscillator back into a normal one. Let's say we put a microscopic signal of some sort into our amplifier and amplify it considerably. Then we stop the process and reinvert the central oscillator so that the signal is attenuated back down to its original strength. The question we now ask is, will we get the oscillator back to the quantum state in which it began? The attenuation process, we've noted, is completely noiseless as long as the heat bath is at temperature zero. Let's assume the heat bath always starts at temperature zero, even for the amplification process as well.

There would be some strange consequences if, in fact, it were possible to get back to the original state. The amplification process we are discussing is irreversible and this may be one of the better illustrations of that fact. The analysis is quite elementary, so I will not go through it in detail. Let me just say that by changing the sign of the frequency parameter you can indeed turn the amplifier into an attenuator with a decay period equal to the amplification period. You can decrease the oscillator excitation noiselessly, but you can never get its state back to the form in which it began. What happens is that the cycle of amplification and attenuation superposes on that state a certain amount of Gaussian noise. If each half of the cycle lasts for a length of time t, the number of quanta of noise added when the heat bath is at zero temperature is just $1 - e^{-2\kappa t}$. For $\kappa t \gg 1$ then each such cycle adds a single quantum of noise. If we carry out the cycle with an excitation $\langle n_\omega \rangle$ in the heat bath oscillators at frequencies near ω we find that the number of noise quanta introduced by the cycle is $(1 + 2\langle n_\omega \rangle)(1 - e^{-2\kappa t})$.

Suppose it were true that we could carry out the cycle of amplification and attenuation and still return the system to precisely the state from which it began. We could put the central oscillator into some interesting quantum state and amplify it until all its variables assumed classical strength. We could then measure them all, including the complementary ones, without disturbing any of them significantly, and finally, after attenuating the signal noiselessly to reestablish exactly the initial state, we would find ourselves in possession of all sorts of information about it that contradicts the uncertainty principle. It is the quantum mechanical nature of the amplification process ultimately that makes it both noisy and irreversible, and thus prevents the occurence of any such miracle.

Our model for the amplifier may seem a bit unrealistic. It depends, after all, quite explicitly on the use of an inverted oscillator, and oscillators of that kind are not available "off-the-shelf". In fact, if we were in possession of one just such oscillator it could solve the world's energy problems -- and that doesn't sound too likely. But there is no need, in fact, for us to find an inverted oscillator in the literal sense. All we need is something that behaves like one over the limited range of variables in which it is actually used. Many systems are available which do approximate that sort of behavior.

Let us consider, for example a system, say an atom, with a large total angular momentum, $j \gg 1$. If the atom has a magnetic moment in the direction of the angular momentum $\vec{J}$ and we place it in a uniform magnetic field it will have $2j + 1$ equally spaced Zeeman levels. As long as the atom is in any of the states with J_z not too far from j, that is near the state of maximum energy, its behavior will quite closely approximate that of an inverted oscillator. The operator $(J_x + iJ_y)/\sqrt{2j}$, that raises the magnetic quantum number J_z by one unit, then plays the role of the annihilation operator a. Its commutator with its hermitian adjoint has eigenvalues appropriately close to unity for the states with J_z near j. We can then identify the angular momentum states $|J_z = j-n\rangle$ with the states $|n\rangle$ of the inverted oscillator. (See Fig. 10).

When this atom is in any state of small n, and is coupled to the radiation field in the usual way, it will begin to emit quanta and descend to states of larger n, in just the way we have described for the inverted oscillator. Of course, the acceleration of the radiation rate won't continue indefinitely. When the quantum number n becomes

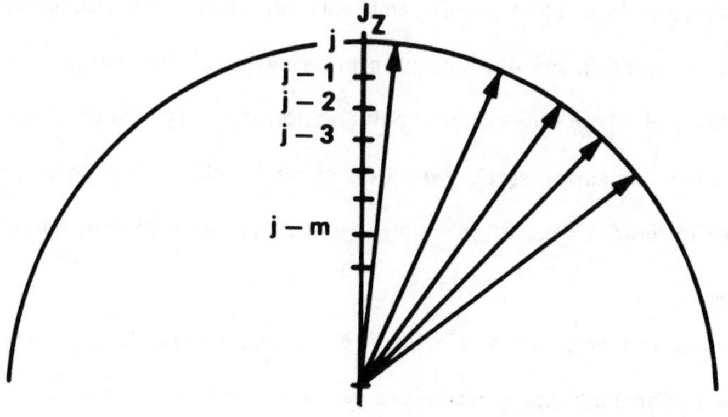

Fig. 10

comparable to j it will no longer increase as rapidly. The rate of radiation reaches a maximum for n = j and then begins to decrease. All of this is only to say that a magnetic moment associated with a large value of j is not really a linear amplifier. It is a nonlinear one. But there is a regime for large values of J_z in which it duplicates quite accurately the behavior of a linear amplifier.

The example of an atom with a huge angular momentum j may seem a bit far-fetched, but there is a context in which we often deal with such systems these days. A single two-level atom is equivalent algebraically to a system of spin 1/2 in a magnetic field. A system of N such atoms with identical couplings to the radiation field is therefore precisely equivalent to the angular momentum model we have just discussed with $j = N/2$. When all or nearly all of the N atoms are initally in their excited states the emission process that ensues is superfluorescent in character. The initial phase of superfluorescence is one of linear amplification. It continues long enough in practice to generate fields of essentially classical strength that are easily observed with large-scale laboratory equipment. The inverted oscillator model, in fact, is just an idealization of the way in which the superfluorescent radiation process begins.

V. Specification of Photon Polarization States

One of the subjects we want to discuss, you may remember, is a scheme for superluminal communication involving photon polarizations. That proposal gave a special role to states of plane and circular polarization, but any two distinguishable pairs of orthogonal polarization states would do equally well . It will be of some help therefore to have in hand a general procedure for dealing with photon polarization states of all sorts.

For a light wave that propagates in a fixed direction, the polarization vectors are defined in a two-dimensional space transverse to that direction. Complex vectors of unit length in that two-dimensional space define a two-parameter family of elliptical polarization states; any two orthogonal vectors of that sort are eligible to be a pair of basis states. Transformations from one pair of basis states to another must then be unitary matrix transformations in two dimensions. They are thus a two-parameter subgroup of the group we call SU(2), the special unitary group in two dimensions.

There is a well-known correspondence between the transformation of SU(2) and real rotations in three dimensions which gives us a simple way of picturing the various polarization states. They can be put in one-to-one correspondence with the points on a unit sphere, so that the transformations of SU(2) simply rotate them into one another. This way of dealing with all of the polarizations at once was invented, in fact, by Poincaré[13]), in 1892, somewhat before the discovery of spinors by Cartan.

Figure 11 then is a picture of the Poincaré sphere. We have taken the north and south poles of the sphere to represent the two orthogonal states of circular polarization and have accordingly labelled them for the state $|+>$ of helicity one and for the state $|->$, of helicity minus one. Of course you can equally well label them L and R for left and right circular polarizations -- if you can remember which is which. An arbitrary

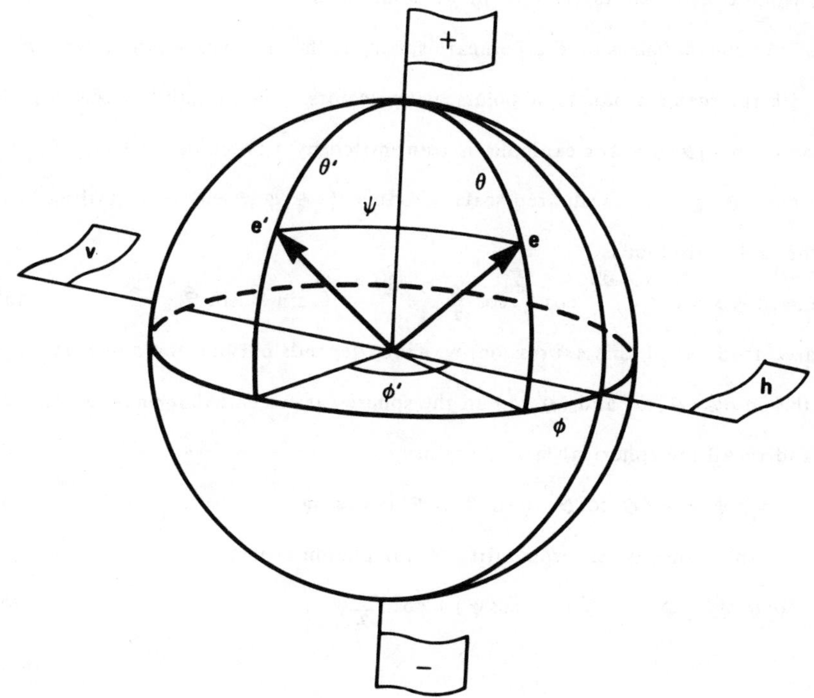

Fig. 11

polarization state, which we can call $|\theta, \varphi>$, can then be written as

$$|\theta,\varphi> = e^{\frac{i\varphi}{2}} \cos\frac{\theta}{2}\ |+> + e^{-\frac{i\varphi}{2}} \sin\frac{\theta}{2}\ |-> . \tag{53}$$

where the angles ϑ and φ are limited by

$$0 \le \vartheta \le \pi \quad , \; -\pi \le \varphi < \pi . \tag{54}$$

Each point on the surface of the sphere thus defines a unique state of elliptical polarization. Antipodal points on the sphere always represent orthogonal polarization states.

Let us take $|h>$ and $|v>$ to represent the usual horizontal and vertical states of plane polarization, so that the circular polarization states can be written as

$$|\pm> = \frac{1}{\sqrt{2}}(|h> \pm i\ |v>) \tag{55}$$

Then we find that in the $|\theta, \varphi>$ scheme these states are

$$|h> = |\tfrac{\pi}{2}, 0> \quad , \quad |v> = |\tfrac{\pi}{2}, -\pi> , \tag{56}$$

We have indicated their locations too on the Poincaré sphere in Fig. 11. They lie on the equator, which consists exclusively of plane polarizations.

One of the conveniences of the Poincaré sphere is that it gives a simple way of dealing with the scalar products of polarization vectors. The probability that a photon with polarization $|\theta, \varphi>$, for example, is transmitted by a filter that selects polarizations $|\theta', \varphi'>$ is the squared scalar product $|<\theta', \varphi'|\theta, \varphi>|^2$. According to Eq.(55) the scalar product is

$$<\theta', \varphi'|\theta, \varphi> = e^{\frac{i}{2}(\varphi - \varphi')} \cos\frac{\theta'}{2} \cos\frac{\theta}{2} + e^{-\frac{i}{2}(\varphi - \varphi')} \sin\frac{\theta'}{2} \sin\frac{\theta}{2} . \tag{57}$$

The squared modulus of this expression, we find, depends only on the angle between the radii to the points (θ, φ) and (θ', φ') on the sphere. If we call that angle ψ , as in Fig. 11, and recall the spherical law of cosines,

$$\cos\psi = \cos\theta \cos\theta' + \sin\theta \sin\theta' \cos(\varphi - \varphi') , \tag{58}$$

we find that the transmission probability of our photon is just

$$|<\theta', \varphi'|\theta, \varphi>|^2 = \frac{1}{2}(1 + \cos\psi) = \cos^2 \frac{1}{2}\psi . \tag{59}$$

VI. Measuring Photon Polarizations

If we are to communicate by means of photon polarizations we must be prepared to answer questions like this: A photon goes through a filter that transmits with 100% efficiency the polarization $|\theta', \varphi'>$; what is the probability that such a transmitted photon really had the initial polarization $|\theta, \varphi>$? That is a question that can only be posed probabilistically because you cannot go back and verify what the state was. The possibility of communication, on the other hand, depends critically on the determination of those probabilities. How do we find them? We will presently show that Bayes' theorem gives a convenient way of determining such <u>a posteriori</u> probabilities, but we don't need all of that generality quite yet. Let's assume we have some arbitrary beam of photons. We place in the beam a filter that transmits only the polarization $|\theta', \varphi'>$. When a photon is transmitted, we can obviously say that the probability that its initial polarization was $|\theta, \varphi\rangle$ is proportional to the probability that a photon of polarization $|\theta, \varphi\rangle$ is transmitted by a filter that transmits polarizations $|\theta', \varphi'\rangle$. That probability, in other words, is proportional to the squared scalar product of Eq.(59), and has the angular dependence $\cos^2(\psi/2)$. The $\cos^2(\psi/2)$ distribution is spread out quite smoothly all over the sphere but does convey some information. When you make a measurement on one photon, what you find is a new way of weighting whatever information you had initially about where its polarization vector might have been pointing. The new weighting multiplies whatever distribution you knew of before by this factor $\cos^2(\psi/2)$ -- if you knew nothing before, if the a priori distribution were uniform, then the final distribution over the surface of the sphere would just be a constant times $\cos^2(\psi/2)$.

If you were to attempt to perform the Einstein-Podolsky-Rosen experiment we described initially, and to use it to communicate via measurements made on individual photon polarizations, you would quickly find you cannot distinguish between circular polarizations and linear polarizations. In fact you can't distinguish between any pairs of antipodal points on the Poincaré sphere. The angles ψ for any pair of antipodal points are naturally supplementary, so the sum of the two probabilities is proportional to

$$\cos^2(\psi/2) + \cos^2\left(\frac{\pi-\psi}{2}\right) = \cos^2(\psi/2) + \sin^2(\psi/2)$$
$$= 1 ,$$

and that is constant -- and quite independent of which pair of orthogonal polarization states you've chosen. So you are stuck! One polarization measurement on a single photon can't ever provide the information desired.

We have the alternative suggestion, however, that we use a "laser gain tube" to multiply the number of photons to be processed. What such a laser gain tube does, presumably, is just to amplify in similar ways photons that are in any of the possible polarization states. Those polarization states can all be considered to be superpositions of one pair of basis states. Then everything the laser gain tube does can be represented by the action of two identical amplifiers, each of which is fed by the appropriate polarization component.

An appropriate sort of amplifier-detector arrangement is shown in Fig. 12. The quarter-wave plate converts circular polarizations into plane polarizations and the Nicol or Wollaston prism separates the latter into two beams, each of which is sent into its own amplifier and emerges in considerably strengthened form. In effect then, one of those amplifiers amplifies the right-handed circularly polarized component of the incident beam and the other the left-handed component. The two output beams are so strong we may describe them classically and can measure them to our heart's content without disturbing them in any way.

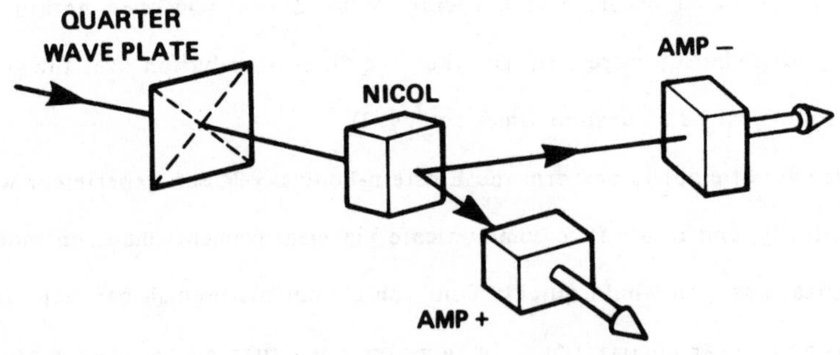

Fig. 12

Let us agree then that we are going to carry out our calculations by using circularly polarized basis states. Let $a_+^\dagger$ be a creation operator for the state of helicity one and $a_-^\dagger$ the creation operator for the state of helicity minus one. Then a state with n_+ photons of the first variety and n_- of the second can be generated from the vacuum state $|n_+ = 0, n_- = 0> \equiv |0,0>$ by applying the appropriate creation operators,

$$|n_+ , n_-> = \frac{(a_+^\dagger)^{n_+}}{\sqrt{n_+!}} \frac{(a_-^\dagger)^{n_-}}{\sqrt{n_-!}} |0, 0> . \tag{60}$$

We want still to be able to deal with photons in arbitrary polarization states. For a photon in a polarization state $|\theta, \varphi>$ we can define the creation and annihilation operators

$$a^\dagger_{\theta\varphi} = e^{\frac{i\varphi}{2}} \cos\frac{\theta}{2}\, a_+^\dagger + e^{-\frac{i\varphi}{2}} \sin\frac{\theta}{2}\, a_-^\dagger , \tag{61}$$

$$a_{\theta\varphi} = e^{-\frac{i\varphi}{2}} \cos\frac{\theta}{2}\, a_+ + e^{\frac{i\varphi}{2}} \sin\frac{\theta}{2}\, a_- , \tag{62}$$

so that a single-photon state $|\theta,\varphi>$ can be written as

$$|\theta , \varphi> = a^\dagger_{\theta\varphi} | 0, 0 > . \tag{63}$$

The initial density operator that represents this state is just this vector multiplied by its dual,

$$\rho(0) = a^{+}_{\theta\varphi} |0,0\rangle\langle 0,0| a_{\theta\varphi} \ . \tag{64}$$

Now I am a little embarrassed because the amplifiers we are discussing are transient amplifiers, and it would be nicer, of course, to carry out this experiment with continuously operating linear amplifiers. But it's a bit more complicated, as you know, to construct models of CW amplifiers, and not really very essential for conceptual purposes. The time dependent scheme we are discussing, in fact, suits the practical needs -- of a gedanken experiment -- quite well. One can adjust the various initial times and amplification times and the like so that one does secure an appropriately strong output signal from a single photon arriving at time zero. It's a slightly touchy thing because these amplifiers are devices which give you an output signal even if you have no photon there at time zero. You have got to be pretty careful about coordinating everything, but there is nothing difficult about that -- at least, in principle.

VII. Use of the Compound Amplifier

We can deal with our pair of amplifiers very much as we dealt with a single one before. If the photon incident upon our detector is in the state $|\theta,\varphi\rangle$, the initial density operator for the two amplifiers is given by Eq. (64). The final reduced density operator for the two amplifiers can then be written as a two-mode P-representation that has the general form

$$\rho(t) = \int P(\gamma_+, \gamma_-, t|\theta,\varphi)\ |\gamma_+\rangle\langle\gamma_+|\ |\gamma_-\rangle\langle\gamma_-|\ d^2\gamma_+ d^2\gamma_- \ . \tag{65}$$

Evaluating the function P precisely is now, for the most part, a repetition of the calculation described earlier for the single mode case. The only new points to be observed are that the function depends on two variables $\gamma_\pm$ and that the initial density operator given by Eq.(64) contains off-diagonal terms in the quantum numbers

$n_{\pm}$. The effect of such terms, as we've noted, can easily be found by the same generating function devices that we used for the diagonal ones.

The function P that we find in this way is

$$P(\gamma_+, \gamma_-, t\,|\,\theta, \varphi) = \frac{1}{\pi^2 \eta^2}\{1 - \frac{|U|^2}{\eta} + \frac{|U|^2}{\eta^2}|\gamma_+ e^{-\frac{i\varphi}{2}} \cos\frac{\theta}{2} + \gamma_- e^{\frac{i\varphi}{2}} \sin\frac{\theta}{2}|^2\} \cdot \exp\{-\frac{|\gamma_+|^2 + |\gamma_-|^2}{\eta}\}, \tag{66}$$

where the functions U and η for the amplifier are those defined earlier and approximated by Eqs. (41) and (42) respectively. We can express this result a bit more simply by imagining the two amplified output fields to be superposed and then defining a polarization vector for the superposed fields. An appropriate unit polarization vector is

$$|\hat{e}_{\gamma_+\gamma_-}\rangle = \frac{1}{\sqrt{|\gamma_+|^2 + |\gamma_-|^2}}\{\gamma_+|+\rangle + \gamma_-|-\rangle\}. \tag{67}$$

We can define the intensity associated with any polarization state $|\theta', \varphi'\rangle$ of the superposed output fields as

$$I(\theta', \varphi') = \text{Trace}\{\rho(t)\, a^\dagger_{\theta'\varphi'}\, a_{\theta'\varphi'}\} \tag{68}$$

$$= \int P(\gamma_+, \gamma_-, t\,|\,\theta, \varphi)\, |\gamma_+ e^{-\frac{i\varphi'}{2}} \cos\frac{\theta'}{2} + \gamma_- e^{\frac{i\varphi'}{2}} \sin\frac{\theta'}{2}|^2\, d^2\gamma_+\, d^2\gamma_-$$

$$= \int |\langle \theta', \varphi' | \hat{e}_{\gamma_+\gamma_-}\rangle|^2 (|\gamma_+|^2 + |\gamma_-|^2)\, P(\gamma_+, \gamma_-, t\,|\,\theta, \varphi)\, d^2\gamma_+ d^2\gamma_- . \tag{69}$$

We can use the vector $|\hat{e}_{\gamma_+\gamma_-}\rangle$ furthermore to write the expression for the function P a bit more compactly as

$$P(\gamma_+, \gamma_-, t\,|\,\theta, \varphi) = \frac{1}{\pi^2 \eta^2}\left\{\frac{|U|^2}{\eta^2} |\langle \hat{e}_{\gamma_+\gamma_-} | \theta\varphi\rangle|^2 (|\gamma_+|^2 + |\gamma_-|^2) - \frac{1}{\eta}\right\} \exp\{-\frac{|\gamma_+|^2 + |\gamma_-|^2}{\eta}\}. \tag{70}$$

The Gaussian integral for the polarization dependence of the amplified intensity leads to the simple zero-temperature result

$$I(\theta', \varphi') = |U(t)|^2 \{1 + |\langle \theta', \varphi' | \theta, \varphi\rangle|^2\} - 1 \tag{71}$$

$$= e^{2\kappa t}\{1 + |\langle \theta', \varphi' | \theta, \varphi\rangle|^2\} - 1 . \tag{72}$$

The compound amplifier , we see, does indeed "remember" the initial polarization state of the photon in the classical output it generates. While the output field follows the distribution of Eq.(70) and is therefore highly random, it is indeed polarized, on the average in the direction $|\theta, \varphi\rangle$. The fact that we can measure the polarization of a classical field as precisely as we like may seem to favor the EPR communication scheme, so we had better say a bit more about polarizations.

To define the (θ , φ) polarization of the amplified beam we must compare the intensity I(θ, φ) with the intensity for the orthogonal polarization state. Let $\bar{\theta} = \pi - \theta$ and $\bar{\varphi} = \varphi \pm \pi$ be the polar coordinates of the antipodal point on the Poincaré sphere. Then the polarization in the (θ, φ) direction is

$$p\ (\theta,\varphi) = \frac{I(\theta,\varphi) - I(\bar{\theta},\bar{\varphi})}{I(\theta,\varphi) + I(\bar{\theta},\bar{\varphi})} \tag{73}$$

$$= \frac{|U(t)|^2}{3|U(t)|^2 - 2} \tag{74}$$

$$= \frac{1}{3 - 2e^{-2Kt}} \ . \tag{75}$$

The last of these expressions shows that though the compound amplifier does remember the initial photon polarization, it doesn't have the clearest of memories. The strong field that emerges for $\kappa t >> 1$ only has polarization 1/3. That happens because of the amplified noise output both amplifiers generate even with no initial photon.

The amplified intensity given by Eq.(71) has a fully polarized component $|U(t)|^2\ |\langle\theta',\varphi'|\theta,\varphi\rangle|^2$, which we may consider as the amplified signal due to the incident photon. But it also contains an unpolarized component $|U(t)|^2 - 1$ which represents the noise output, contributed equally by the two amplifiers. The amplification of the noise keeps pace with the signal, and in the long run there is usually even somewhat more noise than signal.

The compound amplifier that is part of our detector scheme reproduces quite accurately, I believe, what should be the action of a "laser gain tube" on the incident photons. But it is not at all clear that such action deserves to be called "cloning". Both the compound amplifier and the laser gain tube are bound to generate more or less random numbers of photons in two orthogonal polarization states. There is no alternative, clearly, to suffering the presence of two varieties of photons, if we are to be able to analyze polarization states with any generality.

In a note inspired by the EPR communication scheme, Wootters and Zurek[14)] have shown that it is not possible, by using a single amplifier, to clone photons of arbitrary polarization. Their analysis takes the definition of cloning quite literally, requiring all photons to be identical, and places the further restriction that the initially pure one-photon state always remain pure. It is not related, therefore, to the action of any real amplifier, let alone the pair of them necessary for the measurement of polarizations. That a pair of analyzing systems is sufficient for the unbiased analysis of polarizations has been pointed out by Mandel[15] who discussed a detector consisting of two atoms.

VIII. Superluminal Communication?

Let me now return to our superluminal communication problem. Having led you through through several areas of quantum optics, I would like to persuade you that we can in fact dismiss the "laser gain tube" scheme very quickly. I will nonetheless go on to analyze it a bit further as it is interesting to see in somewhat more detail how and why it doesn't work -- and there are some more practical uses for the analysis.

Let us try to describe some devices which boil the communication problem down to its barest essentials. Let's say we have two devices, they could even be different states of the same device, that produce two varieties of signal. These are wave packets being sent off to a distant receiver. You can see them in Fig. 13. One variety I will call a dot and the other one a dash (Morse code). Our distant observer listens with whatever equipment he has and measures some property X of those wave packets. He hopes, on

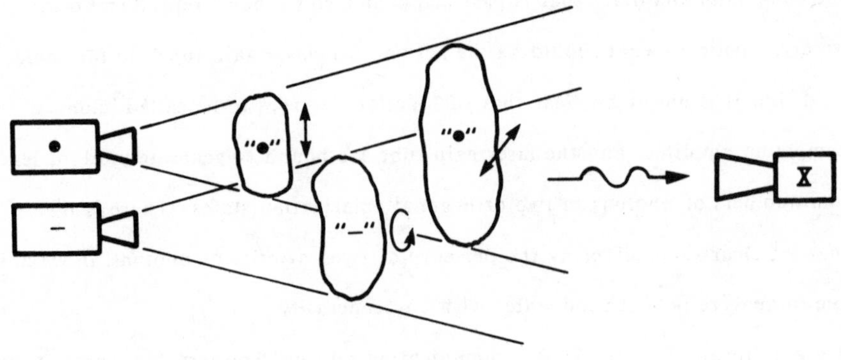

Fig. 13

the basis of a measurement of X, to determine which variety of wave packet it is that he has received; that is whether it is a dot or a dash. He must choose X to be a quantity which makes the clearest distinction between the two. But he must inevitably face the problem, given the observation X = x, what is the probability that the signal is a dot? And given the observation X = x, what is the probability that it is a dash? Now here I do not know of any alternative to the use of Bayes' theorem or more sophisticated decision theoretical notions.

What are the dashes and dots in our communication model? Well, the dashes might be represented by circularly polarized photon packets and the dots by linearly polarized packets. We want, if possible, to distinguish clearly between them. But we must carefully note one thing more. In the class of linearly polarized packets that represent dots we have no control over which of two polarizations is sent. Some of those dots are going to be vertically polarized, and an equal number, on the average, are going to be horizontally polarized. In the class of dashes some of the packets are going to be right-handed circularly polarized, and some are going to be left-handed circularly polarized.

To phrase the decision problem more formally our signaling devices have to be described statistically. Each of our signal generators puts the field into a state described by a particular density operator. The detection problem is to make measurements that offer a clear distinction between those two density operators. By measuring X, or whatever it is, in each wave packet, the detector device asks, in effect: do you represent the density operator for a dot, or the density operator for a dash?

We can phrase the mathematical part of the problem as follows: Let the probability that a given packet represents a dash or a dot be $p(-)$ and $p(\cdot)$, respectively. Let the probability that the detector observes the value x if the input is a dash be $P(x|-)$. Then the joint probability that a packet represents a dash and the detector observes x is

$$P(x,-) = P(x|-)\,p(-)\,. \tag{76}$$

Bayes' theorem asserts that this joint probability can be also be written as the product

$$P(x,-) = P(-|x)p(x)\,, \tag{77}$$

in which $p(x)$ is the probability that the detector registers x no matter which sort of packet arrives, and $P(-|x)$ is the probability, given the measurement x, that the incident packet represents a dash. The probability p(x) is evidently given by

$$p(x) = P(x|-)\,p(-) + P(x|\cdot)\,p(\cdot)\,, \tag{78}$$

while the probability we seek, the probability that the incident packet represents a dash, is evidently

$$P(-|x) = \frac{P(x|-)\,p(-)}{P(x|-)\,p(-) + P(x|\cdot)\,p(\cdot)}\,. \tag{79}$$

An analogous expression of course holds for $P(\cdot|x)$.

But, now that we have created a'l this machinery, we have to face the fundamental difficulty that if we look at the ensemble of wave-packets that represent dots, we will find them unpolarized. What is the density matrix that represents an unpolarized beam? It is one half of the unit matrix; the density operator is just one half of the unit operator. Now let's take the ensemble that represents dashes. We've said they are circularly polarized, but in fact they are circularly unpolarized; they have no net polarization either. What is the density matrix that the dash generator turns out? That is once again one half of the unit matrix. So we are asking our poor distant friend, whatever listening device he may be using, to make a distinction between things which are absolutely identical. The two density operators he must recognize individually are identical twins. This is a classic example of a distinction without a difference. The probabilities that result must satisfy the identity

$$P(x|-) = P(x|\bullet) \tag{80}$$

and thus

$$P(-|x) = p(-)\,. \tag{81}$$

The a posteriori probability, in other words, is equal to the raw a priori probability. Measuring x, whatever it may be, doesn't add to our knowledge of the message at all.

That difficulty is bound to frustrate our distant friend and in the meantime it can't be too encouraging to the theorists who insist on examining the detection problem microscopically rather than surveying it from the more global view point. Our argument assures us that whatever useful possibilities we seem to find microscopically are going to have to cancel out, and that it is just not possible to use this scheme as a means of communication. Having said that, I want nevertheless to look a bit further at the scheme because I think it is interesting -- most particularly as an illustration of what amplifiers do. We'll find our calculations do have practical applications -- in other experimental contexts.

Let us go back to the pair of amplifiers we are using to amplify each photon that is received and see what additional information we can elicit from them. What we do is let the amplifiers amplify for several gain periods, so that the field intensity is blown way up. We are then talking about classical outputs. We have no embarrassment about measuring those field amplitudes, they can be just as strong as we like; we can measure them without disturbing anything.

In this strong-field limit we have

$$\frac{1}{n(t)} = \frac{1}{e^{2\kappa t} - 1} \longrightarrow 0$$

and

$$\frac{|U(t)|^2}{n(t)} = \frac{e^{2\kappa t}}{e^{2\kappa t} - 1} \longrightarrow 1 .$$

When the output fields are that strong, the P-distribution for their amplitudes becomes, in effect, a classical probability density; Eq.(70) then reduces to the form

$$P(\gamma_+, \gamma_-, t | \theta, \varphi) = \frac{1}{\pi^2 n^2} |< \hat{e}_{\gamma_+ \gamma_-} | \theta \varphi >|^2 (|\gamma_+|^2 + |\gamma_-|^2) \exp\{- \frac{|\gamma_+|^2 + |\gamma_-|^2}{n}\} . \qquad (82)$$

We can use this expression now to determine what classical measurements made on the amplified fields can tell us, for any one incident photon, about the original state of its polarization.

In any given detection process we insert a photon in the compound system and let the two amplifiers blow up their initial fields. Then we observe a pair of classical output field amplitudes γ_+ and γ_-, that are governed by the probability distribution of Eq.(82). Once we have measured any such pair of amplitudes we can ask the question; what is the probability that the initial photon had a polarization $|\theta, \varphi>$? The answer is, in fact, staring at us in Eq.(82). It is the very same weighting that we would have arrived at by making a measurement of a single photon with a single polarizer. We are now no longer talking simply about the superposition of different polarization states. The amplification process, combined with the corruption of the information by noise has brought us back, however, to exactly the same position we were in before. The compound amplifier, as far as polarization is concerned, has contrived to tell us nothing at all new because of the noise it has added. We can only make the same inferences about initial polarization states that we made without it. In fact, from that standpoint the compound amplifier is no better than a single polarizer.

IX. Interference Experiments and Schrödinger's Cat

Since we have some experience now at using amplifiers in pairs there are all kinds of interesting games we can play. One is to perform Young's classic double-pinhole interference experiment with an amplifier placed behind each of the pinholes -- as in Fig. 14.

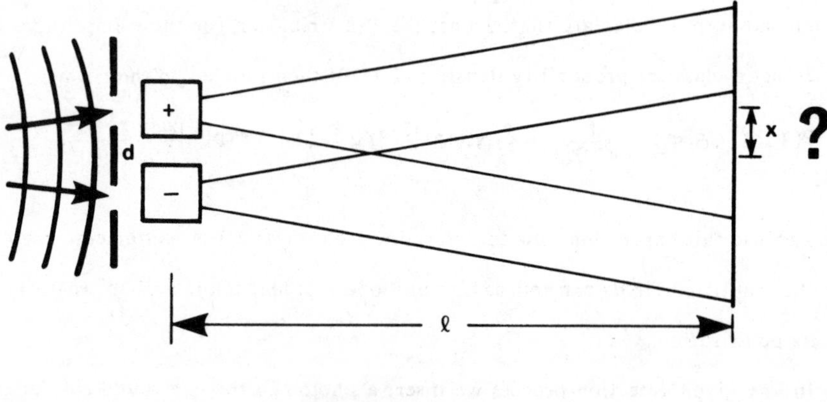

Fig. 14

Conceivably we could use such a scheme to determine which pinhole any given photon has really passed through. Let's agree that we are going to look only at cases in which a single photon packet falls symmetrically on the two pinholes. That means that we are starting our amplifiers out in a pure initial state that is a superposition of the two states in which a photon enters one and the other remains in its empty or "ground" state. We can use the plus and minus signs now just to label the two amplifiers. Those signs need no longer refer to polarizations at all. And with that one change we can use the same calculations we carried out before.

Now what shall we expect to see when the amplified fields are superposed by projecting them onto the distant screen? To set up some straw men let me list three bad guesses. Bad guess number one: someone who just believes in probability theory and no more will say that the photon is going to go through one hole or the other. It will have a probability 1/2 of going through the upper hole and if it does that, the upper amplifier is going to produce a blast while the lower one will produce nothing. Of course it may happen the other way around but we will never see interference fringes. -- That is what he would say, however benightedly.

Bad guess number two might be made by somebody who has studied Young's experiment as it is usually described, and takes the classical view of it. He will note that the photon wave enters those two holes coherently and is strengthened symmetrically by passage along both those routes. The fields projected by the two amplifiers, he will say, should simply show Young's interference fringes on the screen. In saying that, he too is of course overlooking something quite important.

This may not be an exhaustive list of bad guesses; naivety can take almost as many forms as sophistication. But a naively sophisticated view we should not overlook would be this. The photon begins in a well defined initial quantum state. You may remember the Schrödinger cat example -- the situation is not unlike that. Here we have a well defined initial superposition state. The photon may alternatively, but coherently, enter either of the two amplifiers. If their action has the elegant simplicity Schrödinger assumed his diabolical cat-killing machine to have, we should then find, as their output, a quantum mechanical superposition of two amplified fields.

Now what is the meaning of such a quantum mechanical superposition of two quantum states containing highly amplified fields? The two output states involved are orthogonal -- and quite dissimilar. They can easily be distinguished from one another classically. A quantum mechanical superposition of those states regards the two amplified outputs as alternatives. That feature is quite characteristic of superpositions of states in quantum mechanical Hilbert space. When you superpose two orthogonal states, $|A\rangle$ and $|B\rangle$, in Hilbert space you are saying in a certain sense, "either $|A\rangle$ or $|B\rangle$, but not both". So, you may just see in this third, slightly more sophisticated view of the amplification problem, the ghost of Schrödinger's cat.

Here we indeed have a device in which we can trace what happens all the way from the quantum to the classical domain. It permits us, at last, to answer the question of whether we will see the quantum mechanical superposition of two macroscopic states that Schrödinger talks about. That state, if it were to occur, would be characterized by a strong output from either one amplifier or the other -- but not both. The product of the two classical output field strengths would always be zero and thus there would be no interference fringes.

What is the correct way to treat the interference problem? When the photon impinges on the two pinholes, we can think of the two amplifiers, labelled + and -, as beginning in the pure state

$$|\psi\rangle = \frac{1}{\sqrt{2}}\{\,|1\rangle_+\,|0\rangle_- + |0\rangle_+\,|1\rangle_-\,\}\,, \tag{83}$$

in which $|0\rangle_\pm$ and $|1\rangle_\pm$ are the $n_\pm = 0$ and $n_\pm = 1$ states, respectively. This state corresponds precisely to the plane-polarization state $|h\rangle = |\frac{\pi}{2},0\rangle$ for a single photon in the analysis we have just described, and so the results of that calculation may be taken over directly. If $\gamma_\pm$ are the complex amplitudes of the classical fields generated by the two amplifiers, we can write the probability distribution for those amplitudes, according to Eq.(82), as

$$P(\gamma_+,\gamma_-,t|\,\psi) = P(\gamma_+,\gamma_-,t|\,\tfrac{\pi}{2},0)$$

$$= \frac{1}{2\pi\,n^3}\,|\gamma_+ + \gamma_-|^2 \exp\{-\frac{|\gamma_+|^2+|\gamma_-|^2}{n}\}\,. \tag{84}$$

This result shows us immediately that the two amplifiers will have positively correlated outputs; they will, to some degree, tend to radiate coherently, and thus to create fringe patterns on the screen.

How do we calculate the intensity of the light on the screen? We can assume that the fields emitted by the two amplifiers are nearly parallel plane waves, and then try to recall the geometric approximations that we used so often as students. When the pinholes are a distance d apart, the screen is at a distance ℓ, and we observe at a point x on the screen, as opposed to the central point, (See Fig.14.), the waves arriving from the two amplifiers undergo shifts of phase by $\pm \pi dx/\lambda \ell$.

When a single photon enters the system and the output fields of the amplifiers have amplitudes γ_+ and γ_-, the intensity on the screen will be proportional to

$$I(x, \gamma_+, \gamma_-) = |\gamma_+ \exp(i\pi dx/\lambda\ell) + \gamma_- \exp(-i\pi dx/\lambda\ell)|^2 . \tag{85}$$

We should emphasize that this is an intensity distribution for an entire interference pattern, not a probability distribution for the appearance of a single photon on the screen. A single photon arriving at the front end of our system produces an intense, classical interference pattern on the distant screen. (In fact we would even find an interference pattern there when no photon arrives.)

To the extent that the amplitudes γ_+ and γ_- are more or less random variables, the interference pattern of Eq.(85) will also have some random features. It will always contain parallel intensity fringes, unless one or the other of $\gamma_\pm$ happens to vanish, but the fringes will shift in position from one repetition of the experiment to another. The fringes, furthermore, will usually not have the strong contrast typical of Young's experiment. The intensity has no zeroes unless $|\gamma_+| = |\gamma_-|$.

When we repeat the experiment many times, the random fringe system will have the average intensity

$$\begin{aligned} I(x) &= \int |\gamma_+ e^{\frac{i\pi dx}{\lambda\ell}} + \gamma_- e^{-\frac{i\pi dx}{\lambda\ell}}|^2 P(\gamma_+, \gamma_-, t|\psi)\, d^2\gamma_+\, d^2\gamma_- \\ &= 2\ (1 + \cos^2 \frac{\pi d x}{\lambda \ell}) . \end{aligned} \tag{86}$$

Taking the average, in other words, leaves us with Young's fringe pattern set against a constant background intensity. That constant background, of course, is the intensity due to spontaneous emission noise. It's just the average intensity we'd find for the random fringes generated when no photon enters the system. The average visibility of the fringes when a photon does enter is

$$\mathcal{V}(x) = \frac{I_{max.} - I_{min.}}{I_{max.} + I_{min.}} = \frac{1}{3} . \tag{87}$$

This result corresponds, loosely speaking, to a mixture of all three of the wrong guesses we listed earlier -- and probably some others as well. In any case, it does not correspond very closely at all to the guess Schrödinger would have made by means of the same reasoning he used for his cat. The final state with an amplified wave coming from the upper amplifier and none from the lower one could correspond, according to Schrödinger's picture, to finding the cat alive, and the opposite configuration to finding the cat dead. The actual state of the amplified fields, however, is far from being Schrödinger's superposition of those two states. It isn't any pure state at all; in the coherent state representation it is a Gaussian mixture with an enormous variance. In the n-quantum state representation it is a quasi-diagonal mixture with a vast dispersion too. The error in Schrödinger's argument is its omission of the effects of noise -- and the noise is unavoidable; it's there for intrinsically quantum mechanical reasons.

It is occasionally said that quantum mechanics deals only with averages taken over ensembles of experiments, but here we have an explicit counter-example. The outputs of our amplifiers, and the appearance of the fringes on the screen vary greatly from one repetition of the experiment to another, but they are all observable individually -- and we have no trouble measuring them without disturbing them in any way. In any one repetition, for example, the amplifiers may give us the two amplitudes γ_+ and γ_- The visibility of the set of fringes that result is then

$$\mathcal{V}(\gamma_+, \gamma_-) = \frac{2|\gamma_+||\gamma_-|}{|\gamma_+|^2 + |\gamma_-|^2}. \tag{88}$$

This expression only takes on the value one, expressing strong fringe contrast, when the two amplitudes $\gamma_\pm$ are equal in modulus.

The question that always fascinates us about Young's experiment is: which pinhole did the photon really go through? We can, in fact, say something about that in the present version of the experiment. If it should turn out in one repetition of the experiment that $|\gamma_+|$ is far larger than $|\gamma_-|$, we would have a certain indication that the photon went through the upper pinhole and amplifier rather than the lower ones. Of course we can only make such a statement on a probabilistic basis. The way to make it is to use Bayes' theorem again to define the probability $P(+\,|\,\gamma_+, \gamma_-)$ that the photon passed through the upper pinhole, given that the two field amplitudes are $\gamma_\pm$ The result, you can easily see, is

$$P(+|\gamma_+, \gamma_-) = \frac{|\gamma_+|^2}{|\gamma_+|^2 + |\gamma_-|^2}. \tag{89}$$

Let's abbreviate the two probabilities $P(\pm|\gamma_+,\gamma_-)$ as $p_\pm$, so that we have

$$p_\pm = \frac{|\gamma_\pm|^2}{|\gamma_+|^2 + |\gamma_-|^2}. \tag{90}$$

The product of these probabilities, according to Eq.(89), is proportional to the square of the fringe visibility,

$$p_+ p_- = \frac{1}{4} \mathcal{V}^2$$

An alternative way of phrasing the same relation is to write

$$(p_+ - p_-)^2 = \left\{\frac{|\gamma_+|^2 - |\gamma_-|^2}{|\gamma_+|^2 + |\gamma_-|^2}\right\}^2 = 1 - \mathcal{V}^2. \tag{91}$$

Both of these relations show that when we have any degree of certainty about which way the photon went, i.e. $p_\pm = 1$, the fringe contrast $\mathcal{V}$ goes to zero. It is only when we have no inkling of which way the photon went, i.e. $p_+ = p_- = \frac{1}{2}$, that we can see fringes with strong contrast, $\mathcal{V} = 1$. Complementarity, in short, has won again. The incident photon can behave either as a particle or a wave, but never exhibits both extremes of behavior at once. What one usually sees is neither the one behavior nor the other, and Eqs.(91) and (92) describe, a whole continuum of possible compromises.

Young's experiment could be a bit difficult, in practice, to carry out with two amplifiers. The pinholes, after all, must be quite close together and the space available for the amplifiers is rather cramped. There are other interference experiments, however, in which the geometry is less constraining. Plenty of room would be available for the amplifiers in the two arms of an interferometer, for example, and one might entertain the hope of using their outputs to determine which arm a photon actually entered. A suggestion of just such an experiment, using a single amplifier in one arm of a Mach-Zehnder interferometer has been made recently by Gozzini[16]. That arrangement, with a second amplifier added in the other arm, is shown in Fig. 15. Placing the photodetectors C_1, C_2 and C_3 in the positions shown, and letting the mirror M_+ be slightly transparent make it possible to carry out several interesting experiments.

If the geometry is such that the two interferometer paths are precisely symmetric, and if the amplifiers were absent, then the detector C_1 would register photons while C_2 would detect none. The two waves reaching the latter would interfere destructively.

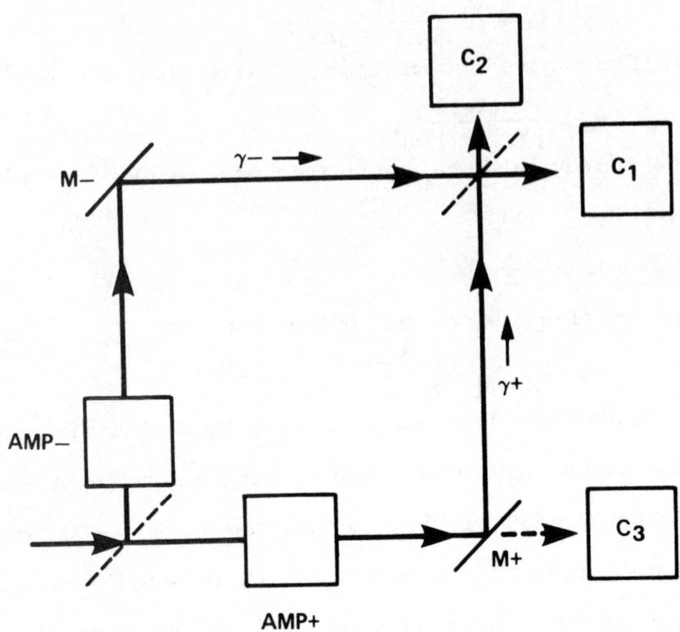

Fig. 15

When the amplifiers are present we may assume that their outputs are described in precisely the same way as in Young's experiment. In other words, the two output fields have random amplitudes γ_+ and γ_- that are governed by the probability distribution in Eq.(84).

The total field amplitude at the counter C_1, we may assume, is proportional to the symmetric sum $\frac{1}{\sqrt{2}}(\gamma_1 + \gamma_2)$ while that at the counter C_2 is the antisymmetric sum $\frac{1}{\sqrt{2}}(\gamma_1 - \gamma_2)$. Then the average counting rates at the two counters will be

$$I_{C_1} = \{\frac{1}{2}|\gamma_+ + \gamma_-|^2\}_{av.} = 2\,n(t) \tag{93}$$

$$I_{C_2} = \{\frac{1}{2}|\gamma_+ - \gamma_-|^2\}_{av.} = n(t)\;. \tag{94}$$

The effects of spontaneous emission noise are again evident in these results; they provide a non-vanishing intensity for C_2 and contribute 2/3 of the total intensity recorded by the two detectors.

Making the mirror M_+ slightly transparent and placing the detector C_3 behind it offers an interesting, if slightly sneaky, way of trying to determine which of the two interferometer paths the incident photon actually took. If the amplifiers were noiseless, the detection of any light by C_3 would mean the initial photon penetrated the first beam splitter and set out on the route we've labelled +. The amplitude γ_- would then be zero and the counters C_1 and C_2 would receive equal amplitudes $\frac{1}{\sqrt{2}}\gamma_+$. The coincidence rate of C_1 and C_3, in other words, would be equal to the coincidence rate of C_2 and C_3. Deviations from that prediction, it has been suggested, might then mean that there was still some interference involving a pilot wave or other mysterious goings on in the half of the interferometer that the photon avoided.

Well, that prediction is indeed false, but it's because the amplifiers are anything but noiseless. A signal recorded by C_3 doesn't tell us uniquely which path the original photon took. The actual C_1-C_3 coincidence rate is proportional to

$$\{\tfrac{1}{2}|\gamma_+ + \gamma_-|^2\,|\gamma_+|^2\}_{av.} = 4n^2\ , \tag{95}$$

while the C_2-C_3 rate is proportional to

$$\{\tfrac{1}{2}|\gamma_+ - \gamma_-|^2\,|\gamma_+|^2\}_{av.} = 2n^2\ . \tag{96}$$

The two rates are not equal; the C_1-C_3 rate is twice the C_2-C_3 rate. That shouldn't be too difficult a result to verify experimentally.

I would like to thank Philip Pearle for bringing the superluminal communication problem to my attention and Professor E. R. Pike and his staff for considerable help in editing the manuscript. Partial support for this work has been provided by the Department of Energy under Contract DE-AC02-76ER03064 and by the Office of Naval Research under Contract N0014-85-K-0724.

References

1. A. Einstein, B. Podolsky and N. Rosen, Physical Review 47 (1935) 777.
2. N. Herbert, Found. Phys. 12, (1982) 1171.
3. E. Schrödinger, Naturwiss, 23 (1935) 807, 823, 844.
4. R. J. Glauber in Quantum Optics, Proceedings of the Enrico Fermi International School of Physics, Course 42, R. J. Glauber, ed., (Academic Press, New York, 1969) p. 32.

5. R. J. Glauber, Phys. Lett. 21 (1966) 650.

6. R. J. Glauber, Phys. Rev. 131 (1963) 2766.

7. V. Weisskopf and E. Wigner, Zeits. Phys. 63 (1930) 54.

8. R. J. Glauber V. I. Man'ko, J.E.T.P. 87 (1984) 790
Soviet Physics, JETP 60 (1984) 450.

9. I have presented descriptions of this model for an amplifier at several conference University of Texas Workshop on Irreversible Processes in Quantum Mechanics and Quantum Optics, San Antonio, March 14-18, 1982.

R. J. Glauber in Group Theoretical Methods in Physics, Proceedings of the International Seminar, Zvenigorod, November 24-26, 1982. Vol. II (Nauka Publishers, Moscow 1983) p. 165.

See also, Group Theoretical Methods in Physics, Vol. I. (Harwood Academic Publishers, 1985) p. 137.

R. J. Glauber, in Proceedings of the VI International School on Coherent Optics, Ustron, Poland, September 19-26, 1985 (in press).

The inverted oscillator has also been used, in connection with a laser model, and with a different coupling, by

F. Schwabl and W. Thirring, Ergeb. d. Exact. Naturwiss. 36, 219 (1964).

10. K. E. Cahill and R. J. Glauber, Phys. Rev. 177 (1969) 1882, in particular Eqs.(3.21) and 3.22) p. 1887.

11. See, for example, B. R. Mollow and R. J. Glauber, Phys. Rev. 160 (1967) 1076, in particular the appendix, p. 1096.

12. C. Caves, Phys. Rev. D23 (1981) 1693.

13. M. Born and E. Wolf, Principles of Optics, 6th ed. (Pergamon Press, Oxford, New York, 1980) p.31.

14. W. K. Wootters and W. H. Zurek, Nature 299, (1982) 802.

15. L. Mandel, Nature 304, (1983) 188.

16. A. Gozzini in The Wave-Particle Dualism , eds. S. Diner et al. (D. Reidel Publishing Co., Dordrecht, 1984) p. 129.

INDEX